TEACHER'S BOOK

Anne Joshua

MA, Dip Ed (Syd); MSc (Oxon)

Contents

Introduction

Dear Teachers

Thank you for purchasing *Enrich-e-matics 3rd Edition Teacher's Book.*
This book accompanies the series of six student books *Enrich-e-matics 3rd Edition Books 1* to *6*.

Enrich-e-matics 3rd Edition Teacher's Book **contains:**

- an introduction and a teacher's ideas section with helpful guidelines for teaching gifted students
- four screening tests for students in Years 1 to 6. For Years 3 to 6, one of the tests is multiple choice to reflect the format of common tests such as the Basic Skills and AIM tests. Altogether, 24 tests are included to help to identify talented students. New tests can be given throughout the year to monitor the progress and the mathematical development of students
- answers to the 24 screening tests with a suggested marking scheme
- grid paper and diagram masters
- answers and fully worked solutions for most of the exercises in *Enrich-e-matics 3rd Edition Books 1* to *6*, plus a guide to teaching techniques for many of the problems.

The screening tests

The purpose of these tests is to identify mathematically promising students.

These tests contain questions that:

- test and extend mathematical skills
- require insight, ability and logical thought
- require creativity.

Who should take the tests?

These tests are for students who show mathematical ability and are nominated by parents and/or teachers. The questions are too difficult for students of average or below-average ability, so it is not advisable to give the tests to a whole class. To help both parents and teachers with the nominations, see 'Characteristics of the mathematically gifted student' on page 126 of this book.

How to administer the screening tests

- Ask the student to write his or her name and class at the top of the page.
- Explain to the students that it is not expected that they will be able to do every question. They should do their best, but should not worry if they are unable to answer a question.
- Please do not help or interpret the question for the student.
- Each test has an allocated period of time written on it.

For Year 1 tests the teacher should read each question aloud and repeat it to ensure understanding. The meaning of certain words may need to be explained. In Year 2, students should be able to read the questions on their own. However, if a student needs help reading a word or a question, the teacher should help. Wait until almost everyone has finished the question before reading the next one. Some questions will take longer to answer than others.

Interpreting the test results

It is very important to build up a profile in order to identify a student as gifted in mathematics. Observing and listening to children applying mathematical skills and concepts in the classroom is extremely important. Though many other criteria need to be considered, these tests do provide one objective measure of a student's problem-solving ability. Students' work should be continually monitored as some children's ability may develop at a later stage. Discussing children's ability and progress with their parents is most valuable.

If a child lacks concentration or is not motivated to do his or her best, the test will have little significance. A child's performance in class is most important and if there are inconsistencies between teacher assessment and observation and the result of a test, the child should be re-tested on another occasion.

There are four tests for each year included in this book so that a new test can be given, and a student's progress and mathematical development can be monitored, at different times throughout the year.

Answers and a suggested marking scheme are included for all tests. Teachers should mark leniently, but consistently, and should award part marks where applicable. Each test is scored out of 100 marks.

Since these tests aim to identify more able students, they are difficult, and any student whose score is 50 per cent or above could be considered for a 'gifted and talented' program. At least two tests should be given before a child is included in, or excluded from, a program.

A student who scores more than 75 per cent on a test has exceptional ability in mathematics and needs a modified curriculum and an enrichment program. *Enrich-e-matics 3rd Edition* provides an excellent source of material for such a program.

Enrich-e-matics 3rd Edition student books

The *Enrich-e-matics 3rd Edition* student books are a series of six books designed to develop and enrich students' problem-solving skills for able students aged 6 to 13. *Enrich-e-matics 3rd Edition* deepens students' mathematical concepts and encourages flexibility of thinking along with willingness to tackle challenging and fascinating problems. The series is much more than a collection of puzzles and difficult problems. The exercises are graded, concepts and strategies are developed throughout the series to provide for a systematic development of problem-solving and mathematical ability.

Enrich-e-matics 3rd Edition is a complete package of material which may be used to supplement and be integrated into a school's mathematics program from Year 1 to Year 7.

Each book consists of 88 pages of challenging problems. The books are designed for students in overlapping age ranges, as set out in the following table.

	Age range
Book 1	6 to 8
Book 2	7 to 9
Book 3	8 to 10
Book 4	9 to 11
Book 5	10 to 12
Book 6	11 to 13

Even in young children there is a great spread of ability in mathematics. Parents and teachers find that some children love playing with numbers or they look for patterns, and are able to generalise and exhibit great talent in spatial ability. These talented children yearn to be challenged and can get very frustrated and bored in the ordinary classroom. These books have been designed to cater for their needs, in an interesting and fun way.

Success leads to greater confidence, and talented students too, need appropriate material so that each individual has balanced experiences of challenge, success and failure. To develop a positive attitude towards mathematics, students should experience success in completing the activities. It is advisable, therefore, that students should start with an appropriate book to build up important techniques.

A student identified as gifted in Year 3 may need to work through Book 2 first to build up important techniques and to gain confidence.

Young children's reading skills may not be as advanced as their mathematical ability. Parents or teachers may need to read the instructions with the children to ensure understanding of the activities, especially for Book 1. Responding orally to some of the activities may be helpful.

It is important that students develop a positive attitude towards mathematics through experiencing success in completing the activities. However, success may not be immediate and students should be encouraged not to 'give up' when doing the problems from the books. Taking risks is an important part of learning. If a student is not ready for a topic, like Time, or 3-Dimensional Space, then that topic could be left and revisited at a later date.

These books are also an excellent resource for enrichment and problem solving for slightly older children of average ability in mathematics. Whereas Book 3 is recommended for able 8 to 10-year-olds, it is also suitable for students of average ability in the 9 to 11 year age group. The age of students using a particular book in the series may vary considerably according to their ability and previous experience. For example, while an exceptionally gifted 10-year-old may be working through Book 4 or 5, another 10-year-old in the same class may find Book 3 both demanding and enjoyable. When starting to work from the series, it is advisable that students be given problems that build up important techniques from the earlier books.

Many of the activities suggest the use of equipment or concrete materials to solve the problems. Young children need 'hands-on' experience with concrete materials to build up their understanding of mathematical concepts.

Aim

Enrich-e-matics is designed to meet the needs of talented students by providing appropriate challenging material that:

- provides a ready source of challenging problems for enrichment and extension
- reinforces concepts and skills in the curriculum
- develops students' problem-solving strategies and extends their mathematical insight, ability and logical thought
- develops their ability to work mathematically
- provides a useful supplement to the mathematics curriculum
- provides an opportunity for students to experience the joy of problem solving
- motivates students to learn mathematics by making learning fun
- is valuable for teachers and trainee teachers who wish to teach mathematics in a pleasurable and captivating way, so that the creative and intellectual potential of able students may be fully realised
- prepares students for mathematics competitions.

Content

All the outcomes of the primary mathematics curriculum are catered for, with a balance of challenging activities covering a wide range of skill areas.

The exercises and activities have been arranged into mathematics strands—**Number, Patterns & Algebra, Chance & Data, Measurement, Space** and **Working Mathematically**—with hundreds of new problems added. This allows students and teachers to work systematically through a number of similar problems, focusing on one area of mathematics. Although some of the activities should be done in order, most can be selected in any order.

Enrich-e-matics 3rd Edition contains many challenging problems on topics that broaden the curriculum. These include:

- investigations and open-ended problems
- creative and recreational topics and number games, including 'Find my rule', 'Knight's tour', 'Magic squares', multiplication short cuts and mathematical curiosities
- problem solving with the use of a calculator
- topics that highlight the aesthetic aspect of mathematics—triangular numbers and their applications and the Fibonacci sequence
- 'competition-style' problems
- topics that are not usually covered in the maths syllabus such as clock arithmetic, Goldbach's conjecture, and number theory
- true or false and multiple choice questions, that can test, in depth, students' understanding of a topic.

The answers for all problems are included in a removable section in the centre of each student book.

enrich-e-matics TEST

3rd EDITION

Time: 30 minutes

YEAR 1
TEST 1

Name: ____________________

Class: ____________________

1 Fill in the blank cards so that the sum of the two cards makes a total of 10.

a 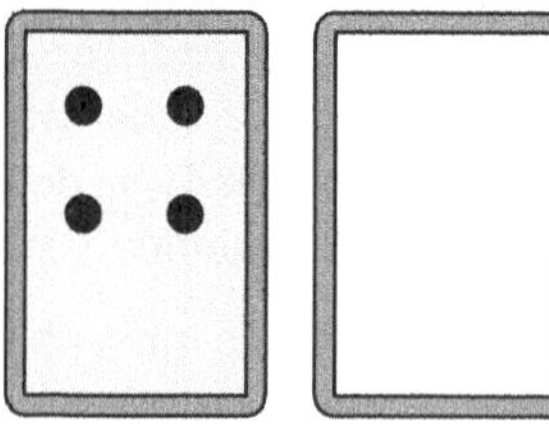

b 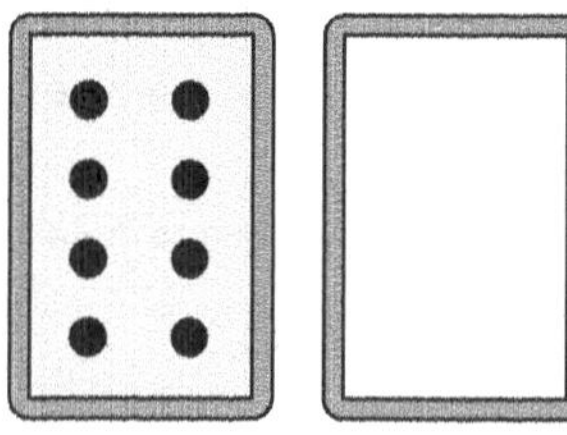

2 Write down how many small squares are in each shape.

 ______

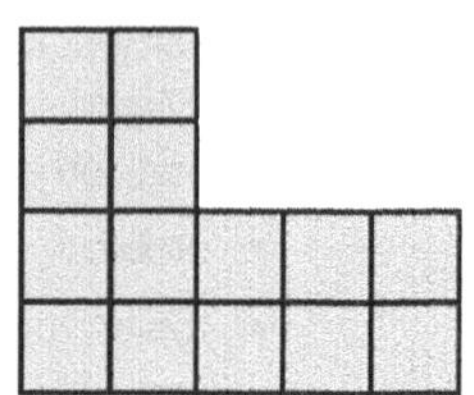 ______

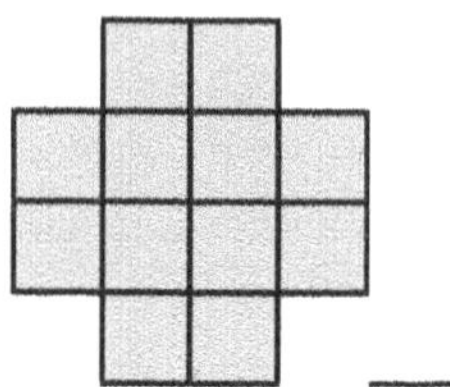 ______

3 Draw the two missing shapes in these patterns.

a 

b 

4 Draw the 5th row of circles.

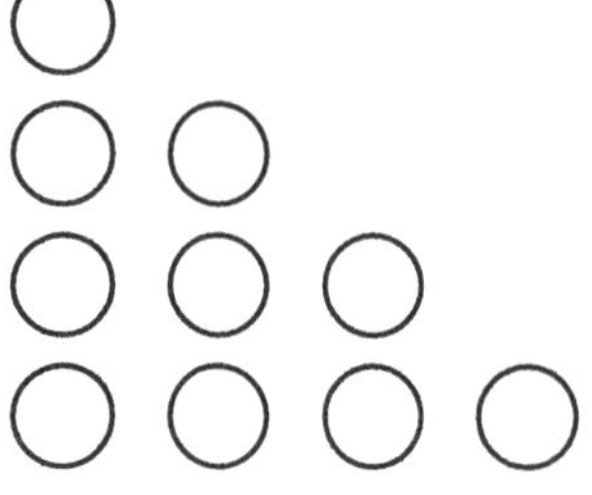

1st row

2nd row

3rd row

4th row

____________________ 5th row

5 Copy this figure very carefully onto the larger squared paper.

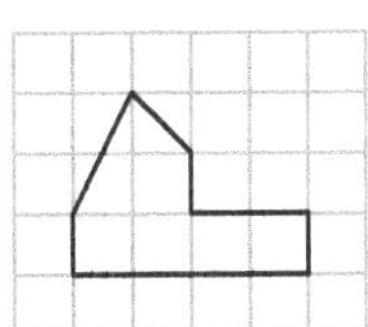

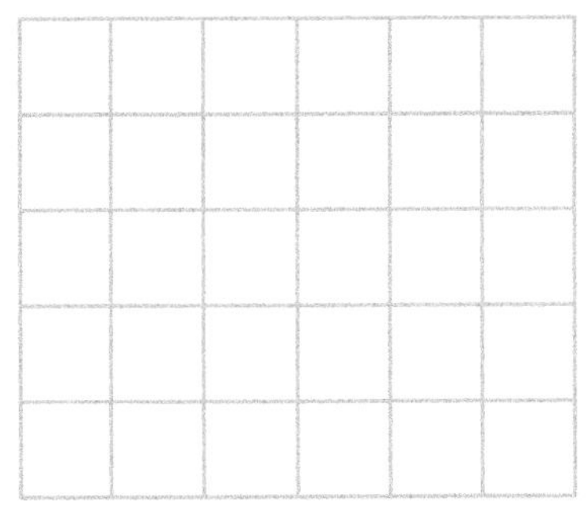

6 Find the next two numbers in the following pattern.

a	1,	3,	5,	7,	______,	______
b	2,	5,	8,	11,	______,	______
c	20,	30,	40,	50,	______,	______
d	7,	17,	27,	37,	______,	______

7 Write the missing numbers in these patterns.

a (10) (11) ( ) ( ) (14) ( )

b (12) (10) ( ) (6) ( ) (2)

c ( ) ( ) ( ) (13) (12) (11)

d (21) ( ) ( ) (15) ( ) (11) (9)

8 Write the missing number in the box.

3 + ☐ = 8

9 Place the numbers 2, 3 and 4 in the circles so that the sum of the numbers in each line is 10.

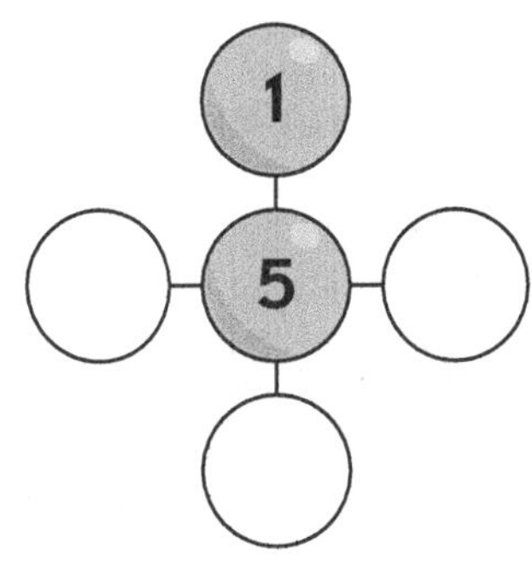

10 Draw the shapes that must be added in this grid to make the pattern complete.

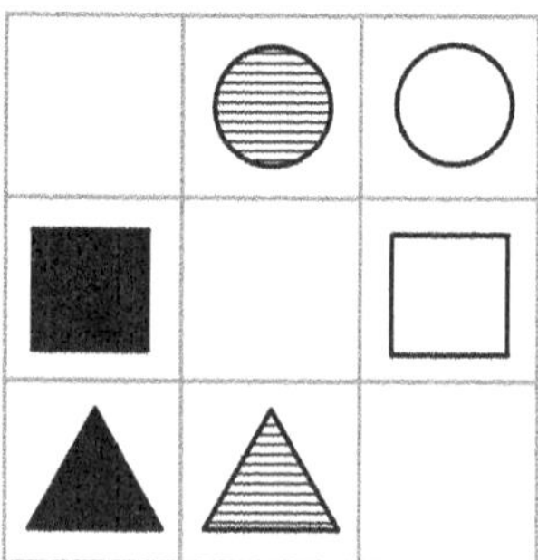

11 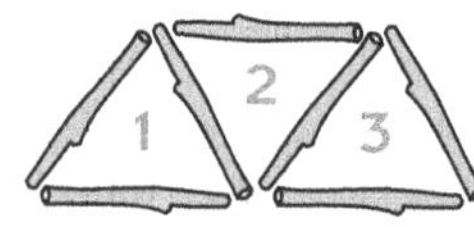 7 sticks are used to make these 3 triangles.

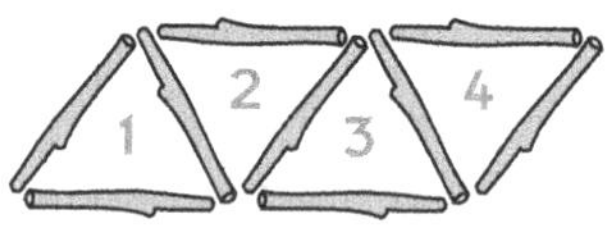 9 sticks are used to make these 4 triangles.

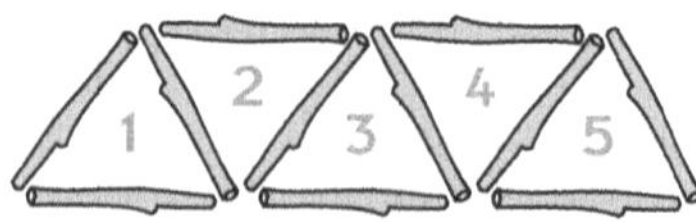 11 sticks are used to make these 5 triangles.

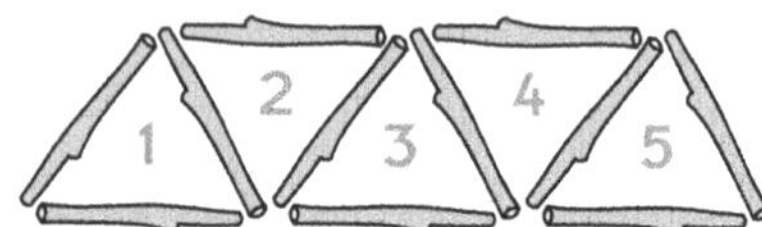

Add 2 more triangles to this diagram so that it will have 7 triangles.

How many sticks are used to make 7 triangles? ________

12 Square tiles are used to make each shape and the shapes form a pattern.

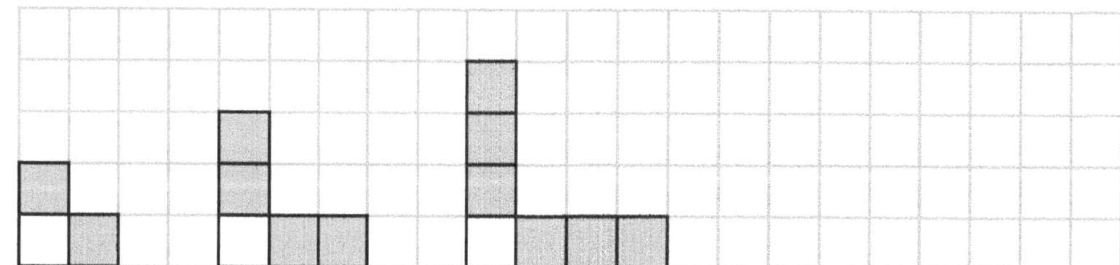

Draw the next shape in the pattern above.

13 If ◆ = 3 ice creams,

Then ◆◆ = 6 ice creams and ◆◆◆ = 9 ice creams.

Work out how many ice creams the diamonds below are worth.

a  = ________ ice creams

b ◆◆◆◆◆  = ________ ice creams

14 **a** If it is 8 a.m. now, what time will it be in 3 hours? ______________

b If it is 8 a.m. now, what time was it 3 hours ago? ______________

c Today is Tuesday. What day will it be in 8 days' time? ______________

15 **a** How many 50-cent coins do I need to make $2? ______________

b If 2 stamps cost $1, what is the price of 4 stamps? ______________

16 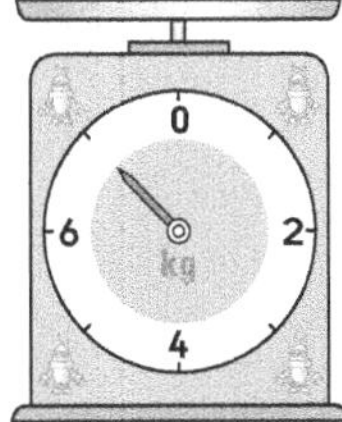

This scale shows 7 kg.
What does the scale show in each of the following?

a 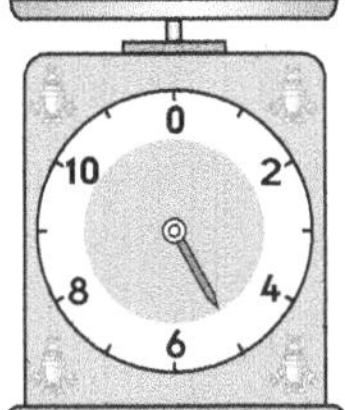

_______ kg

b 

_______ kg

17 Find the mass of 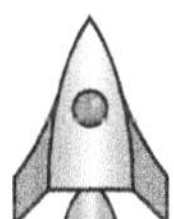 using the scale shown.

______________ kg

18 7 and 4 have a difference of 3, since 7 – 4 = 3
10 and 7 have a difference of 3, since 10 – 7 = 3

Write down some other numbers that have a difference of 3.

a ◯ – ◯ = 3 **b** ◯ – ◯ = 3

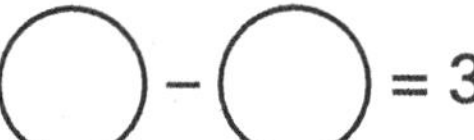

enrich-e-matics 

3rd EDITION

Time: 30 minutes

YEAR 1
TEST 2

Name: ______________________

Class: ______________________

1 How many times does the letter **X** appear in this diagram?

____

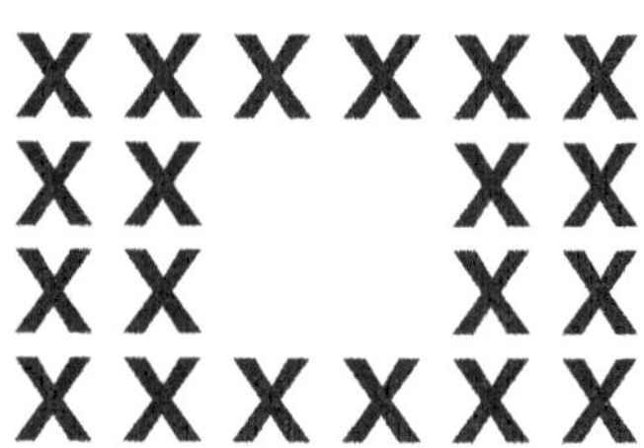

2 Continue this pattern of shapes for one more figure.

   ____________

3 Draw the next three shapes in each pattern.

4 Draw the shapes that must be added to complete the pattern in this square.

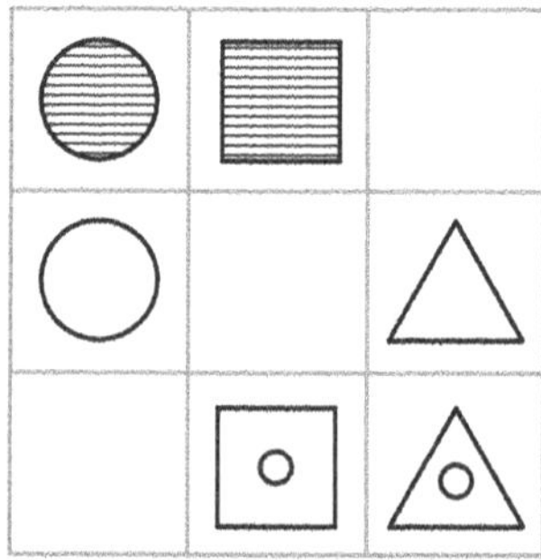

5 Place the numbers 3, 4 and 5 in the circles, so that the sum of the numbers in each line is 11.

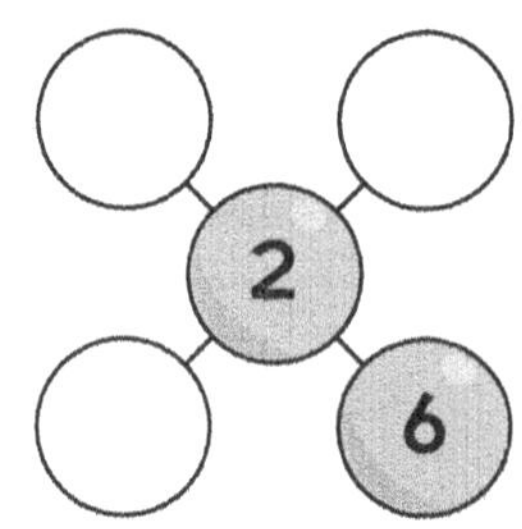

6 The numbers in each set of circles below follow a pattern.
The arrows in the first circle of each set show how the numbers should be read.

What numbers are missing from the last two circles in each set?

a

10 8 12

6 4

b

8 9 7

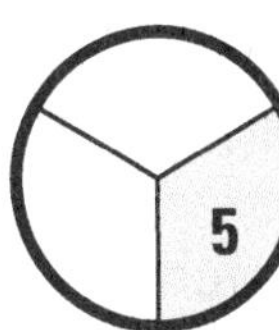

7 Write the missing number in each box.

a ☐ + 2 = 6

b ☐ − 3 = 4

8 If three circles balance one square

and six circles balance two squares ,

then how many ◯s are needed to balance ☐☐☐?

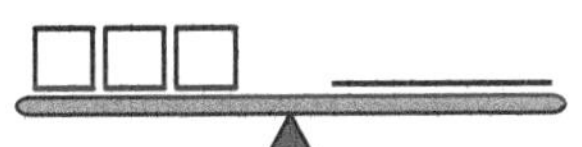

______ circles

9 Place numbers in the squares so that the sum of the numbers in all directions adds to 15.

		8
9	5	1
2	7	

10 In these flags only two of the squares are coloured in.

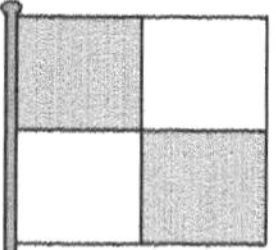
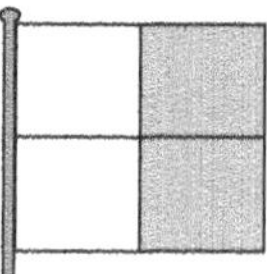

Can you make four other *different* flags by colouring in only two squares on the flags below?

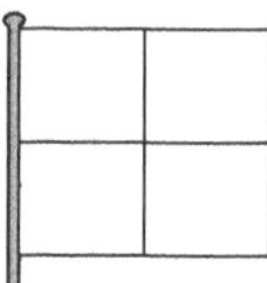
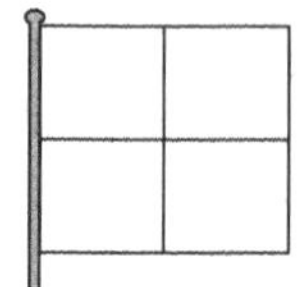
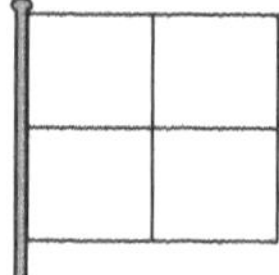
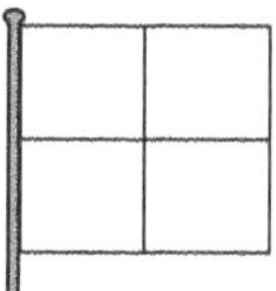

11 **a** Two sisters share $12 so that each girl gets the same amount.

How much does each girl get? ______________

b Three sisters share $12 so that each girl gets the same amount.

How much does each person get? ______________

c If six children share $12, how much does each person get? ______________

12 Look at each picture carefully and fill in the empty square to complete the pattern.

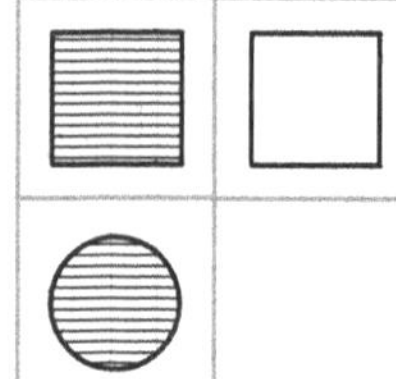
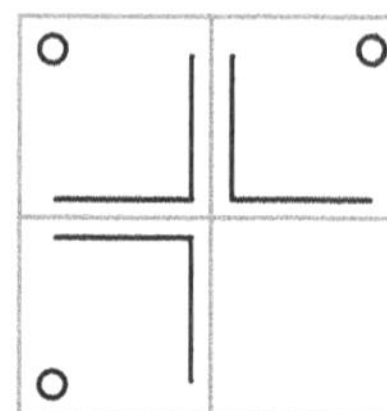
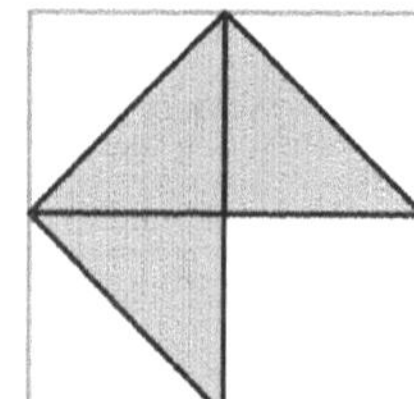
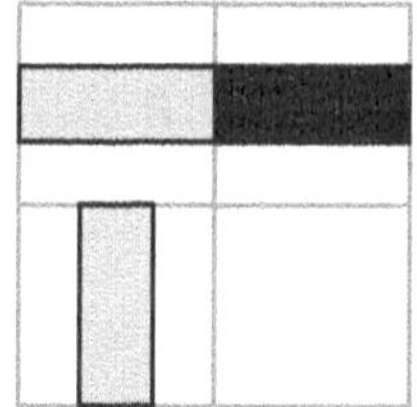

13 Find the mass of  using the scale.

____________ kg

14 Melissa is stacking square tiles. Draw the next diagram in her pattern.

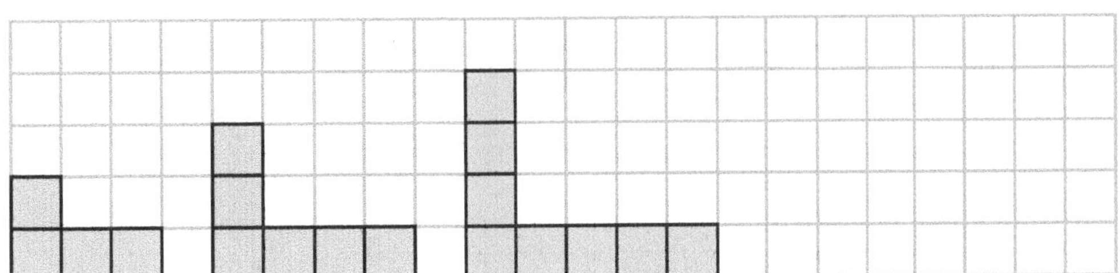

15 The two pieces on the right will fit together to make shape R.

Find which two pieces will fit together to make the shape on the right. Show how they will fit.

a

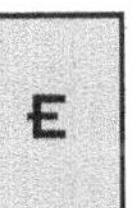

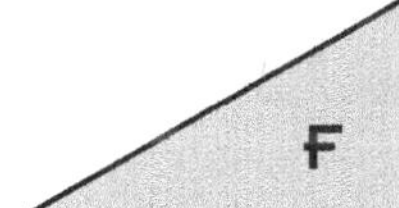

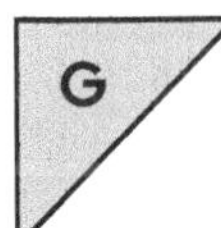

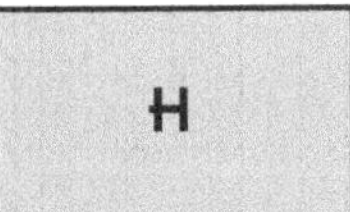

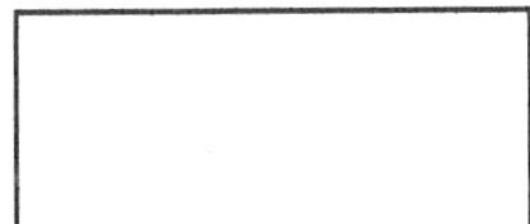

b

U V T W

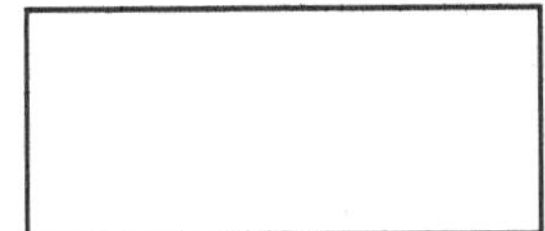

enrich-e-matics 3rd EDITION TEST

Time: 30 minutes

YEAR 1 TEST 3

Name: ____________________

Class: ____________________

1 Write down the next two terms in each pattern.

a 9, 12, 15, 18, ______, ______

b 30, 25, 20, 15, ______, ______

c 1, 6, 11, 16, ______, ______

d A B C A B C A ______ ______

2 Draw the next shape for each pattern below.

a

b

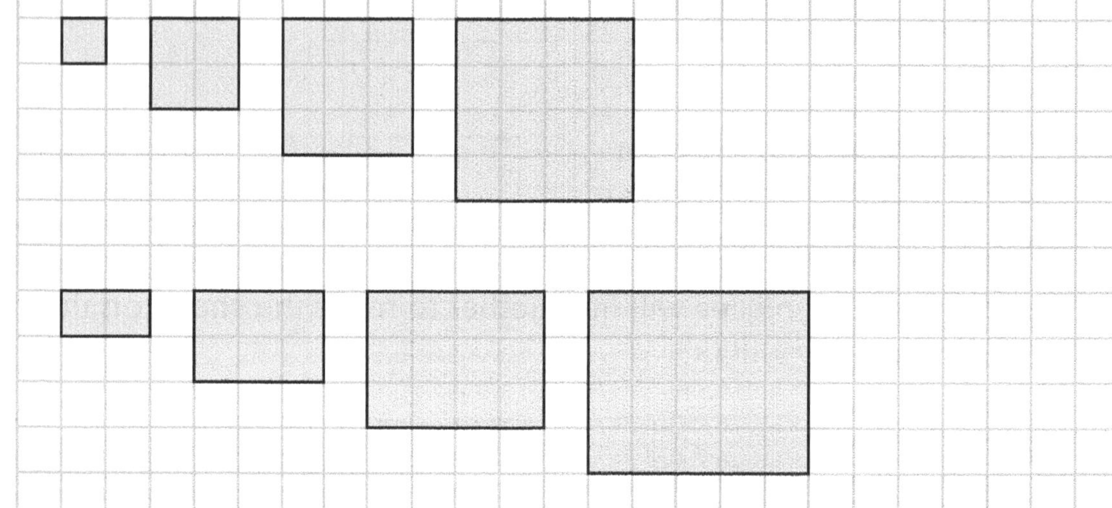

3 Count the number of sides of each shape and write it under the shape.

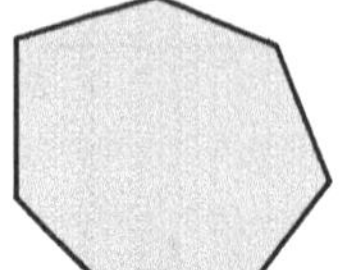 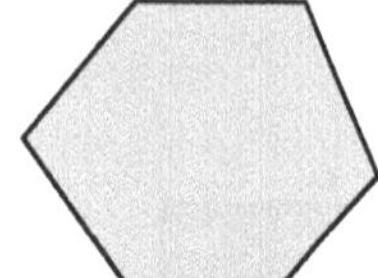 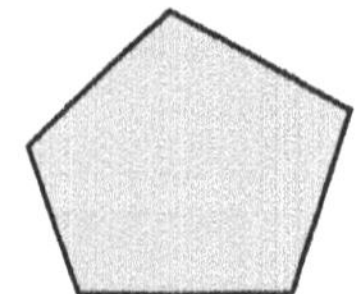 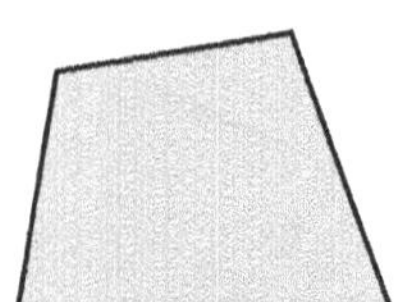

______ ______ ______ ______ ______

Now draw the next shape in this pattern.

4 Find the pattern and complete the missing numbers.

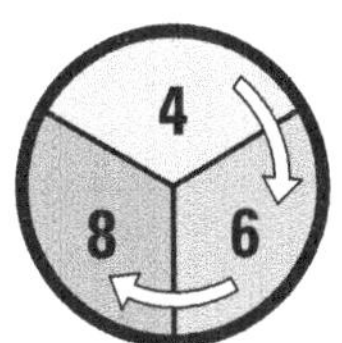

5 Try to discover the pattern in each of the sequences below.
Complete the empty squares.

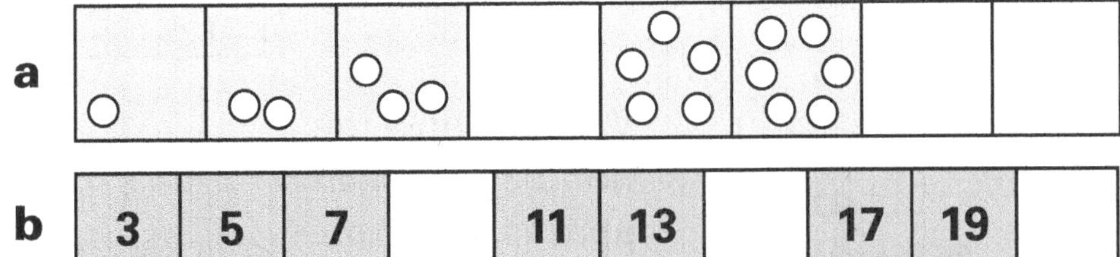

b

3	5	7		11	13		17	19	

6 Figure A changes to B and figure C changes to D in these logic puzzles.
Look at the changes carefully.
Draw the shape that E changes to in each question below.

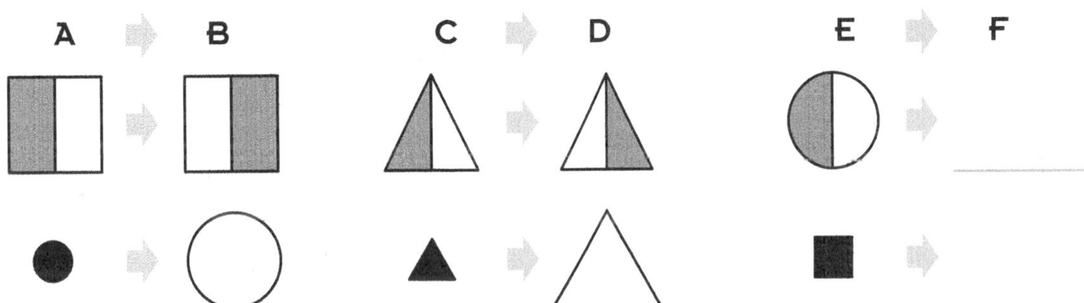

7 I can see three children and one dog. How many legs in all? ________

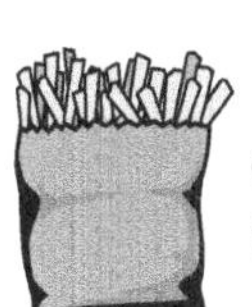
Chips
60c

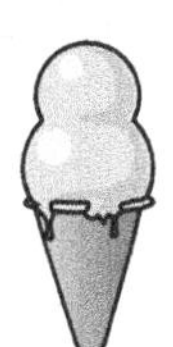
Ice cream
40c

Apple
30c

a At this tuckshop an apple costs 30c.
Work out how much two apples cost. ________

b What did I buy at the tuckshop if I spent exactly $1?

____

9 Half of each of these shapes has been shaded.

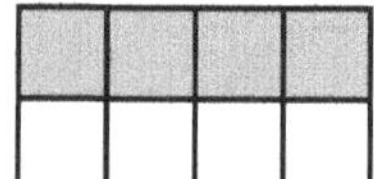
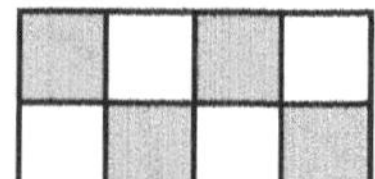
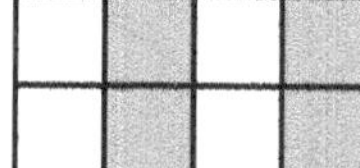

Can you shade half of these shapes?

a 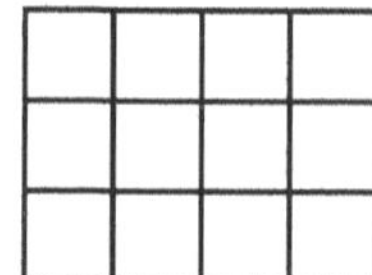

b 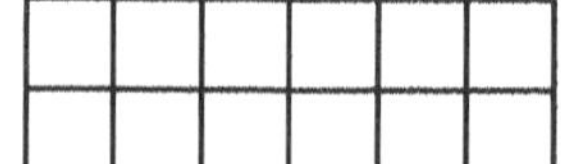

c 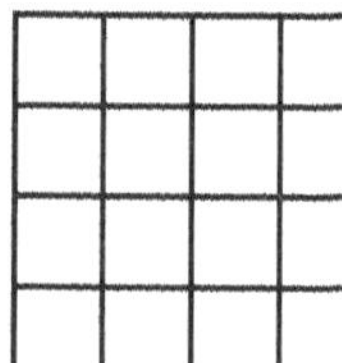

10 We can show the number 25 on an abacus like this.

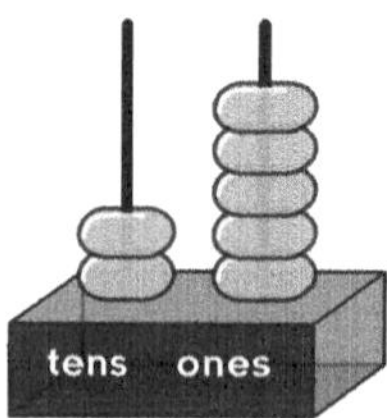

We can show the number 41 on an abacus like this.

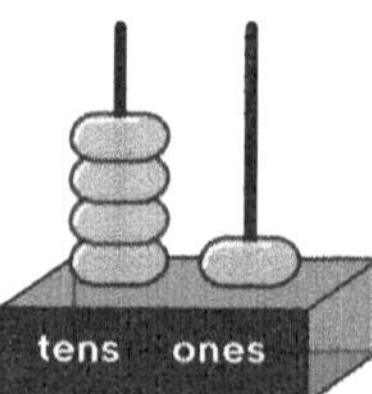

a What number is shown on each abacus below?

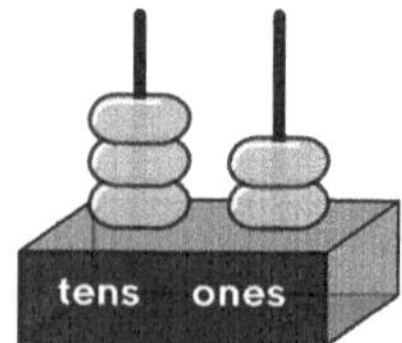

____

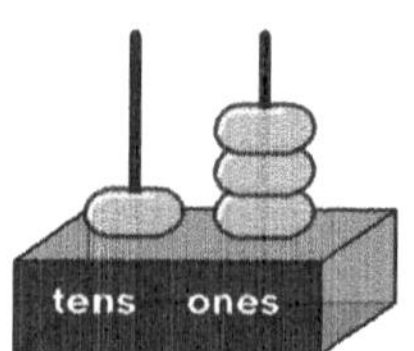

____

b What is the largest number you can make using five beads? Show this on the abacus.

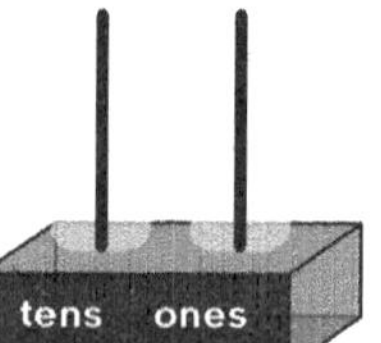

c What is the smallest number you can make using five beads? Show this on the abacus.

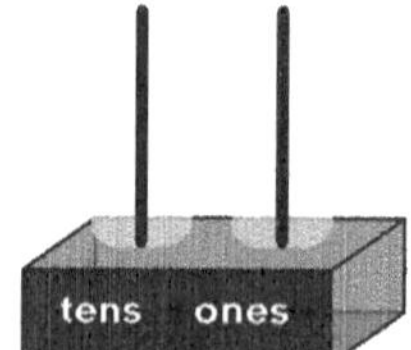

11 The length of this rocket is 3 cm.

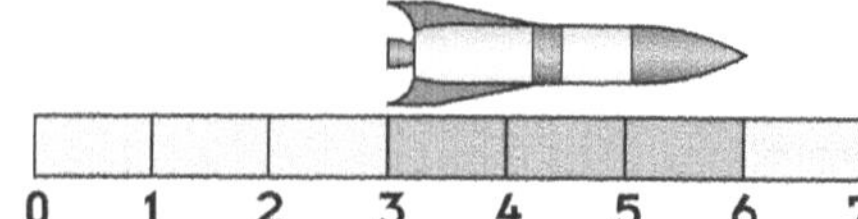

Find the length of this rocket.

_______ cm

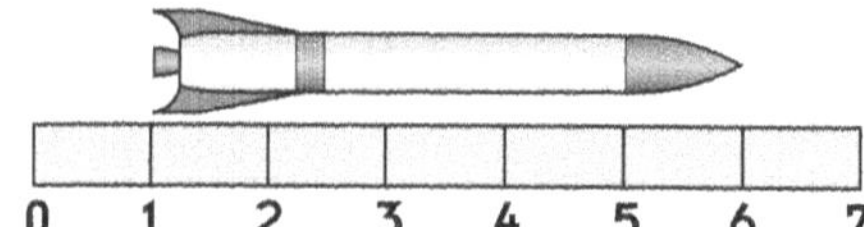

12 Check the pattern in the first square by adding the numbers in the shaded squares across and down. Now complete the other squares.

4	3	7
7	2	9
11	5	16

4	5	9
2	6	
6		

7		8
	3	
9	4	13

13 We need 5 matches to make a 'nine'.

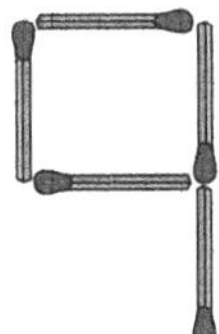

a Draw 3 more 'nines' exactly like the one above.
How many matches are needed to make these 4 'nines'? ________

b How many 'nines' can we make if we use 15 matches? ________
Hint: use your diagram above.

14 Nick bought square tiles like this ☐.

He could cut one in half to make ◺ and ◹.

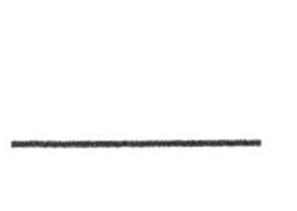 This shape is made from one tile as it is two halves.

How many square tiles did he buy to cover each shape below?

a 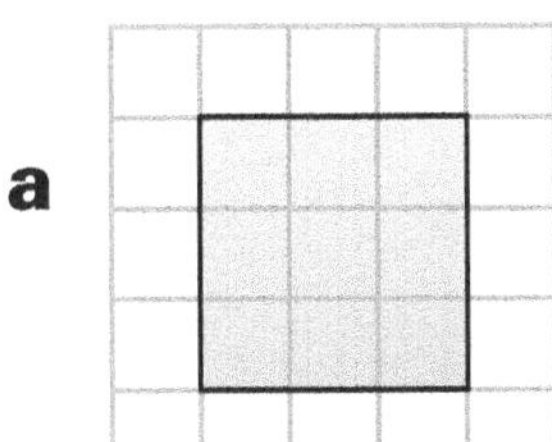 ______ **b** 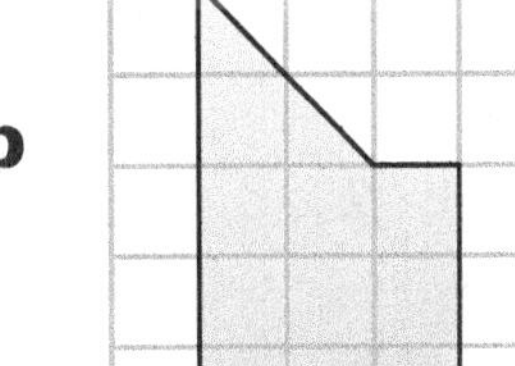 ______

15 Numbered beads are placed on red, green and yellow coloured spikes on a ring in a pattern as shown.

On which coloured spike will 20 be placed?

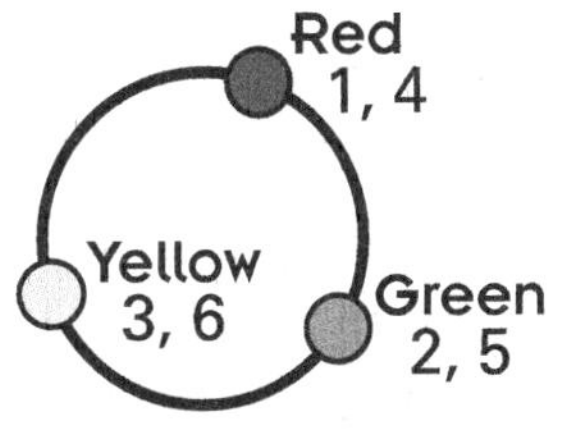

____

enrich-e-matics TEST

3rd EDITION

Time: 30 minutes

YEAR 1
TEST 4

Name: ____________________

Class: ____________________

1 Write the missing number in each box.

a When I add 5 to a number the answer is 11.
What number am I thinking of?

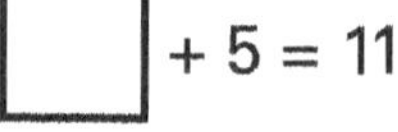

b When I take 4 from a number the answer is 9.
What number am I thinking of?

☐ – 4 = 9

2 Draw the next figure in this pattern of shapes.

3 Write down the missing number in each number pattern below.

a 11, 15, 19, 23, ____, 31

b 22, 19, 16, ____, 10, 7

4 There is one mistake in each of these number patterns. Find the mistake, draw a neat line across it and write the correct answer above it.

a 19, 18, 17, 16, 16, 14, 13

b 6, 10, 15, 20, 25, 30, 35

c 80, 70, 65, 50, 40, 30, 20

5 Anna had eight balls as shown.

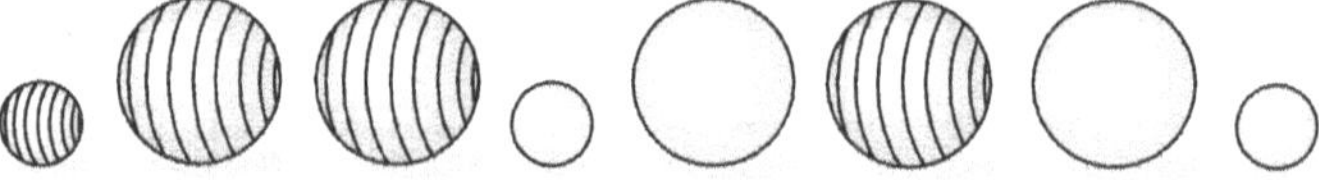

Colour the bubble next to the correct answer.

◯ **A** Half the balls are big. ◯ **B** Half the balls have stripes.

6 Place the numbers 1, 2, 3, 4 and 5 in the spaces, so that each vertical and horizontal set of three numbers is a true number sentence.

7	−	☐	=	6
−				+
☐				☐
=				=
☐	+	☐	=	8

7 An ice cream costs 60c and a packet of chips costs $1.

Ice cream
60c

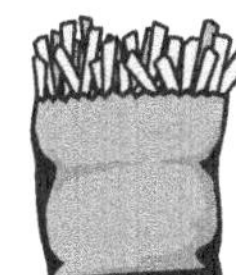

Chips
$1.00

a How much do 2 ice creams cost? __________

b How much do 3 packets of chips cost? __________

c How much do 2 ice creams and 3 packets of chips cost? __________

8 Half of each of these shapes has been shaded.

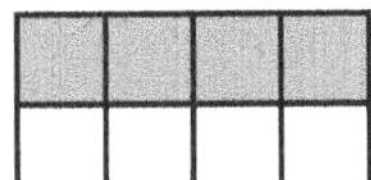 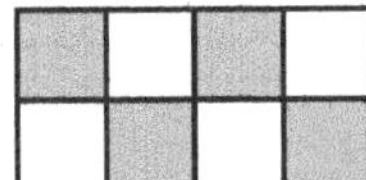 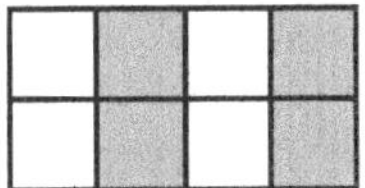

Colour the bubble(s) next to the square(s) in which half the figure has been shaded.

A 

B

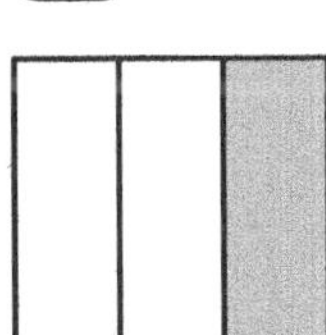

C

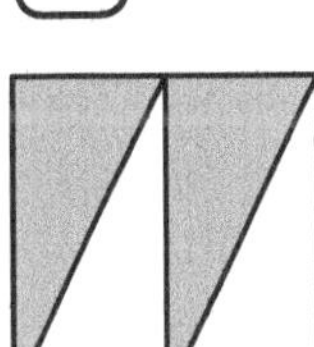

D

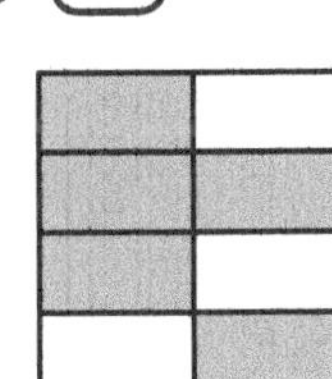

9 I can see three children and two dogs.

How many legs in all? __________

10 The scale on the right shows ________ kg.

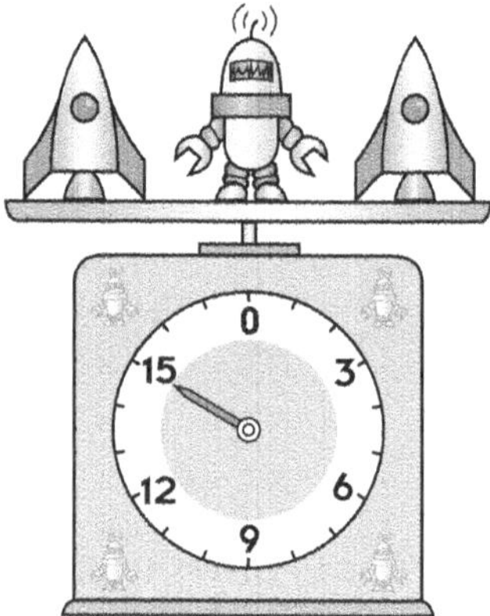

The mass of 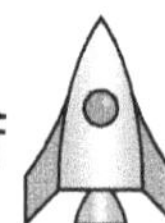 is 4 kg.

Find the mass of 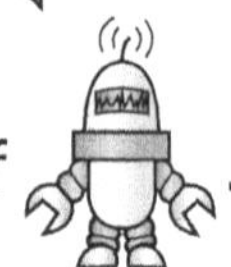. ________ kg

11 In this pictograph [book] stands for 2 books read.

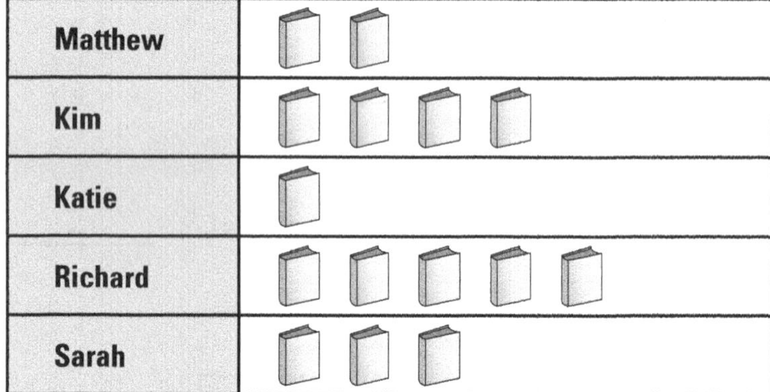

a How many books did Katie read? ________

b How many books did Matthew read? ________

c Who read the most books? ________

d How many more books did Richard read than Sarah? ________

12 The sum of the numbers in the two big circles next to one another is the number in the small circle between them.

Work out which numbers belong in the empty circles.

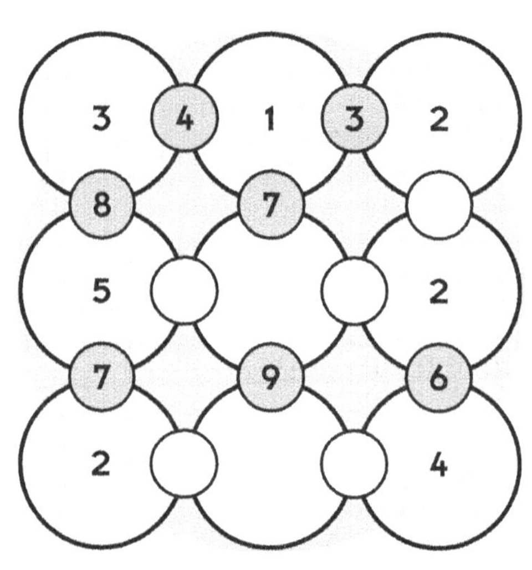

13 If numbers are written in columns in the pattern shown, name the letter of the column in which the number 13 appears.

A	B	C
1	2	3
4	5	6
7	8	

____

14 Jason walks from his home to school each day. He can take *six different* shortest routes along the lines drawn.

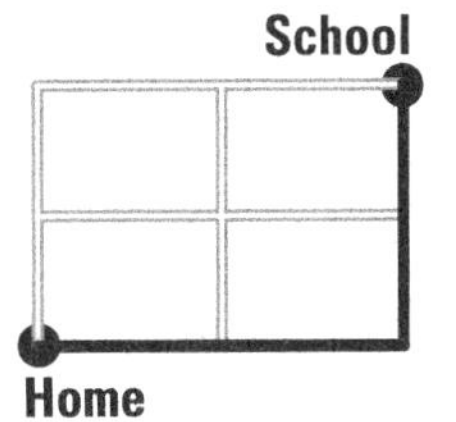

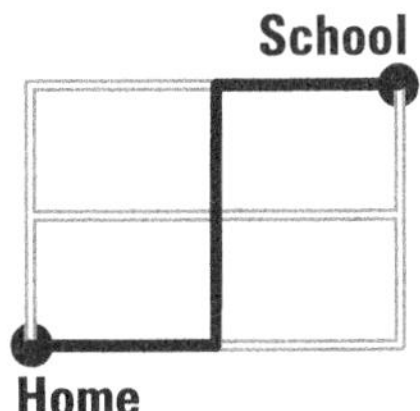

Two of them are drawn for you.

Can you draw the other four in the boxes below?

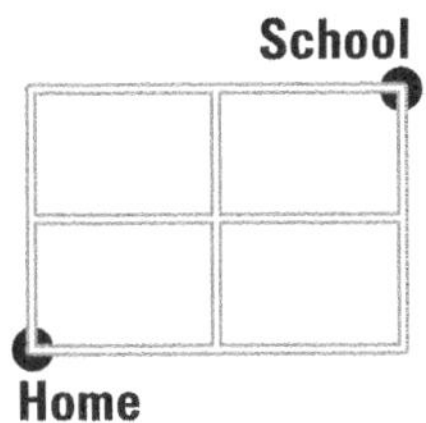

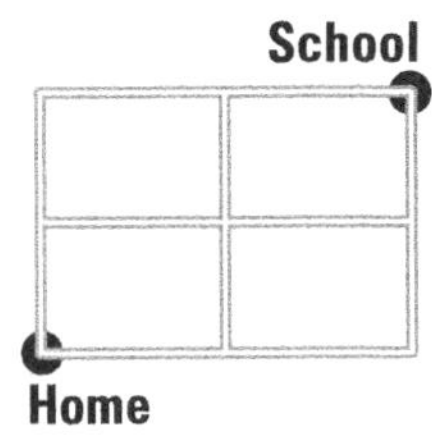

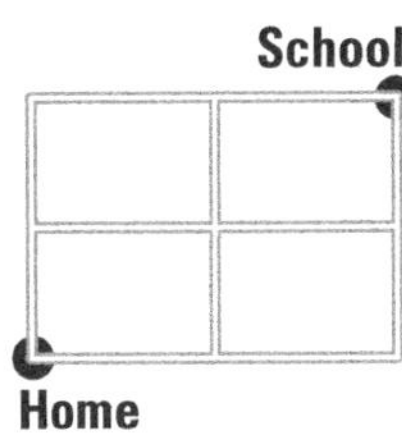

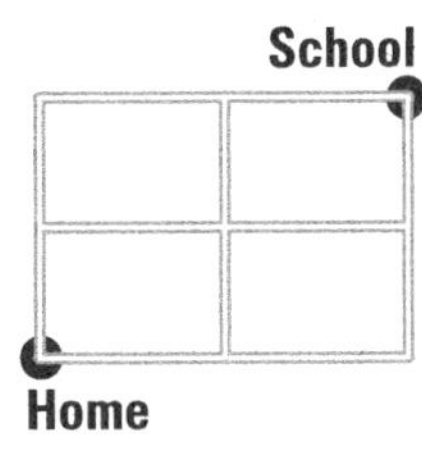

15 If each shape has the given value, work out the value of the third balance.

If □ 5 and

, then

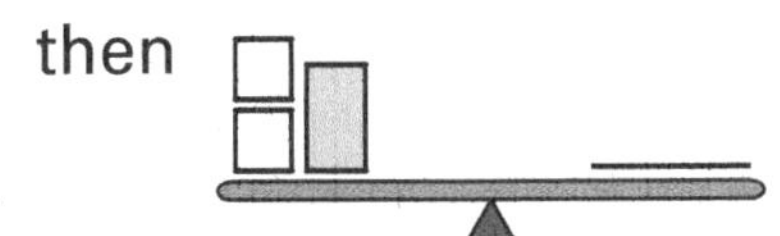

_____.

16 If it is 10:15 a.m. now:

a what will the time be in 45 minutes? ____________

b what was the time 45 minutes ago? ____________

enrich-e-matics 3rd EDITION TEST

Time: 30 minutes

YEAR 2 TEST 1

Name: ____________________

Class: ____________________

1 Write down the next two numbers each pattern.

a 3, 7, 11, 15, _______, _______

b 22, 19, 16, 13, _______, _______

c 1, 3, 6, 10, 15, _______, _______

d 1, 3, 7, 9, 13, 15, _______, _______

2 Square tiles are used to make each shape and the shapes form a pattern.

Draw the next shape in the pattern.

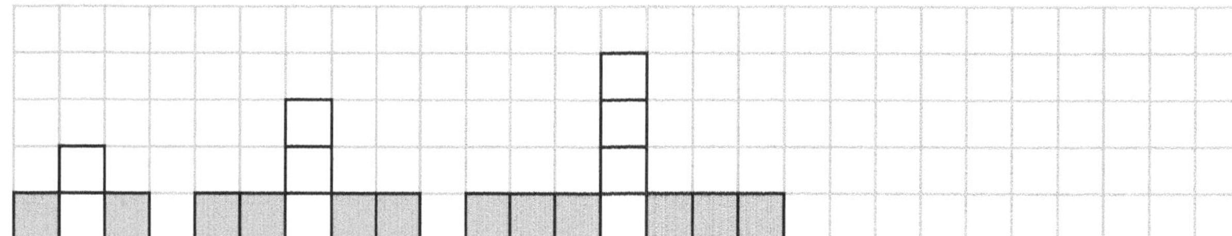

3 Draw the two missing shapes in these patterns.

4 Write down how many small squares are in this shape. _________

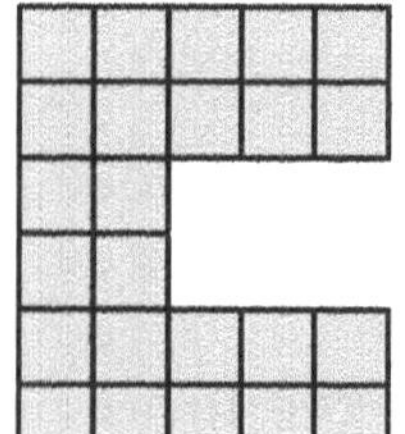

5 The sum of three numbers is 14.
Two of them are 3 and 9.
What is the third number?

3 + 9 + △ = 14

6 Fill in the spaces to complete the patterns below.

a 

b 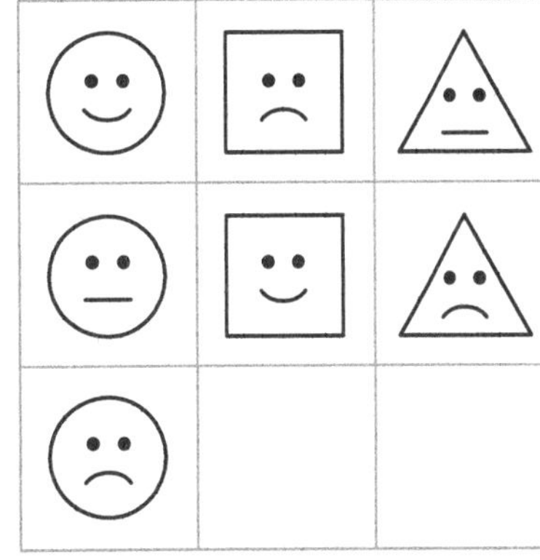

7 **a** Joseph is at school from 9 a.m. till 3 p.m.

How many hours is that? __________

b How many minutes are there
from 8:40 a.m. to 9:00 a.m. the same morning? __________

c Today is Tuesday. What day will it be in 15 days' time?

____

8 If ◆ = 5 raffle tickets, work out the number of raffle tickets sold by each class.

Class A ◆◆◆ __________

Class B ◆◆◆◆◆ __________

9 If I can see 3 cars, 2 bicycles and 1 tricycle, how many wheels in all can I see? Fill in the gaps to find your answer.

3 cars have __________ wheels

2 bicycles have __________ wheels

1 tricycle has __________ wheels

Total __________ wheels

10 **a** From a medium pizza Joseph can cut 3 equal slices for 3 children.
How many medium pizzas does
Joseph need if 12 children have to be served? ________

b If Joseph had 6 medium pizzas,
how many children could he serve? ________

11 Carefully draw the 5th and 6th rows in this pattern of circles.

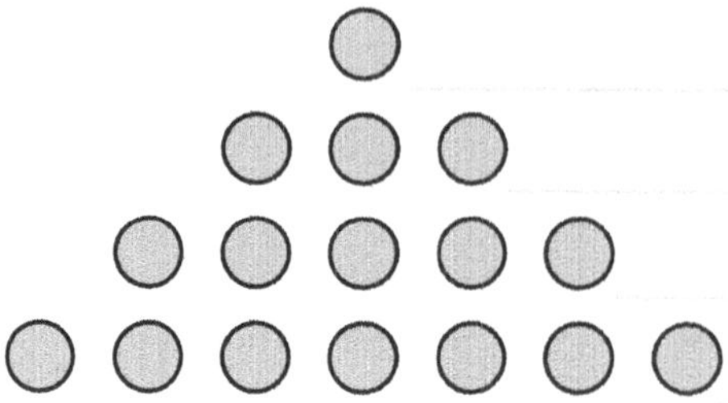

1st row
2nd row
3rd row
4th row
5th row
6th row

Write down how many circles there are in the:

1st row ______ 2nd row ______ 3rd row ______
4th row ______ 5th row ______ 6th row ______

Have you noticed a pattern? Check your work carefully.
Work out how many circles there will be in the 8th row. ________

12

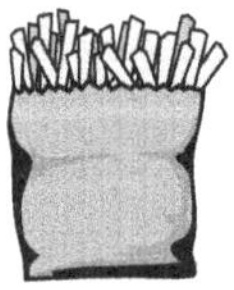

Chips
80c

Ice cream
60c

Apple
30c

a At the tuckshop a packet of chips costs 80c.
Work out how much two packets of chips cost. ________

b Work out how much an ice cream and an apple cost. ________

c How much change would I get from $1
if I bought an ice cream and an apple? ________

d Two ice creams cost exactly the same as ______ apples.

13 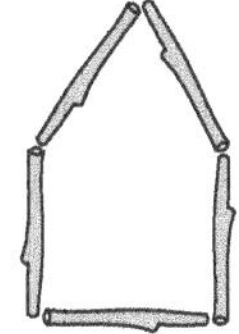 Five sticks are used to make one house.

 Nine sticks are used to make two houses.

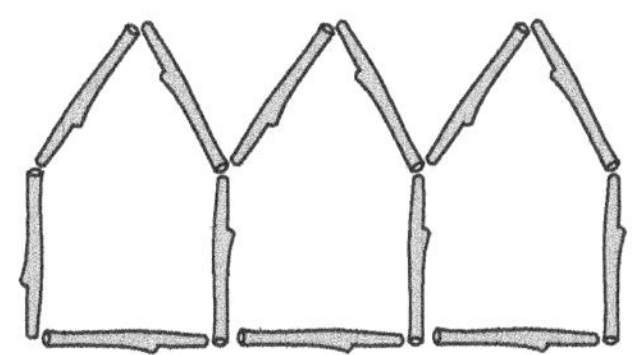 Thirteen sticks are used to make three houses.

a Draw the picture to show four houses.

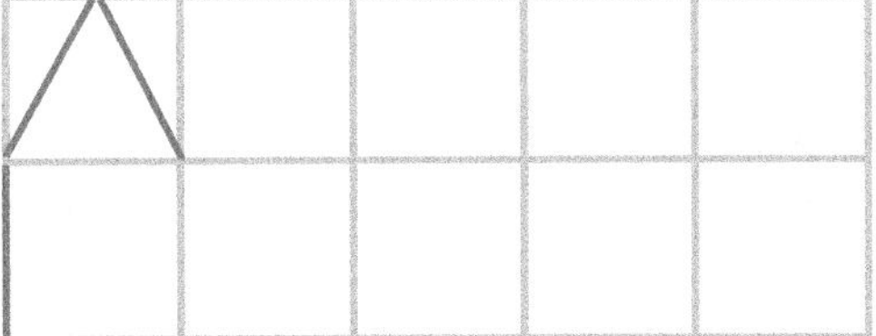

b How many sticks are used to make four houses? ________

c How many sticks are used to make five houses? ________

14 We can pour 4 glasses of juice from 1 litre of lemon drink.

a How many glasses of juice can we pour from 3 litres? ________

b How many litres of lemon drink did I have, if 20 glasses of juice were poured? ________

This table may help you work out the answer.

	Number of litres	Number of glasses
	1	4
a	3	
b		20

15 From the two balances given, work out how many □s you need on the right side of the third balance.

If 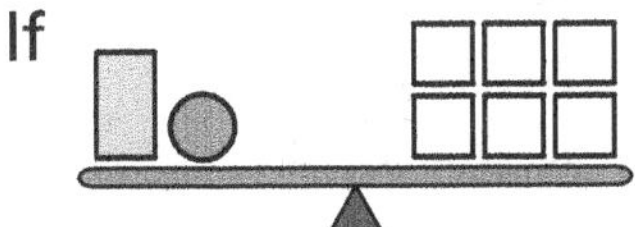 and 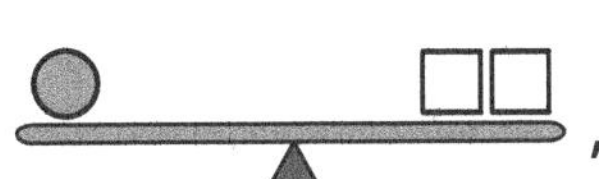, then 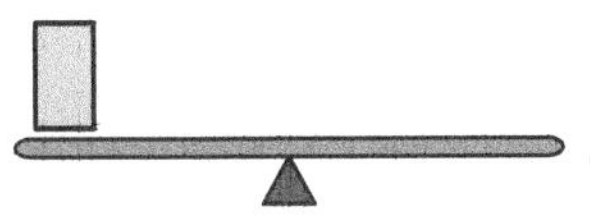.

enrich-e-matics 3rd EDITION TEST

Time: 30 minutes

YEAR 2
TEST 2

Name: ______________________

Class: ______________________

1 Place the numbers 2, 3 and 5 in the circles so that the sum of the numbers in each line is 10.

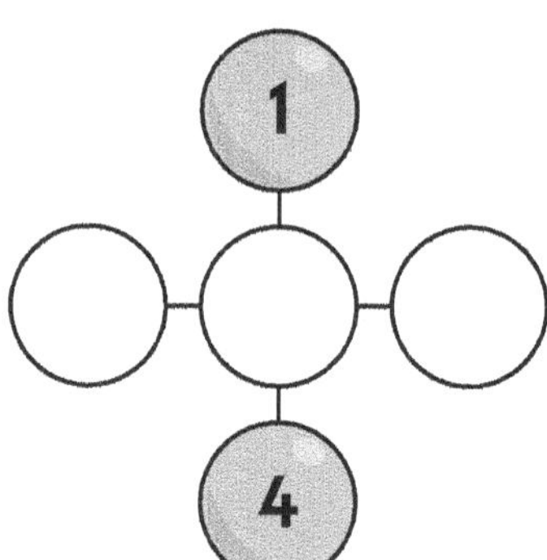

2 The two shapes represent different numbers. What is the value of each shape?

○ + ○ + ○ = 12   ○ – □ = 1

3 Complete the missing numbers in the patterns below.

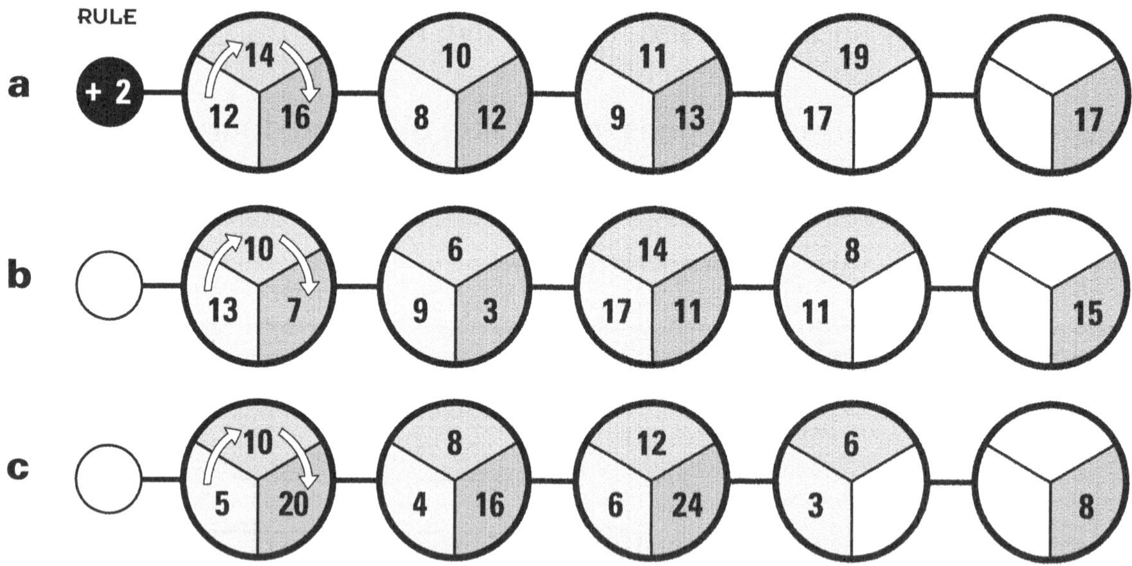

4 How many times does the letter **X** appear in this diagram?

____

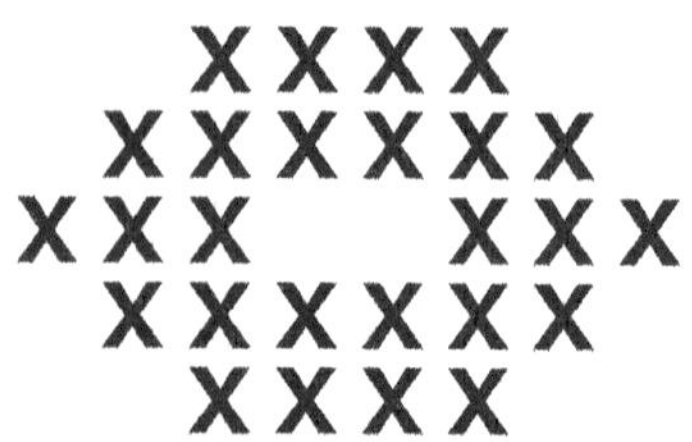

5 Nick is stacking square tiles. Draw the next diagram in his pattern.

6 Place the numbers 3, 4 and 5 in the circles so that each side of the triangle adds up to 9.

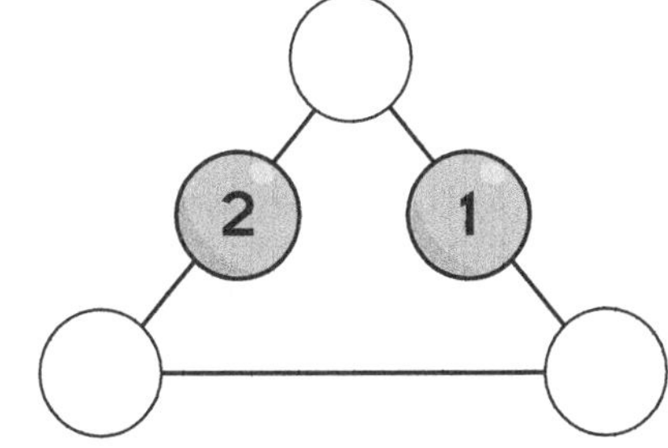

7 We need six matches to make a 'six'.

a Draw a diagram to show four 'sixes'.

b How many matches are needed to make four 'sixes'? ________

c How many 'sixes' could be made with 18 matches? ________

8 Draw the answers on the balances.

If  :

a then how many ○s are needed to balance two □s?

b how many □s are needed to balance 12 ○s?

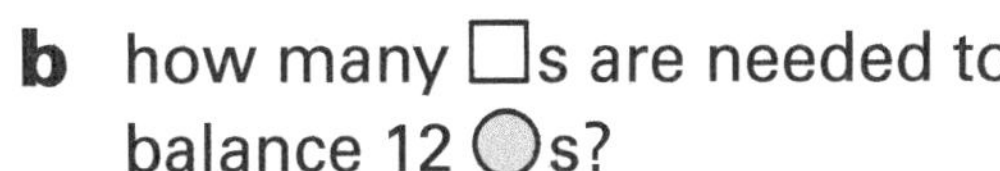

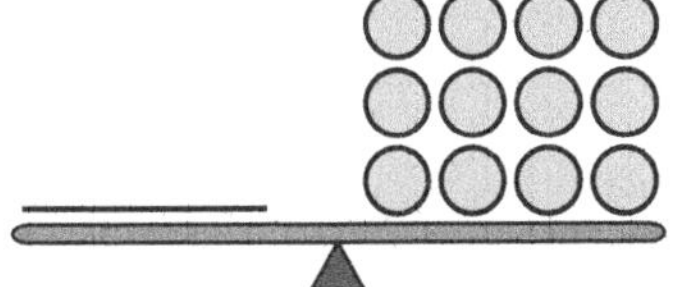

9 In this pictograph 🍎 stands for 3 pieces of fruit a person ate in a day.

David	🍎 🍎
Kurt	🍎 🍎
Chris	🍎 🍎 🍎 🍎
Harry	🍎
Michelle	🍎 🍎 🍎

a How many pieces of fruit did Kurt eat? __________

b Who ate the most fruit? __________

c Who ate nine pieces of fruit? __________

d How many more pieces did David eat than Harry? __________

10 If I have a large number of 5c, 10c and 20c coins, in how many different ways can I make up exactly 35 cents? __________

What coins would you use? 10c, 5c, 5c, 5c, 5c, 5c is one possibility.

____

____

____

11 Two numbers have a sum of 9 and a difference of 3.

Find the two numbers.

$\square + \triangle = 9$ $\square - \triangle = 3$

Check your answer carefully.

12 Place numbers in the squares so that the sum of the numbers in every row (→), column (↓) and diagonal (↘ or ↙) is 21.

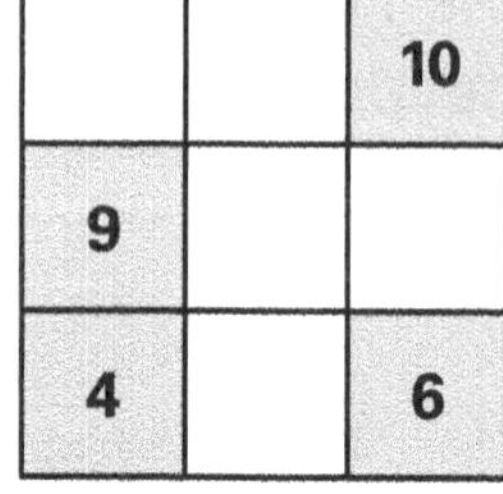

		10
9		
4		6

13 **a** Mrs Young is at work from 8 a.m. until 6 p.m. How many hours is this? ____________

b My bus ride from home to school is from 8:20 a.m. until 8:45 a.m. How long is the trip? ____________

14 If you add the number in the triangle to twice the number in the square, you will get the number in the circle.

The first two examples are done for you. Find the other missing numbers.

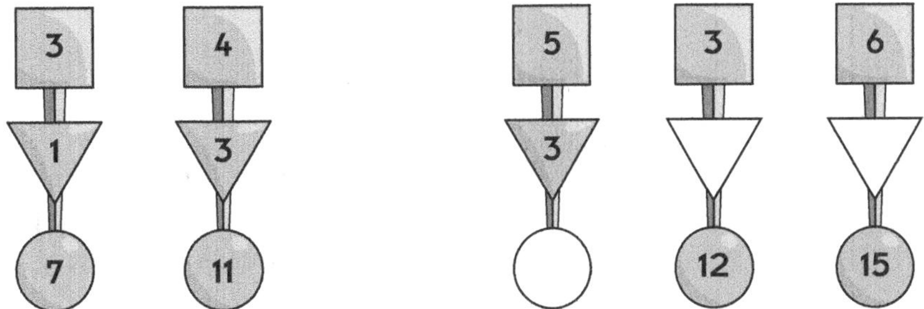

15 Using these scales:

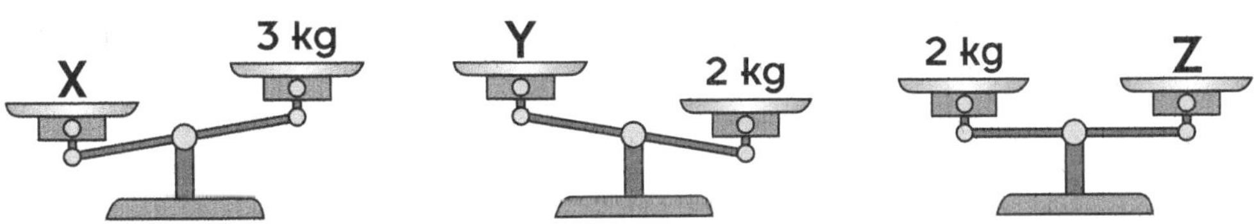

a work out which item (X, Y or Z) has a mass of 2 kg ________

b arrange the items (X, Y and Z) from the lightest to the heaviest.

_______ _______ _______

16 Three tiles are used to make two steps. 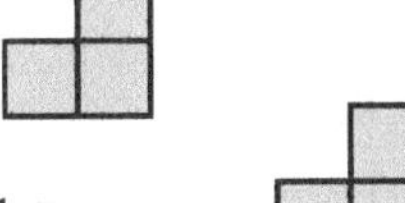

Six tiles were used to make these three steps.

a Add tiles to this shape to make four steps. How many tiles are used altogether to make these four steps? ________

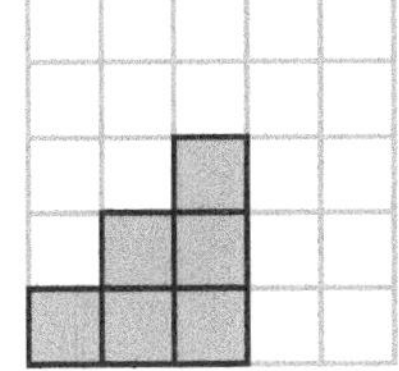

b Continue the pattern and add tiles to the diagram above to use 15 tiles. How many steps are there now? ________

enrich-e-matics 3rd EDITION TEST

Time: 30 minutes

YEAR 2 TEST 3

Name: ____________________

Class: ____________________

1 Write down the next two numbers in each pattern.

a 29, 25, 21, 17, ______, ______

b 66, 55, 44, 33, ______, ______

c 45, 55, 50, 60, 55, 65, 60, ______, ______

2 Complete the missing numbers in the patterns below.

a

9 4 14 19

b

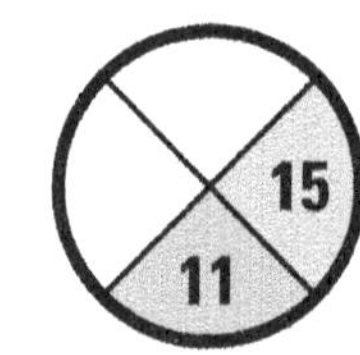

15 11

3 Two balances are given. Draw how many squares you need on the right side of the third balance.

If

and

, then ______ .

4 Numbers in the shaded square are added across and down. Complete these two squares.

5	3	8
8	7	15
13	10	23

6	7	13
5	8	
11		26

7		11
	5	12
14	9	23

5 When Lee looks straight down from directly above a cube 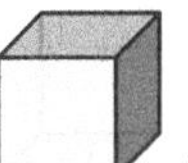,

she sees □.

When she looks down on this figure 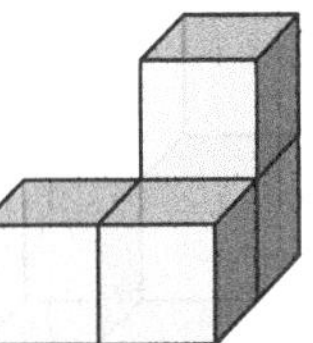, she sees 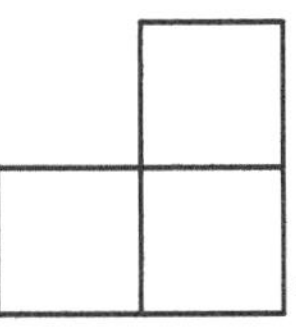.

Draw what she can see when she looks down on each of these shapes.

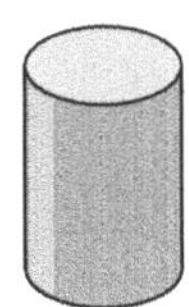 She sees

____

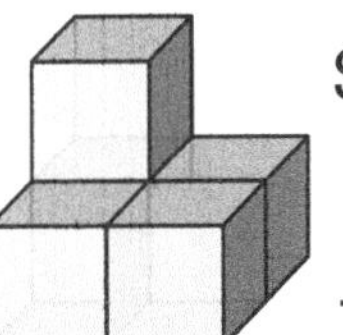 She sees

____

6 In this diagram the number in the square is the sum of the numbers in the circles next to it.

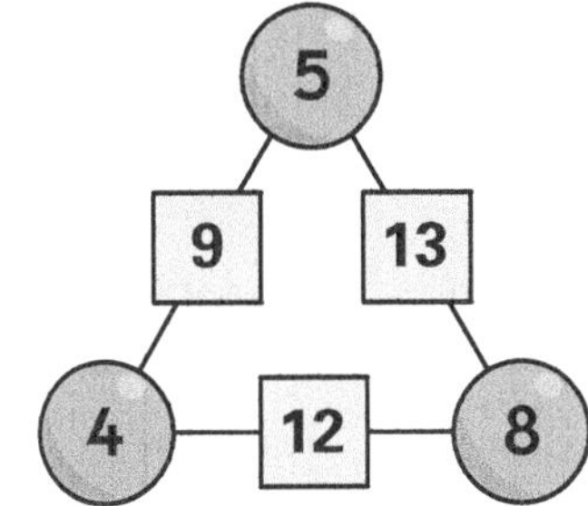

a Write the missing numbers in the squares.

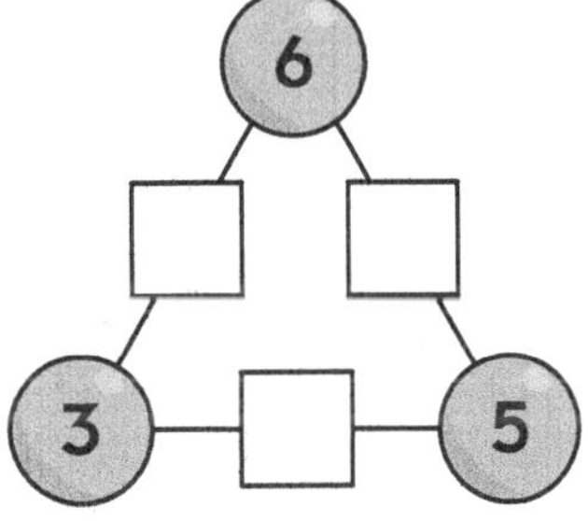

b Find the numbers that go in the circles if the number in the square is the sum of the numbers in the circles.

i

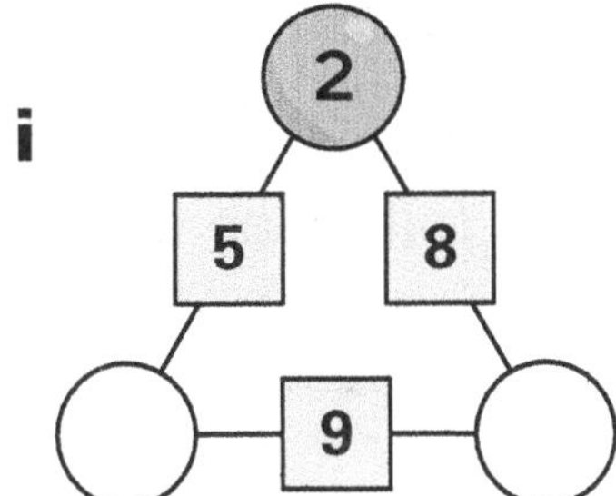

ii

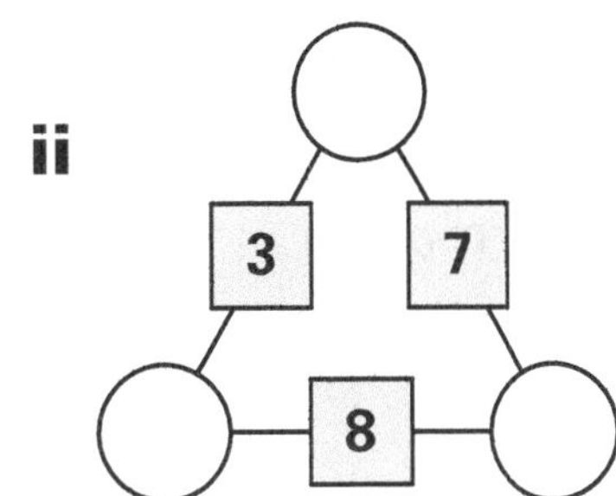

7 Find the temperature shown on the thermometer.

Look at the scale carefully first.

_________ °C

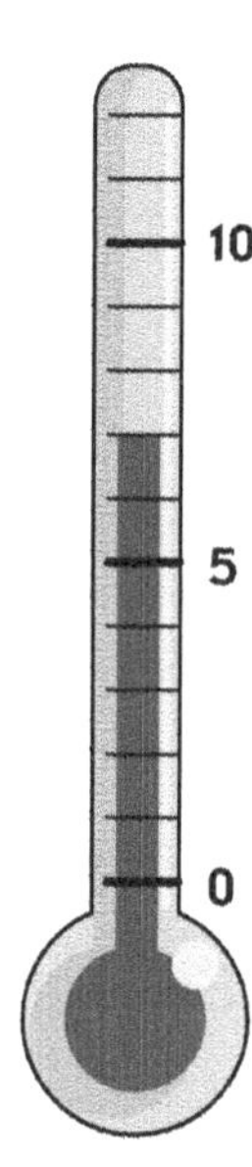

8 In this table:

sunny	cloudy	windy	rainy

	week 1	week 2	week 3	week 4
Monday				
Tuesday				
Wednesday				
Thursday				
Friday				
Saturday				
Sunday				

In **week 1** there were 3 sunny days, 2 windy days and 2 cloudy days. Use the information in this table to answer the following questions.

a Which week had the most sunny days? ______________________

b Which week had the most rainy days? ______________________

c During which week did it not rain? ______________________

9

Chips
80c

Ice cream
40c

Apple
60c

a At this tuckshop how much do three ice creams cost? __________

b What did I buy at the tuckshop if I spent exactly $1.20?

____________________________ or ____________________________

c What did I buy at the tuckshop if I spent exactly $2?

____________________________ or ____________________________

10 Find how many sweets Amy brought to school.
It was less than 14. It was more than 8.

a The possibilities are ______________________________

b She could divide them into two equal shares.

Now the possibilities are ______________________________

c She could divide them into three equal shares, as well.

Now the possibility is ______________________________

11 Find which two pieces will fit together to make the square on the right, and draw the diagram.

a

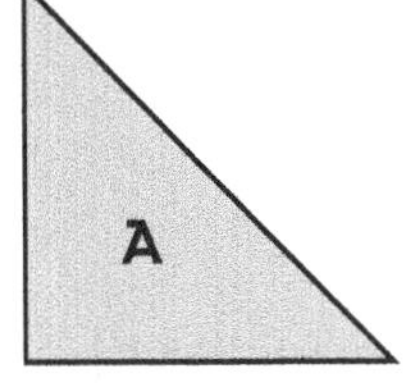

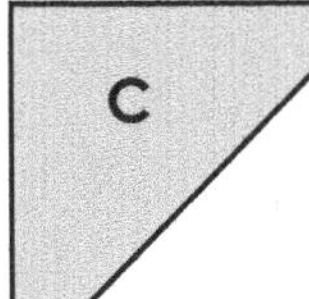

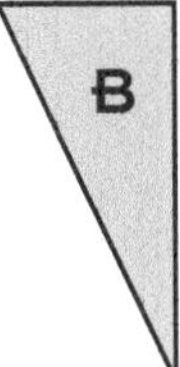

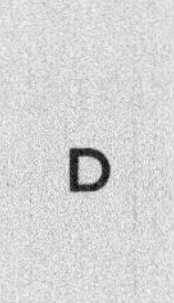

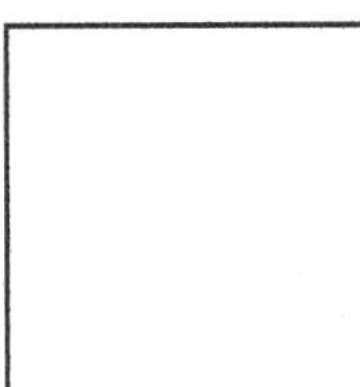

b

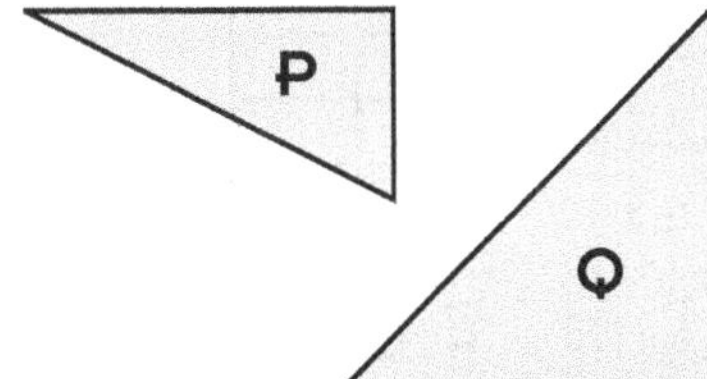

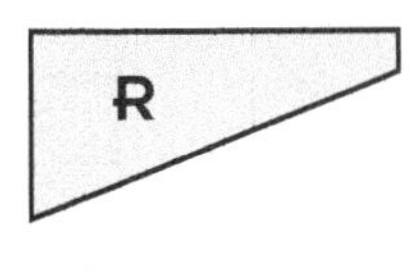

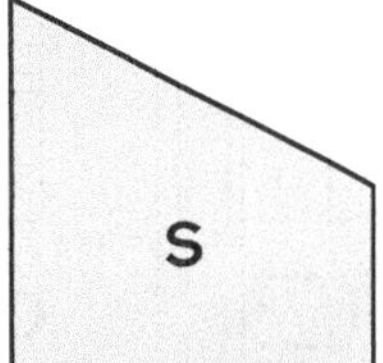

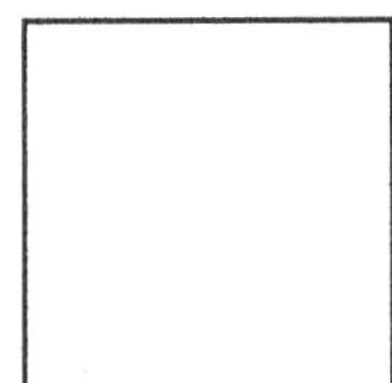

enrich-e-matics TEST

3rd EDITION

Time: 30 minutes

YEAR 2
TEST 4

Name: ____________________

Class: ____________________

1 Write down the next two numbers in each pattern.

a 2, 8, 14, 20, ______, ______

b 32, 27, 22, 17, ______, ______

c 1, 2, 4, 7, 11, 16, ______, ______

2 Continue this pattern for one more figure.

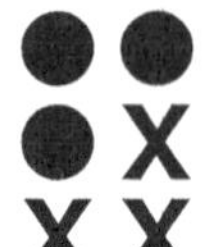

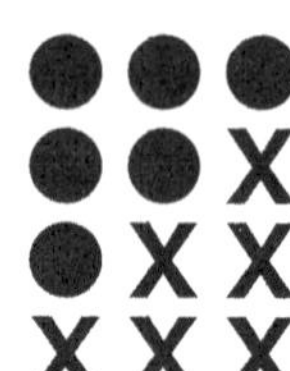

____

3 Place the digits 2, 3, 4, 5, 6 and 7 in the empty circles so that the sum of each line is 10.

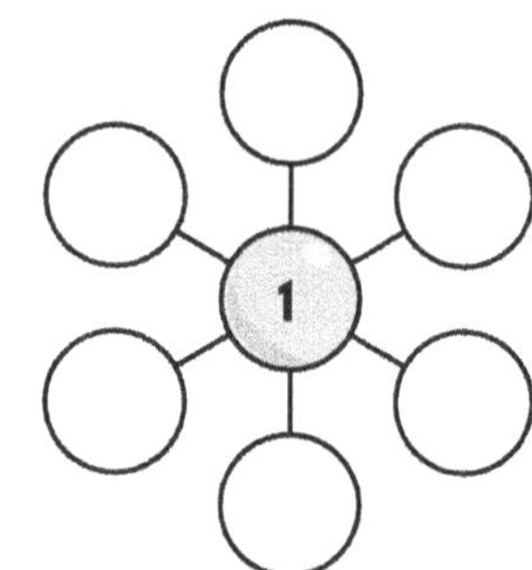

4 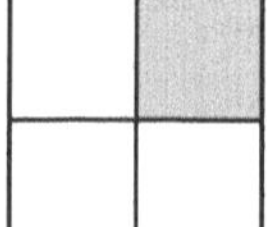

In this square one quarter or one part in four has been shaded.

Colour the bubbles next to the squares that have one quarter shaded.

A ◯

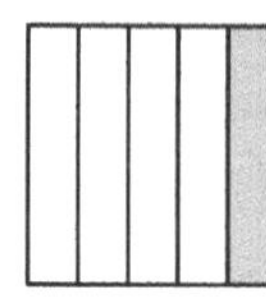
B ◯

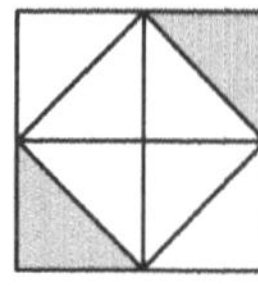
C ◯

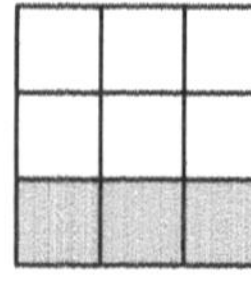
D ◯

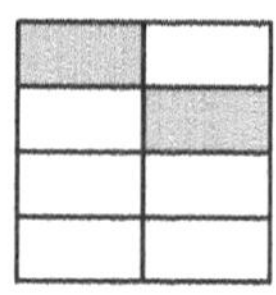
E ◯

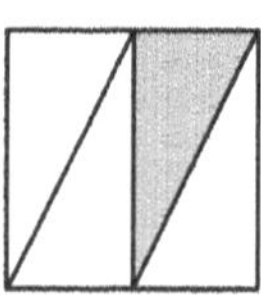
F ◯

5 Andrew had eight balls as shown.

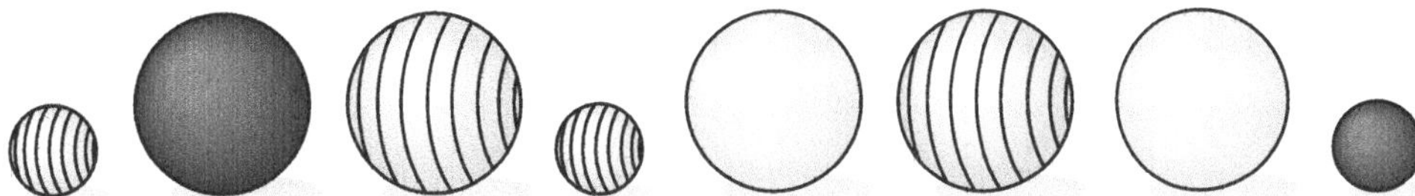

Colour in the bubble next to the true statements.

A ⬭ Half the balls are big.

B ⬭ Half the balls have stripes.

C ⬭ One quarter of the balls are black.

6 Complete the following.

a

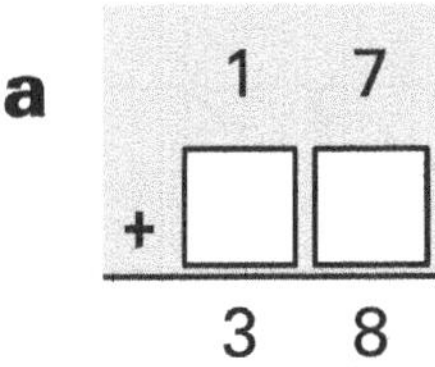

b

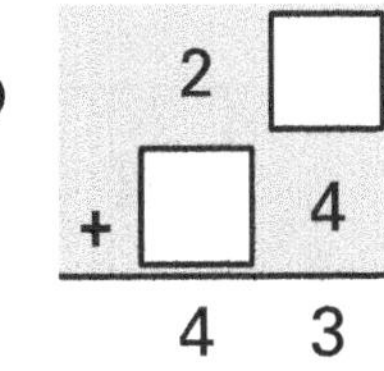

c

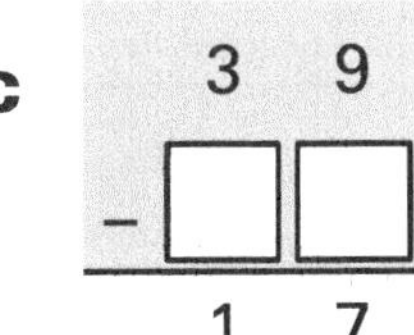

d

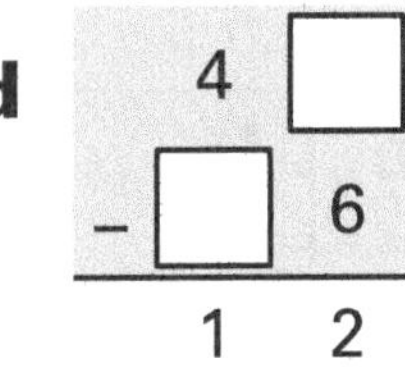

7 Numbers are written in columns in the pattern shown.

a Continue the pattern and complete this table.

b Name the letter of the column in which the number 25 would appear.

____

A	B	C	D	E
1		2		3
	4		5	
6		7		8
	9		10	

8 Two balances are given.
Work out how many □s you need on the right side of the third balance.

If

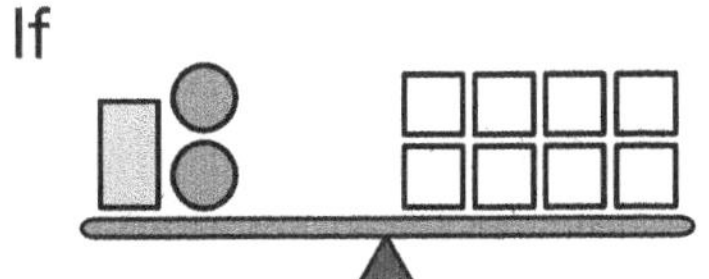

and

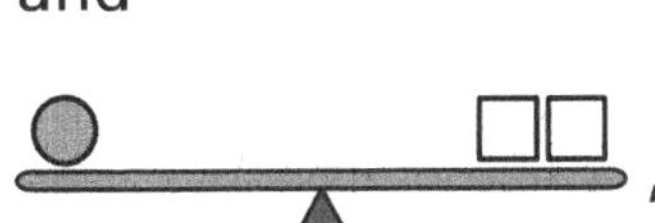,

then

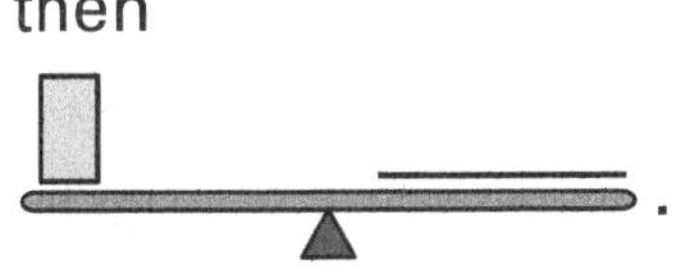.

9 Each of these number patterns has a mistake in it.

Find the mistake, draw a neat line across it and write the correct answer above it.

a 1, 4, 7, 11, 13, 16, 19

b 35, 31, 27, 23, 18, 15, 11

c 4, 8, 14, 16, 20, 24, 28

10 Try to discover the pattern in each of the sequences below then complete the empty squares.

a

b

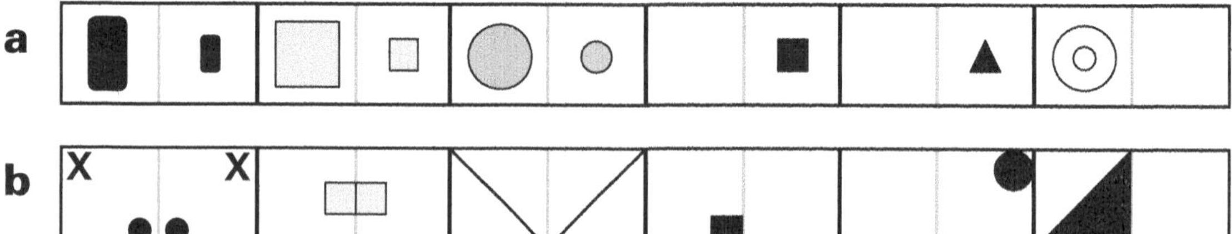

11 If the mass of [rocket] is 6 kg,

find the mass of [robot] using this scale.

_________ kg

12 Find the pattern and complete the missing numbers.

2 ⇩ 8	4 ⇩ 10	8 ⇩ 14	15 ⇩ 21	1 ⇩	⇩ 9

13 Faye was on a train for 4 hours. Her train trip finished at 3:30 p.m.

What time did her trip start? _____________

14 When the numbers in the two big circles next to one another are *multiplied*, the answer is the number in the small circle drawn between them.

Check the rule carefully, and fill in the empty circles.

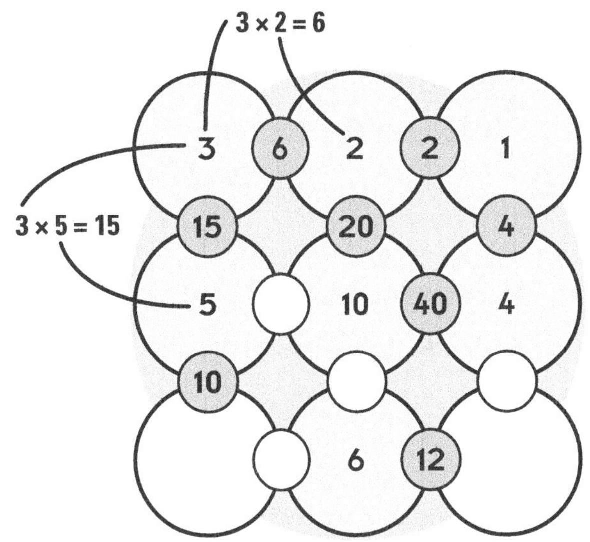

15

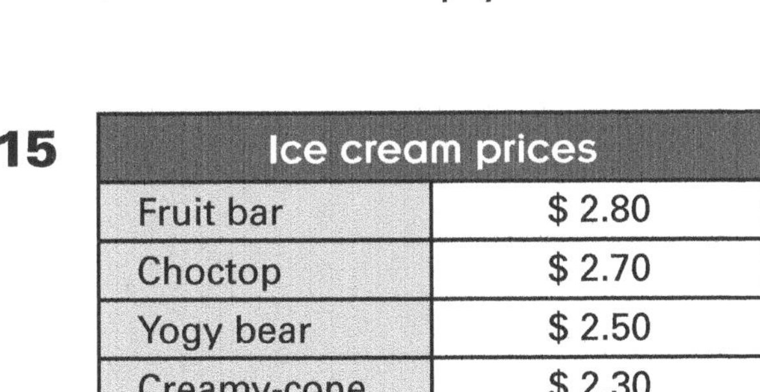

Ice cream prices	
Fruit bar	$ 2.80
Choctop	$ 2.70
Yogy bear	$ 2.50
Creamy-cone	$ 2.30

Alex spends exactly $5 on two ice creams.

Which two ice creams can he buy? ______________ ______________

16 This set of 22 figures can be grouped in many ways.

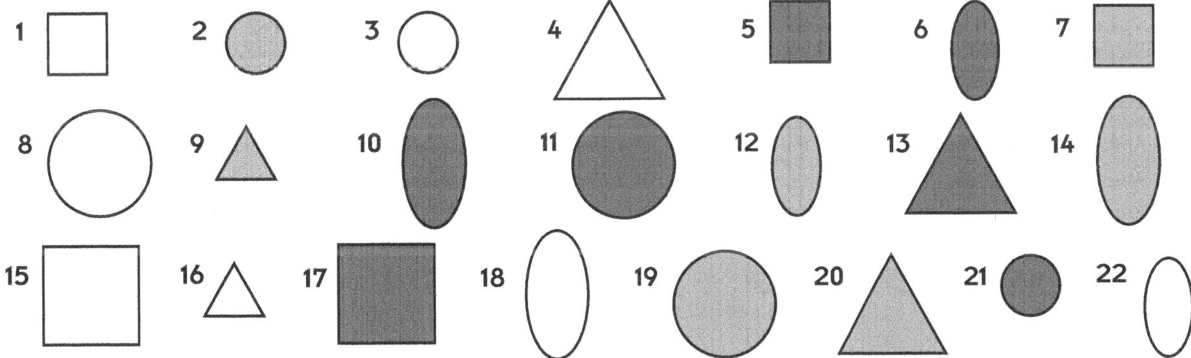

One way is to group all the white small shapes: 1, 3, 16, 22.

Another way is to group all oval shapes even though some are shaded, some are large and some are small: 6, 10, 12, 14, 18 and 22.

Complete other shapes that belong to each of these groups.

a 2, 3, 11, 21, _______, _______

b 2, 9, 14, 19, _______, _______, _______

enrich-e-matics TEST

3rd EDITION

Time: 45 minutes

YEAR 3
TEST 1

Name: ______________________

Class: ______________________

1 In each line the shapes follow a pattern. Draw the next two shapes

a ○★□★○★□★○ ___ ___

b △△□△●△△□△●△ ___ ___

2 Write down the next three numbers in each line that will continue the pattern.

a 3, 10, 17, 24, ______________

b 3, 6, 10, 15, ______________

c 6, 9, 8, 11, 10, 13, 12, ______________

3 Which of the four numbered figures is needed to complete the pattern?
Write the number in the square.

1 2 3 4

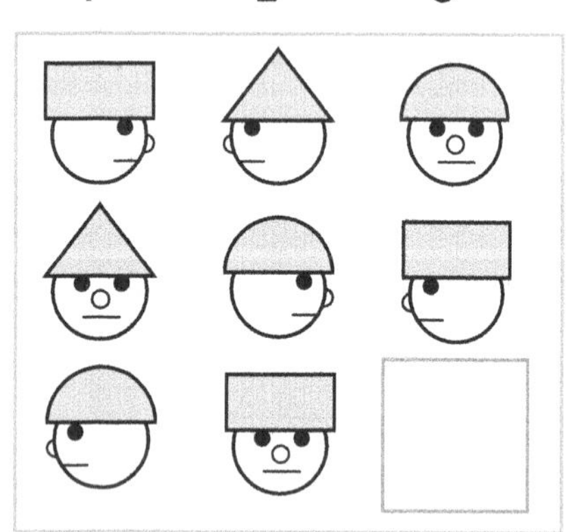

4 How much longer is one rocket than the other? Measurements are in cm.

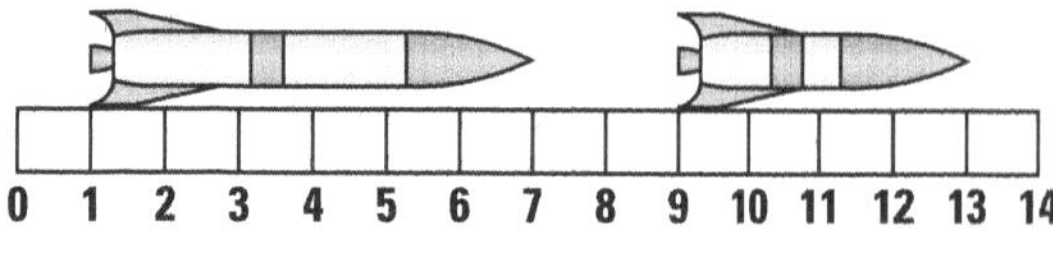

____

5 Two balances are given. You have to work out what mass you need on the right side of the third balance.

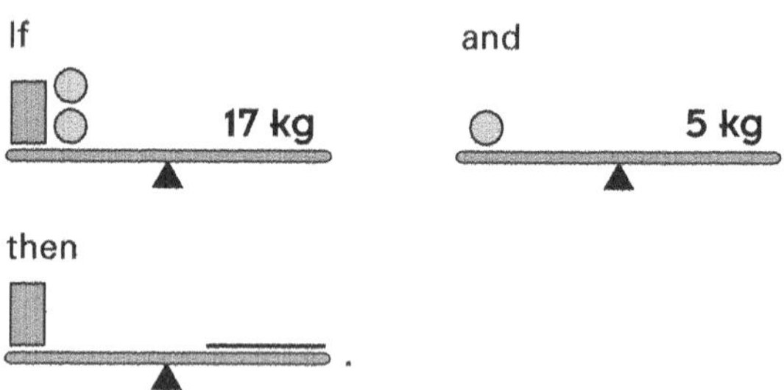

6 There are triangles of three different sizes in this square:

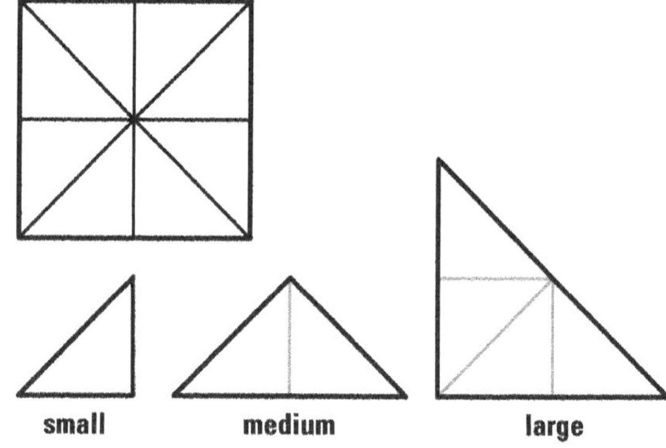

a How many small triangles can you count in the square? ________

b How many medium triangles can you count? ________

c How many large triangles can you count? ________

d How many triangles can you count in total? ________

7 In a magic square, the numbers in any row, column or diagonal line have the same sum. Find the missing numbers in this magic square.

		5
2		6
		1

8 Two numbers have a sum of 8, and a difference of 2.

□ + △ = 8

My numbers are:

9 Complete the pattern in this figure by drawing the correct shapes in the empty squares.

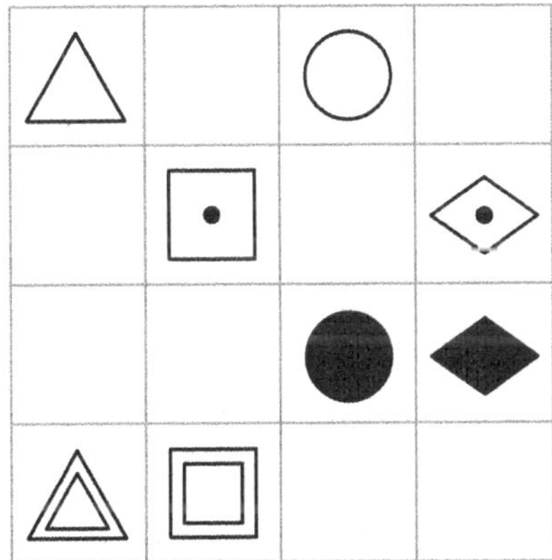

10 How many squares made up this rectangle before it was torn? ________

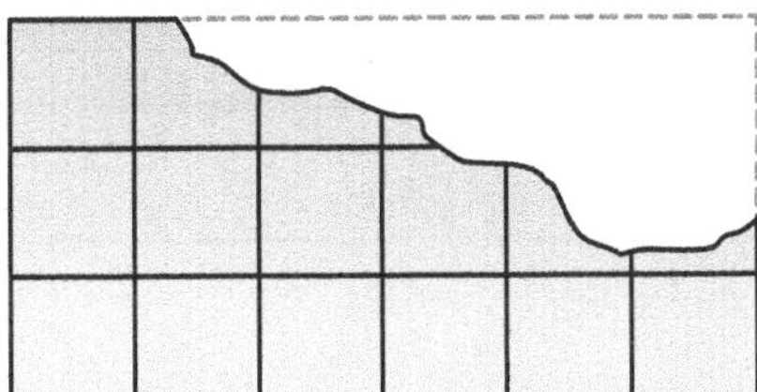

11 Arrange the numbers 3, 4, 5 and 6 in the other circles of the diagram so that the sum of each line is 13.

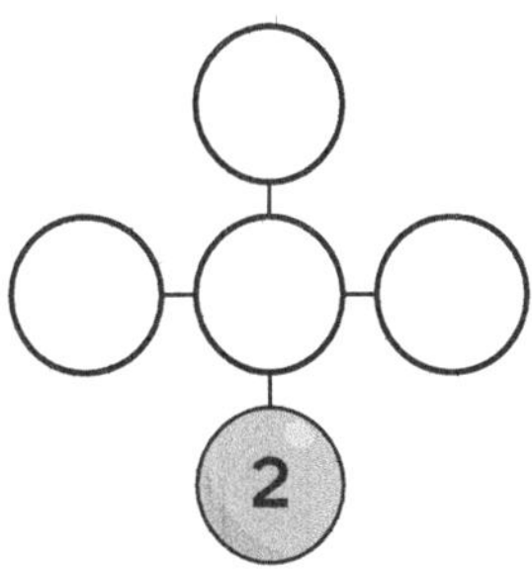

12 Toddler Daniel is playing with yellow, red and blue building blocks.

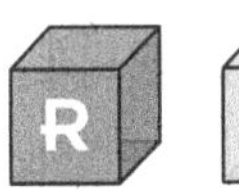

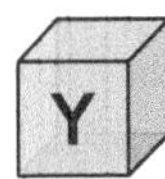

How many different two-block towers can he build? One possibility is shown. Draw the others.

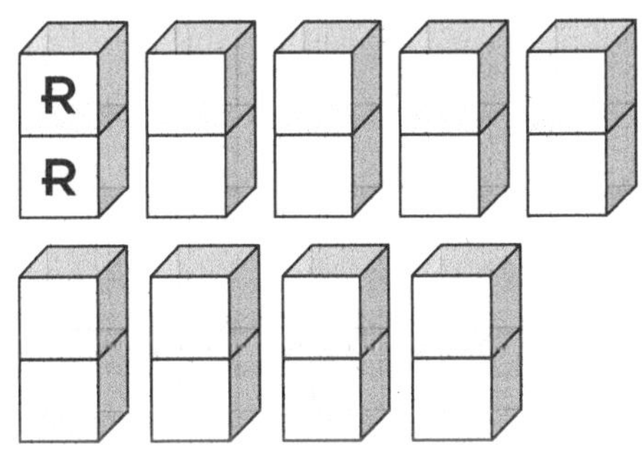

13 Calculate the number of:

a minutes in 3 hours. ________________

b seconds in 2 minutes. ________________

14 a Two computer disks cost $6.00.

i Find the cost of 1 disk. __________

ii Find the cost of 5 disks. __________

b A group of 6 girls went to the movies.
Altogether they paid $33.
How much did each girl pay? ________

15 Melissa bought 2 chocolate frogs and received 20 cents change from $1.
How much did each chocolate frog cost?

____

16 Jane and Kevin share 12 kg of potatoes so that Jane gets 2 kg more than Kevin. Using 'guess and check' and this table, work out the two shares.

Jane's share	Kevin's share

Check that these two amounts add to 12 kg.

17 Allan put a box on a shelf that is 120 cm long. The box is 32.5 cm long.
How many more boxes of the same size can he put on the rest of the shelf?

____

18 How many small cubes are there in this shape? ________

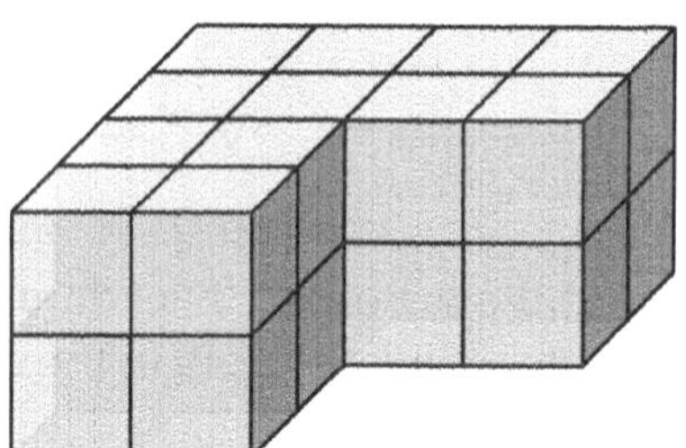

19 Here is a number line. Show the position of each number given on the number line.

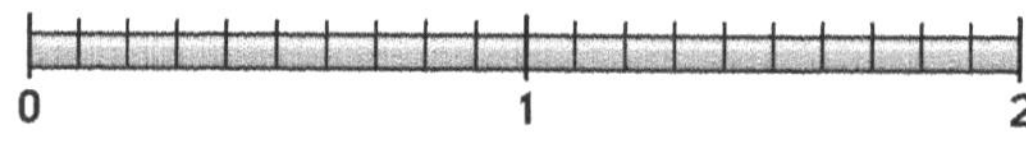

a $1\frac{1}{10}$ **b** $\frac{7}{10}$

20 Four bananas are being weighed on this scale.

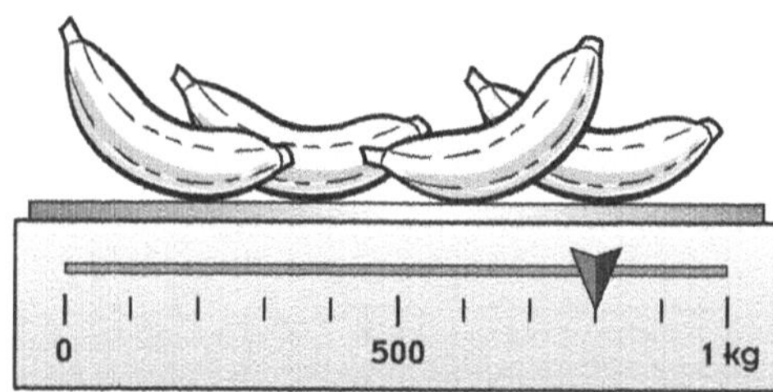

Find the mass of 1 banana. ________

21 Each row of /s has three more /s than the row immediately above it, as shown.

Row 1 /
Row 2 ////
Row 3 ///////
Row 4 ____________
Row 5 ____________

a Complete the pattern for rows 4 and 5.

b How many /s will there be in the sixth row? ________

c How many /s are there altogether in the first five rows?

____

22 This diagram shows how many black, grey and white marbles are on a tray. Complete the statements below.

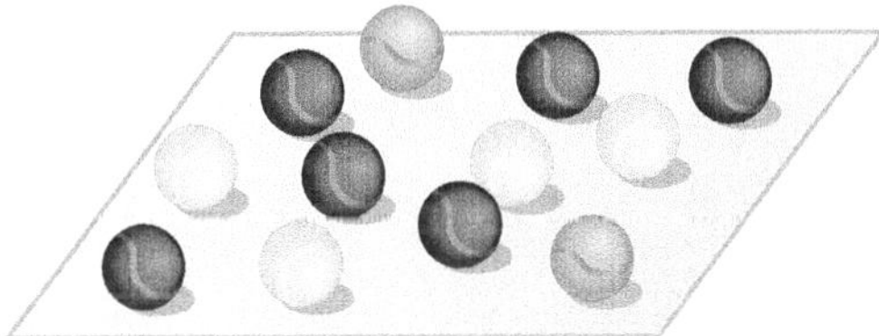

a Half the marbles are ____________

b You choose a marble from the tray without looking. If the chance is 1 in 6 of selecting this colour, it is

____

23 From a large pizza, Mario can cut 7 equal slices for 7 children.

a If Mario had 5 large pizzas, how many children could he serve with an equal slice? ________

b How many large pizzas does Mario need if 28 children have to be served?

____

24 Wendy drew this shape and found that the distance all around it is 10 units.

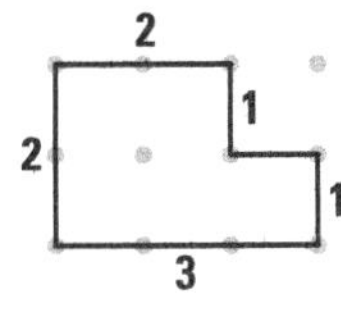

a Find the distance around these other three shapes.

i

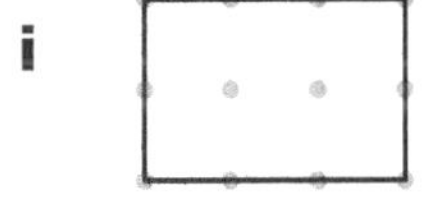

units ________

ii

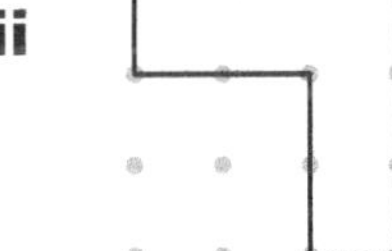

units ________

iii

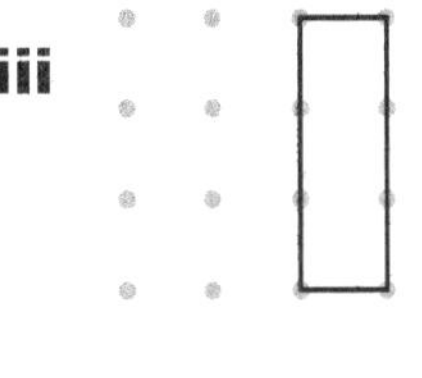

units ________

b Wendy found that the distance around one of these shapes is greater than the others. Colour in this shape.

enrich-e-matics TEST

3rd EDITION

Time: 45 minutes

YEAR 3 TEST 2

Name: ____________________

Class: ____________________

1 Place numbers in the squares so that the total of each line equals the number in the circle for each line.

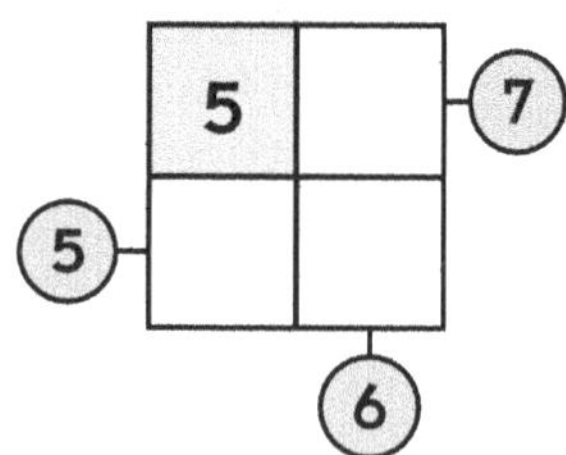

2 If 5 apples is $\frac{1}{3}$ of all the apples that I have, find how many apples I have.

____

Use the diagram to help you.

3 This rectangular shape is divided into three equal parts. If the value of the shaded parts is $14, what is the value of the whole shape? ________

4 Two numbers have a sum of 15 and a difference of 7. Find the two numbers.

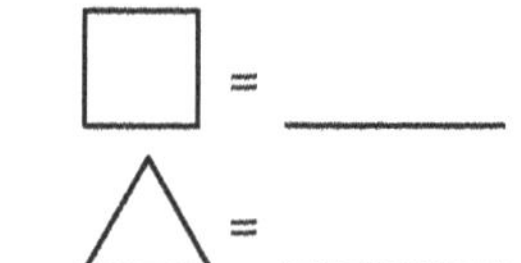

□ = ________

△ = ________

5 Find the temperature shown on the thermometer in °C.

First look at the scale carefully.

________ °C

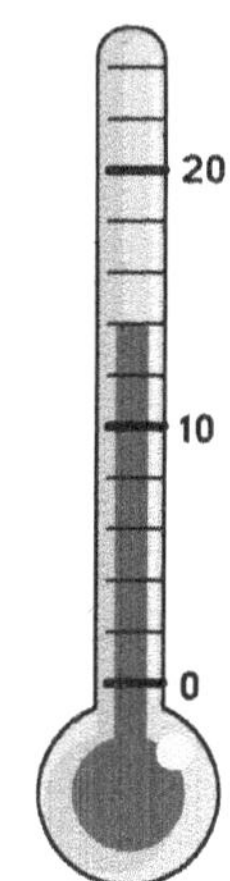

6 Continue this pattern of shapes.

7 In this pictograph, 🍎 = 6 pieces of fruit that a person ate in a week.

David	🍎 🍎
Jessica	🍎 🍎
Helen	🍎 🍎 🍎 🍎
Joan	🍎
Wasim	🍎 🍎 🍎

a How many pieces of fruit did Jessica eat? ________

b Who ate the most pieces of fruit?

____

c Who ate 18 pieces of fruit? ________

d How many more pieces did Wasim eat than Joan? ________

e How many pieces of fruit did these children eat altogether? ________

8 Find the number of little squares needed to balance the circle.

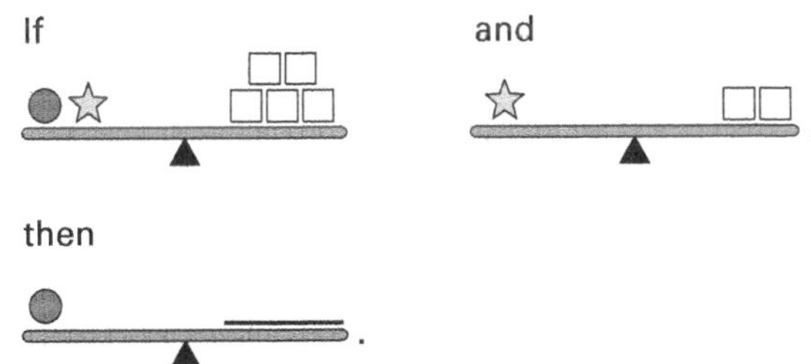

9 Jamie can catch the school bus at the following times in the morning:

7:49 a.m. 8:11 a.m. 8:23 a.m.

If Jamie arrived 4 minutes late for the 7:49 a.m. bus, how long must he wait for the next bus?

____

10 Greg drove to Queenscliff to visit his grandparents. He left home at 7:30 a.m.

a If the trip lasts 1 hour and 45 minutes, what time will he arrive?

____

b The following week when the new freeway opened, Greg left at the same time, and arrived at his grandparents' place at 8:55 a.m. How long did the trip take him this time?

____

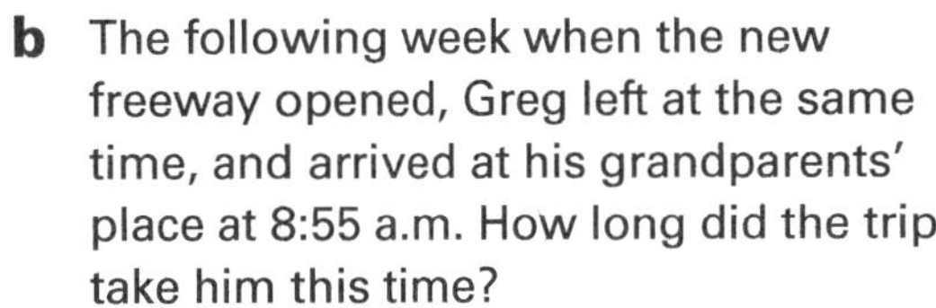

11

Year 3 Sport Stall SALE

3 tennis balls for	$2.40
2 cricket balls for	$3.00
1 basketball for	$4.00
5 golf balls for	$6.00

a Write down the price of:

i 1 tennis ball. ________

ii 1 cricket ball. ________

iii 1 golf ball. ________

b If Georgina bought 1 cricket ball and 1 tennis ball:

i how much did she spend? ________

ii how much change did she get from $5? ________

12 A box of 6 mangoes costs $12.30.

What is the price of one mango? ________

13 Jenna weighed herself and her dog. Jenna weighs twice as much as her dog and together they have a mass of 60 kg.

What is Jenna's mass? ________________

14 Alex has 24 marbles of the same size in a bag.

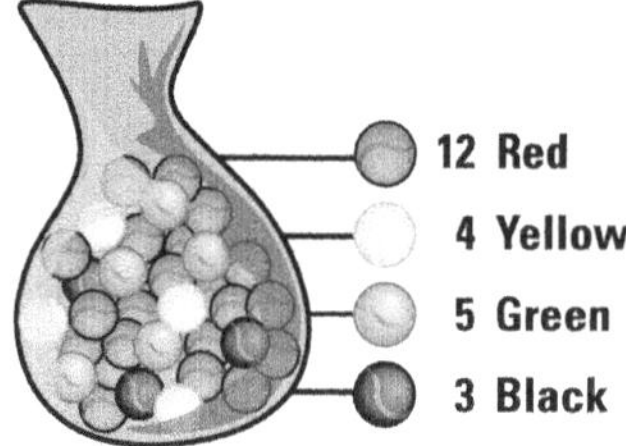

Without looking he takes one marble out of this bag. If there is 1 chance in 2 that Alex picks this colour, what colour did he pick?

____

15 What is the largest 4-digit even number that uses each of the digits 5, 7, 8 and 9 exactly once?

____

16 Sarah is 98.3 centimetres tall. When she was born she was 50.9 centimetres.

How many centimetres has she grown since she was born? ________

17 In this shape each shaded square has an area of 1 cm^2.

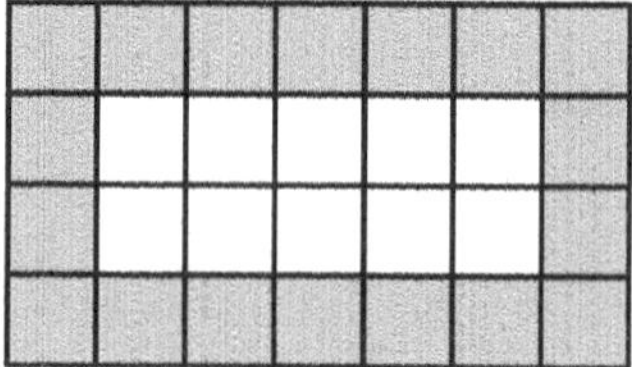

What is the area of the white rectangle in the middle? ________________

18 On these scales all measurements are in kilograms.

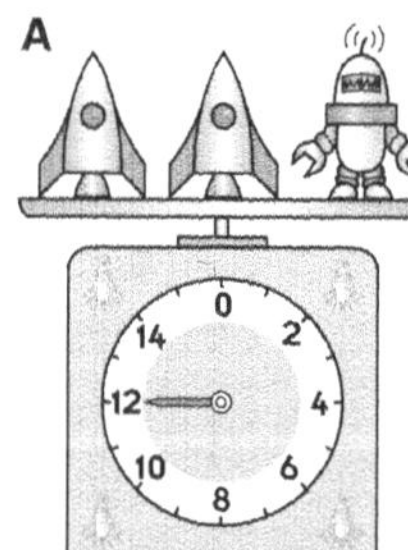

a Write down the mass of the three objects on scale **A**. ___________

b Write down the mass of the two objects on scale **B**. ___________

c Find the mass of 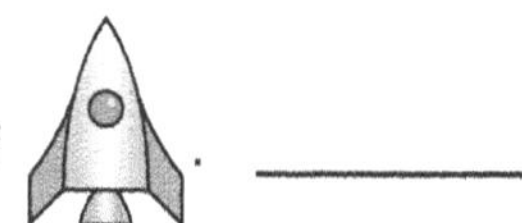. ___________

19 Jamie built a fence around his garden. He only measured some of the sides.

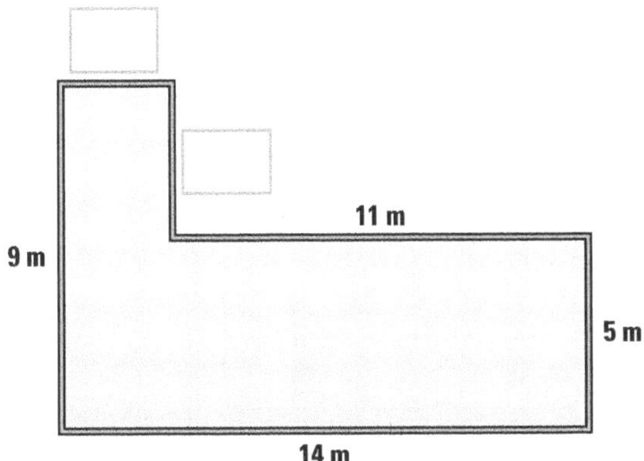

a Write the lengths of the missing sides in the boxes.

b Find the length of the fence. ________

20 If you have a piece of paper and fold it as shown, draw what the result would look like if the paper was opened out after a hole was punched in it.

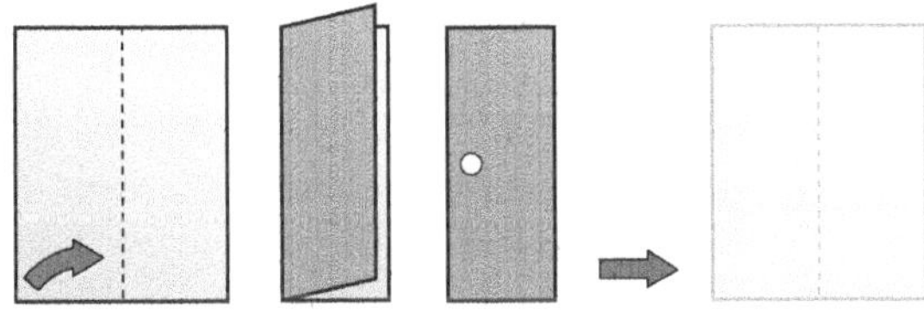

21 Six chocolate frogs cost the same as 4 ice blocks.

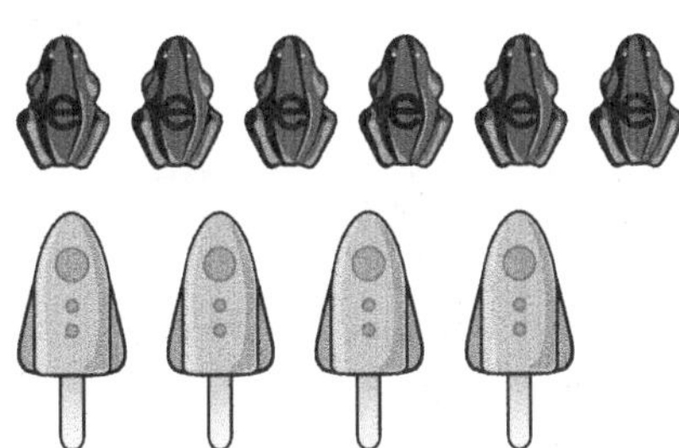

How many chocolate frogs cost the same as:

a 2 ice blocks? ________

b 10 ice blocks? ________

22 By looking at all the balance scales shown, number the item which is the lightest (1) and heaviest (4).

______ ______ ______ ______

23 Gina tosses a coin twice that has two sides, a head and a tail.

a List all the possible ways her coin can land.

____________ ____________

____________ ____________

b What is the chance she will get two tails? ____________

24 In the expression □★△, the symbol ★ means 'Double the first number and to the result add the second number'.

For example:

4★3 $= 2 \times 4 + 3$

$= 8 + 3$

$= 11$

Find the value of 5★4. ____________

enrich-e-matics TEST

3rd EDITION

Time: 45 minutes

YEAR 3 TEST 3

Name: ______________________

Class: ______________________

1 Use all the digits 8, 3, 4.

a Complete these additions.

i ☐☐ + ☐ = 51

ii ☐☐ + ☐ = 42

b Complete these subtractions.

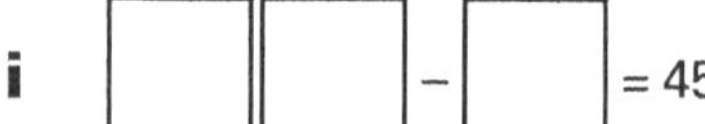

i ☐☐ − ☐ = 45

ii ☐☐ − ☐ = 79

2 If I can buy 6 pens for $1.80,

a find the cost of 1 pen. ____________

b find the cost of 5 pens. ____________

3 Complete this pattern of shapes to fill the grid.

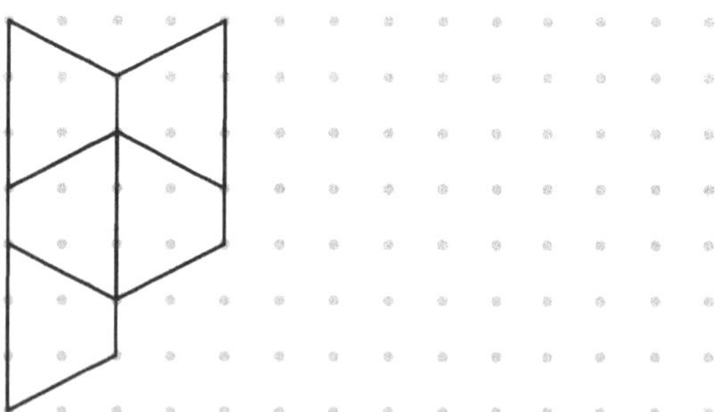

4 Two numbers have a sum of 11 and a difference of 7. Find the two numbers.

☐ + ○ = 11 ☐ = ______

☐ − ○ = 7 ○ = ______

5 Shade $\frac{3}{5}$ of this large rectangle.

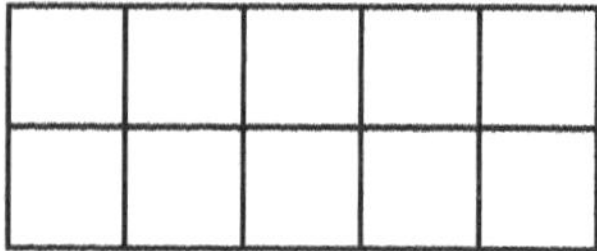

6 Shade 0·3 of this large rectangle.

7 If this clock is 30 minutes slow, find the correct time.

____

8 What is the date 15 days before 9 June?

____

9 Draw the missing figure that will complete this pattern of cats.

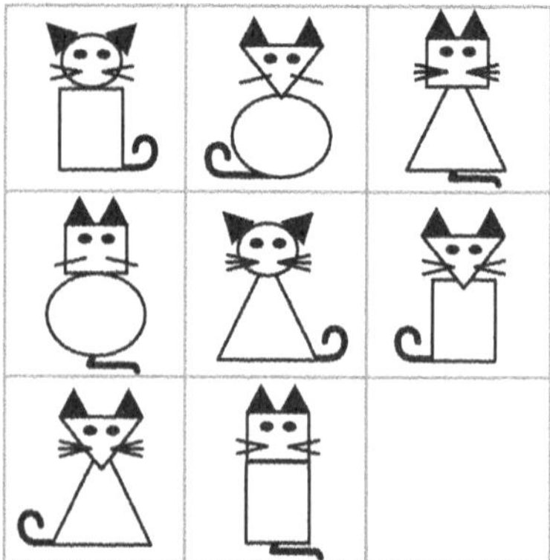

10 Michael has put 3 crosses on each grid. Add one more cross so that the 4 crosses form a square.

11 The Babylonian cuneiform (wedge writing) numerals used two symbols only, **V** and **<**.

V represented 1 and **<** represented 10; therefore:

VVV = 3

<<V = 21

a What does **<VVVVV** represent?

______

b How would the number 32 be written as a cuneiform numeral?

____

12 If it takes three minutes to cut a log once, how long would it take to cut this log into 4 equal lengths?

Hint: show the cuts.

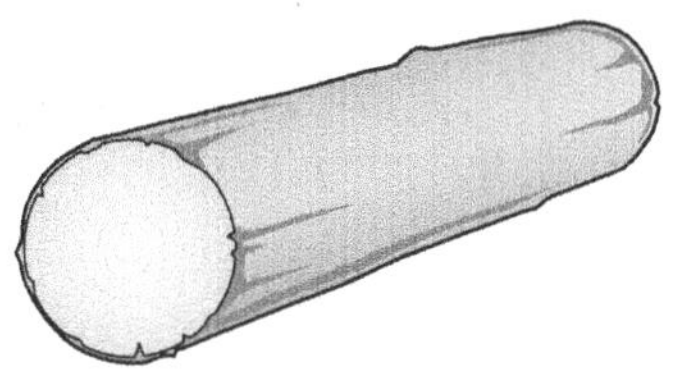

____

13 a Write down the lengths of the missing sides on the figure below.

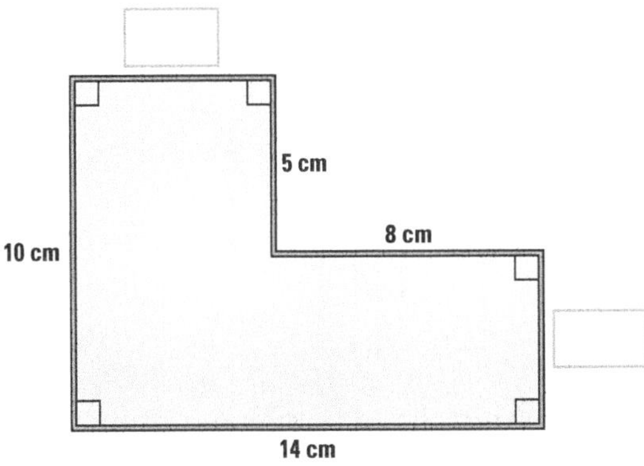

b Find the perimeter of this shape.

____

14 Using the two balances given, work out the value of the square.

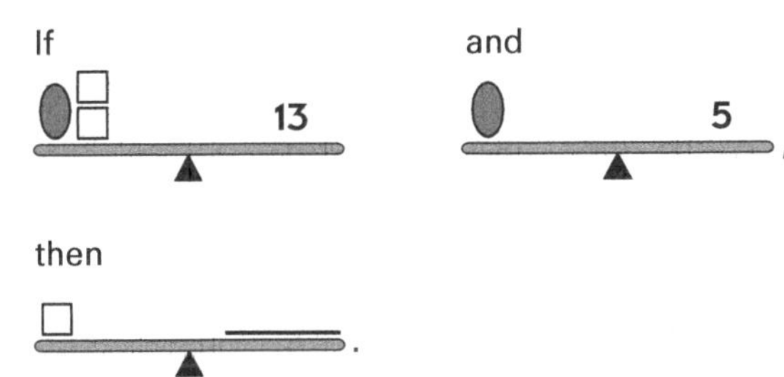

15 If Gary is at school every Monday from 7:30 a.m. till 6 p.m., how long does he spend at school on Mondays?

____

16 Each figure represents a fraction.

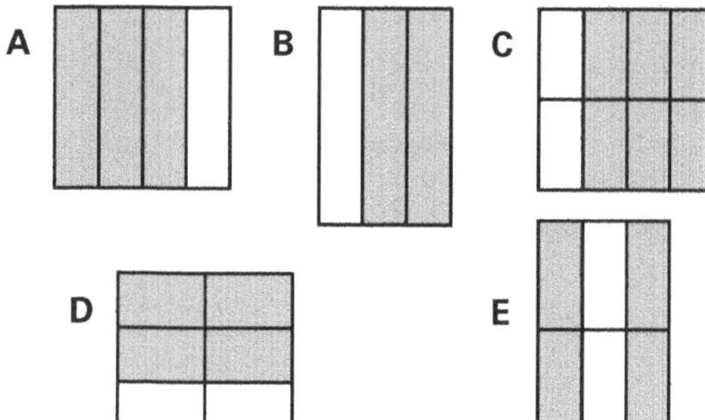

a Which three figures represent the same fraction?

____

b Which figures represent $\frac{3}{4}$? ________

17 **a** Two pens cost 88 cents.
Find the cost of 5 pens. ____________

b If 2 ice creams cost $4.20, find the cost of 3 ice creams.

____

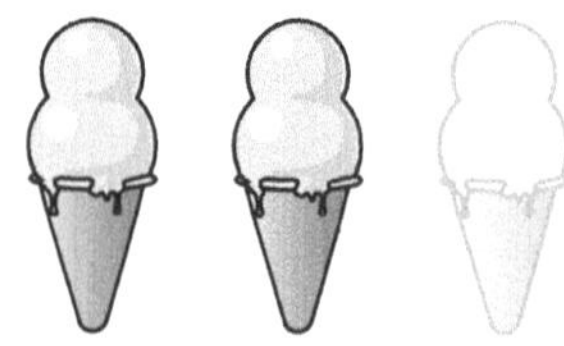

18 **a** Write down the number that is 10 more than 5299.

____

b Write down the number that is 1000 less than 34 872.

____

19 A teacher marks 10 of her students' tests every half an hour.

a If she has 22 students in her class, how long will she take to mark their tests?

____

b If it takes her one hour and 24 minutes to mark all her students' tests, how many are in her class? ________

20 There are two red marbles in each bag. The rest are marbles of different colours.

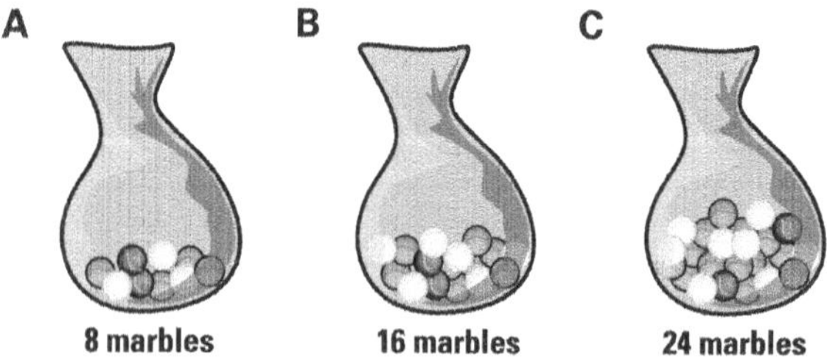

a Without looking in the bag, if you pick a marble out of one of the bags, which bag would give you the greatest chance of picking the red marble?

____

b What is this chance? ________

21 Which of these does not show a line of symmetry? ________

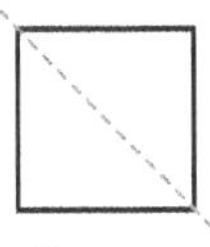
A

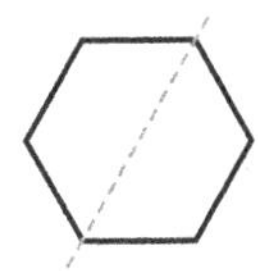
B

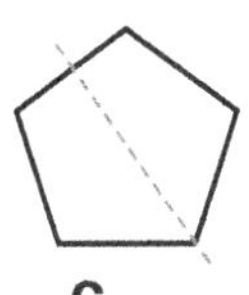
C

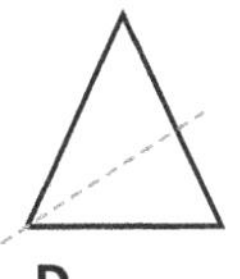
D

22 The graph shows the number of sandwiches sold each day of a week at Melbourne City Primary School.

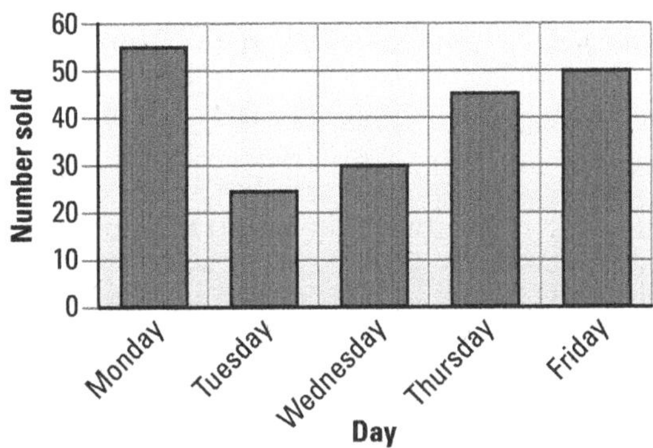

a How many more sandwiches were sold on Monday than on Tuesday?

____

b How many sandwiches did the school sell that week? ________

23 25 x 13 is more than 24 x 13.

How much more? Try to answer this without working out the multiplication.

____

24 How many cubes are in this box? 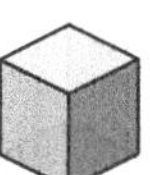

____

Be careful when you count the cubes, as there are more than you can actually see.

25 Two darts are thrown at the target. Assume that each dart lands within one of the rings or within the bullseye.

How many different point totals are possible? Remember to work systematically.

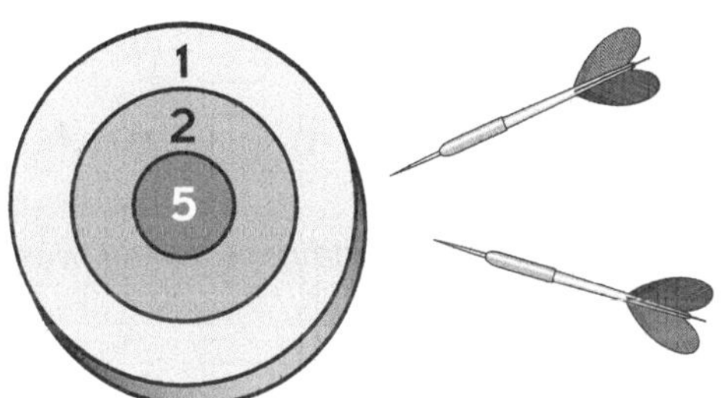

________ ________ ________

________ ________ ________

enrich-e-matics TEST

3rd EDITION

Time: 45 minutes

YEAR 3 TEST 4

Name:

Class:

Multiple choice

Colour in the bubble next to the answer.

1 One more than 399 is:

A 3991 B 401
C 400 D 1399

2 What does the digit 5 stand for in 3592?

A 5 x 100 B 5 x 10
C 5 x 1000 D 5 x 1

3 How many halves ($\frac{1}{2}$) are there in $5\frac{1}{2}$?

A 10 B 11
C 9 D 21

4 If tomorrow is Thursday, what day was it 9 days ago?

A Wednesday B Tuesday
C Monday D Sunday

5 Suppose I collected $7.00 worth of 20 cent coins. How many coins have I saved?

A 70 B 14
C 140 D 35

6 What fraction of the figure is shaded?

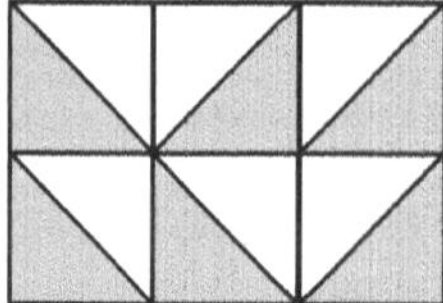

A $\frac{1}{4}$ B $\frac{2}{3}$ C $\frac{1}{2}$ D $\frac{3}{8}$

7 The number of cents in 3 dollars and 8 cents is:

A 38 B 3008
C 380 D 308

8 $6 \times 5 = 3 \times \square$

Find the missing number in the box.

A 30 B 20
C 2 D 10

9 If $\frac{1}{4}$ of a number is 3, find the number.

A 16 B 9
C 6 D 12

10 Which of the following sentences is *not* true?

A 217 < 172 B 564 < 654
C 342 > 243 D 316 > 163

11 Apples are sold at 3 for $2. Mrs Kemp needs 12 apples. How much does she need to pay for them?

A $4 B $8
C $9 D $6

12 The value of the shaded part is $9. What is the value of the whole shape?

A $12 B $15
C $3 D $18

13 Alex is shorter than Jason, but taller than Peter. Bob is taller than Peter, but shorter than Alex. Which is the correct order when the boys are placed from shortest to tallest?

A ⬭ Alex, Bob, Peter, Jason

B ⬭ Alex, Peter, Bob, Jason

C ⬭ Peter, Bob, Alex, Jason

D ⬭ Peter, Alex, Bob, Jason

14 Amanda saves 75c of her pocket money each week. How many weeks will it take her to save $5 to buy a present for her friend?

A ⬭ 7 weeks B ⬭ 5 weeks

C ⬭ 12 weeks D ⬭ 6 weeks

15 It takes you 15 minutes to have a bath each night. What is the total time that you would have spent having a bath during the past week if you had a bath each night?

A ⬭ 1 hour 30 minutes

B ⬭ 1 hour 45 minutes

C ⬭ 95 minutes

D ⬭ 1 hour 35 minutes

16 George correctly subtracted 5 tens and 7 tenths from 2731. What was his answer?

A ⬭ 2681·7 B ⬭ 2780·3

C ⬭ 2680·7 D ⬭ 2680·3

17 A bus travels 110 kilometres in 2 hours. At this speed, how many kilometres would the bus travel in 1 hour?

A ⬭ 55 B ⬭ 108

C ⬭ 220 D ⬭ 50

18 Judy can make 6 ribbons in 2 minutes. At this rate, find the time taken to make 15 ribbons.

A ⬭ 8 min. B ⬭ 11 min.

C ⬭ 6 min. D ⬭ 5 min.

19 Which thermometer shows 10°C?

A ⬭ B ⬭ C ⬭ D ⬭

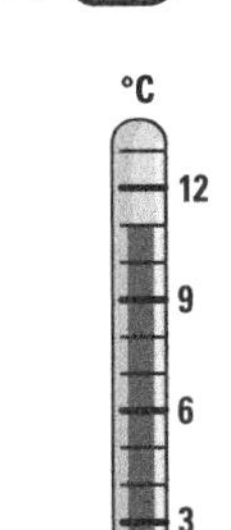

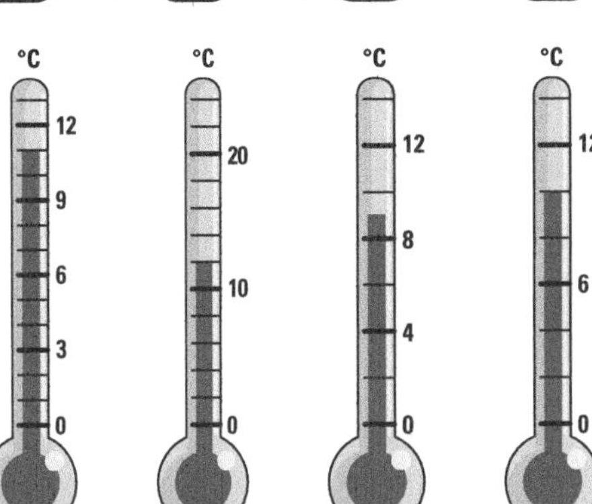

20 Paul bought 3 oranges for 45c each. He gave the shopkeeper $2. How much change should he get?

A ◯ $1.35 B ◯ $1.25

C ◯ 65c D ◯ 75c

21 A sheet of paper is folded in half and then a hole is punched as shown.

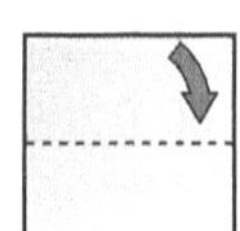

If the paper is unfolded, it will appear as:

A ◯ B ◯ C ◯ D ◯

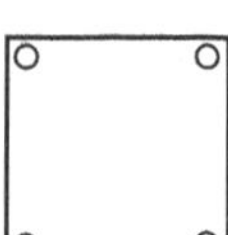

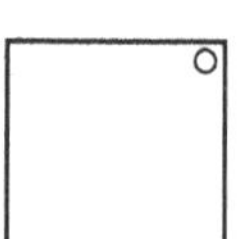

22 Dena built this solid from sugar cubes. If each cube has a mass of 3 g, what is the mass of this solid?

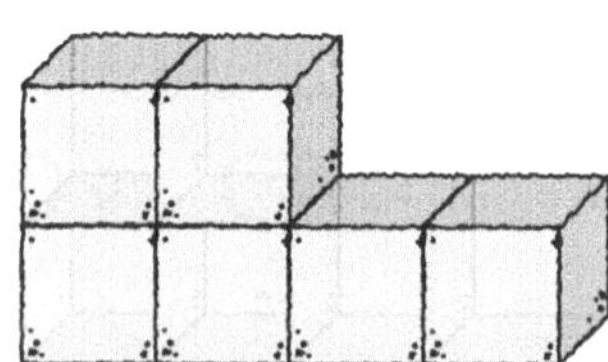

A ◯ 36 g B ◯ 18 g

C ◯ 12 g D ◯ 27 g

23 Work out the pattern and find the next shape.

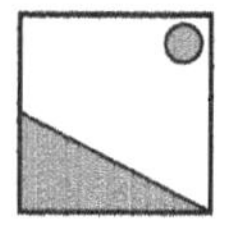

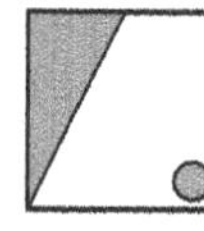

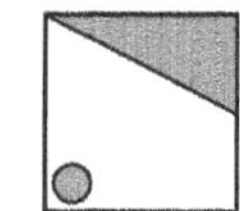

A ◯ B ◯ C ◯ D ◯

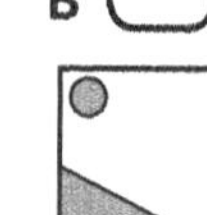

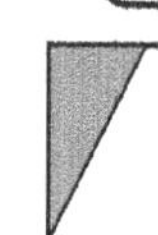

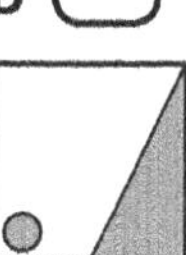

24 I am thinking of a number that is greater than 20 but less than 30. If it is even and divisible by 3, find my number.

A ◯ 21 B ◯ 27

C ◯ 30 D ◯ 24

25 Yesterday in Melbourne it was 24 degrees at 3 p.m. If today it is 9 degrees colder at 8 a.m., then:

A ◯ 17 hours later, the temperature is 15 degrees.

B ◯ 17 hours later, the temperature is 33 degrees.

C ◯ 15 hours later, the temperature is 15 degrees.

D ◯ 15 hours later, the temperature is 5 degrees.

26 What is the maximum number of cards measuring 3 cm by 4 cm that can be cut from a piece of cardboard 9 cm by 8 cm?

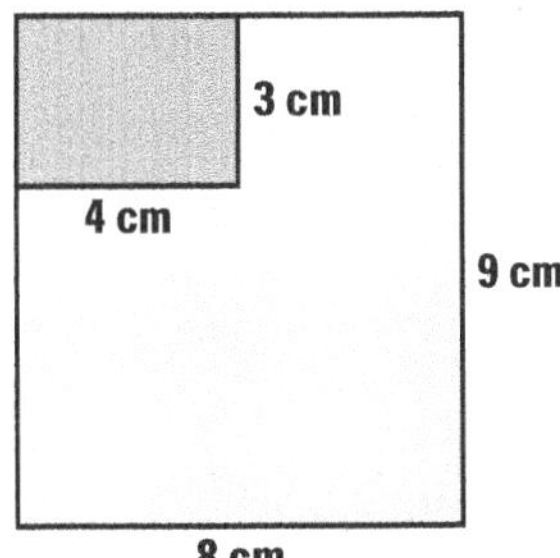

A ⬭ 6 B ⬭ 12 C ⬭ 8 D ⬭ 9

27 These shapes are growing in a certain pattern. Carefully study how each side is growing and select the next shape in the pattern.

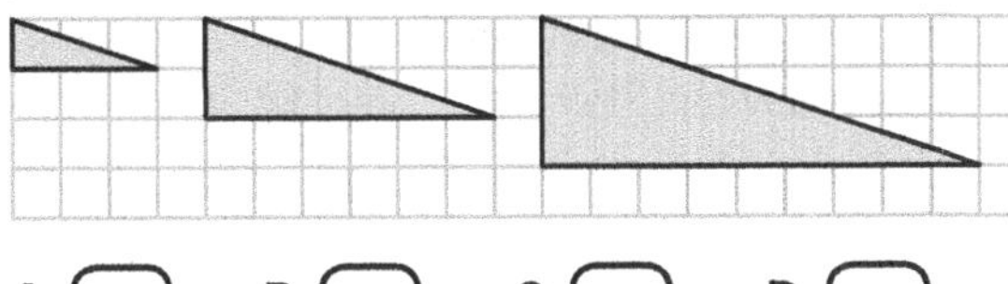

A ⬭ B ⬭ C ⬭ D ⬭

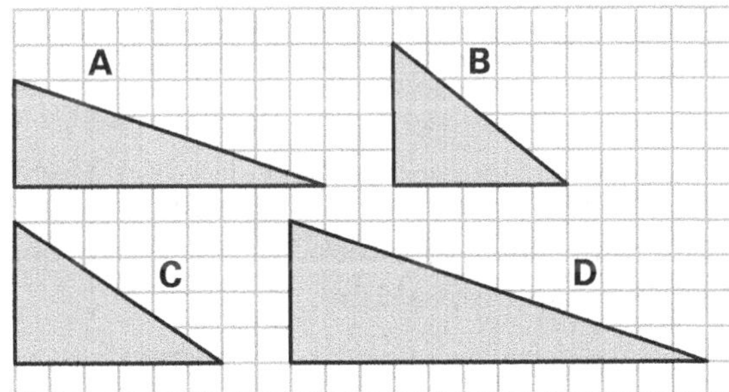

28 This is Kelly's clock.

Which clock shows the correct time if Kelly's clock is 10 minutes slow?

A ⬭

B ⬭

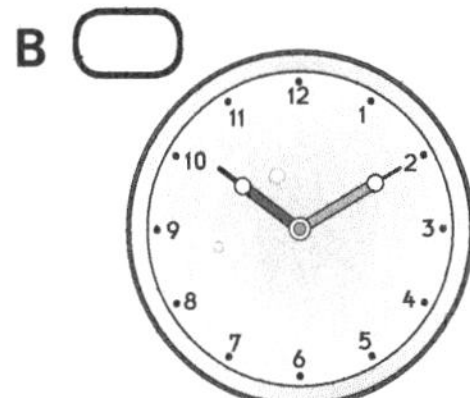

C ⬭

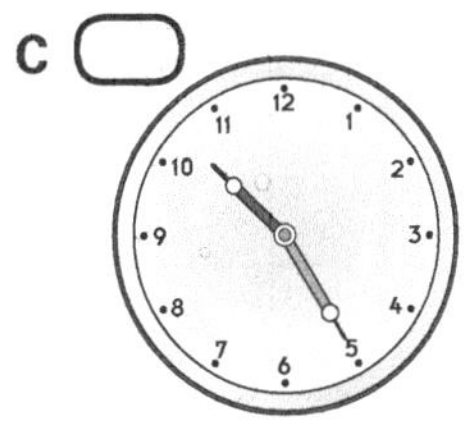

D ⬭

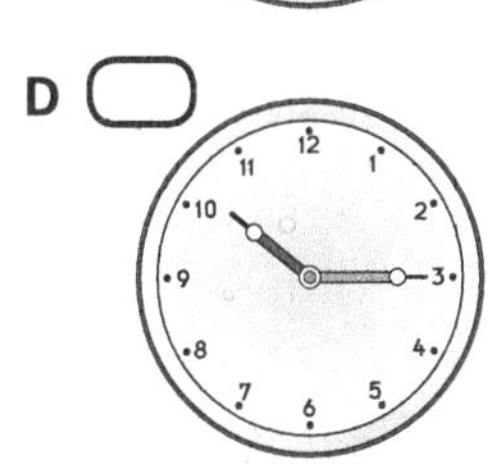

29 A library card number is made up of 8 digits. The sum of any 3 digits next to each other is always 16.
Which digit will replace ◆ ?

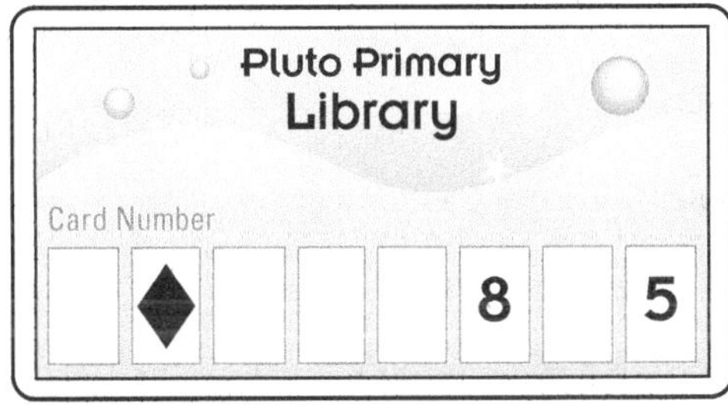

A ⬭ 8 B ⬭ 5
C ⬭ 3 D ⬭ 11

30 Jamie has one 10-cent coin, one 20-cent coin and one 50-cent coin. How many different amounts of money can he make?

A ⬭ 3 B ⬭ 6
C ⬭ 7 D ⬭ 8

enrich-e-matics 3rd EDITION TEST

Time: 45 minutes

YEAR 4 TEST 1

Name: ____________________

Class: ____________________

1 Place numbers in the squares so that the total of each line equals the number in the circle for each line.

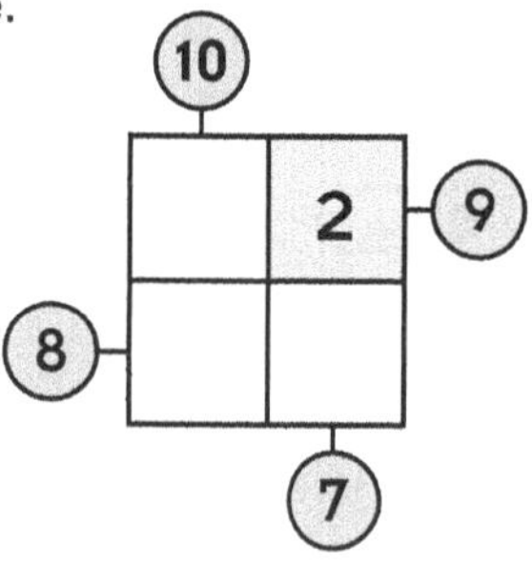

2 A group of 6 people won $24,000 in a lottery. How much should each person receive? ____________

3 Four lemons cost 84 cents.

a Find the cost of 1 lemon. ____________

b Find the cost of 5 lemons. ____________

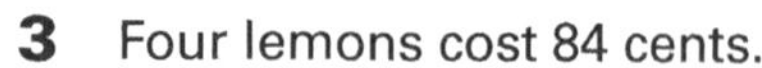

4 Using the two balances given, work out the value of the square.

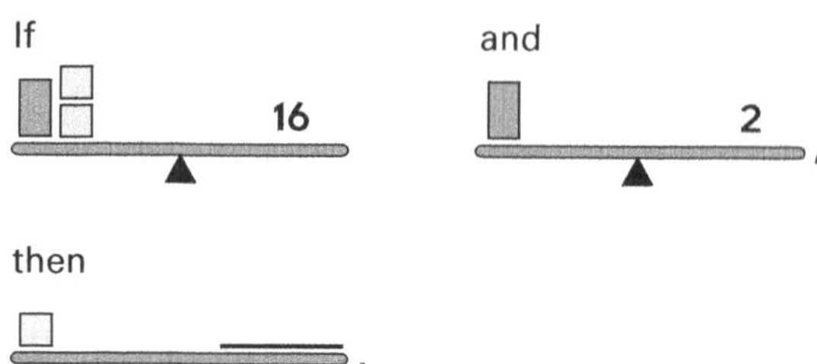

5 Complete this square with the pattern.

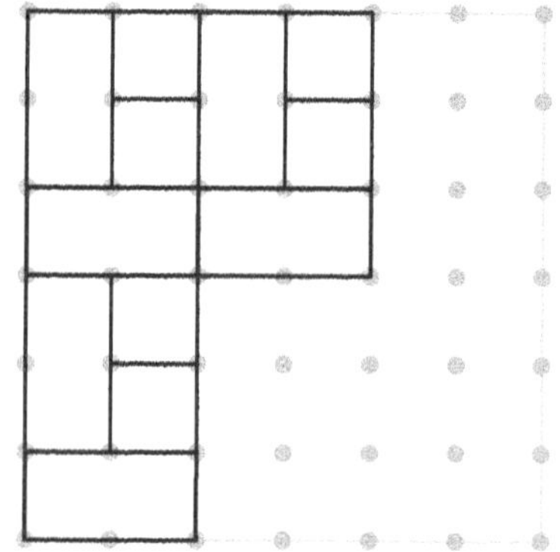

6 Diane entered a shop with $10. She bought a cup of tea for $2.50 and a cheese sandwich. She was given $4.10 change.

How much did the cheese sandwich cost?

____

7 The figure below is cut out around the thick outer lines and folded on the thin inner lines to form a cube.

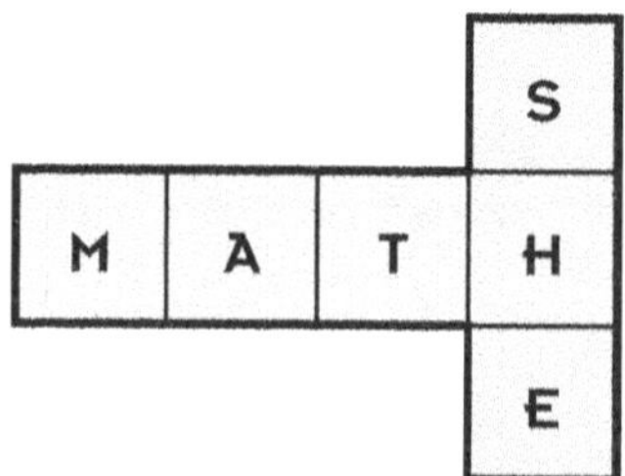

Which letter on the face of the cube will be opposite the:

a H ________ **b** E ________

8 A machine produces 4 bolts each minute.

a How many bolts would be produced in 2 hours? ________

b How long would it take to produce 200 bolts? ________

9 One comic book costs 80 cents, a balloon 40 cents, and an icy pole 50 cents.

Comic book
80c

Balloon
40c

Icy pole
50c

a How much will Jeremy spend if he buys 2 balloons and 1 comic book?

____

b How much will Amy spend if she buys 3 balloons and 3 icy poles?

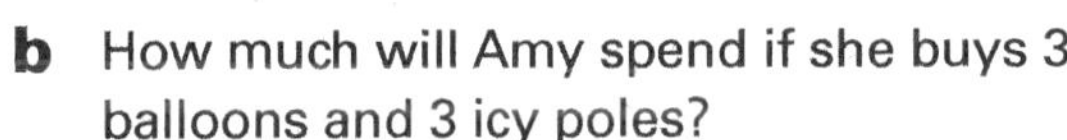

____

c How much change will Amy get from $5 if she buys 3 balloons and 3 icy poles?

____

d What items can Sharon buy for exactly $2?

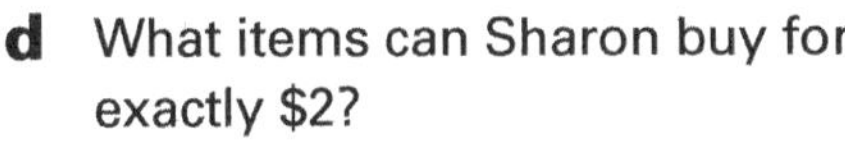

____

____

e If Joe wants to buy 3 items for exactly $2, what should he buy?

____

____

10 The sum of two numbers is 10 and their product is 16. Find the two numbers.

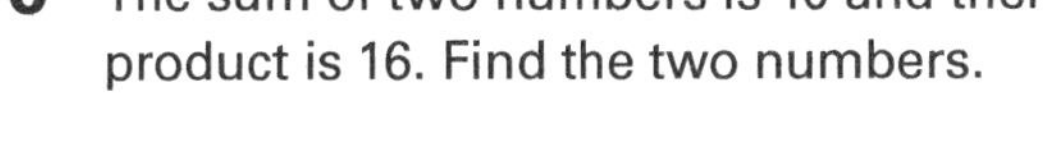

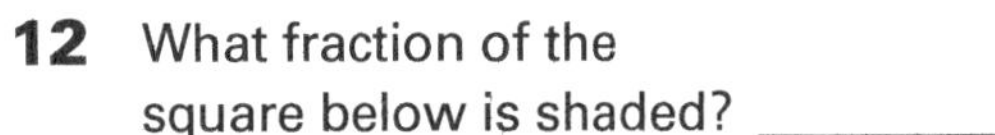

11 In a choir there are twice as many girls as boys. If there are 24 students in the choir, how many are girls? ________

12 What fraction of the square below is shaded? ________

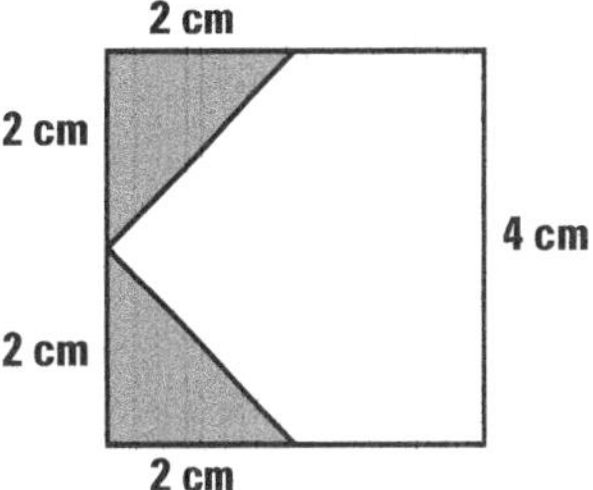

13 At my school there are 7 tennis teams in the intermediate division.
How many matches must there be for all teams to play each other?

____

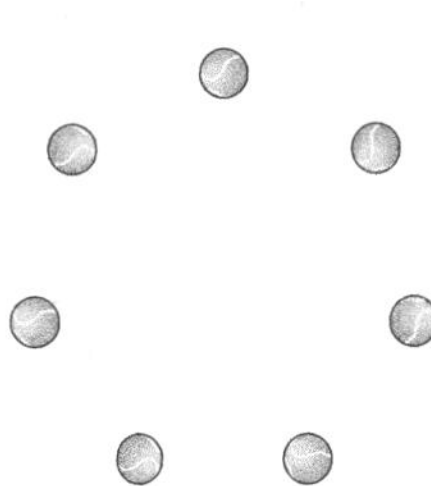

14 If you made a spinner like this, and:

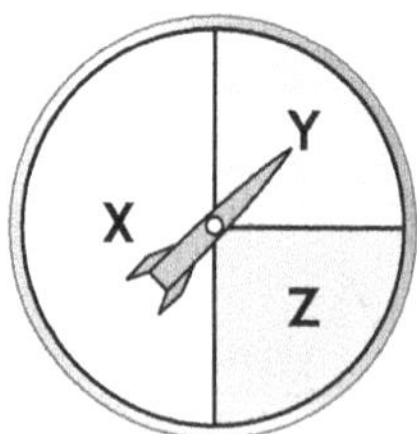

a if you made 20 spins, how many times do you predict that it would land in X?

____

b if the spinner landed 6 times in Y, predict how many times the person spun the spinner altogether.

____

15 In the following sequence there are 9 numbers:

6
66
666
6666
66666
666666
6666666
66666666
666666666

What is the tens digit of the sum of these 9 numbers? ________

16 Gary takes 20 equal paces to walk from A to B in his garden. How many such paces will he take to walk once around his garden?

____

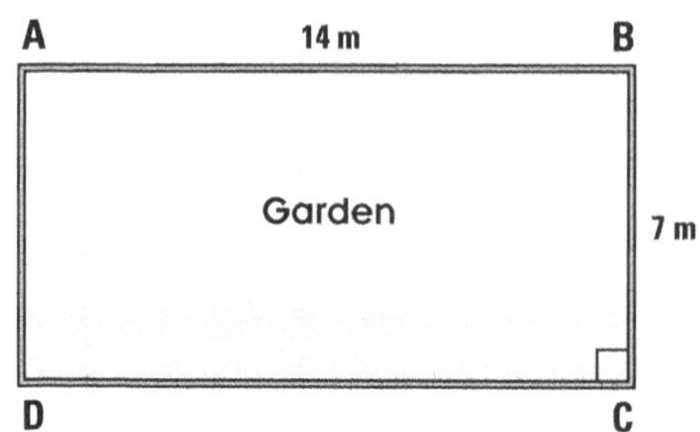

17 Mr White gives $36 to his two sons. The elder son receives three times as much as the younger son. How much does each son receive?

younger son ________

elder son ________

18 Anthea flipped this shape over the dotted line and drew its reflection. Draw what the shape will look like after the flip.

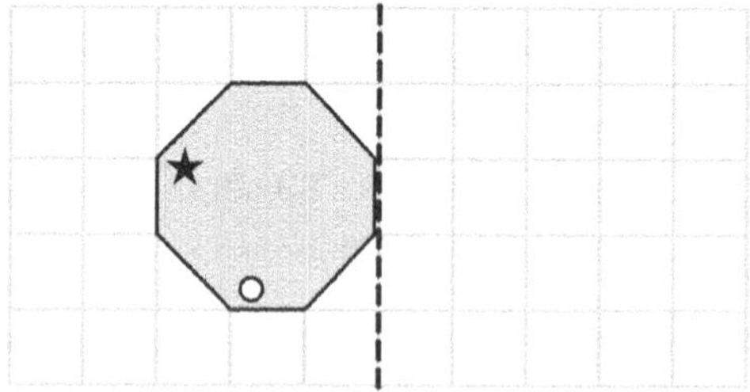

19 Shaun has 5 coins in his wallet: a 5c, a 10c, two 20c coins and a 50c coin.

If he takes out a coin without looking, what is the chance that it is a 20c coin?

____

20 This solid shape is made up of a number of centicubes.

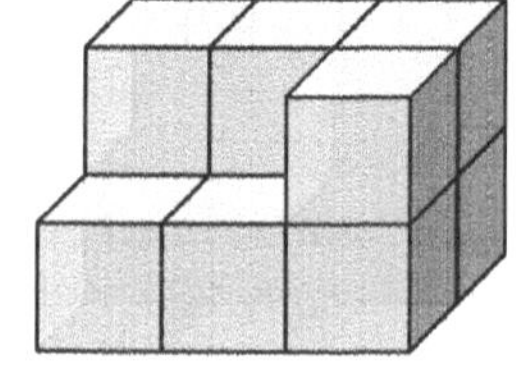

a How many centicubes were needed to build the shape above?

____

b How many more centicubes would need to be added to make a cube of side 3 cm? ________

21 Australia is divided into three time zones:

A Eastern Standard Time
B Central Standard Time
C Western Standard Time.

Complete the following time zone grid.

Western Standard Time	Central Standard Time	Eastern Standard Time
02:00	03:30	04:00
07:00	08:30	09:00
	09:00	09:30
10:15		12:15
	09:40	
		14:15

22 When Linda looks down from directly above:

- a cube, she sees.

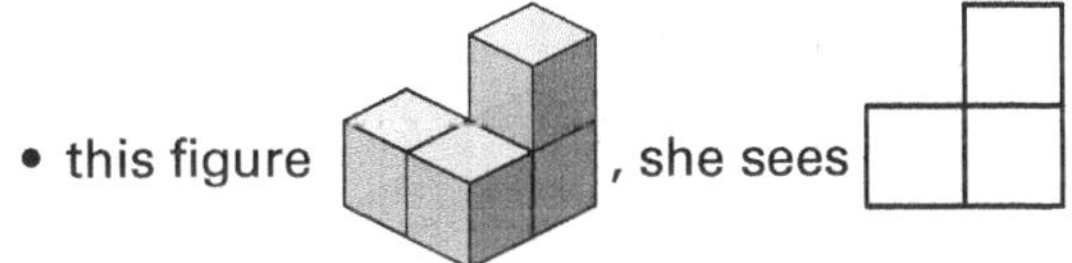

- this figure, she sees.

Draw what she can see when she looks straight down on each of these shapes.

a

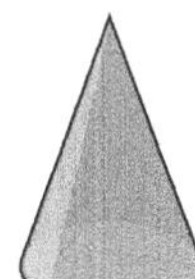

b

__________ __________

23 On Monday morning, Frank had a certain mass of potatoes. He sold half, then sold another 4 kg, so he ended up with 10 kg of potatoes. How many kilograms of potatoes did Frank have at the beginning of the day?

____

24 The diagrams in this pattern are made up of matches.

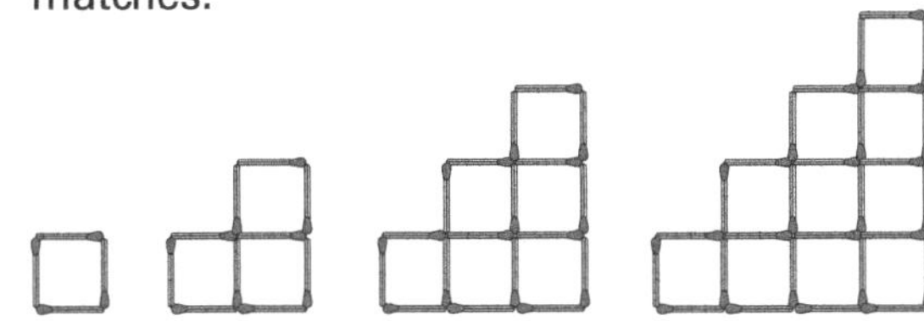

a Draw the fifth diagram.

b How many squares are there in the fifth diagram? ________

c How many matches are there in the fifth diagram? ________

enrich-e-matics 3rd EDITION TEST

Time: 45 minutes

YEAR 4 TEST 2

Name: ______________________

Class: ______________________

1 Daniel spent $5.85. How much change would he get from $20?

____

2 Paul and his sister, Martine, collect marbles. Martine has 3 more marbles than Paul, and together they have 19. How many marbles does each person have?

Paul	Martine

3 Find the missing number in the box.

(3 × ☐ – 2) = 19

4 A plane due at the airport at 4:00 p.m. arrived 25 minutes early.

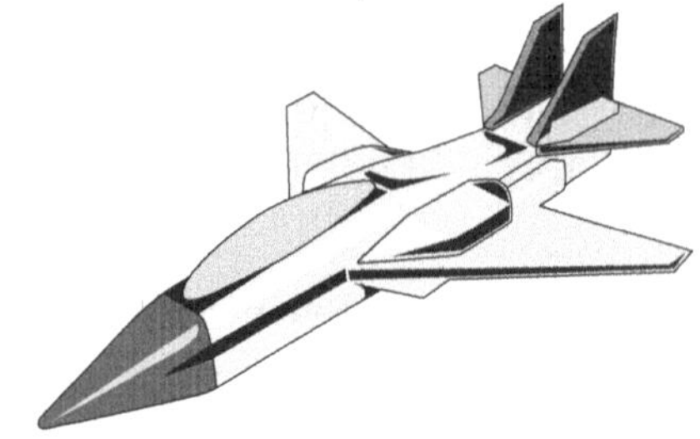

At what time did it arrive? __________

5 How many minutes in two and a quarter hours? __________

6 Write the next two numbers in each pattern.

a 93, 87, 81, 75, 69, ______ , ______

b 2, 4, 7, 11, 16, ______ , ______

c 1, 2, 4, 8, ______ , ______

7 There is a rule connecting the numbers in the top row with the numbers in the second row. Find the rule and work out the missing numbers.

2	3	7	5	9	4	
⇩	⇩	⇩	⇩	⇩	⇩	⇩
3	5	13	9	17		11

8 How many cubes make up this solid figure?

1 cube

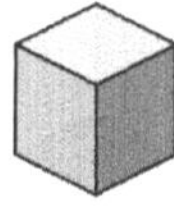

____

9 If

and ★●●● □□ :

a how many ★s are needed to balance two □s?

b how many ★s are needed to balance one □?

10 Three rockmelons were cut into quarters and then each quarter was cut in half.

How many pieces of rockmelon were there? ____________

11 I am thinking of two numbers. When I add them the sum is 9, and when I multiply them the product is 20.

Find my two numbers.

□ + △ = 9 □ = ______

□ × △ = 20 △ = ______

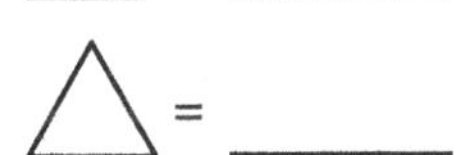

12 A tap drips every $2\frac{1}{2}$ seconds.

How many times will it drip in 1 minute?

____

13 If Paul is fifth from the start of a straight line, and third from the end of the line, how many are in the line?

Draw a diagram to show Paul's position.

Start ———————————— End

14 Daniel has more than 25 but fewer than 40 marbles. When he counts the marbles by threes, there is 1 left over. When he counts by fives, there are 2 left over.

How many marbles does Daniel have?

____

15 Mr Berger gives $24 to his two sons. If the elder son receives three times as much as the younger son, how much does each son receive?

elder ________ younger ________

16 If a car travels 12 km in 10 minutes, find how far it travels in 1 hour.

____

17 If seven friends sent birthday cards to one another, how many cards did they send altogether in a year?

Use the seven points below to help you work out the answer.

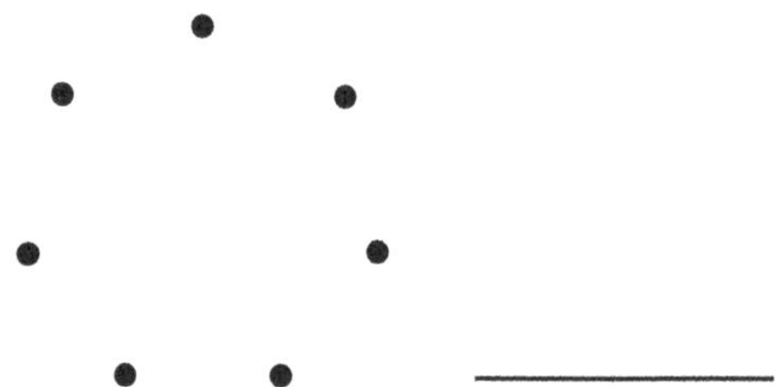

____

18 A block of cheese is cut in half as shown and separated.

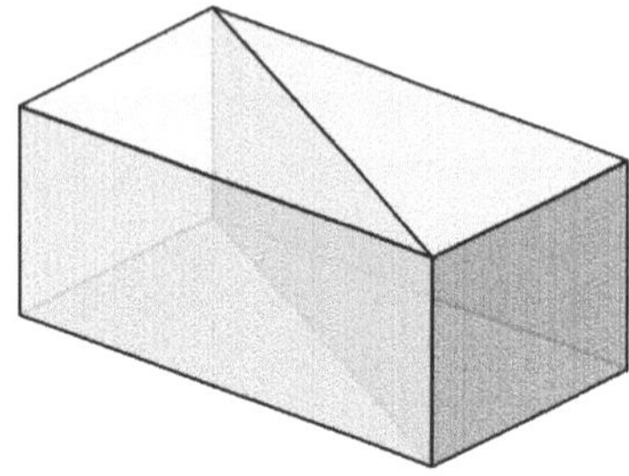

a What is the total number of edges on these two pieces? ________

b What is the total number of corners on these two pieces? ________

19 Mayan numerals consisted of dots and strokes representing the numbers up to 20.

4 was ●●●●

5 was ———

6 was ● ———

17 was ●● ≡ (5 + 5 + 5 + 2)

a Write the number 13 as a Mayan numeral. ______________

b What does ●●●● ——— mean? __________

20 This pattern of rectangles was constructed from matches. If this pattern continued, how many matches would be needed to build the 4th and 5th rectangles?

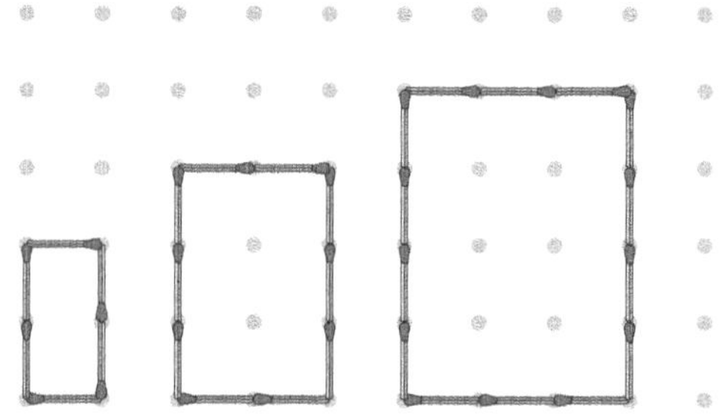

Rectangle	Number of matches
1	6
2	10
3	14
4	
5	

21 The following is the amount of money raised at the school fete by each year level.

Year 1	\$60
Year 2	\$50
Year 3	\$70
Year 4	\$80
Year 5	\$85
Year 6	\$95

Money raised at the fete

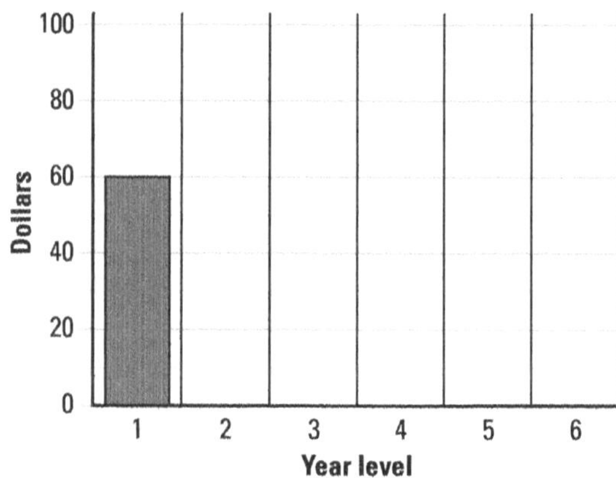

22 I can buy 3 pens for $2.40.

a How many can I buy for $3.20?

____

b How much will 7 pens cost?

____

23 In Roland's Year 4 class of 24 students in total, there are 6 students with black hair, 12 with brown hair, 4 with blond hair and 2 with red hair. If a student is picked at random (without looking):

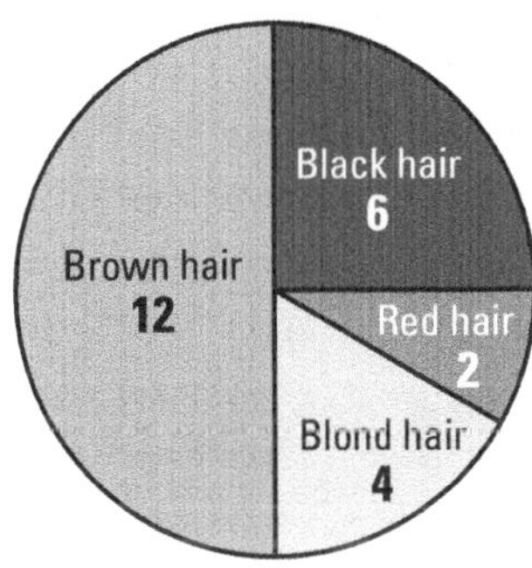

a what is the chance that the person picked has black hair?

____

b what is the colour of the student's hair if there is 1 chance in 2 of that colour being picked?

____

24 a What is the mass of the four objects on this scale? ________

b If the mass of 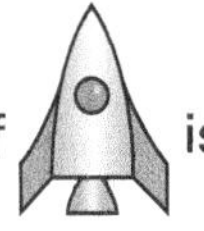 is 5 kg,

find the mass of . ________

25 Robbie has to number 45 seats for a concert. He sticks digits from 0 to 9 on the backs of seats, so seat 25 will need one digit 2 and one digit 5.

a How many digit 3s will he need?

____

b How many digits will he need altogether? ________

enrich-e-matics TEST

3rd EDITION

Time: 45 minutes

YEAR 4
TEST 3

Name: ____________________

Class: ____________________

1 How much more than 37 420 is 43 420?

____

2 Three books cost $39.

a Find the cost of one book. ________

b Find the cost of 7 of these books.

____

3 A plane due at the airport at 3:20 p.m. arrived 35 minutes early.

What time did it arrive? __________

4 How much longer is 1 metre than 35 cm?

____

5 A ticket to a school concert costs $5.50. Jay bought two tickets. How much change should Jay get from $20?

____

6 Shade $\frac{1}{3}$ of this shape.

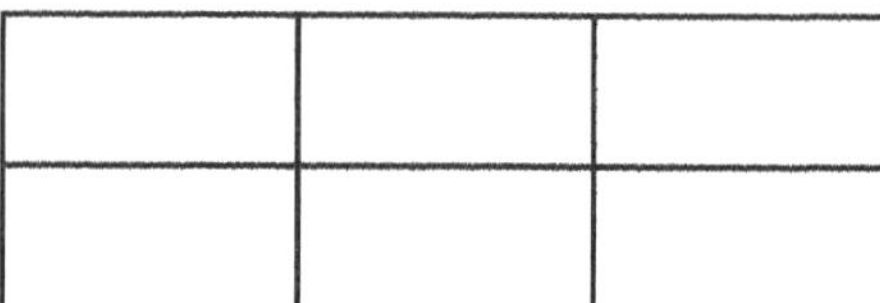

7 I think of a number, double it, add 8 to the result and my answer is 22.

My number is ________

8 Place these numbers in order from smallest to largest.

2·9, 2·09, 2·99, 2·92

________ ________ ________ ________

9 Emma is facing East. She turns clockwise through an angle of 90°.

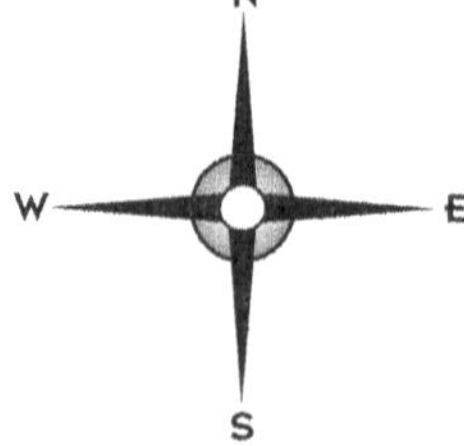

Which way is she now facing? __________

10 Douglas has a mass of 84 kg. If he loses 500 g per week, what will Douglas's mass be after 10 weeks?

____

11 I am thinking of a two-digit number. If the sum of the digits of my number is 9, find the:

a smallest possible value of the even number I am thinking of.

____

b largest possible value of the even number I am thinking of.

____

12 Five boys swim a 50 metre race. Their times are recorded below.

Lane	Name	Time (seconds)
1	Jason	87·4
2	Leon	91·2
3	Robert	87·9
4	Gabor	84·3
5	Nick	89·3

a Who finished last? _________________

b Who finished second? _______________

c How many seconds faster was Gabor than Jason? _______________

13 The Ho family drink 3 litres of milk in 4 days.

a How many litres will they drink in 24 days? _________

b How many days will it take the Ho family to drink 12 litres of milk?

____

14 In this addition some of the digits are missing. Write your answer clearly in the boxes.

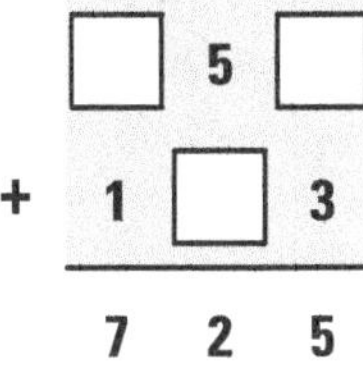

15 If we count by 4s starting with 1, we obtain the sequence 1, 5, 9, 13, ...

1 is the first term,

5 is the second term,

9 is the third term and so on.

Find:

a the 10th term. _________

b which term is 81. _________

16 A container holds 2·5 litres of orange juice. How many 250 mL glasses can be filled from this container?

17 Daniella has made this shape by stacking some centicubes together.

a How many cubes did Daniella use to make this shape? ________

b How many more cubes does Daniella need to add to this shape to form a cube of side 3 cm? ________

18 One table seats 4 people. Two tables in a row seat 6 people.

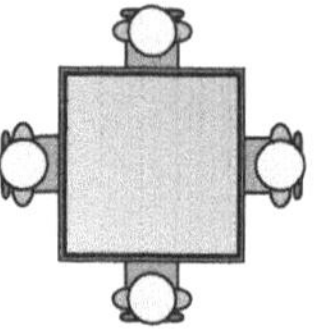

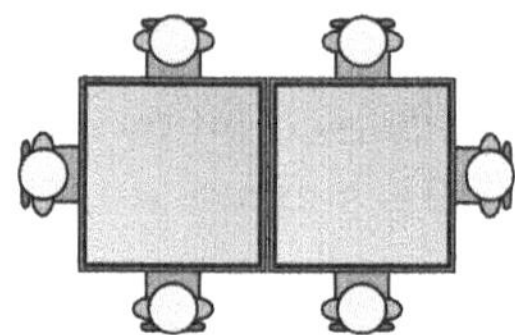

a Draw 3 tables in a row and find the number of people that can be seated.

b Look for a pattern and work out how many tables in a row are needed to seat 12 people? ________

19 This is Richard's clock.

On each clock face below show the time required:

a at the same time as shown on Richard's clock.

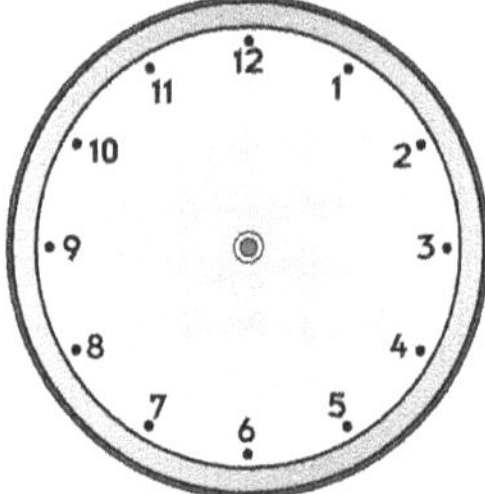

b 45 minutes before the time shown on Richard's clock.

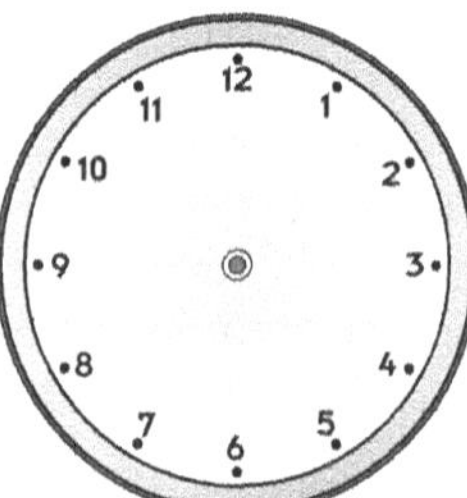

c 90 minutes after the time shown on Richard's clock.

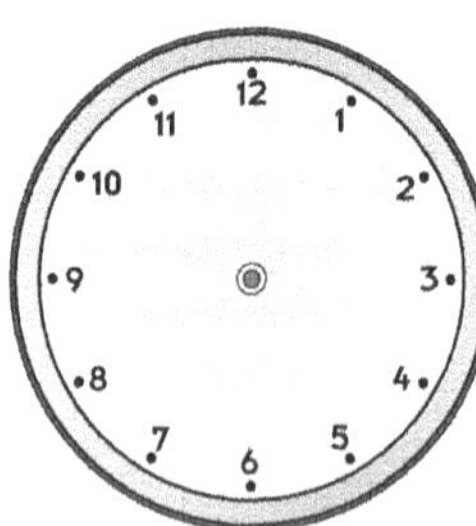

20 Perimeter is the distance around the outside of a shape.
This rectangle has a perimeter of 12 units.

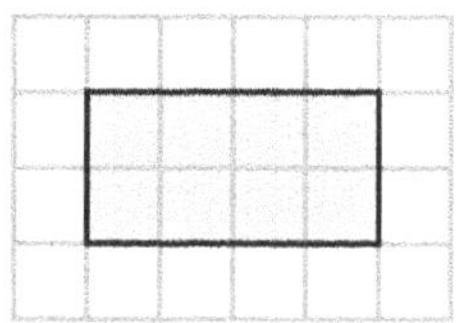

On the grid paper below draw two different rectangles each with a perimeter of 10 units.

21 If [square] is 5 kg, estimate the mass of the other shapes.

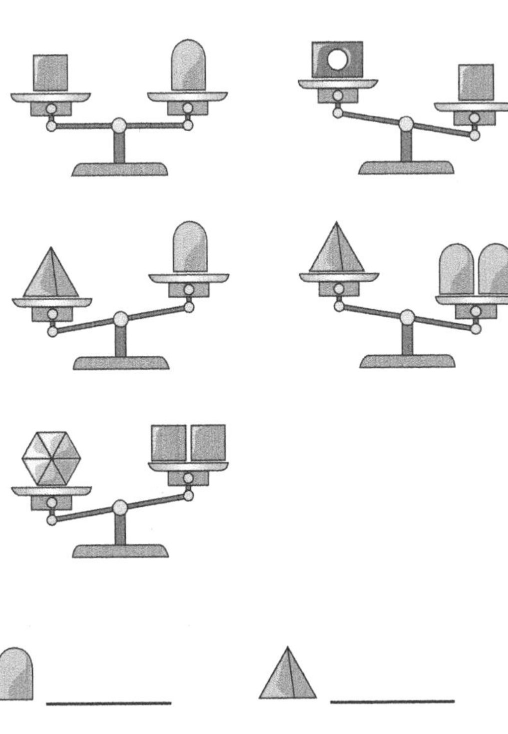

[dome] ________ [triangle] ________

[rectangle with circle] ________ [hexagon] ________

22 Consecutive numbers are numbers that follow one another. Examples of three consecutive numbers are 3, 4, 5 and 11, 12, 13.
Find three consecutive numbers for which the sum is 63.

____

23 Here is a picture of three people.

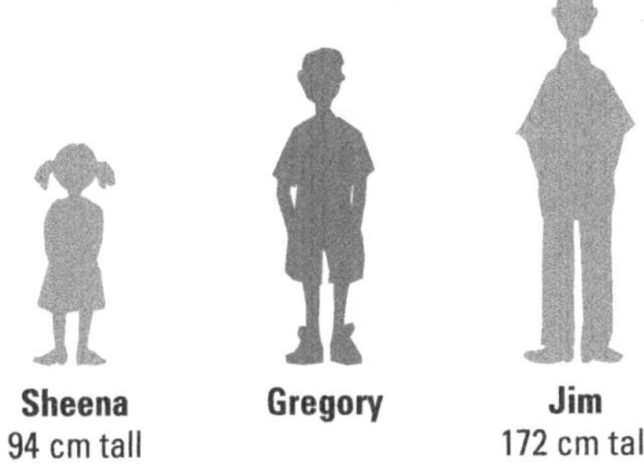

a Gregory's height is halfway between Sheena's height and Jim's height. Calculate Gregory's height.

____

b If Sheena's mother, Helena, is twice as tall as Sheena, calculate her height.

____

24 Seven cycle riders passed our house.
Some were riding bicycles with 2 wheels and some were riding tricycles with 3 wheels. I counted 19 wheels.

How many bicycles and how many tricycles were there?

Number of bicycles	Number of tricycles	Total wheels

enrich-e-matics 3rd EDITION TEST

Time: 45 minutes

YEAR 4
TEST 4

Name: ____________________

Class: ____________________

Multiple choice

Colour in the bubble next to the answer.

1 Which sentence is true?

A ◯ 625 > 526 B ◯ 345 > 435

C ◯ 564 < 554 D ◯ 874 < 784

2 $12 \times 8 = 6 \times \square$

Find the missing number in the box.

A ◯ 96 B ◯ 4

C ◯ 2 D ◯ 16

3 $32 = \square \times 5 - 3$

Find the missing number in the box.

A ◯ 16 B ◯ 35

C ◯ 9 D ◯ 7

4 24 ◯ 8 ◯ 5 = 8

The missing signs are:

A ◯ + and ÷ B ◯ ÷ and +

C ◯ − and ÷ D ◯ ÷ and −

5 The sector graph shows how Rita spends 24 hours of each day. How many hours does she spend at school?

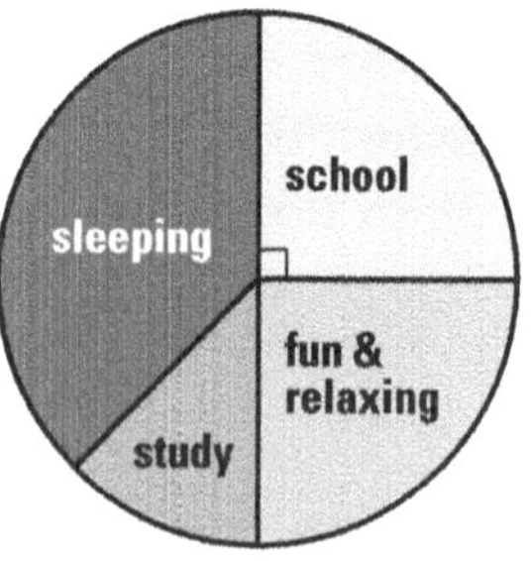

A ◯ 6 hours B ◯ 4 hours

C ◯ 8 hours D ◯ 5 hours

6 What is the real time if my watch, which is 3 minutes slow, is showing 8:45 a.m.?

A ◯ 8:42 a.m. B ◯ 8:75 a.m.

C ◯ 8:48 a.m. D ◯ 11:45 a.m.

7 If a $2\frac{1}{4}$ hour movie starts at 9:50 a.m., it should finish at:

A ◯ 12:05 a.m. B ◯ 11:05 a.m.

C ◯ 12:05 p.m. D ◯ 1:05 p.m.

8 Elizabeth spent two hours at the gym. She spent the first 30 minutes doing warm-up exercises.
What fraction of the time spent at the gym was she doing warm-ups?

A ◯ $\frac{1}{5}$ B ◯ $\frac{3}{20}$

C ◯ $\frac{1}{2}$ D ◯ $\frac{1}{4}$

9 $\square < 19$

 > 8

One of these numerals below will make *both* these sentences true. Choose the correct one.

A ◯ 21   B ◯ 5

C ◯ 15   D ◯ 28

10 Mindy can make 75 ribbons in 30 minutes. At this rate, find the time taken to make 100 ribbons.

A ◯ 40 minutes   B ◯ 50 minutes

C ◯ 25 minutes   D ◯ 80 minutes

11 Which one of these gives the largest answer?

A ◯ one quarter of 20

B ◯ one third of 9

C ◯ one tenth of 45

D ◯ one fifth of 24

12 What fraction of the square is shaded?

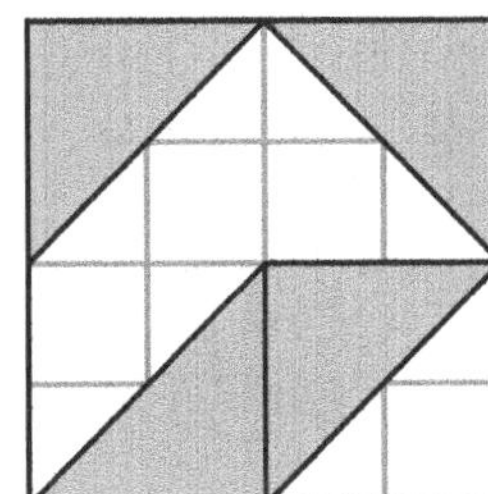

A ◯ $\frac{2}{3}$ B ◯ $\frac{1}{2}$

C ◯ $\frac{7}{16}$ D ◯ $\frac{5}{8}$

13 If 3 kg of jelly beans cost $12, what would 2 kg of jelly beans cost?

A ◯ $24   B ◯ $4

C ◯ $9   D ◯ $8

14 If $1\frac{1}{2}$ kg of biscuits cost $6, how much will 500 g of biscuits cost?

A ◯ $4   B ◯ $2

C ◯ $3   D ◯ $5

15 Carly works for 6 hours in order to earn $24. If she is paid the same rate per hour, find how many hours she will have to work in order to earn $44.

A ◯ 20 hours   B ◯ 11 hours

C ◯ 9 hours   D ◯ $10\frac{1}{2}$ hours

16 A class of 28 students record their favourite pets:

3 students like birds
12 students like dogs
7 students like cats
6 students like horses.

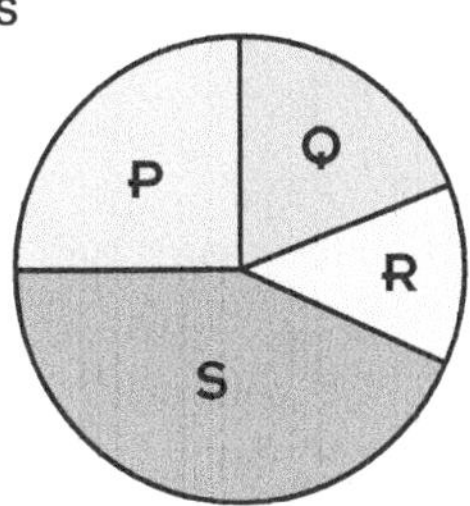

Which part of the pie chart shows the students who like cats?

A ◯ P   B ◯ Q

C ◯ R   D ◯ S

17 What is the maximum number of cards measuring 3 cm by 5 cm that can be cut from a piece of cardboard 20 cm by 6 cm?

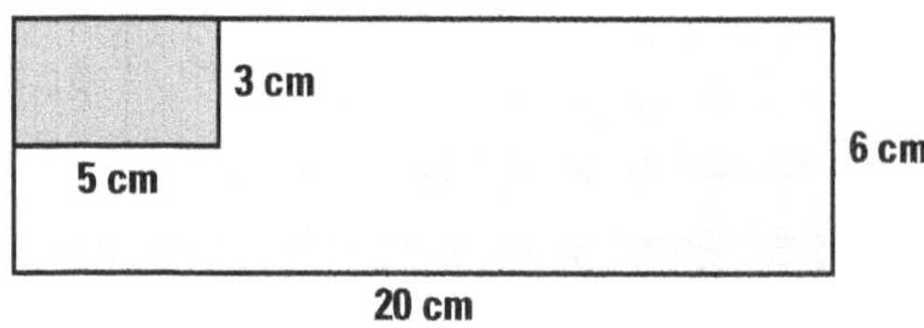

A 12 B 15
C 8 D 4

18 In space arithmetic:

8 is written as ●ΛΛ
13 is written as ●●Λ
15 is written as ●●ΛΛΛ

How would 10 be written in space arithmetic?

A ●ΛΛΛ B ΛΛΛΛΛ
C ●● D ●ΛΛΛΛ

19 After a cataract operation, Mrs Kavanagh has to put 2 drops in her eyes every 8 hours. How long will a bottle containing 96 drops last her?

A 6 days B 48 days
C 16 days D 8 days

20 How long will it take Kevin to walk 200 metres, if he walks 40 metres in 30 seconds?

A 5 minutes B 150 seconds
C $1\frac{1}{2}$ minutes D 5 seconds

21 At 7 p.m. the temperature was 17°C. By midnight the temperature had fallen 6°C, and between midnight and 7 a.m. it rose 8°C. Which thermometer shows the correct temperature at 7 a.m.?

A B C D

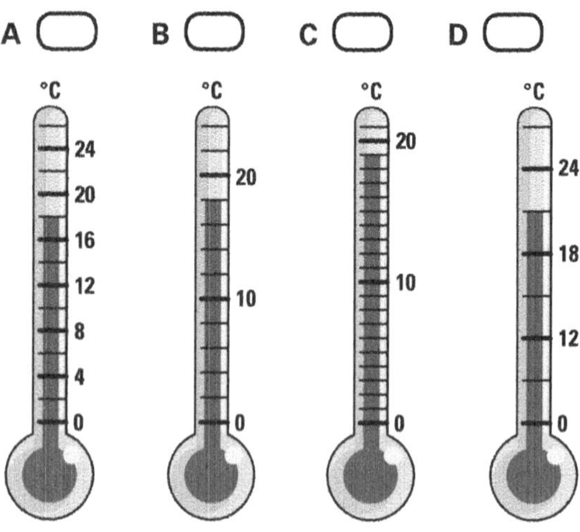

22 Suppose you have 4 black, 6 brown and 2 navy socks in your drawer. They are exactly the same quality. The room is dark.

What is the least number of socks you must take out of your drawer to be certain that you have a pair of the same colour?

A 2 B 3
C 4 D 5

23 The perimeter of this rectangle is 50 m. Find its width.

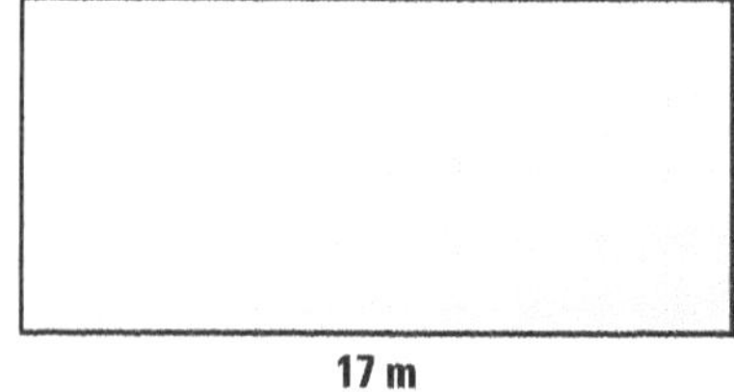

A 13 m B 33 m
C 16 m D 8 m

24 Barbara has two 10-cent and two 50-cent coins. How many different amounts of money can she make if she wants to use at least one (one or more) of each type of coin?

A ⬭ 8 B ⬭ 4
C ⬭ 10 D ⬭ 5

25 Which whole number should be written in the box to make a fraction whose value is between 1 and 2?

A ⬭ 9 B ⬭ 14
C ⬭ 6 D ⬭ 10

26 If I draw a dot on each face of a cube, there will be 6 dots.

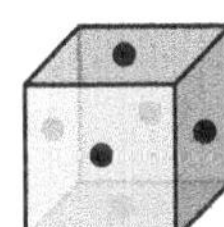

If 2 cubes are glued together and a dot is drawn on each square face that is showing, there will be 10 dots.

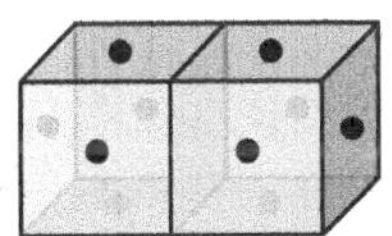

If you made the shape below by sticking cubes together and placed a dot on each square face, how many dots would you count?

A ⬭ 7
B ⬭ 11
C ⬭ 14
D ⬭ 16

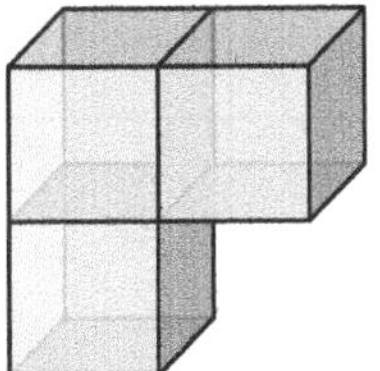

27 On Friday a 40-litre petrol tank was $\frac{1}{2}$ full.

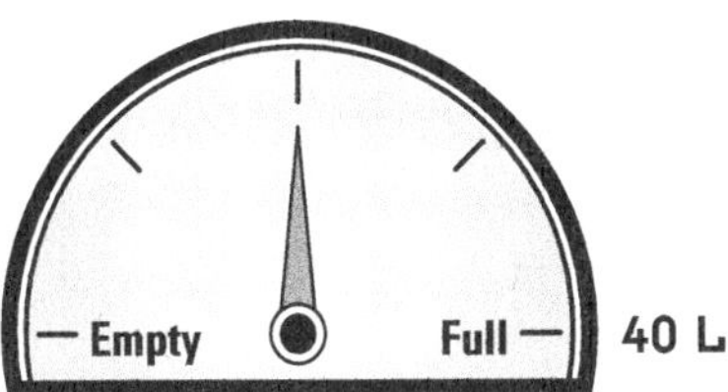

On Saturday 4 litres of petrol were used.
On Sunday the tank was filled till it was $\frac{3}{4}$ full.
How many litres of petrol were added on Sunday?

A ⬭ 14 L B ⬭ $22\frac{1}{2}$ L
C ⬭ 24 L D ⬭ 30 L

28 Michael worked from 9 a.m. to 9 p.m. He was paid $20 per hour for the hours 9 a.m. till 5 p.m. and $30 per hour for the rest of the time. How much was Michael paid?

A ⬭ $50 B ⬭ $240
C ⬭ $190 D ⬭ $280

enrich-e-matics TEST

3rd EDITION

Time: 45 minutes

YEAR 5 TEST 1

Name: ______________________

Class: ______________________

1 What is the value of each of the following:

a 98 + 98 + 98 + 98 + 98 + 98 + 98 + 98 + 98 + 98 =

____

b 10 + 11 + 12 + 13 + 14 + 15 + 16 + 17 + 18 + 19 + 20 + 19 + 18 + 17 + 16 + 15 + 14 + 13 + 12 + 11 + 10 =

____

2 If

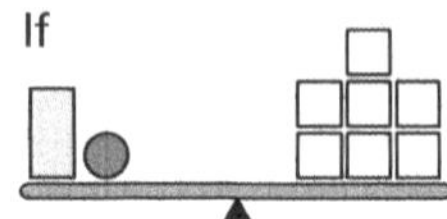

and

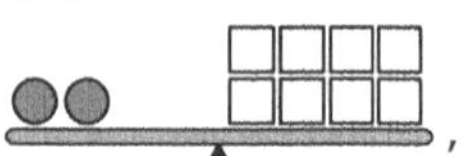,

then

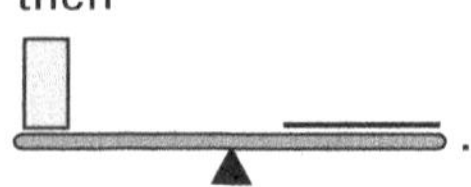.

3 Five darts landed on the dartboard shown. Each dart scores 3, 5 or 7 points.

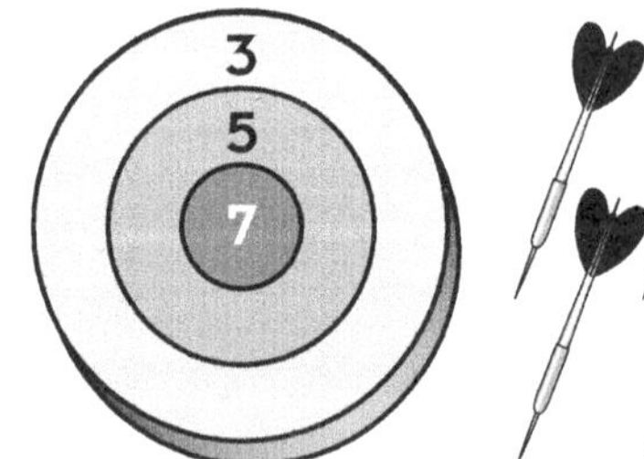

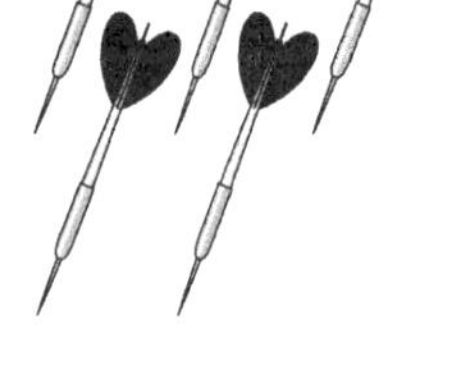

One way for 5 darts to score a total of 25 points is shown in the table.

Can you find two other different ways for 5 darts to score a total of 25 points?

3	5	7
1	3	1

4 Maureen is making ribbons. From 2 metres of material, she can cut 5 ribbons.

a How many metres did she have if she cut 20 ribbons? __________

b How many ribbons did she cut from 6 metres of material? __________

5 **a** If you buy 6 pens at $1.25 each, find the change from a $20 note.

____

b How many 45-cent postage stamps can I buy if I have $9.00?

____

c Which is the better buy?

Colour in the bubble next to the answer.

A ◯ 4 pens costing $2.76

B ◯ 3 pens costing $1.80?

Explain your reason carefully by giving the price of one pen (or 12 pens).

____

____

____

____

6 1 m × 1 m concrete slabs are laid around a large rectangular swimming pool.

How many slabs are needed altogether to completely surround the pool?

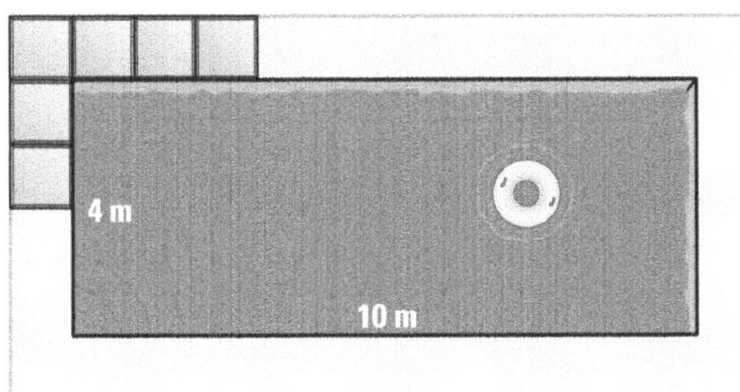

____

7 **a** Write down three different fractions (not equivalent) between $\frac{1}{4}$ and $\frac{1}{2}$.

________ ________ ________

b Find the fraction which is exactly halfway between $\frac{1}{4}$ and $\frac{1}{2}$.

____

8 Calculate the area of the shaded triangle.

(Each square represents 1 cm².)

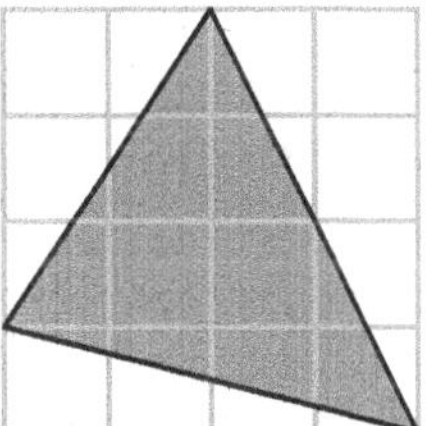

____

9 Pauline has more than 25 marbles but fewer than 50. When the marbles are counted by threes, there is one left over. When they are counted by fives, there are two left over.

How many marbles does Pauline have?

____

10 Suppose all the counting numbers are written in columns in the pattern shown.

A	B	C	D	E	F	G
1		2		3		4
	5		6		7	
8		9		10		11
	12		13		14	
15		16		17		18

a Continue the pattern for three more rows.

b Name the letter of the column in which the number:

i 35 appears. ________

ii 71 appears. ________

11 **a** How many:

faces ________,

edges ________ and

vertices ________ has a cube?

b If this cube was truncated as shown, how many:

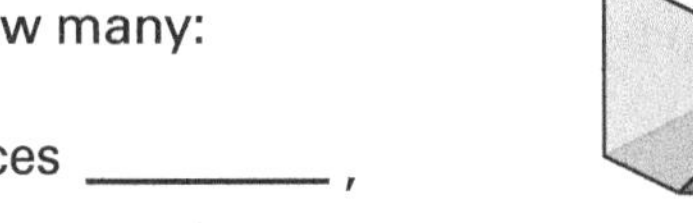

faces ________,

edges ________ and

vertices ________ would it have?

12 This pie graph shows how a student, Michael, spends his time at school on Tuesdays.

If Michael spends 1 hour 20 minutes playing sport, how long does he spend:

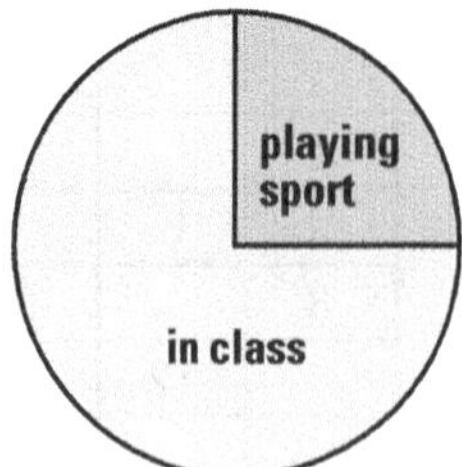

a in class? ____________

b at school on Tuesdays? ____________

13 If you made a spinner like this, and:

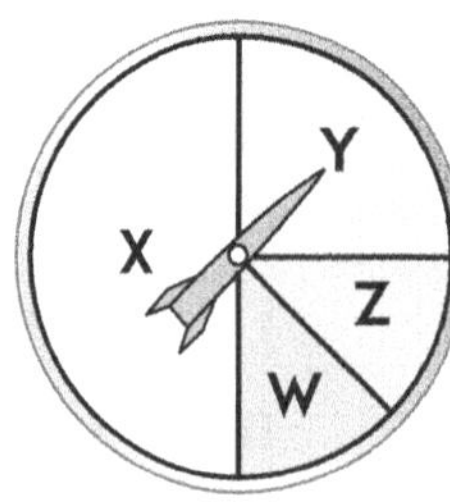

a you made 16 spins, how many times do you predict that it would land in W?

____

b the spinner landed 7 times in Y, predict how many times the spinner was spun altogether?

____

14 Here is a hexagon. Three of its diagonals have been drawn using broken lines.

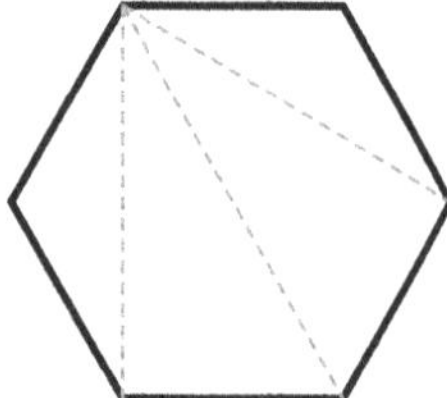

How many diagonals does a hexagon have altogether? ________

15 Imran had two maths tests this term. He got 82 in the first test and 73 in the second test.

What mark should he get in the third test so that his average for the three tests is 80?

____

16 PQRS is a square. A and B are mid-points of PQ and SR respectively.

Find what fraction of the square is shaded.

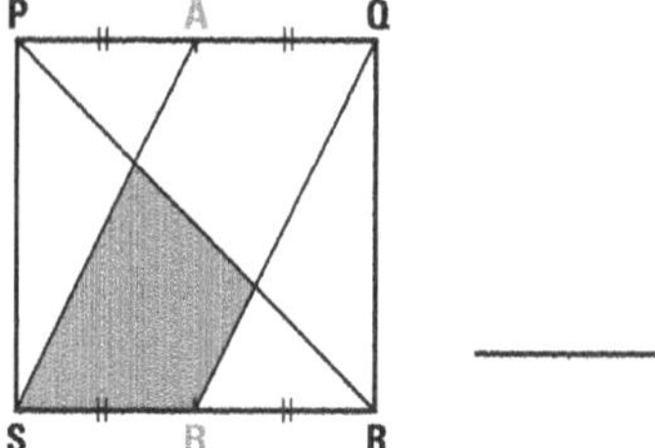

____

17 Given the birth date of the following people, number them in order from youngest to oldest.

The first three have been done for you.

	Name	Day	Month	Year
3	Kate	25	May	1966
	Phillip	7	August	1952
	George	12	February	1949
	Elizabeth	7	April	1952
2	Aaron	30	Dec.	1976
1	Peter	26	January	1980
	Sylvia	5	June	1947
	Mary	8	Sept.	1947

18 On a map, a length of 3 centimetres represents an actual distance of 12 kilometres. How many kilometres apart are two towns which are drawn as $2\frac{1}{2}$ centimetres apart on the map?

____

19 Three different points are marked on each of two parallel lines. Find the number of different triangles that can be formed using three of the six points.

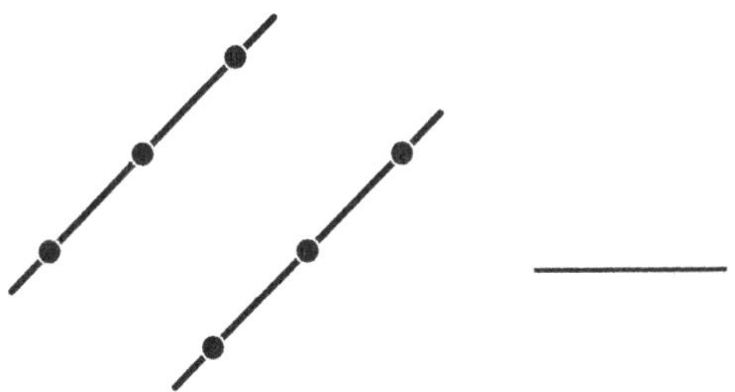

____

20 Look at these shapes and answer the questions that follow.

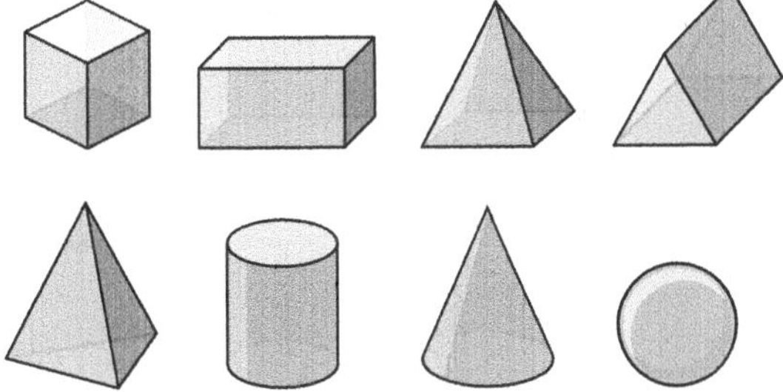

a Circle the solid which has 5 faces, 8 edges and 5 corners.

b Colour in the solid which has 3 faces, 2 edges and no corners.

c Name two solids which, when sliced in a certain way, will have a circle as a face. Draw a line through these solids to show the slicing you would use.

____

____

21 Shopping centre parking fees

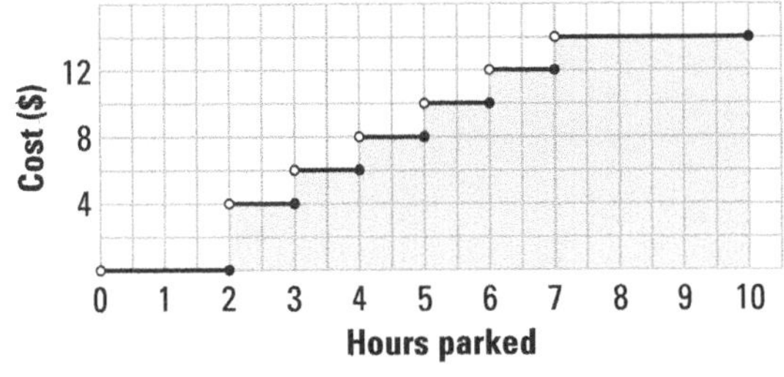

a Find the cost of parking for $3\frac{1}{2}$ hours.

____

b What is the maximum time a car can be parked for $10? ________

22 What are the values of A and B in each of the following if any three numbers next to each other have a sum of 16?

a

	2		5					A	B		

A ________ B ________

b

7		A	B	4				5		

A ________ B ________

23 Sharon walks along the edges of a rectangular pool from A to B to C to D, a distance of 42 metres. Nicky walks along the edges of the same pool starting at B and going to C to D to A, a distance of 33 metres.

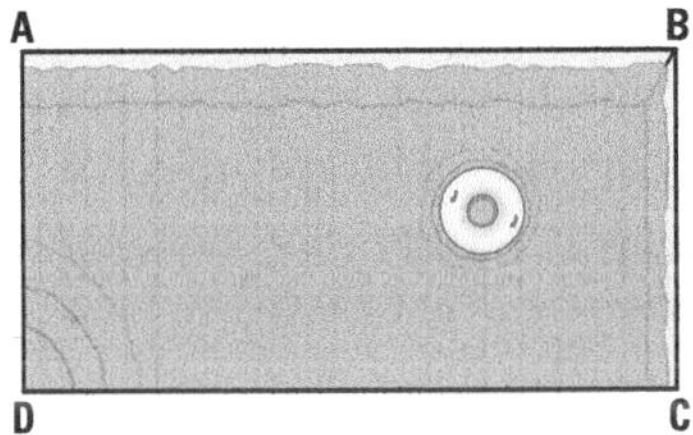

What is the perimeter of the pool in metres? ________

24 Each person tells a secret to three others in an hour's time and each one of these three tells three others, in an hour's time. If this continues:

a how many people will know the secret in 2 hours' time? (Remember, one person starts telling the secret in the beginning.) ________

b how long will it take for 40 people to know the secret?

____

enrich-e-matics TEST

3rd EDITION

Time: 45 minutes

YEAR 5 TEST 2

Name: ____________________

Class: ____________________

1 The sum of two numbers is 16 and their product is 48. Find the two numbers.

__________ __________

2 A book has 68 pages. How many times is the 5 digit used in numbering the pages?

____

3 Diane entered a shop with $20. She bought an orange juice for $3.50 and two cheese sandwiches. She was given $8.10 change.

How much did one cheese sandwich cost?

____

4 In a class of 28 students, there are six more boys than girls.

How many girls are there in the class?

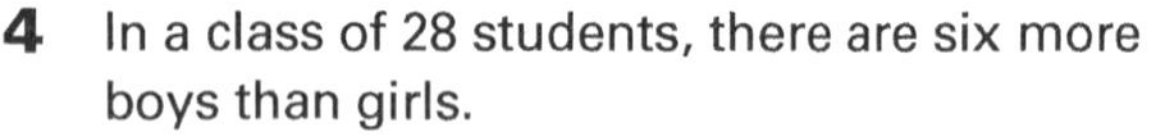

Boys	Girls

5 If the area of a square is 64 cm^2, find the perimeter of this square.

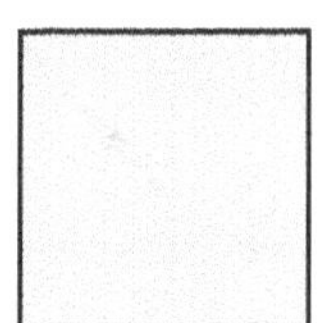

____

6 **a** Write down the missing lengths of the sides of the figure below.

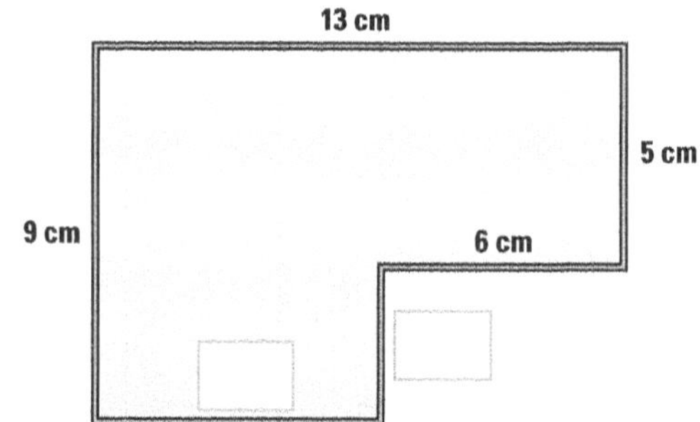

b Find the perimeter of this figure.

____

c Find the area of this figure.

____

7 Find the value of $5 + 55 + 555 + 5555 + 55555 + 555555$.

____

8 **a** Look at this pattern and continue it for two more lines.

$2^2 - 1 = 3 = 1 \times 3$

$3^2 - 1 = 8 = 2 \times 4$

$4^2 - 1 = 15 = 3 \times 5$

____

____

b Find the value of $99^2 - 1$ using the pattern above.

____

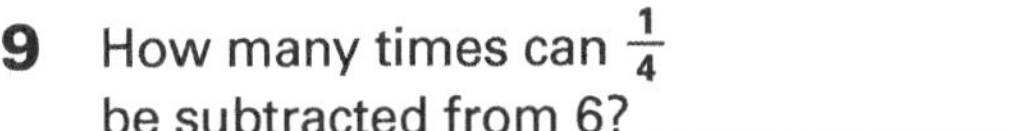

9 How many times can $\frac{1}{4}$ be subtracted from 6? ____________

10 To send a parcel, it costs $2.50 for the first 300 grams and 20 cents for each additional 10 grams. What is the cost of sending a parcel that has a mass of 450 grams?

____

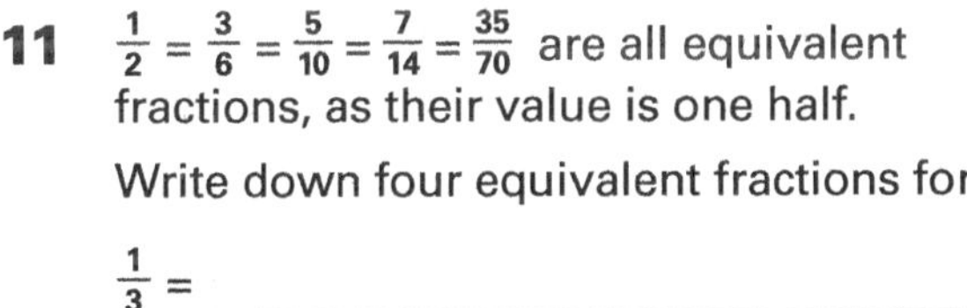

11 $\frac{1}{2} = \frac{3}{6} = \frac{5}{10} = \frac{7}{14} = \frac{35}{70}$ are all equivalent fractions, as their value is one half.

Write down four equivalent fractions for

$\frac{1}{3} =$ ______ ______ ______ ______

12 What is the next number in the pattern?

$4\frac{7}{12}$, $4\frac{2}{12}$, $3\frac{3}{4}$, $3\frac{4}{12}$, ________

13 A plane flew 410 km in 25 minutes. At this rate, how many km would the plane fly in:

a 5 minutes? ________

b 1 hour? ________

14 This shape is made up of 3 cm cubes.

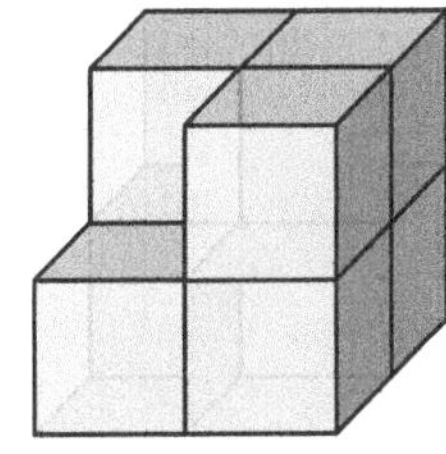

a What is the volume of each cube? ________

b Find the total volume of this shape.

____

15 Mr Green gives $480 to his two sons.

a If the elder son receives twice as much as the younger son, how much does each son receive?

younger ________ elder ________

b If the elder son receives three times as much as the younger son, how much does each son receive?

younger ________ elder ________

16 If 18 men can do a piece of work in 10 days:

a in how many days will 6 men do the same work? ________

b in how many days will 36 men do the same work? ________

c and the work must be completed in 2 days, how many men are required?

____

17 Michael spent $\frac{1}{4}$ of his money on a book and found that $\frac{1}{3}$ of the remainder was $15.

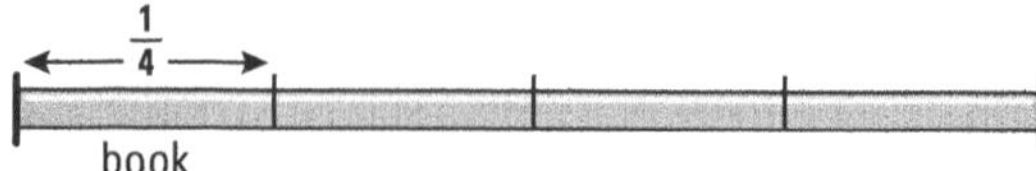

How much did he have at first? ________

18 E is the mid-point of AB.
F is the mid-point of BC.
ABCD is a square.

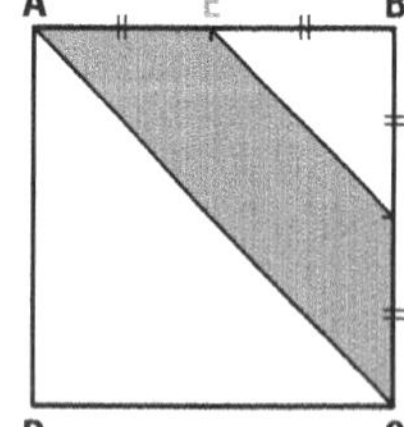
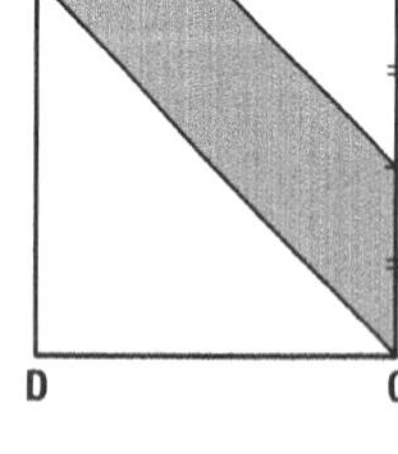

What fraction of the square is shaded? ________

19 In this spinning wheel, the area of number 3 is the same as the area of number 4, while the area of number 5 is 4 times the area of number 4.

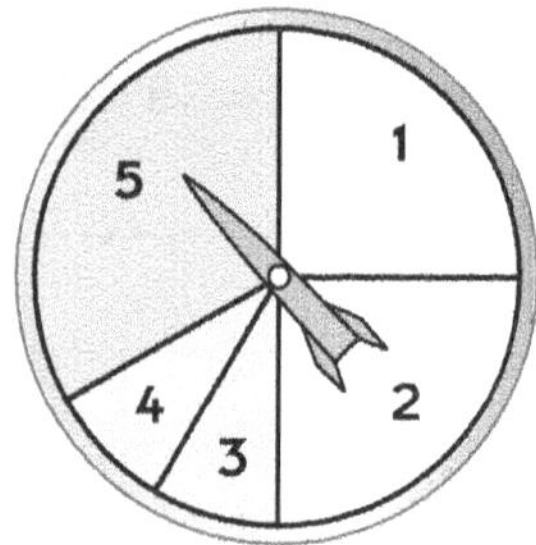

a What fraction of the area of the circle is the area of 3? ____________

b What is the probability of spinning 1? ________________

c What is the probability of spinning 5? ________________

d What is the probability of spinning an even number? ________

20 A man plants three rows of corn. Each row is 5 m long. The plants in each row are spaced 50 cm apart. If the first and last plants in each row are 25 cm from the ends of the rows, what is the total number of plants?

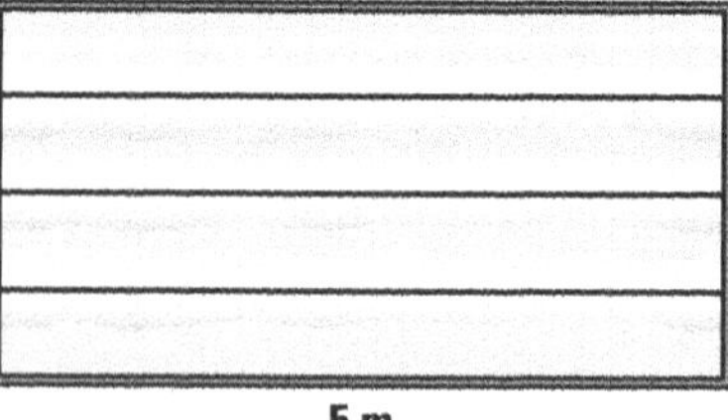

____

21 All the shapes drawn are squares. Find the area of each white square (with side x cm).

Hint: find the length of A first.

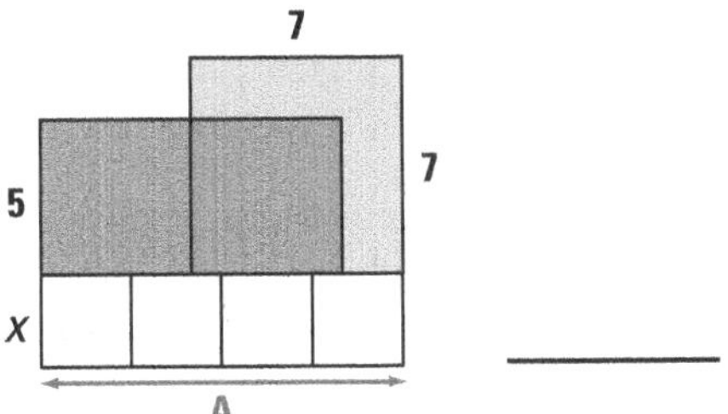

____

22 If a block of chocolate is divided into 5 equal pieces, each piece would be 12 grams heavier than if the block of chocolate was divided into 6 equal pieces.

What was the mass of the block of chocolate? ________

23 3, 4, 7, 1, 8, 9, ...

______ ______ ______ ______ ______

Continue this chain by adding the last two numbers together each time, and writing down only the unit digit.

Write down the next five numbers.

24 Bell A rings every 6 minutes, bell B rings every 10 minutes and bell C rings every 9 minutes. If the three bells have just rung together, how many minutes until they next ring together?

____

25 In the addition problem below, each letter represents a digit, and different letters represent different digits. Work out the sum and the value of each letter.

Can you find two different solutions?

```
  A B C D
    A B C
      A B
+       A
---------
  1 9 9 3
```

A = ______ ______

B = ______ ______

C = ______ ______

D = ______ ______

26 A large cube is built of 64 white building blocks. After painting the surface of this large cube red, it is taken apart.

How many cubes will have:

a 3 red faces? ________

b 1 red face? ________

c no red face? ________

enrich-e-matics TEST

3rd EDITION

Time: 45 minutes

YEAR 5
TEST 3

Name: ______________________

Class: ______________________

1 Sonya left home at 9:45 a.m. and came home at 3:20 p.m.

How long was she away from home?

____

2 What fraction of the figure is shaded?

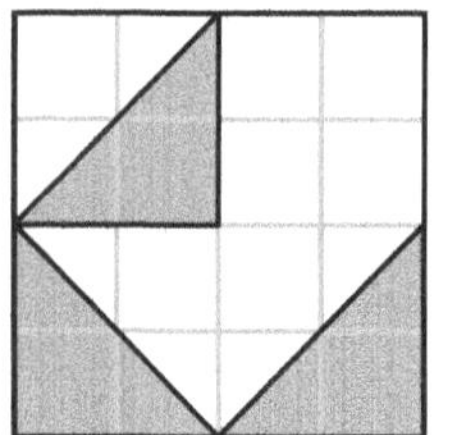

____

3 Write down the digit that should be written in the box to make a fraction whose value is 2.

____

4 The perimeter of a square is 24 m.

What is the area of the square? __________

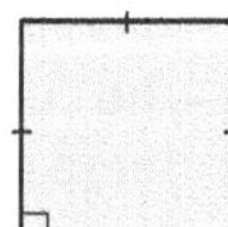

5 If you have a piece of paper and fold it as shown, draw what the result would look like if the paper was opened out after a corner and a piece were cut out.

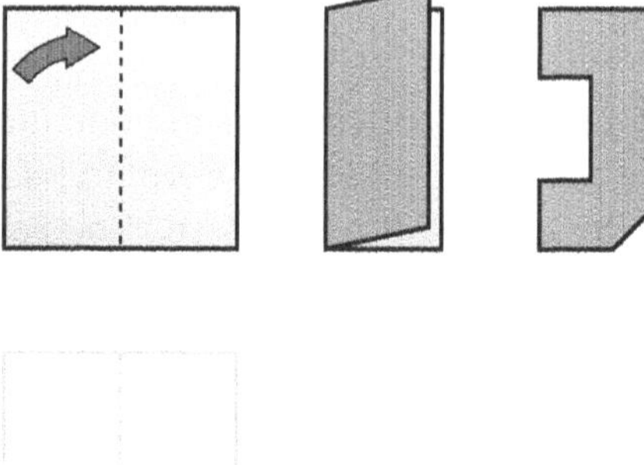

6 If this tank with a capacity of 60 L was empty and David put in 45 L, draw what the petrol gauge will show.

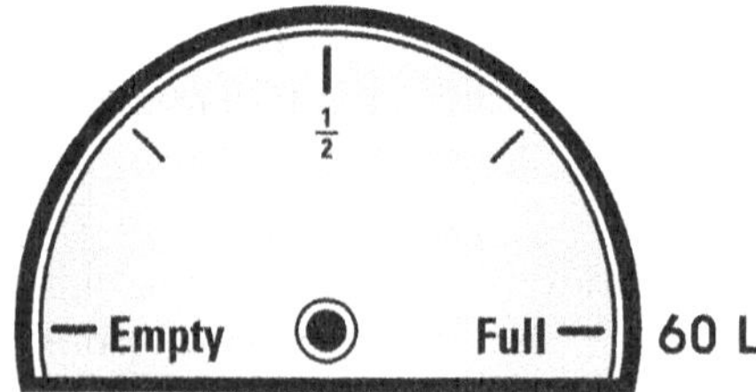

7 I bought an apple and a drink for $1.70, while James bought three apples and a drink for $3.20.

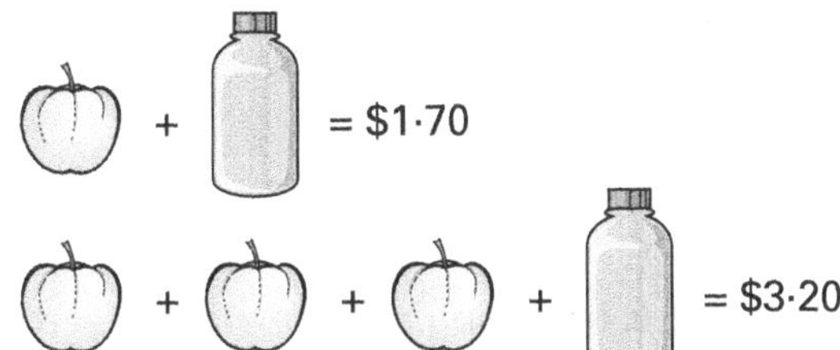

a Use the diagram to work out the cost of two apples. __________

b How much did one apple cost?

____

8 If

14

and

6,

then

__________ .

9

Fare rates	Melb. – Syd.	Melb. – Bris.	Syd. – Bris.
Economy Class by plane	$180	$259	$190
Business Class by plane	$240	$330	$250
First Class by plane	$270	$390	$280
Economy Class by train	$70	$120	$80
First Class by train	$95	$130	$100
Bus	$50	$75	$55

a Find the cost of an Economy Class plane fare from Melbourne to Brisbane.

____

b Find the difference between two Business Class and two Economy Class fares for a plane trip from Melbourne to Sydney. __________

10 Cards measuring 4 cm by 3 cm are cut from a piece of cardboard 12 cm by 9 cm.

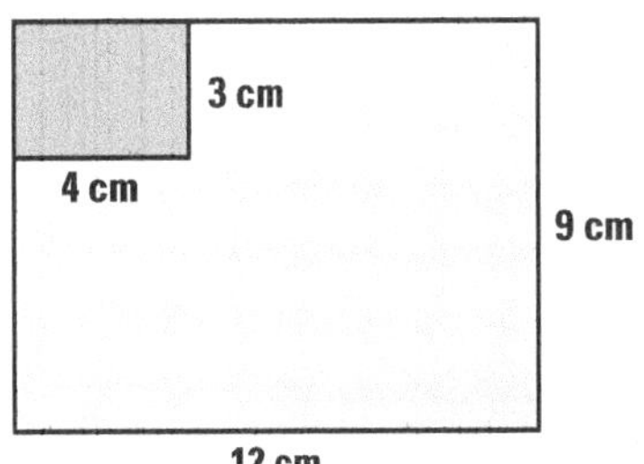

a How many cards can be cut from the top row shown? __________

b What is the maximum number of cards that can be cut from the cardboard?

____

11 A floor tiler uses 4 grey tiles to every 3 white tiles.

Grey tiles	White tiles	Total tiles
4	3	7
20		
		42

a If 20 grey tiles are used, how many white tiles are there? __________

Hint: use the table above to help you.

b If there are 42 tiles altogether, how many grey tiles are used?

____

12 Find the temperature shown on the thermometer in °C.

First look at the scale carefully.

__________ °C

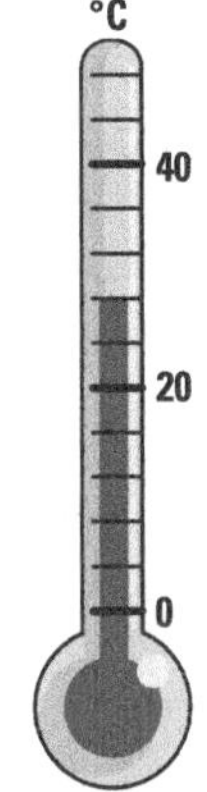

13 All integers are arranged in six columns as shown.

a Continue the pattern above for the next two lines.

A	B	C	D	E	F
	1	2	3	4	5
10	9	8	7	6	
	11	12	13	14	15
20	19	18	17	16	
	21	22	23	24	25

b In which column will 51 be? ________

c In which column will 77 be? ________

14 Write down two numbers less than 50 that can be divided by 5 and 2 with a remainder of 1.

____________ ____________

15 Find the area of the shape below if □ is 1 cm^2.

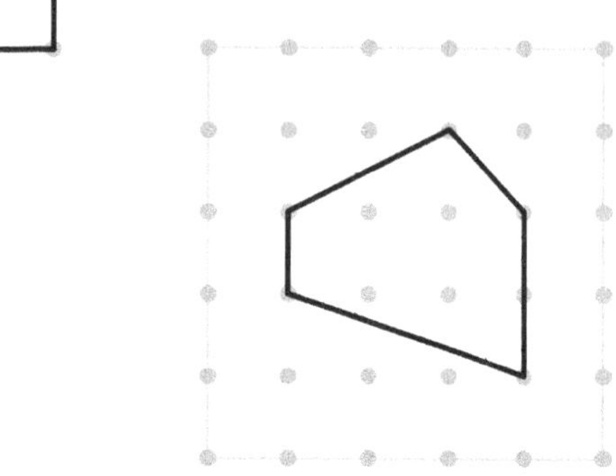

Area = ____________

16 If it takes 3 men 12 days to do a piece of work, how long would it take 9 men working at the same rate?

____

17 One bell rings every 6 minutes, another rings every 8 minutes. If the bells have just rung together, how many minutes until they next ring together?

____

18 A car is travelling at a constant speed of 60 km/hr, that is, 60 kilometres in one hour.

a How far does it travel in 2 hours?

____

b How far does it travel in $\frac{1}{2}$ hour?

____

c If it travels 210 km, how long does it travel for? ____________

19

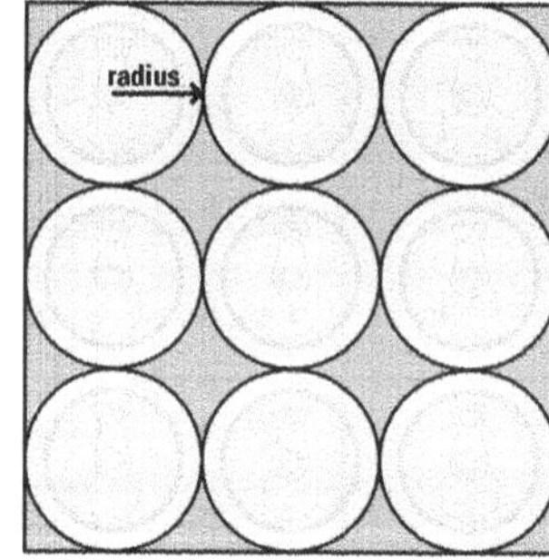

a If the radius of each circle is 2 cm, find the area of the square.

____

b The area of this square is 81 cm^2. Find the radius of each circle.

____

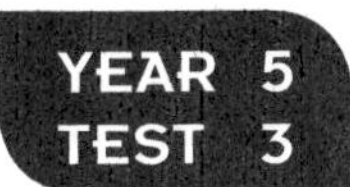

20 One day Anna bought apples at 3 for 75c and sold all of them at 2 for 80c.

a If Anna bought and sold 6 apples, how much profit did she make?

____

b If she made a profit of $18, how many apples did she sell that day?

____

21 Five cards with numbers on them are in a bag.

Without looking, Rachel takes one card out of the bag. What is the chance that the number on the card:

a will be even? ________

b will be prime? ________

c will be divisible by 3? ________

d will have 5 as a factor? ________

22

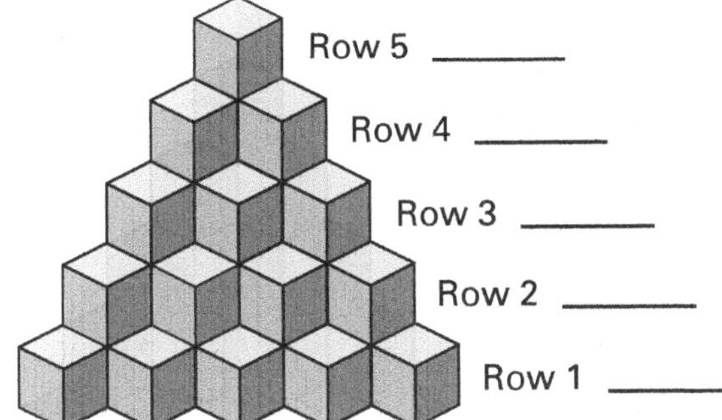

Row 5 ______

Row 4 ______

Row 3 ______

Row 2 ______

Row 1 ______

a Can you count the cubes in each row? (Start at the top row.)

b How many cubes in all? ________

23 In a maths quiz of 10 problems, 5 points are given for each correct answer and 2 points are deducted for each incorrect answer. If Kevin did all 10 problems and scored 29 points, how many correct answers did he have? ________

This table may be useful.

Number correct (5 each)	Number incorrect (– 2 each)	Total score

24 David and Danny bought some toy models. David bought 3 cars and 1 aeroplane for $35, while Danny bought 1 car and 3 aeroplanes for $41. How much was each toy model?

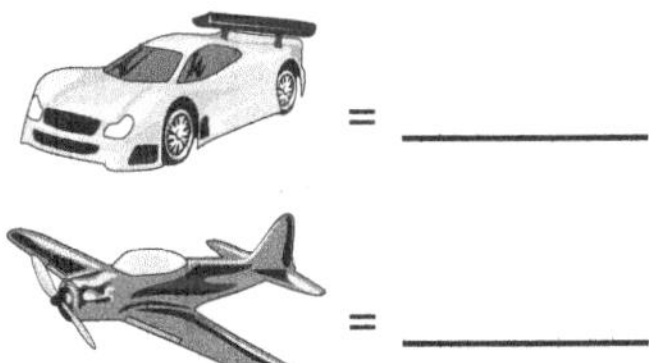

= ________

= ________

25 Two whole numbers when multiplied together give an answer of 1000. Neither of the two numbers contains any zeros.

What are the two numbers?

__________ __________

enrich-e-matics TEST
3rd EDITION

Time: 45 minutes

YEAR 5
TEST 4

Name:

Class:

Multiple choice

Colour in the bubble next to the answer.

1 $6\frac{3}{5}$ is the same as:

A ◯ $\frac{23}{3}$ B ◯ $\frac{33}{3}$

C ◯ $\frac{23}{5}$ D ◯ $\frac{33}{5}$

2 Jerry wakes up at 6:50 a.m., takes 30 minutes to get dressed, 10 minutes to eat breakfast and 15 minutes to travel to school. At what time does he get to school?

A ◯ 7:45 a.m. B ◯ 7:35 a.m.

C ◯ 7:55 a.m. D ◯ 7:05 a.m.

3 Which one of the following has the largest answer?

A ◯ $\frac{1}{5} \times 15$ B ◯ $\frac{1}{3} \times 6$

C ◯ the number of quarters in 1

D ◯ the number of halves in 3

4 How many quarters ($\frac{1}{4}$) are there in $2\frac{1}{2}$?

A ◯ 4 B ◯ 8

C ◯ 10 D ◯ 12

5 1563 rounded to the nearest 100 would be:

A ◯ 2000 B ◯ 1500

C ◯ 1600 D ◯ 1560

6 Which of these sentences is *not* true?

A ◯ $4 \times 5 > 13 + 5$

B ◯ $12 - 8 < 5 - 2$

C ◯ $3 \times 5 = 5 \times 3$

D ◯ $12 \div 2 = \frac{1}{2} \times 12$

7 To work out how many pieces of string each $\frac{1}{8}$ m long can be cut from a roll $1\frac{3}{4}$ m long, which one of the following needs to be worked out?

A ◯ $\frac{7}{4} \div \frac{1}{8}$ B ◯ $\frac{7}{4} - \frac{1}{8}$

C ◯ $\frac{7}{4} \times \frac{1}{8}$ D ◯ $\frac{1}{8} \div \frac{7}{4}$

8 In space arithmetic:

9 is written as ●Λ

13 is written as ●ΛΛΛ

16 is written as ●●Λ

How would 6 be written in space arithmetic?

A ◯ ΛΛ B ◯ ΛΛΛΛΛΛ

C ◯ ● D ◯ ΛΛΛ

9 If you multiplied 103 by 270, your answer would be about:

A ◯ 2700 B ◯ 27 000

C ◯ 270 000 D ◯ 2703

10 If $\frac{3}{4}$ of a number is 12, find the number.

A ◯ 16 B ◯ 9

C ◯ 4 D ◯ 48

11 Elizabeth calculates correctly that 25 × 37 is equal to 925. Using this information, what should her answer be for 9250 ÷ 25?

A ⬭ 370 B ⬭ 3·7
C ⬭ 37 D ⬭ 3700

12 Three pencils cost 90c. Which calculation gives the price of 5 pencils?

A ⬭ (90 ÷ 3) ÷ 5c B ⬭ (90 × 3) × 5c
C ⬭ (90 × 3) ÷ 5c D ⬭ (90 ÷ 3) × 5c

13 3 kg 56 g is the same as:

A ⬭ 3056 g B ⬭ 35·6 g
C ⬭ 356 g D ⬭ 3560 g

14 0·1 x 240 is the same as:

A ⬭ 24 × 0·10 B ⬭ 24 × 10
C ⬭ 240 ÷ 10 D ⬭ 240 × 10

15 Which of these fractions is equal to $\frac{3}{4}$?

A ⬭ $\frac{3-1}{4-1}$ B ⬭ $\frac{3\times3}{4\times4}$
C ⬭ $\frac{3+2}{4+2}$ D ⬭ $\frac{3\times2}{4\times2}$

16 How many pieces of ribbon 40 cm long can be cut from a roll $3\frac{3}{5}$ metres long?

A ⬭ 7 B ⬭ 90
C ⬭ 9 D ⬭ 900

17 Which of the following shapes has three axes of symmetry?

A ⬭ equilateral triangle
B ⬭ rhombus
C ⬭ isosceles triangle
D ⬭ rectangle

18 The plan of a house is to be drawn. The actual house is 20 m long and 15 m wide. If the scale that is used is 1 cm = 5 cm, what is the width of the plan?

A ⬭ 5 cm B ⬭ 3 cm
C ⬭ 30 cm D ⬭ 40 cm

19 ☐ > 19

☐ < 8

One of these numerals below will make *both* these sentences false. Choose the correct one.

A ⬭ 21 B ⬭ 5
C ⬭ 1 D ⬭ 13

20 24 ◯ 6 ◯ 3 = 7

The missing signs are:

A ⬭ + and ÷ B ⬭ ÷ and +
C ⬭ – and ÷ D ⬭ ÷ and –

21 If this clock is 10 minutes slow, find the correct time.

A ⬭ 12:35 B ⬭ 12:15
C ⬭ 12:20 D ⬭ 1:35

22 At 7 a.m. the temperature was 7°C. By noon the temperature had risen 6° and by 5 p.m. it had fallen 3°.

Which thermometer shows the correct temperature at 5 p.m.?

A ◯ B ◯ C ◯ D ◯

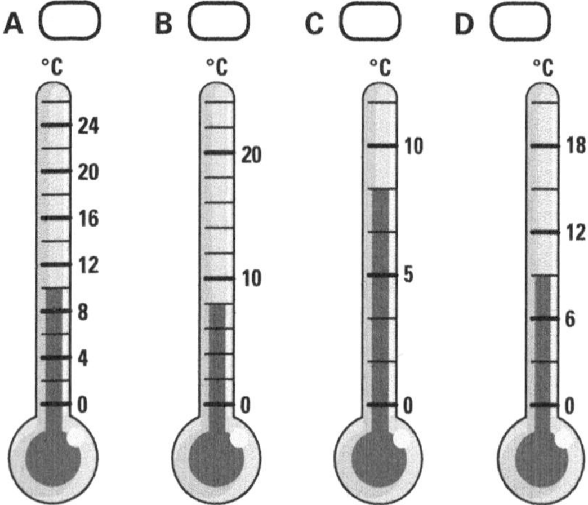

23 This solid block is made up of red cubes. It is then painted black on the outside, even on the base. When the paint is completely dry the pieces are taken apart.

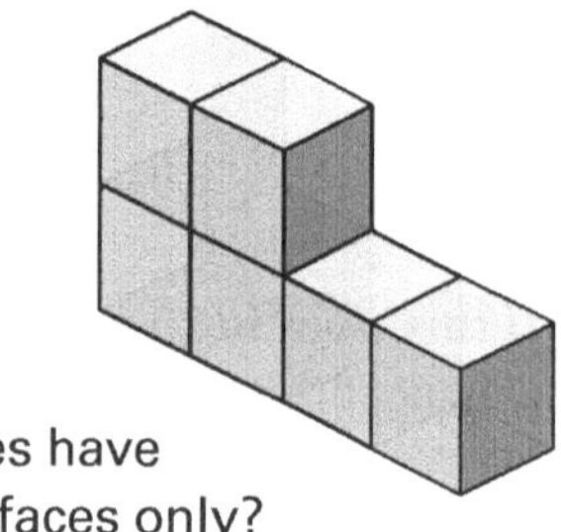

How many pieces have black paint on 4 faces only?

A ◯ 3 B ◯ 4
C ◯ 2 D ◯ 1

24 Which of the following statements is true?

A ◯ A right-angled triangle always has one angle equal to 45°.

B ◯ All three sides of an isosceles triangle are equal.

C ◯ If all three angles of a triangle are equal, it is called an equilateral triangle.

D ◯ A scalene triangle has 2 sides equal.

25 Nicky orders two scoops of ice cream. She can choose from chocolate, vanilla or strawberry and can order different flavours or two flavours the same. How many different combinations can she buy?

You can write out the possibilities here.

A ◯ 10 B ◯ 6
C ◯ 4 D ◯ 9

26 Michelle walks 40 metres in 30 seconds. If she walks at the same speed, how far will she walk in 1 hour?

A ◯ 480 m B ◯ 2·4 km
C ◯ 48 km D ◯ 4·8 km

27 Year 5 students were asked, 'What is your favourite sport?' The pie graph represents their answers.

If 12 students selected cricket, how many Year 5 students are in this survey?

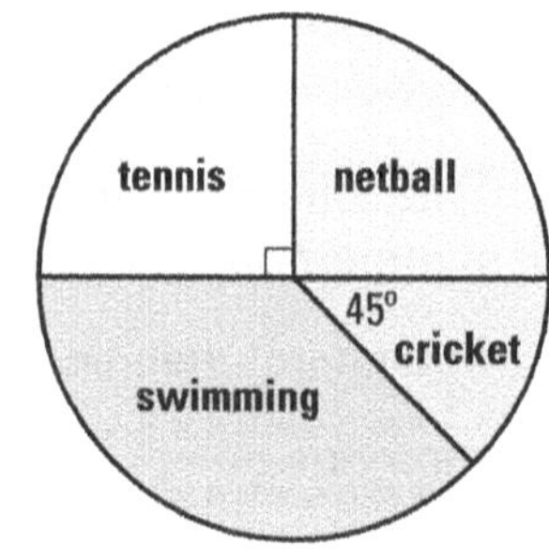

A ◯ 24 B ◯ 48
C ◯ 96 D ◯ 72

28 What is the maximum number of 3 cm cubes that can be placed into a rectangular box 15 cm by 12 cm by 6 cm?

First work out how many cubes will fit on the bottom layer of the box.

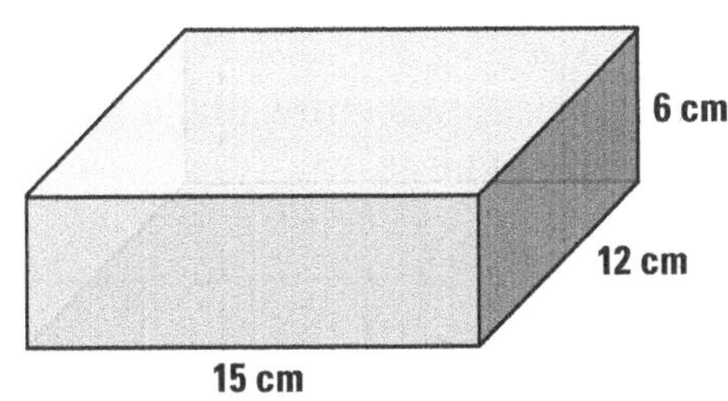

A ⬭ 40 B ⬭ 20
C ⬭ 18 D ⬭ 360

29 What is the volume of this solid, if the side of each cube used to make the solid is 2 cm?

A ⬭ 12 cm^3 B ⬭ 24 cm^3
C ⬭ 96 cm^3 D ⬭ 80 cm^3

30 How many whole numbers between 1 and 100 use the digit 7 at least once?

A ⬭ 18 B ⬭ 19
C ⬭ 20 D ⬭ 21

31 Mr Stern is demonstrating to Year 5 how many different rectangular prisms he can build from 24 cubes. He must use all the cubes. One of the shapes he makes is 2 × 2 × 6. He shows the class that this is the same shape as 2 × 6 × 2. How many other different shapes can he make?

A ⬭ 5 B ⬭ 10
C ⬭ 4 D ⬭ 6

32 A square P with sides 4 cm and a square Q with sides 6 cm are partially overlapping as shown in the diagram.

What is the difference between the shaded area Q and shaded area P?

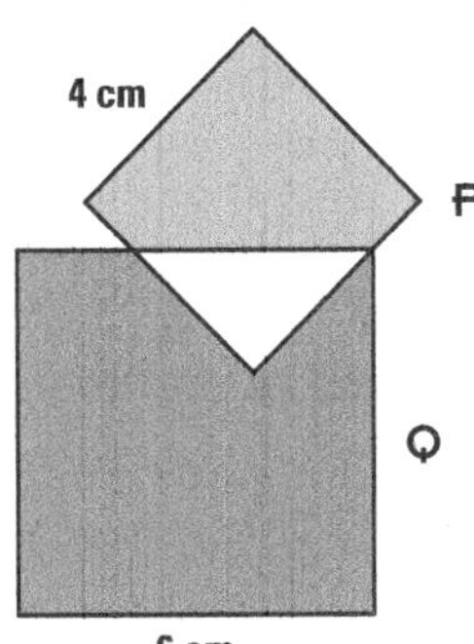

A ⬭ 2cm^2 B ⬭ 4 cm^2
C ⬭ 20 cm^2 D ⬭ 16 cm^2

33 The average sailing speed of a boat is 600 metres every minute, which is 600 m/min. How far can it sail in $\frac{2}{3}$ hour? Give your answer in km.

A ⬭ 2·4 km B ⬭ 24 km
C ⬭ 240 km D ⬭ 4 km

34 A bottle of cough mixture is $\frac{5}{12}$ full. When 75 mL is poured out, the container becomes $\frac{1}{3}$ full. Find the capacity of the container.

A ⬭ 900 mL B ⬭ 1080 mL
C ⬭ 540 mL D ⬭ 1020 mL

35 William paid $36 for 2 kg of fish. When the bones, head, skin and fins were removed, he had 1·5 kg of fish that he could use. What was the price per kilogram of the fish he could use?

A ⬭ $4.50 B ⬭ $24
C ⬭ $22.50 D ⬭ $13.50

enrich-e-matics TEST

3rd EDITION

Time: 45 minutes

YEAR 6 TEST 1

Name: ______________________

Class: ______________________

Please show all necessary working for each question.

1 Evaluate each of the following.

a $\frac{1}{2} + \frac{3}{4} =$ ______________________

b $\frac{1}{2} \times \frac{3}{4} =$ ______________________

c $\frac{3}{4} - \frac{1}{2} =$ ______________________

d $\frac{3}{4} \div \frac{1}{2} =$ ______________________

2 From the numbers in the grid, circle the only number described by these clues.

a All the digits are different.

b The sum of the middle digits is equal to the sum of the first and last digits.

c The thousands digit is the largest.

d None of the digits is odd.

e The units digit is the smallest.

2648		9351		8316
	1537		5713	
4680		6449		6284
	8442		7951	
3174		1627		7624
	1816		1592	
3589		8462		1983

3 Write the next two terms in the following number sequences.

a 55, 48, 41, 34, ________, ________

b 162, 54, 18, 6, ________, ________

c 1, 3, 7, 15 ________, ________

4 Express each of the following as a decimal.

a $\frac{3}{5} =$ ______________________

b 69% = ______________________

c $\frac{71}{100} =$ ______________________

d $\frac{3}{4} =$ ______________________

5 Damien is chasing his younger brother Peter. For every 5 metres that Damien runs, Peter runs 3 metres.

If they are 6 metres apart, how far must Damien run in order to reach Peter?

____

6 The average step taken by Cathy when walking is 0·6 m.

a How far has she walked after 500 steps? ________

b How many steps must she take to walk 3 km? ________

7 How many square tiles of side $1\frac{3}{4}$ cm would be needed to tile a single row 14 cm long?

____

8 If the volume of the cubic box in which 8 balls fit exactly is 64 cm^3, find:

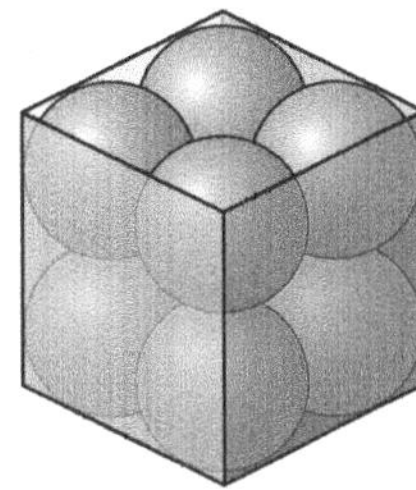

a the length of the side of the box. __________

b the radius of each ball. __________

9 The value of 12! is the product of all the whole numbers from 1 to 12 inclusive:

$12! = 12 \times 11 \times 10 \times 9 \times 8 \times 7 \times 6 \times 5 \times 4 \times 3 \times 2 \times 1$

Find the maximum number of times that 2 will divide into 12! exactly.

____

10 If $\frac{2}{3}$ of a telegraph pole is below the ground and 15 metres is above the ground, how long is the entire pole?

____

11 Find the shaded area.
(Each square represents 1 cm^2.)

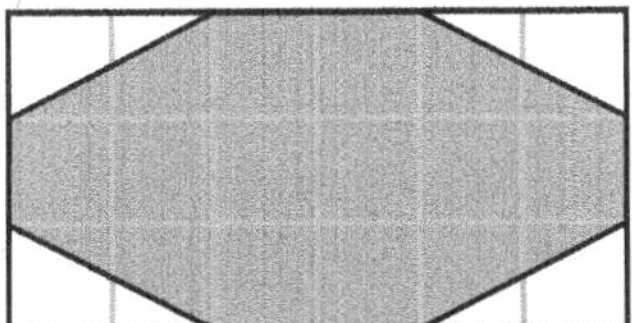

____

12 Rectangular cards 5 cm by 3 cm are cut from a rectangular sheet of cardboard 36 cm by 1 metre.

What is the greatest number of cards that can be cut from the sheet?

____

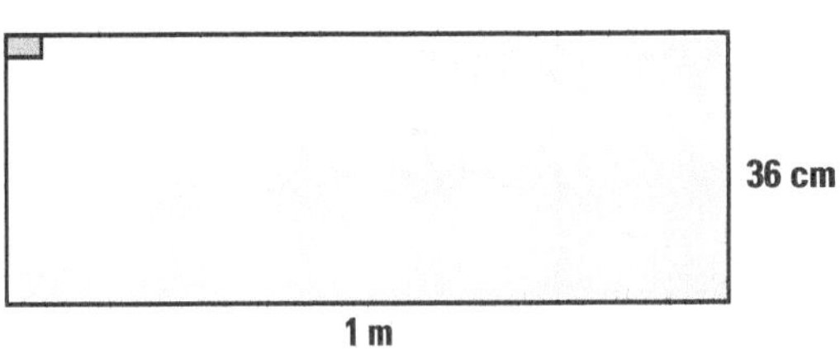

13 Sandra and Joy had a total of $56. If one fifth of Sandra's money is equal to half of Joy's, how much more money does Joy have than Sandra?

Use this diagram to help you.

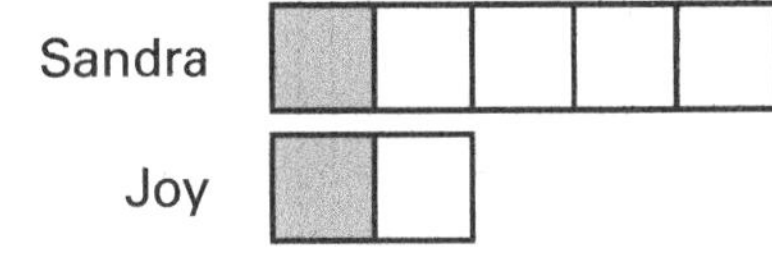

____

14 This solid is made of blue cubes.

a How many cubes were used to make this solid? __________

b The solid is dipped in red paint. Count the number of cubes with 3 red faces.

____

15 At home, Alvin has:

2 choices for entrée—
soup (S) or pasta (P);

3 choices for the main course—
fish (F), chicken (C), vegetarian (V)

2 choices for dessert—
ice-cream (i), cake (c).

How many different three-course meals would it be possible for Alvin to eat?

____

You might like to list the possibilities systematically using the initials.

____

____

____

____

16 A palindromic number is one that reads the same from left to right or from right to left. 787 and 12321 are examples of palindromic numbers. As you can see, the number is unchanged when the digits are reversed.

a How many two-digit palindromic numbers are there?

____

b How many three-digit palindromic numbers are there less than 400?

Hint: listing from smallest to largest might help.

____

17 Colin drives for 100 km at an average speed of 60 km/h and then he drives for 15 km at 45 km/h.

a How long does Colin take for each section of his trip?

____

____

b What was Colin's average speed for the entire trip? __________

18 The age and the hours of sleep of six children A, B, C, D, E and F have been graphed.

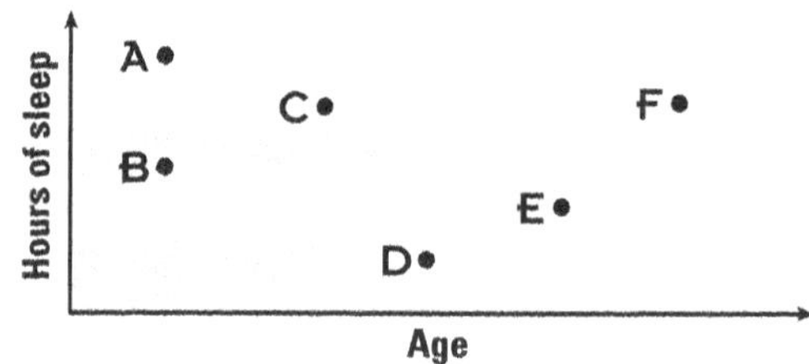

a Which child is the oldest? ________

b Which child slept the most? ________

c Which two children are the same age?

________ ________

d Which two children slept the same number of hours?

________ ________

e Name the child who is younger than C and gets less sleep than C.

________ ________

19 In a scout camp there is enough food for 12 days for 70 boys.

How many boys could be fed at the same rate for the food to last for 14 days?

____

20 A cube of edge 6 cm is built from 1 cm cubes.

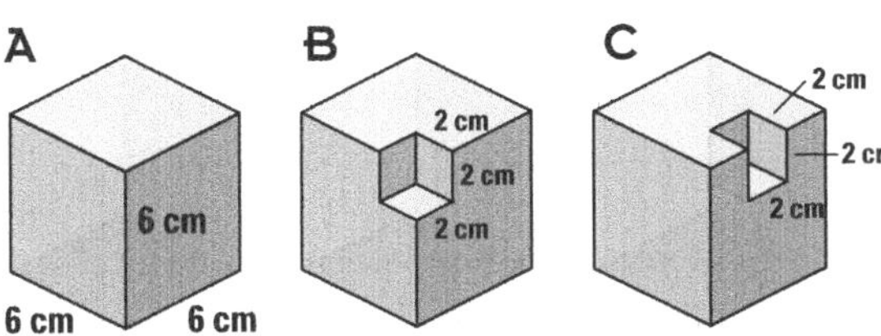

a Find the total surface area of this 6 cm cube (Figure A).

____

Note: the surface area of the cube is the total area of all faces.

b A 2 cm cube is removed from the corner of this cube. Find the total surface area of this solid (Figure B).

____

c Find the total surface area of the solid figure when a 2 cm cube is removed along an edge, but not from a corner (Figure C).

____

21 In a raffle, 200 green tickets and 100 blue tickets were sold.

If 21 tickets are drawn at random, what is the most likely number of each colour?

Green ____________

Blue ____________

22 A pattern of triangles is made from matches, as shown.

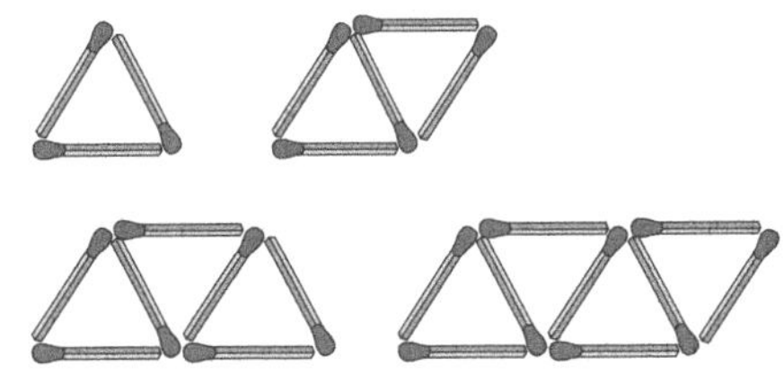

If there are 67 matches used, how many triangles have been formed?

____

23 Show that

$$\frac{1}{2} - \frac{1}{3} = \frac{1}{2 \times 3}$$

____

and using this, find the value of

$$\frac{1}{1 \times 2} + \frac{1}{2 \times 3} + \frac{1}{3 \times 4} + \frac{1}{4 \times 5} + \frac{1}{5 \times 6} =$$ __________

24 Allan could complete a job in 6 hours and David could complete the same job in 4 hours.

If they worked together:

a what fraction of the job could they complete in 1 hour?

____

b how long would they take to complete the job? ________

enrich-e-matics 3rd EDITION TEST

Time: 45 minutes

YEAR 6 TEST 2

Name: ____________________

Class: ____________________

1 A pack of 50 stickers costs \$2.50. Find the cost of 1 sticker.

____

2 Write each as a decimal.

a $2\frac{1}{4}$ __________ **b** $\frac{3}{20}$ __________

3 Write each as a percentage.

a $\frac{74}{100}$ __________ **b** 0·07 __________

4 What fraction is shaded in the figure below?

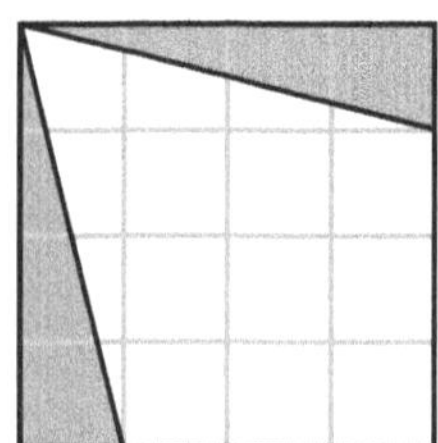

____

5 Work out each of the following.

a $\frac{1}{6} + \frac{3}{4} =$ __________

b $5\frac{1}{2} + 2\frac{1}{4} =$ __________

c $12 \times \frac{2}{3}$ __________

d $13 \div \frac{1}{2}$ __________

e $13\frac{1}{4} - \frac{1}{2}$ __________

6 Arrange these numbers in ascending order (from smallest to largest).

3·6 3·061 3·48 3·099

____

7 How many 600 mm ribbons can be cut from 4·8 m of ribbon?

____

8 It costs a family of 2 adults and 3 children \$28 to go to a show. If a child's ticket costs half of the price of an adult's ticket, find the cost of an adult's ticket.

____

9 How many whole numbers from 10 to 99 inclusive are not divisible by 5?

____

10 Three consecutive odd numbers add up to 45. What are the numbers?

____

11 Find the missing number.

$(1 + \square \times \frac{1}{2}) \div 4 = 1$

12 Baby Jessica has a mass of 5 kg. If she gains an average of 150 g per week:

a what will Jessica's mass be after 20 weeks? ________

b after how many weeks is Jessica's mass 6.5 kg? ________

13 This graph shows the ages of children at a tennis camp.

What percentage of children are older than 13 years? ____________

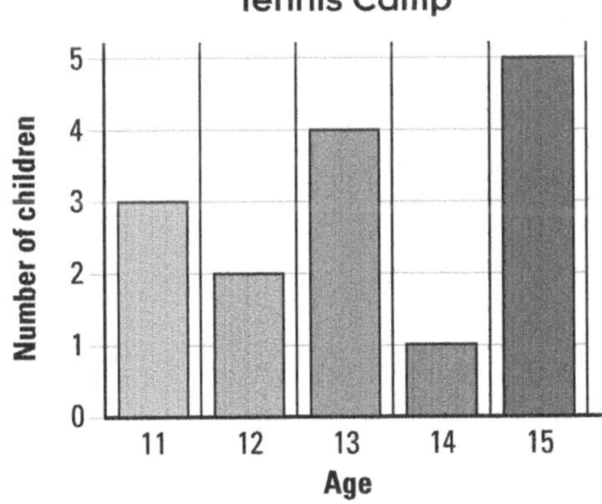

14 Ruth purchased 4 exercise books and 3 novels for $32. If a novel costs 4 times as much as an exercise book, find the cost of each.

Exercise book ____________

Novel ____________

15 How many small cubes were used to make this shape?

____

16 A clock is 10 minutes slow at 8:00 a.m. on Monday, but it is gaining 20 seconds an hour. What time will it show at 8:00 a.m. on Tuesday?

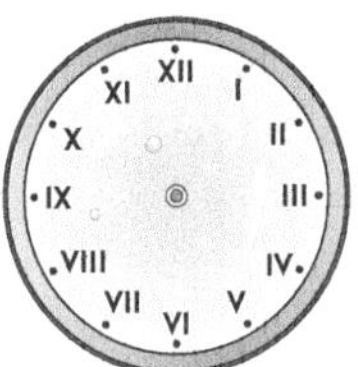

17 

a Complete the table below.

Number of octagons	1	2	3	4	5	6
Number of sides	8					

b Predict the number of sides needed to draw 11 octagons. ________

18 Yuri has two jobs at home. Every 4 days he must clean the fish tank. Every 6 days he must mow the lawn. Last Sunday he did both jobs.

On what day of the week will he next do both jobs? ____________

19 All the shapes drawn are squares.

Find the area of the square with side X cm.

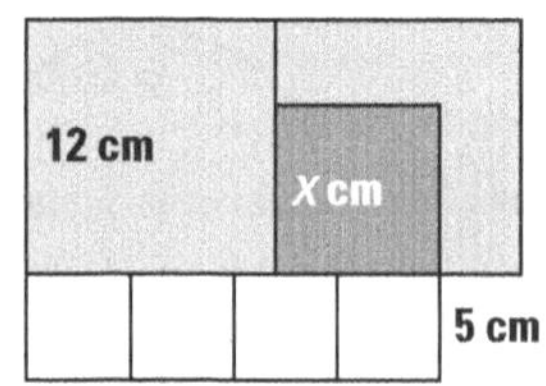

____

20 In a class of 32 students, 18 play tennis and 22 are in the school band. If 16 students both play tennis and are in the band, how many do neither?

____

Putting the information on this Venn diagram may help you.

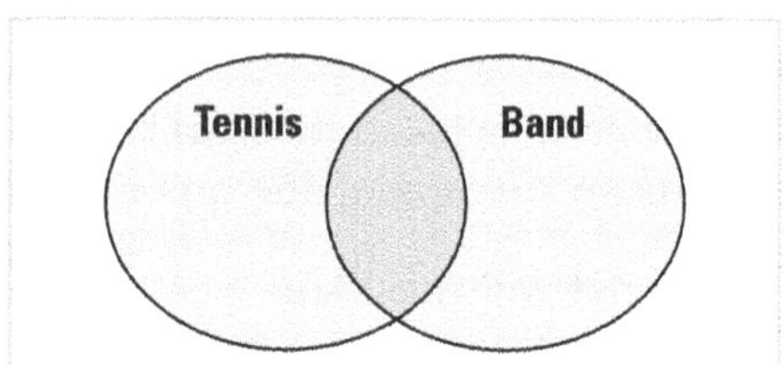

21 A pathway (the shaded region) is built around a rectangular pool, as shown. The pool is 15 metres by 18 metres. The pathway is completely tiled with 1 row of square blocks 3 metres by 3 metres.

How many blocks will be used for this pathway? ________

Draw the blocks on the pathway.

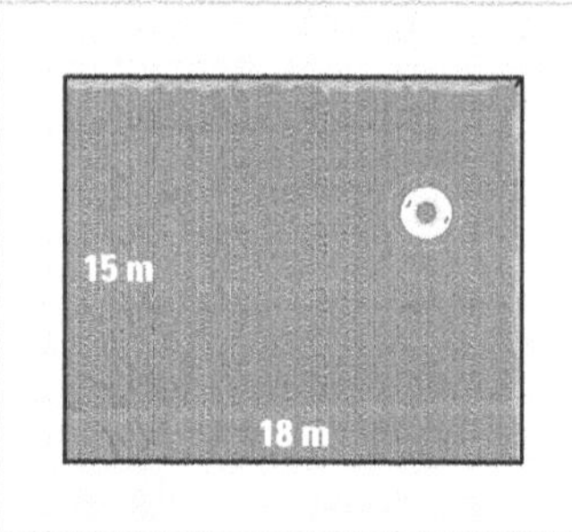

22 **a** How many litres of water are required to fill a pool 12 m long by 4 m wide to a depth of 1·5 m?

____

b If a hose is used to fill this pool and it fills at the rate of 4 litres every 10 seconds, how long will it take to fill?

____

23 The diagram shows a triangular prism. The area of one surface and some information about a triangular face is given.

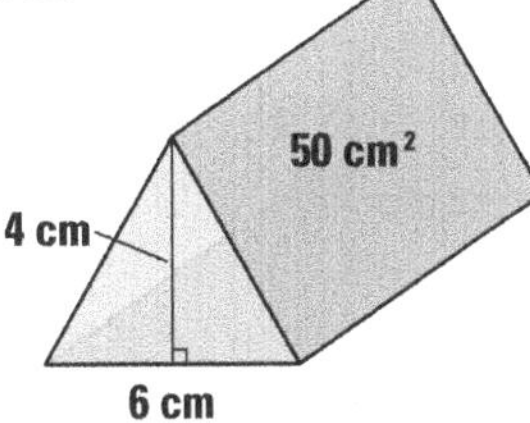

a Find the length of the prism. ____________

b Find the area of the base of the prism. ____________

c Calculate the surface area of the whole prism. ____________

24 A 5 × 3 × 2 solid block was made using interlocking red centicubes.

This block was then entirely covered with yellow cubes so that a new solid was formed.

a Find the dimensions of the new yellow solid. ____________

b Find the number of yellow cubes used to cover the red block.

____

25 Before the metric system, measures of length used in Australia were inches, feet and yards.

12 inches = 1 foot
(so 24 inches = 2 feet)

3 feet = 1 yard

a How many:

i feet in 36 inches? ________

ii inches in 2 yards? ________

iii feet in 3 yards? ________

iv yards in 12 feet? ________

v inches in 5 feet? ________

b Do the following addition very carefully.

	yards	feet	inches
		1	9
+		1	10

c Complete this multiplication.

	yards	feet	inches
	1	2	8
×			4

enrich-e-matics 3rd EDITION TEST

Time: 45 minutes

YEAR 6 TEST 3

Name: ____________________

Class: ____________________

1 Find the value of:

$20 - 19 + 18 - 17 + 16 - 15 + 14 - 13 + 12 - 11 + 10 - 9 + 8 - 7 + 6 - 5 + 4 - 3 + 2 - 1$

____

2 **a** If 16 apples cost $2.96, what is the cost of 20 apples?

____

b James has an annual salary of $25 300. If he receives an increase of 5%, what is his new salary?

____

3 On an international flight, you are allowed to take 20 kg of baggage. If you take more, you must pay $8.50 for each extra kilogram.

a When travelling to Spain, Dennis takes 25 kg of baggage. What will he pay for the excess baggage?

____

b Sheila had to pay $68. How much did her baggage weigh?

____

4 An athlete runs 60 metres in 9 seconds. If he were to maintain the same average speed, how long would he take to run:

a 20 metres? __________

b 100 metres? __________

5 How long will it take a train 1 km long to pass completely through a tunnel 3 km long if the train is moving at 60 km/h?

____

6 To make 24 cornflakes crisps, mix:

6 cups of cornflakes
4 whites of eggs
2 tablespoons of melted butter
40 grams of cocoa
60 grams of nuts
$\frac{3}{4}$ cup of sugar.

Write down the ingredients necessary to make 36 cornflakes crisps.

__________ cups of cornflakes

__________ whites of eggs

__________ tablespoons of melted butter

__________ grams of cocoa

__________ grams of nuts

__________ cup of sugar

7 Suppose all the counting numbers are written in columns in the pattern shown.

A	B	C	D	E
1	2	3	4	5
10	9	8	7	6
11	12	13		

Name the letter of the column in which the number:

a 29 would appear. ________

b 55 would appear. ________

c 72 would appear. ________

8 If ●●△ stands for 11, and

●△△△ stands for 13,

what does ●●●△△ stand for?

____

9 Rebecca has some 10c, 20c and 50c coins in her purse. Altogether she has 14 coins, and the total value of the coins is $3.

If she has more 10c than 20c coins and more 20c than 50c coins, how many 10c coins does she have? ________

10c	20c	50c	Total value

10 I have 6 blouses and 4 skirts in my wardrobe that I can mix and match.

How many different outfits can I wear?

____

11 Car number plates on a small island consist of 1 letter followed by 2 numbers. If only the letters A, B and C are used and only the numbers 7, 8 and 9, what is the total possible number of number plates that can be issued?

____________ possibilities

Complete the tree diagram below to help you.

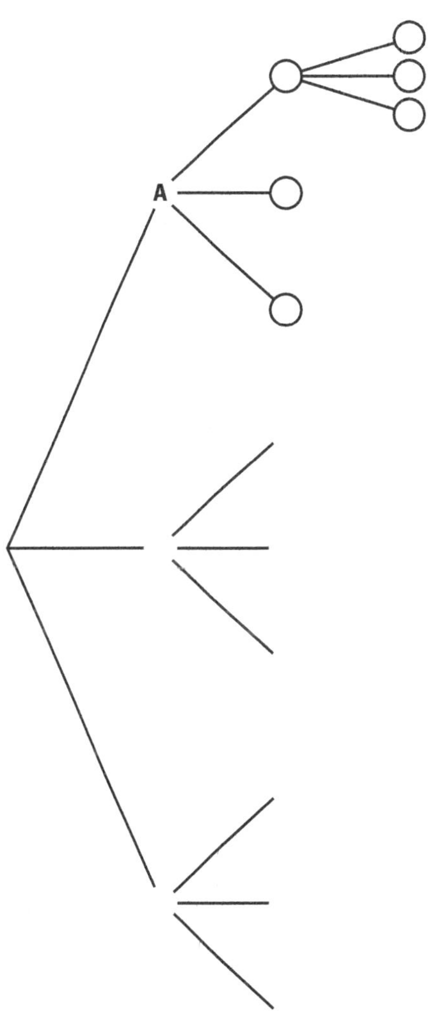

12 This diagram is drawn on 1 cm grid paper.

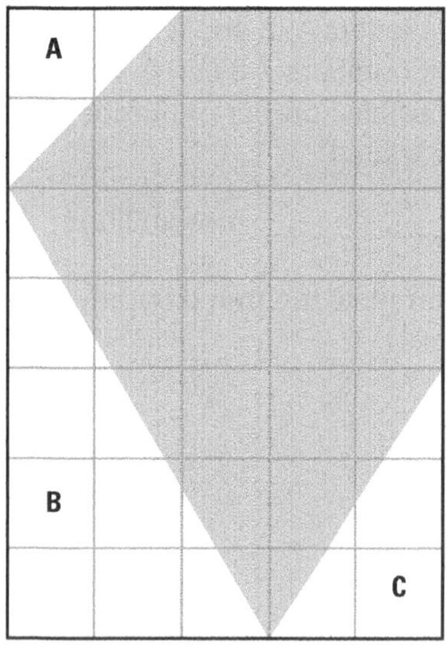

a Find the area of triangles A, B and C.

A ________ B ________

C ________

b Find the area of the shaded figure in square centimetres.

____

13 Two cars are travelling towards each other. One is travelling at 60 km/h and the other at 90 km/h. How far apart are they:

a 1 hour before they meet? ________

b 3 minutes before they meet? ________

14 A motor car travels for 3 hours at 70 km/h and for 2 hours at 80 km/h. Find the total distance covered and the average speed for the whole journey.

distance ____________

average speed ____________

15 The actual measurements of a hall are 16 m by 12 m. Draw a scale drawing of this on your page. Think carefully about the scale you are using and write it down.

16 Three identical cubes are shown. Their faces are coloured blue (B), white (W), red (R), yellow (Y), green (G) and orange (O).

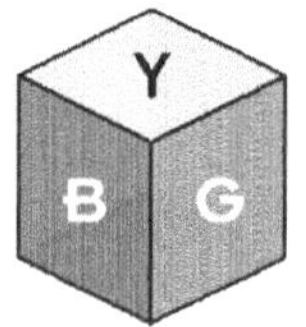

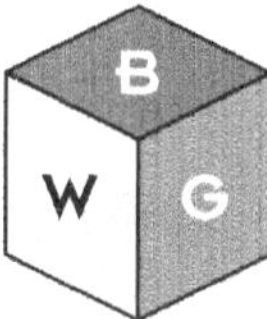

What colour is opposite orange?

____

17 How many small cubes are there in this solid?

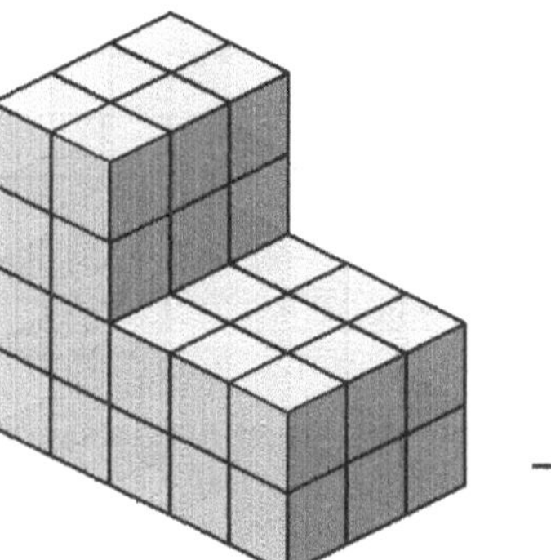

____

18 Approximately 40 bricks are needed for each square metre of a brick wall. The brick wall has two 1-metre square windows.

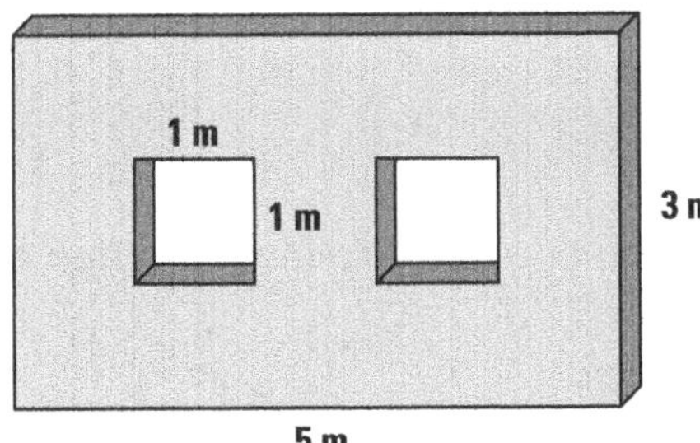

Approximately how many bricks would be needed to build this brick wall?

____

19 A square piece of paper is folded in half vertically. If the resulting figure has a perimeter of 24 cm, what was the side of the original square?

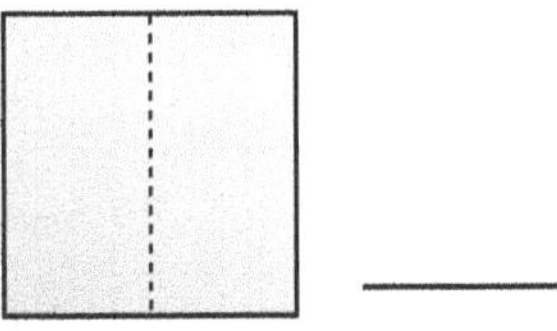

____

20 How many pairs of parallel edges are there on a triangular prism?

____

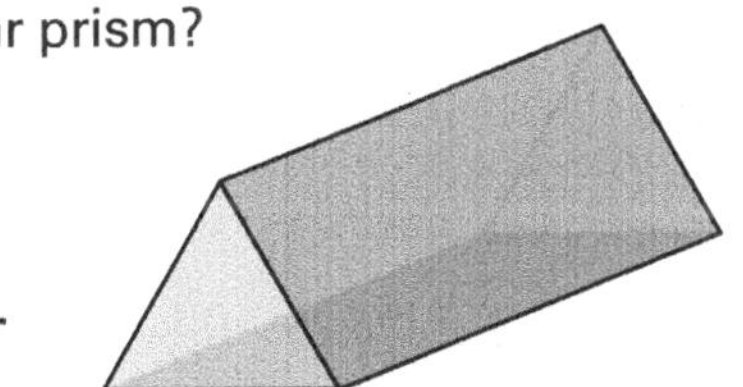

21 The symbol 6! means 6 x 5 x 4 x 3 x 2 x 1.
7! means 7 x 6 x 5 x 4 x 3 x 2 x 1.

Find the value of $\frac{10!}{8!}$

____

Hint: do not evaluate the numerator and the denominator.

22 How many cans can be placed in this cardboard box if they have a diameter of 8 cm and are 12 cm high?

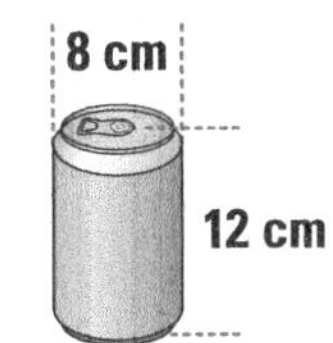

____

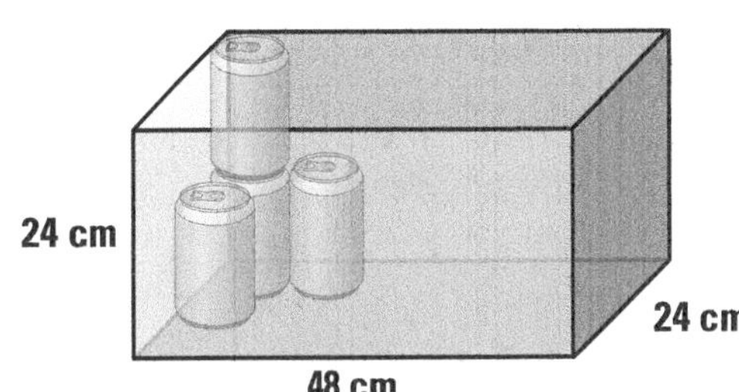

23 Assume that one Australian dollar is worth $\frac{4}{5}$ of one US dollar, so that \$1 (Australian) = US\$0.80.

a What is \$60 Australian worth in US dollars? __________

b What is US\$60 worth in Australian dollars? __________

24 1^2 means 1 × 1; 3^2 means 3 × 3.

If the sum of the squares of the numbers from 1 to 25 is 5525 written as:

$1^2 + 2^2 + 3^2 + 4^2 + \ldots\ldots\ldots\ldots + 25^2 = 5525$,

find

$2^2 + 4^2 + 6^2 + 8^2 + \ldots\ldots\ldots\ldots + 50^2$.

____

Hint: there is a short cut.

enrich-e-matics TEST

3rd EDITION

Time: 45 minutes

YEAR 6 TEST 4

Name: ______________________

Class: ______________________

Multiple choice

Colour in the bubble next to the answer.

1 How many hundredths are there in 0·3?

A ◯ 30 B ◯ 3

C ◯ 0 D ◯ 300

2 When you add two numbers you get 14, and when you find their difference you get 2. What would you get when you multiplied the numbers?

A ◯ 48 B ◯ 28

C ◯ 24 D ◯ 40

3 Select the fraction below that is closest to $\frac{1}{2}$.

A ◯ $\frac{4}{11}$ B ◯ $\frac{8}{13}$

C ◯ $\frac{8}{15}$ D ◯ $\frac{49}{100}$

4 $\frac{1}{10} + \frac{1}{10} + \frac{1}{10} + \frac{1}{10} + \frac{1}{10} + \frac{1}{10} + \frac{1}{10} = \frac{1}{2} + \square \times \frac{1}{10}$

What is the missing number in the box?

A ◯ 2 B ◯ 5

C ◯ 7 D ◯ 1

5 Which of the following numbers is largest?

A ◯ 0·608 B ◯ 0·6008

C ◯ 0·0608 D ◯ 0·680

6 Which one of the following statements is false?

A ◯ $\frac{1}{3} > \frac{1}{4}$ B ◯ $\frac{1}{5} + \frac{1}{4} = \frac{9}{20}$

C ◯ $\frac{1}{3} \div 3 = 1$ D ◯ $3 \div \frac{1}{3} = 9$

7 Which one of the following has the smallest answer?

A ◯ $\frac{1}{5} \times 10$ B ◯ $10 \div \frac{1}{4}$

C ◯ $\frac{2}{3} \times 9$ D ◯ $3 \div \frac{1}{10}$

8 Which whole number should be written in the box to make a fraction whose value is between 3 and 4?

$$\frac{15}{\square}$$

A ◯ 9 B ◯ 3

C ◯ 20 D ◯ 4

9 0·673 is the same as:

A ◯ 6·73 × 10 B ◯ 67·3 ÷ 100

C ◯ 6·73 ÷ 100 D ◯ 673 ÷ 100

10 What percentage of the area of the square is shaded?

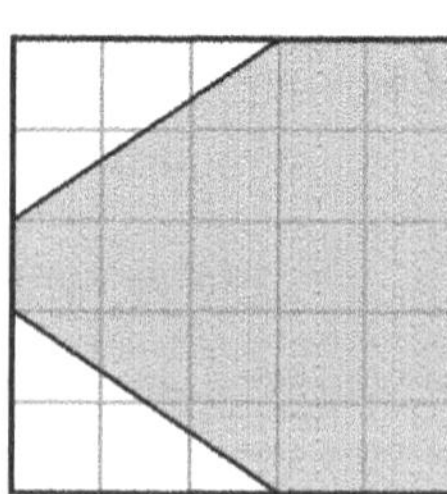

A ◯ 19% B ◯ 76%

C ◯ 80% D ◯ 72%

11

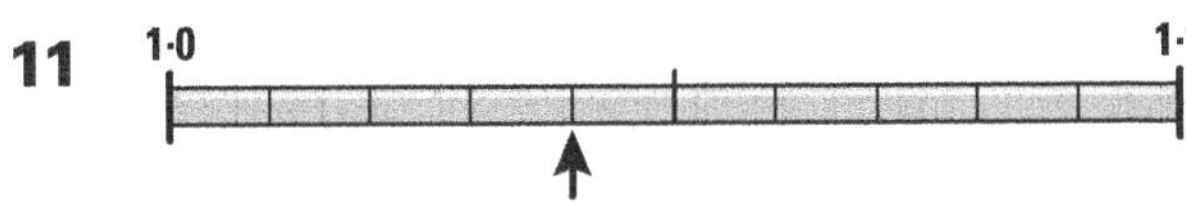

The position on the scale indicated by the arrow is:

A ◯ 1·04 B ◯ 1·004
C ◯ 1·08 D ◯ 1·4

12 The capacity of a tank is 800 litres. It is being filled at a rate of 150 litres per hour.

How long will it take to fill $\frac{3}{4}$ of the tank?

A ◯ 4 h 48 min B ◯ 6 h
C ◯ 5 h 20 min D ◯ 4 h

13 If $2\frac{1}{4}$ kg of grapes cost $6.30, find the cost of 500 g.

A ◯ 70c B ◯ $2.80
C ◯ $1.40 D ◯ $2.40

14 A square, P, is cut off from a piece of cardboard, ABCD.

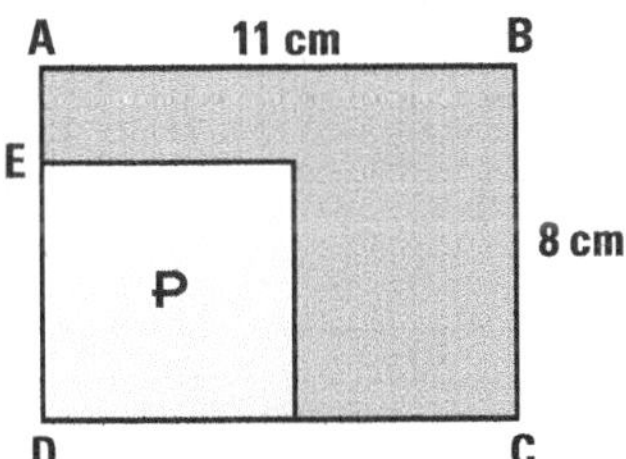

The area of the shaded part is 63 cm². What is the length of DE?

A ◯ 5 cm B ◯ 6 cm
C ◯ 2 cm D ◯ 3 cm

15 If 5 men can paint a house in 8 days, how long would it take 4 men to paint the same house?

A ◯ 9 days B ◯ 10 days
C ◯ $10\frac{2}{3}$ days D ◯ 6 days

16 How many 3 cm cubes can be placed in a box that measures 12 cm by 12 cm and has a height of 9 cm?

A ◯ 16 B ◯ 36
C ◯ 144 D ◯ 48

17 A sugar cube has a mass of about 3 grams. What would be the mass of this stack of sugar cubes?

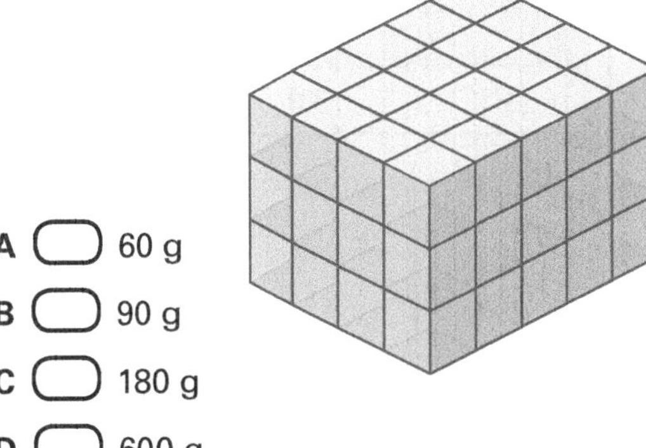

A ◯ 60 g
B ◯ 90 g
C ◯ 180 g
D ◯ 600 g

18 The sector graph shows the fraction of the time that a radio station plays different types of music during broadcasting hours.

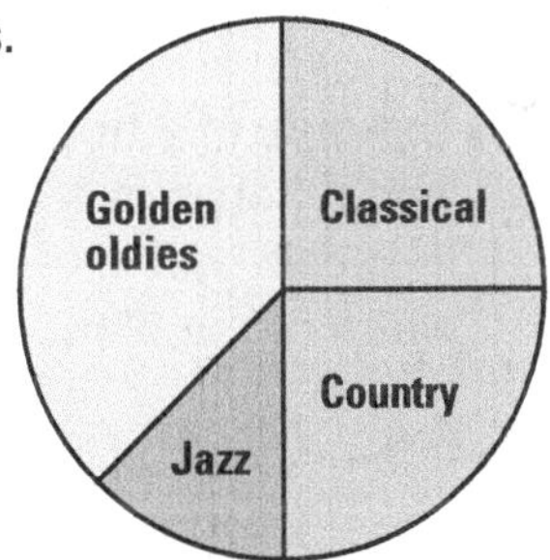

Find the missing data in the table to match the sector graph.

Type of music	Number of hours
Classical	5
Country	5
Golden oldies	
Jazz	3

A ◯ 6 hours B ◯ 7 hours
C ◯ 8 hours D ◯ 2 hours

19 Zoltan is driving his car at 100 km/h along the freeway. How many minutes will he take to travel 40 km?

A ◯ 20 min B ◯ 24 min
C ◯ 16 min D ◯ $\frac{2}{3}$ h

20 The table shows the percentage of some types of household goods sold at a department store in one year.

Television	Washing machines	Fridges	Dish-washers
23%	19%	30%	28%

If 120 fridges were sold, how many dishwashers were sold?

A ◯ 118 B ◯ 112
C ◯ 400 D ◯ 140

21 Refer to the table in question **20**.
If 800 items were sold altogether, how many televisions were sold?

A ◯ 238 B ◯ 230
C ◯ 184 D ◯ 163

22 The plan of a house is to be drawn. The actual house is 16 m long and 12 m wide. If the scale that is used is 3 cm = 4 m, find the width of the plan.

A ◯ 9 cm B ◯ 90 cm
C ◯ 120 cm D ◯ 12 cm

23 Gabrielle has some solid shapes in front of her. If the shape she holds in her hand has 5 faces (one of which is a square), 5 corners and 8 edges, which of the shapes below is she holding?

A ◯ a square prism
B ◯ a square pyramid
C ◯ a triangular prism
D ◯ a triangular pyramid

24 The average mass of 4 girls is 52 kg. One of the girls, who weighs 61 kg, leaves the group. Find the average mass of the 3 girls.

A ◯ 43 kg B ◯ 47 kg
C ◯ 49 kg D ◯ 56·5 kg

25 Lara and Debbie are friends. Lara is taller but is younger than Debbie.

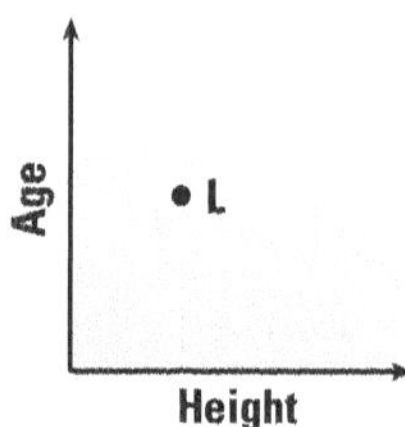

Which graph below could represent Debbie?

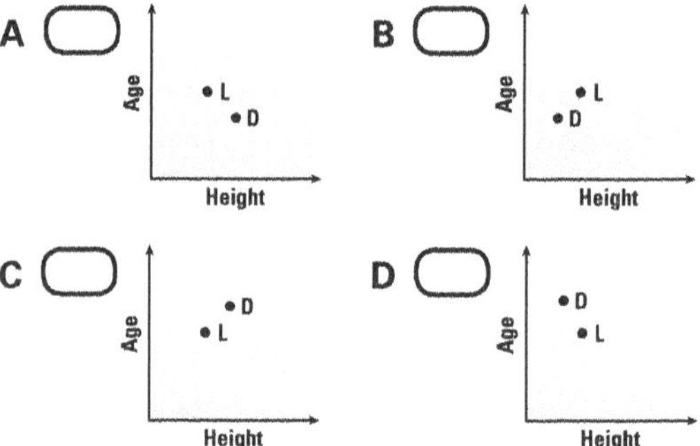

26 The cost price of a computer is $1600. Find the selling price if the dealer makes a profit of 30%.

A ◯ $2080 B ◯ $480
C ◯ $1120 D ◯ $1760

27 Norman sold his bike at a loss of 20%. If the selling price was $160, find the cost price.

A ◯ $200 B ◯ $192
C ◯ $250 D ◯ $180

28 A cube has sides of 0·7 m in length. The volume is:

A ⬭ $3{\cdot}43m^3$ B ⬭ $4{\cdot}9\ m^3$

C ⬭ $0{\cdot}343m^3$ D ⬭ $0{\cdot}49\ m^3$

29 You have to tie a ribbon around this box. If the knot takes 40 cm, how much ribbon will you need?

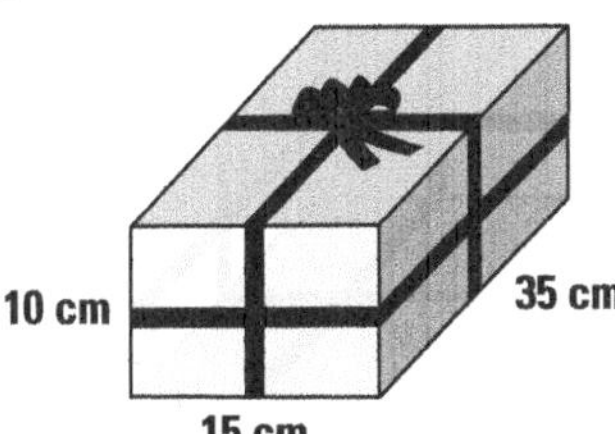

A ⬭ 240 cm B ⬭ 280 cm

C ⬭ 100 cm D ⬭ 160 cm

30 At the school concert $\frac{7}{10}$ of the audience were parents and visitors, $\frac{1}{4}$ were students, and the rest were teachers. If there were 15 teachers, how many people were at the school concert?

A ⬭ 60 B ⬭ 75

C ⬭ 105 D ⬭ 300

31 A certain bacteria doubles in quantity every two hours. At midnight, last night, only two of these bacteria were in a jar. By midnight tonight, the jar will be full. At what time will the jar be half-full?

A ⬭ 12 p.m. B ⬭ noon today

C ⬭ 8 p.m. D ⬭ 10 p.m.

32 Which of the following is the slowest?

A ⬭ 20 km per hour

B ⬭ 200 m per minute

C ⬭ 3 metres per second

D ⬭ 3 km per 10 minutes

33 Mrs Chong took $1\frac{1}{2}$ hours to travel from home to work, a distance of 72 km, in some traffic. On the way home at night she could increase her speed by 12 km/h. By how much would her travelling time decrease?

A ⬭ 12 minutes B ⬭ 48 minutes

C ⬭ 18 minutes D ⬭ 24 minutes

34 If

$3 \times 3 \times 3 \times 3 = 3^4 = 81$

$3 \times 3 \times 3 = 3^3 = 27$

$3 \times 3 = 3^2 = 9$

$3 = 3^1 = 3$

then $3^0 =$

A ⬭ 0 B ⬭ $\frac{1}{3}$

C ⬭ $\frac{1}{9}$ D ⬭ 1

TEST Solutions

YEAR 1 TEST 1 Screening Test 1 (page 6)

Question	Answer	Possible marks
1 a	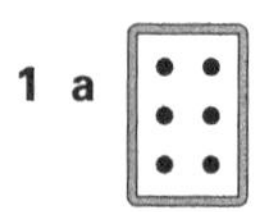	2
b	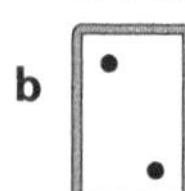	2
2	12, 14, 12	6
3 a	○ ▼	2
b	◆ □	2
4	○ ○ ○ ○ ○ 5th row	2
5	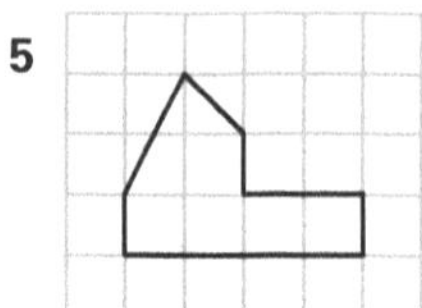	4
6 a	9, 11	4
b	14, 17	4
c	60, 70	4
d	47, 57	4
7 a	10 11 12 13 14 15	3
b	12 10 8 6 4 2	3
c	16 15 14 13 12 11	3
d	21 19 17 15 13 11 9	3
8	3 + **5** = 8	3
9	1, 2 5 3, 4 or 1, 3 5 2, 4	6
10	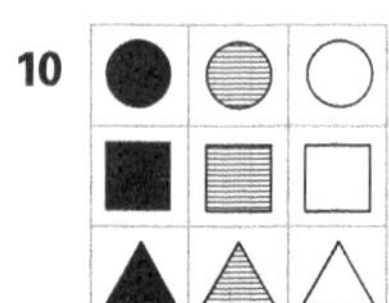	6
11	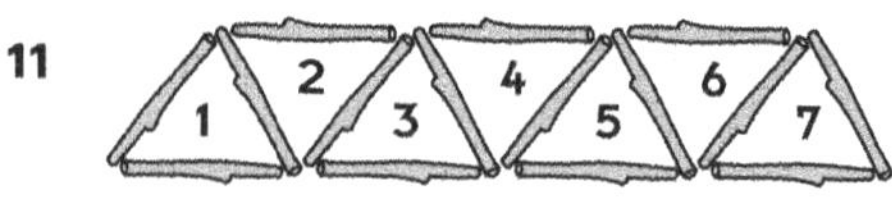	2
	15 sticks	2
12	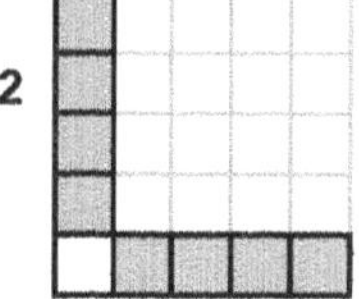	3
13 a	12 ice creams	2
b	15 ice creams	2
14 a	11 a.m.	2
b	5 a.m.	2
c	Wednesday	3
15 a	4	3
b	$2	3
16 a	5 kg	3
b	13 kg	3
17	3 kg	3
18	There are many solutions. The following are some possibilities: 3 – 0 4 – 1 5 – 2 6 – 3 7 – 4 8 – 5 9 – 6 10 – 7 11 – 8 12 – 9 13 – 10 14 – 11 99 – 96 100 – 97	4
	Total	**100**

TEST Solutions

YEAR 1 TEST 2 Screening Test 2 (page 10)

	Possible marks
1 20	
2 	3
3 a ▲●▲	3
b ▮○○	3
4	6
5 a 3, 5, 2, 4, 6 or 3, 4, 2, 5, 6	6
6 a 6, 4, 8; 13, 11, 15	6
b 10, 11, 9; 6, 7, 5	6
7 a **4** + 2 = 6	3
b **7** − 3 + 4	3
8 9 circles	3
9	9
10	12

9

4	3	8
9	5	1
2	7	6

	Possible marks
11 a \$6	3
b \$4	3
c \$2	3
12 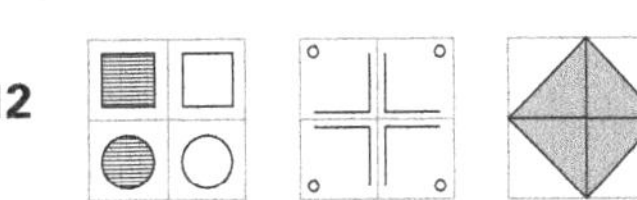	12
13 5 kg	4
14 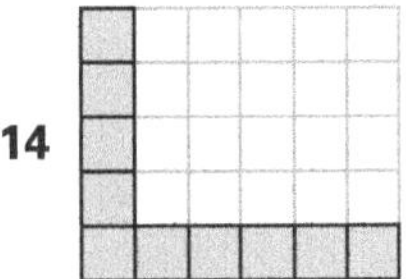	4
15 a E and H	4
b U and W	4
Total	**100**

TEST Solutions

YEAR 1 TEST 3 Screening Test 3 (page 14)

		Possible marks
1 a	21, 24	2
b	10, 5	2
c	21, 26	2
d	B C	2

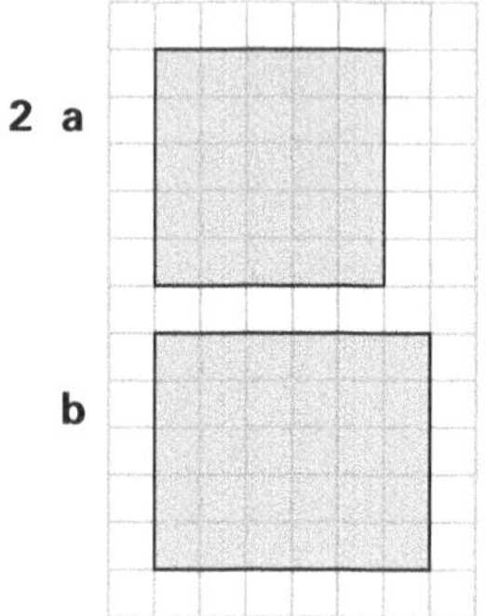

2 a — 3

b — 3

3 7, 6, 5, 4 — 2

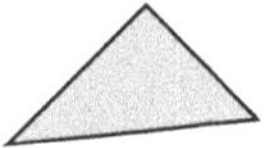

3

4 The rule is add 2 — 6

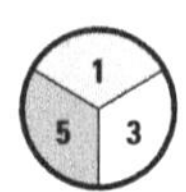

5 a — 3

b — 3

3	5	7	9	11	13	15	17	19	21

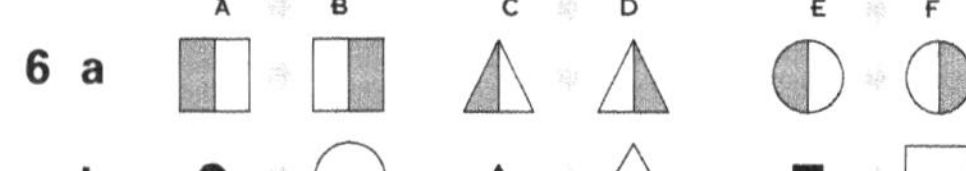

6 a — 3

b — 3

7 6 + 4 = 10 legs — 5

8 a 60c — 3

b For exactly $1, I could buy chips and an ice cream, or two apples and an ice cream. — 4

	Possible marks
9 There are many possible answers. Below are some of the possibilities.	

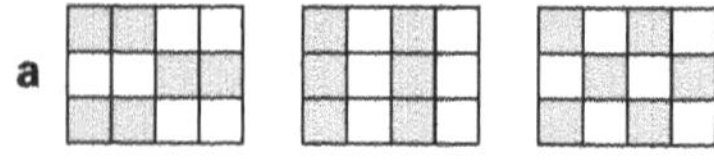

a — 3

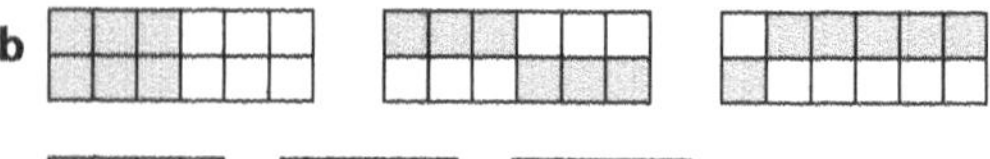

b — 3

c — 3

10 a 32, 13 — 4

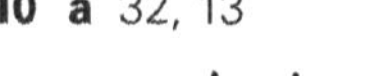

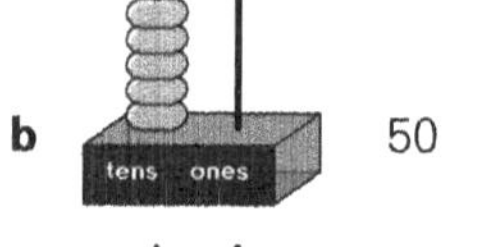

b 50 — 3

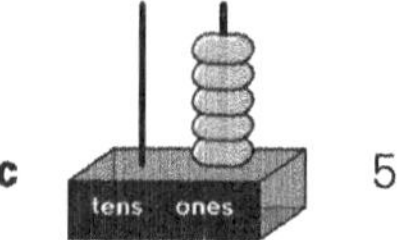

c 5 — 3

11 5 cm — 3

12 — 3

4	5	9
2	6	8
6	11	17

6

7	1	8
2	3	5
9	4	13

13 a — 3

20 matches — 3

b 3 nines — 3

14 a 9 tiles — 3

b 11 tiles — 3

15 Green — 5

Total **100**

TEST Solutions

YEAR 1 TEST 4 Screening Test 4 (page 18)

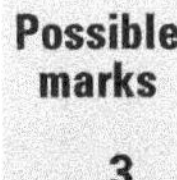

Question	Answer	Possible marks
1 a	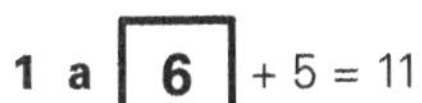 6 + 5 = 11	3
b	13 − 4 = 9	3
2	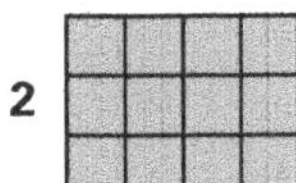	3
3 a	27	3
b	13	3
4 a	19 18 17 16 ~~18~~ 15 14 13	3
b	~~8~~ 5 10 15 20 25 30 35	3
c	80 70 ~~65~~ 60 50 40 30 20	3
5	**B** Half the balls have stripes.	3
6	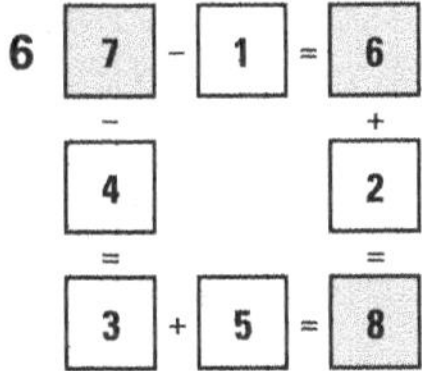 7 − 1 = 6; 7 − 4 = 3; 6 + 2 = 8; 3 + 5 = 8	8
7 a	$1.20	3
b	$3.00	3
c	$4.20	2
8	**A**, **C**	4
9	14 legs	4
10	The scale shows 15 kg.	2
	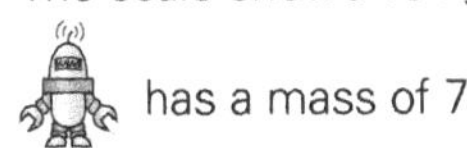 has a mass of 7 kg.	5
11 a	2 books	2
b	4 books	2
c	Richard	2
d	4 more books	2
12	3 (4) 1 (3) 2 / (8) (7) (4) / 5 (11) 6 (8) 2 / (7) (9) (6) / 2 (5) 3 (7) 4	10
13	A	4
14	School / Home (four paths)	8
15	12	4
16 a	11:00 a.m.	4
b	9:30 a.m.	4
	Total	**100**

TEST Solutions

YEAR 2 TEST 1 Screening Test 1 (page 22)

Question	Answer	Possible marks
1 a	19, 23	4
b	10, 7	4
c	21, 28	4
d	19, 21	4
2	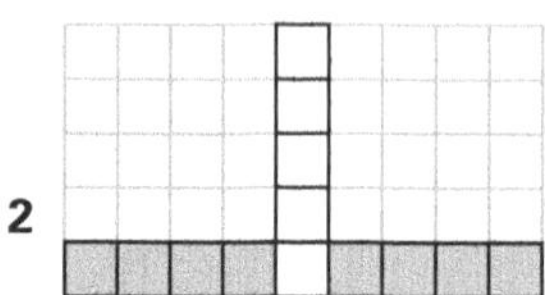	3
3 a		2
b		2
4	24	3
5	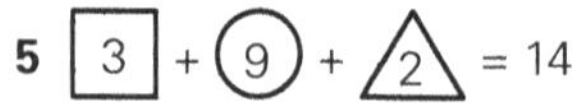	3
6 a		4
b	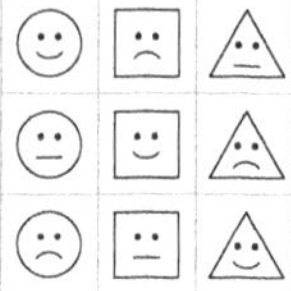	4
7 a	6 hours	2
b	20 minutes	2
c	Wednesday	3
8	15 raffle tickets	2
	25 raffle tickets	2
9	Three cars have 12 wheels.	2
	Two bicycles have 4 wheels.	2
	One tricycle has 3 wheels.	2
	Total 19 wheels	2
10 a	four pizzas	3
b	18 children	3

Question	Answer	Possible marks
11	1st row = 1 2nd row = 3 3rd row = 5 4th row = 7 5th row = 9	2
	6th row = 11	2
	The number of circles in each row is increasing by two and forms a pattern of odd numbers. In the 8th row there will be 15 circles.	3
12 a	$1.60	2
b	90c	3
c	10c	3
d	4 apples	3
13 a		3
b	17 sticks	3
c	21 sticks	3
14 a	12 glasses	4
b	5 litres	4
15		3
	Total	**100**

TeST Solutions

YEAR 2 TEST 2 Screening Test 2 (page 26)

		Possible marks
1	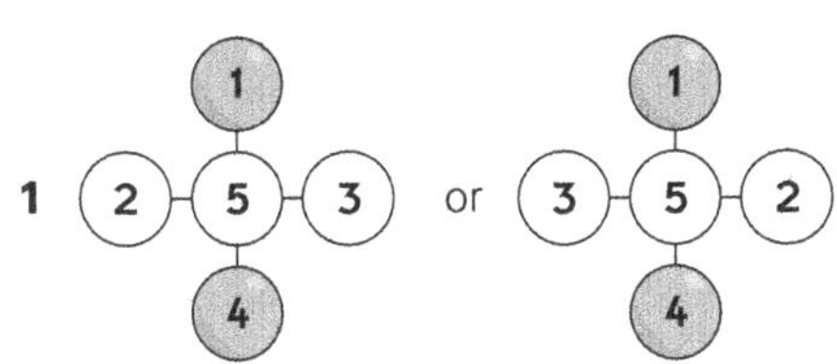	3
2	 = 4 □ = 3	4
3 a	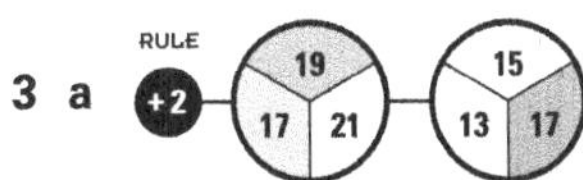	6
b	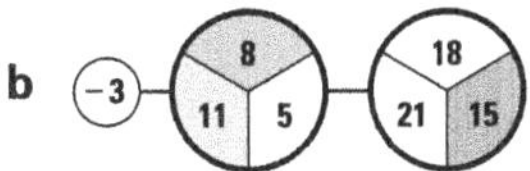	6
c	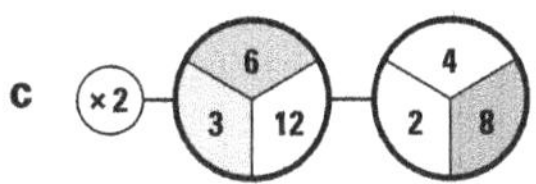	6
4	26	2
5	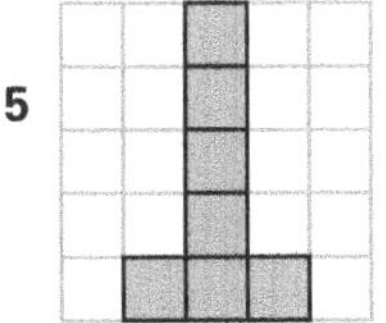	3
6	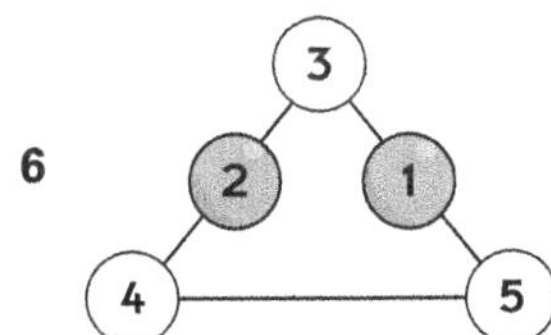	6
7 a	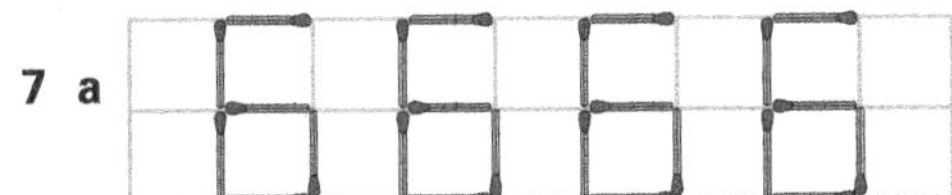	2
b	24 matches	2
c	3 sixes	2
8 a, b	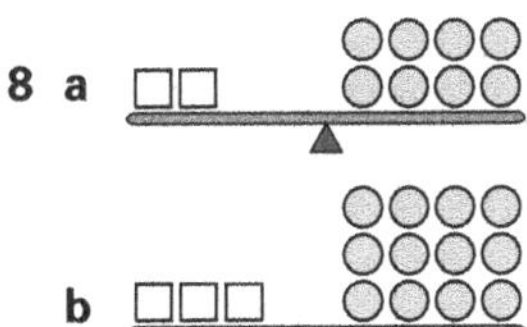	2, 2

		Possible marks
9 a	6	2
b	Chris	2
c	Michelle	2
d	3	2
10	20c, 10c, 5c	2
	10c, 10c, 10c, 5c	2
	10c, 10c, 5c, 5c, 5c	2
	5c, 5c, 5c, 5c, 5c, 5c, 5c	2
11	[6] + △3 = 9 [6] − △3 = 3	4
12	see grid below	10
13 a	10 hours	2
b	25 minutes	2
14	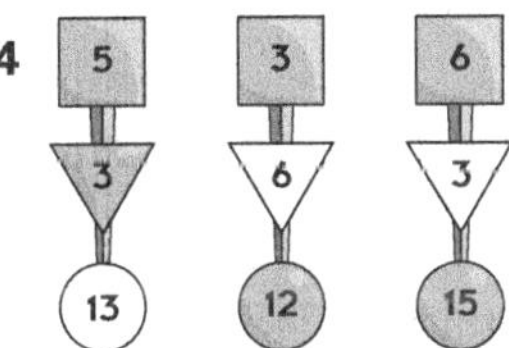	9
15 a	Z	2
b	Y Z X	3
16 a	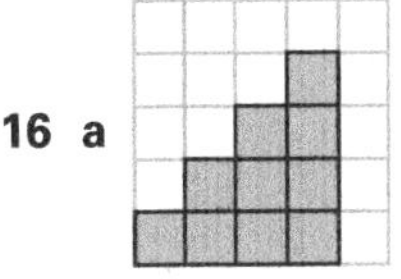	2
	10 tiles are used	2
b	5 steps	2
	Total	**100**

Question 12 grid:

8	3	10
9	7	5
4	11	6

TEST Solutions

YEAR 2 TEST 3 Screening Test 3 (page 30)

		Possible marks
1	**a** 13, 9	4
	b 22, 11	4
	c 70, 65	4

2 a

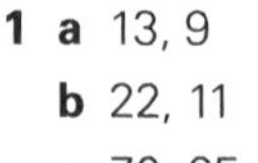

Possible marks: 5

b

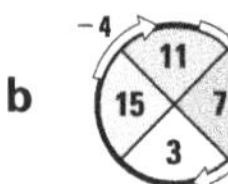

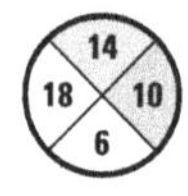

Possible marks: 5

3

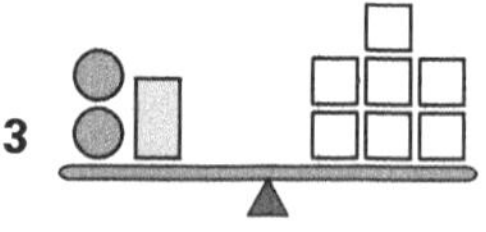

Possible marks: 3

4

6	7	13
5	8	13
11	15	26

Possible marks: 4

7	4	11
7	5	12
14	9	23

Possible marks: 4

5

she sees
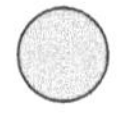

Possible marks: 4

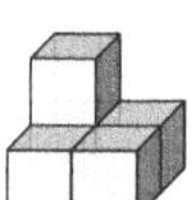
she sees

Possible marks: 4

6 a (triangle: circles 6, 3, 5; squares 9, 11, 8) — Possible marks: 3

b i (triangle: circles 2, 3, 6; squares 5, 8, 9) — Possible marks: 4

ii (triangle: circles 1, 2, 6; squares 3, 7, 8) — Possible marks: 6

		Possible marks
7	7°C	3
8	**a** week 3	3
	b week 4	3
	c week 1	3
9	**a** $1.20	3
	b 3 ice creams *or* chips and an ice cream *or* 2 apples	6
	c 5 ice creams *or* 2 chips and an ice cream *or* 2 apples and 2 ice creams *or* 2 apples and chips (only 2 of 4 answers required)	8
10	**a** 13, 12, 11, 10, 9	3
	b 12, 10	3
	c 12	3

11 a

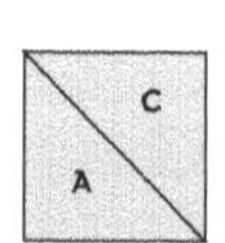

A and C (C will have to be turned) — Possible marks: 4

b

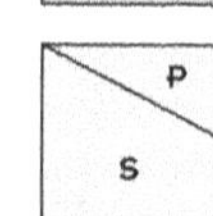

P and S — Possible marks: 4

Total **100**

TEST Solutions

Screening Test 4 (page 34)

	Possible marks
1 a 26, 32	4
b 12, 7	4
c 22, 29	4

2 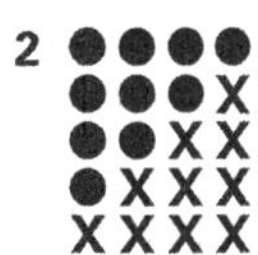 — 4

3 There are many possible answers.
As long as the sum of the numbers on opposite sides of 1 is 9, it is correct.

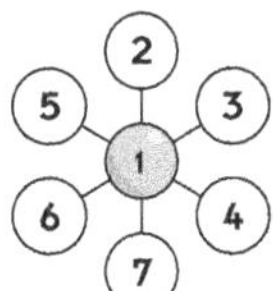

6

4 A C 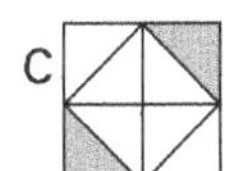 E 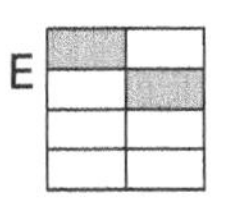 F 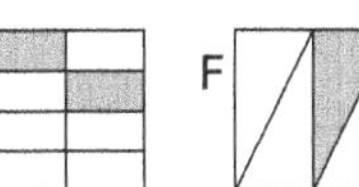 — 4

5 B Half the balls have stripes.
C One quarter of the balls are black. — 4

6 a

$$\begin{array}{r} 1\ 7 \\ +\ \boxed{2}\ \boxed{1} \\ \hline 3\ 8 \end{array}$$

4

b 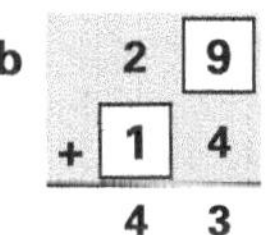

4

c

$$\begin{array}{r} 3\ 9 \\ -\ \boxed{1}\ \boxed{2} \\ \hline 2\ 7 \end{array}$$

4

d

$$\begin{array}{r} 4\ \boxed{8} \\ -\ \boxed{3}\ 6 \\ \hline 1\ 2 \end{array}$$

4

7 a

A	B	C	D	E
1		2		3
	4		5	
6		7		8
	9		10	
11		12		13
	14		15	
16		17		18
	19		20	

3

b D — 3

8  — 3

	Possible marks
9 a 1, 4, 7, ~~11~~ 10, 13, 16, 19	2
b 35, 31, 27, 23, ~~18~~ 19, 15, 11	2
c 4, 8, 12, ~~14~~ 16, 20, 24, 28	2
10	3
	3

11 The mass of [rocket] is 6 kg.
So the mass of [rocket] [rocket] is 12 kg
Since the scale reads 20 kg,
the mass of [robot] is 5 kg. — 4

12 Rule is 'add 6'.

1	2
⇩	⇩
7	9

4

13 11:30 a.m. — 3

14 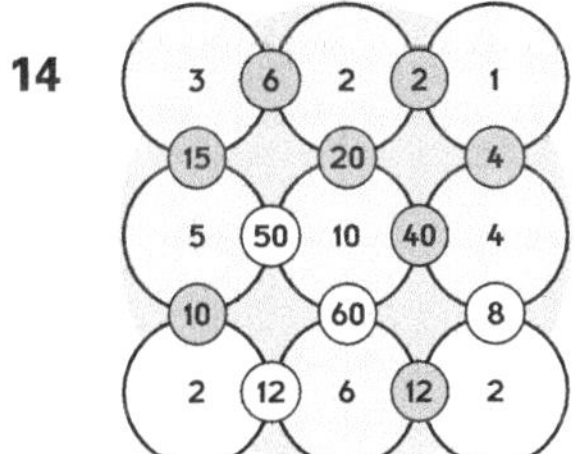

6

15 Choctop and creamy-cone
or 2 Yogy bears — 6

	Possible marks
16 a 8, 19	4
b 7, 12, 20	6
Total	**100**

TEST Solutions

YEAR 3 TEST 1 — Screening Test 1 (page 38)

Question	Answer	Possible marks
1 a		2
b		2
2 a	31, 38, 45	3
b	21, 28, 36	3
c	15, 14, 17	3
3	No. 4	2
4	6 cm – 4 cm = 2 cm	3
5		3
6 a	small triangles = 8	1
b	medium triangles = 4	2
c	large triangles = 4	2
d	total triangles = 16	1
7	7 0 5 / 2 4 6 / 3 8 1	5
8	□ = 5 △ = 3	4
9		4
10	18	2
11	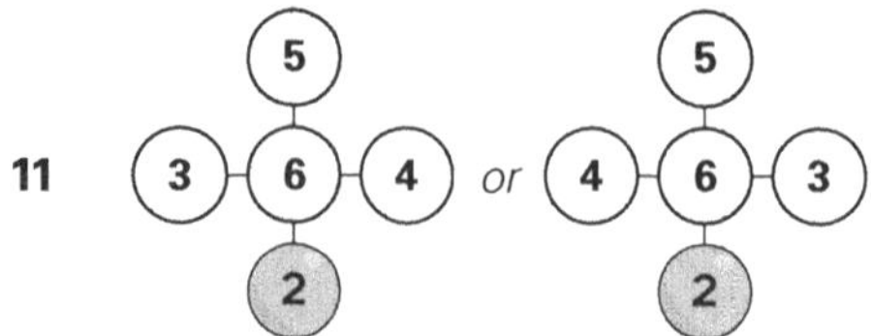	4
12	8 other possibilities: RR, RY, RB, BB, BR, BY, YY, YB, YR	4
13 a	180 minutes	3
b	120 secs	3
14 a	i $3 ii $15	4
b	$5.50	3
15	40 cents each	3
16	Jane's share: 7 kg; Kevin's share: 5 kg	3
17	2	2
18	24 (12 on each layer)	3
19	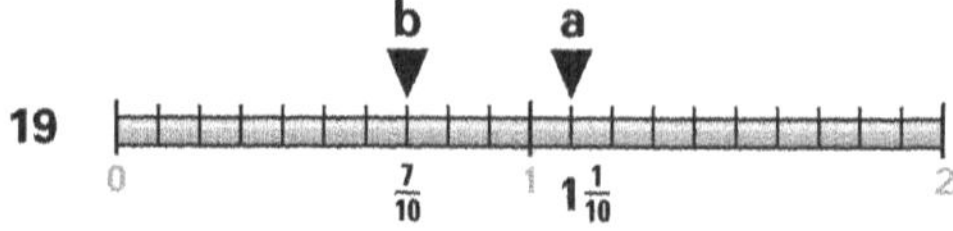	4
20	200 g	4
21 a	Row 4 IIIIIIIIII (10) Row 5 IIIIIIIIIIIII (13)	2
b	16	2
c	35	2
22 a	black	2
b	grey	2
23 a	35 children	2
b	4 pizzas	2
24 a	i 10 units ii 12 units iii 8 units	3
b	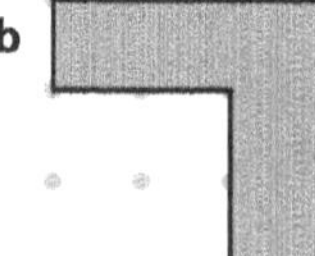	1
	Total	**100**

TEST Solutions

YEAR 3 TEST 2 — Screening Test 2 (page 42)

Question	Answer	Possible marks
1	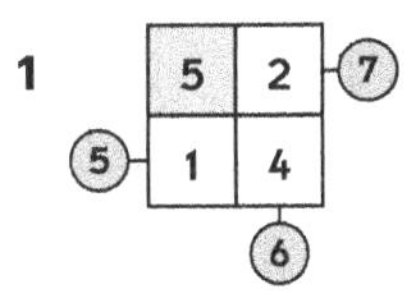	3
2	15	3
3	$21	3
4	□ = 9 △ = 6	4
5	14°C	3
6	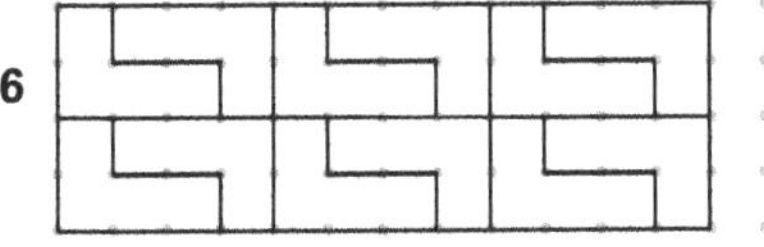	4
7 a	12	2
b	Helen	2
c	Wasim	2
d	12	2
e	72	2
8	□□□	2
9	18 minutes (from 7:53 till 8.11)	3
10 a	9:15 a.m.	3
b	1 hour 25 minutes (He saved 20 minutes.)	3
11 a i	80c	2
ii	$1.50	2
iii	$1.20	2
b i	$2.30	2
ii	$2.70	2
12	$2.05	3
13	40 kg	3
14	Red	3
15	9758	3
16	47.4 cm	3
17	10 cm²	3
18 a	12 kg	1
b	7 kg	1
c	5 kg	2
19 a	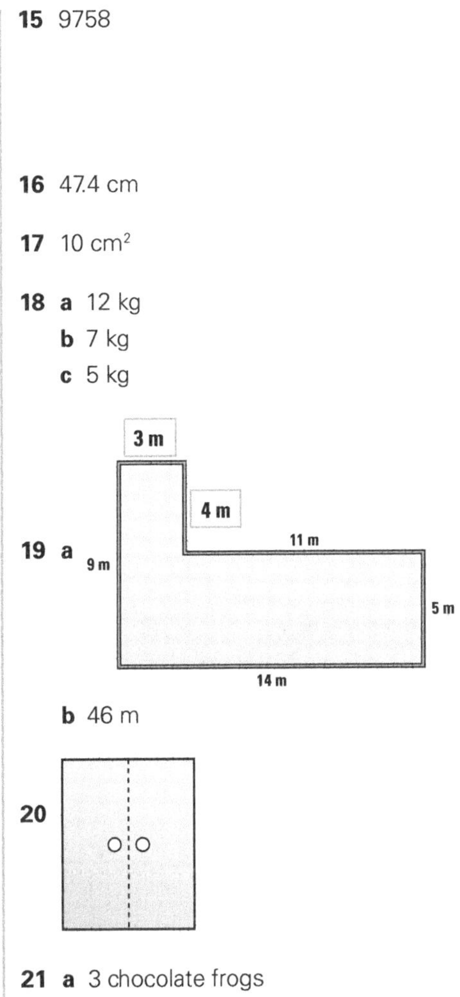	2
b	46 m	3
20	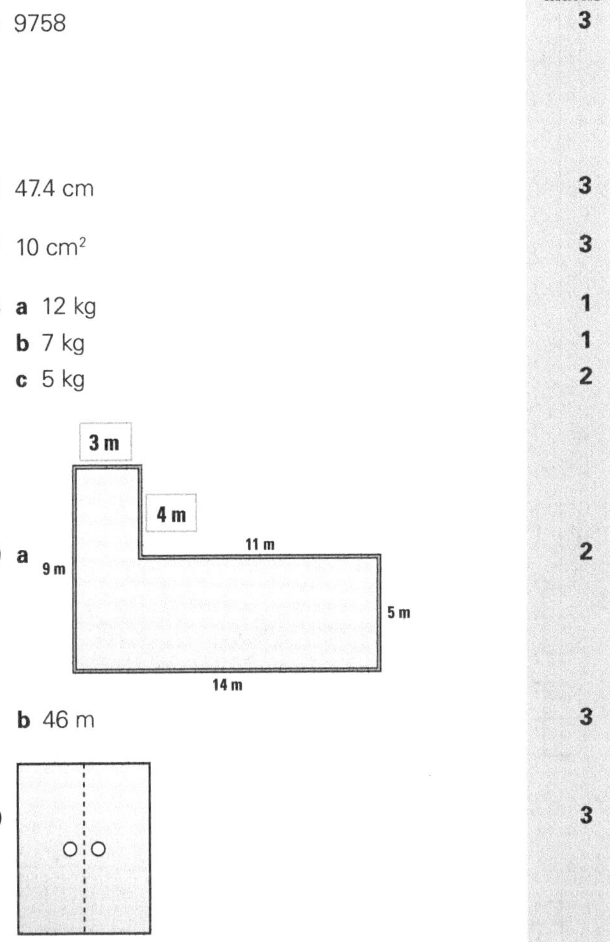	3
21 a	3 chocolate frogs	3
b	15 chocolate frogs	3
22	(1) lightest, (4) heaviest	4
23 a	HH, HT, TH, TT	4
b	1 chance in 4	2
24	14 (= 2 x 5 + 4)	3
	Total	**100**

TEST Solutions

YEAR 3 TEST 3 Screening Test 3 (page 46)

		Possible marks
1 a i	43 + 8 = 51 *or* 48 + 3 = 51	**2**
ii	34 + 8 = 42 *or* 38 + 4 = 42	**2**
b i	48 – 3 = 45	**2**
ii	83 – 4 = 79	**2**
2 a	30c	**2**
b	$1.50	**2**
3	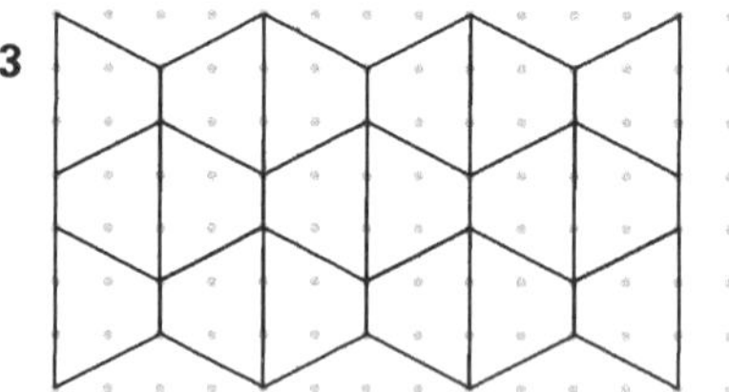	**5**
4	□ = 9 ○ = 2	**4**
5	Shade any 6 of the 10 little squares.	**2**
6	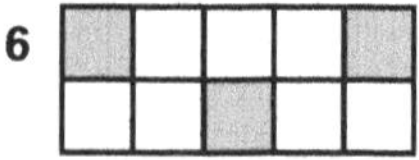 Shade any 3 of the 10 little squares.	**2**
7	4 o'clock	**3**
8	25th May	**3**
9		**3**
10	× × × × (dot grid) × × × ×	**4**
11 a	15	**2**
b	<<<VV	**2**
12	9 minutes (only 3 cuts) 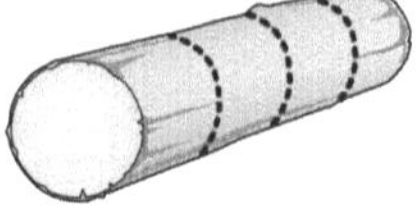	**3**

		Possible marks
13 a	6 cm, 5 cm, 8 cm, 10 cm, 5 cm, 14 cm	**2**
b	48 cm	**3**
14	4	**3**
15	$10\frac{1}{2}$ hours	**3**
16 a	B, D and E ($\frac{2}{3}$)	**3**
b	A and C	**2**
17 a	$2.20	**3**
b	$6.30 (1 ice cream costs $2.10)	**3**
18 a	5309	**2**
b	33 872	**2**
19 a	1 hour 6 minutes	**3**
b	28 students	**3**
20 a	bag A	**3**
b	1 chance in 4	**3**
21	D	**2**
22 a	30 (55 – 25)	**3**
b	205 (55 + 25 + 30 + 45 + 50 = 205)	**3**
23	13(25 lots of 13 – 24 lots of 13 = 1 lot of 13)	**3**
24	24 (4 × 3 × 2 = 24 or 12 in each layer)	**3**
25	1 + 1 = 2, 2 + 2 = 4, 1 + 2 = 3, 2 + 5 = 7, 1 + 5 = 6, 5 + 5 = 10	**3**
	Total	**100**

TEST Solutions

YEAR 3 TEST 4 Screening Test 4 (page 50)

	Possible marks
1 C	3
2 A	3
3 B	3
4 B	3
5 D	3
6 C	3
7 D	3
8 D	3
9 D	3
10 A	3
11 B	3
12 A	3
13 C	3
14 A	3
15 B	3
16 D	3
17 A	3
18 D	3
19 D	3
20 C	3
21 D	4
22 B	4
23 A	4
24 D	4
25 A	4
26 A	4
27 D	4
28 A	4
29 B	4
30 C	4
Total	**100**

TEST Solutions

YEAR 4 TEST 1 Screening Test 1 (page 54)

	Answer	Possible marks
1	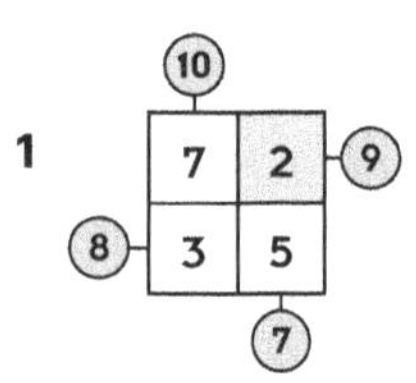	**3**
2	$4000	**2**
3	**a** 21c	**2**
	b $1.05	**2**
4		**3**
5	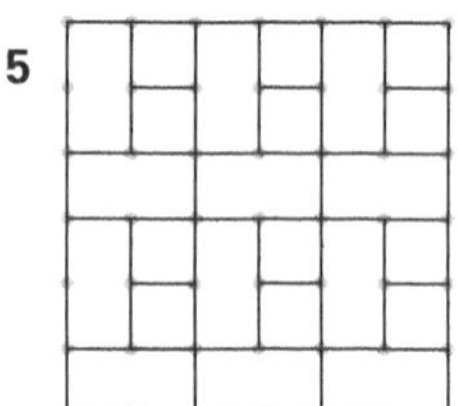	**4**
6	$3.40 (she spent $5.90)	**3**
7	**a** A	**2**
	b S	**2**
8	**a** 480 bolts (60 × 4 = 240 in 1 hour)	**3**
	b 50 minutes	**3**
9	**a** $1.60	**3**
	b $2.70 (3 × 40c + 3 × 50c = $1.20 $1.20 + $1.50 = $2.70)	**2**
	c $2.30	**2**
	d 5 balloons *or* 3 balloons and 1 comic book *or* 4 icy poles *or* 2 comic books and 1 balloon	**3**
	e 2 comic books and 1 balloon	**3**
10	□ = 8 △ = 2	**4**
11	16	**2**
12	$\frac{1}{4}$	**2**
13	21  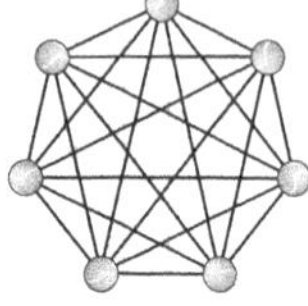 There are 21 lines joining these 7 points.	**4**
14	**a** 10	**2**
	b 24	**2**
	or Each team plays 6 others, however a match is between 2 teams.	

		Possible marks
15	3 Adding 9 lots of 6 gives 54. Carry 5 to the 10s. Adding 8 lots of 6 and carrying 5 gives 53.	**3**
16	P = 60 paces (BC = 10 paces)	**4**
17	younger son = $9 elder son = $27	**4**
18		**4**
19	2 chances out of 5	**3**
20	**a** 10	**2**
	b 17 (27 – 10 = 17)	**3**
21	(see table below)	**3**
22	**a** she sees	**2**
	b  she sees 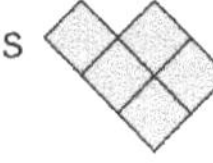	**3**
23	28 kg (Do this question working backwards.)	**3**
24	**a**	**3**
	b 15 squares	**2**
	c 40 matches	**3**
	Total	**100**

Question 21:

WST	CST	EST
07:30	09:00	09:30
10:15	11:45	12:15
08:10	09:40	10:10
12:15	13:45	14:15

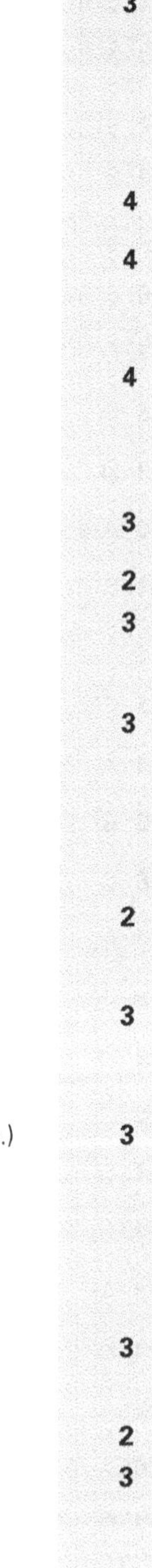

TeST Solutions

YEAR 4 TEST 2
Screening Test 2 (page 58)

Possible marks

1 $14.15 — 2

2 — 3

Paul	Martine
8	11

3 □ = 7 — 3

4 3:35 p.m. — 2

5 135 minutes — 2

6 **a** 63, 57 — 2
b 22, 29 — 2
c 16, 32 — 2

7 Rule is: double top number then subtract 1. — 4

4	6
⇩	⇩
7	11

8 24 (= 4 × 3 × 2) — 3

9 **a** 10 stars — 3
b 5 stars — 2

10 24 — 2

11 □ = 5 and △ = 4 *or*
□ = 4 and △ = 5 — 4

12 24 — 3

13 7 — 3

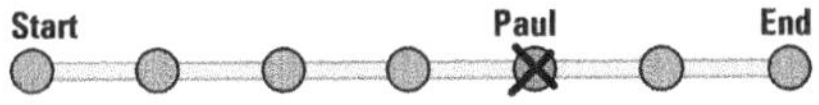

14 37 (Listing possibilities and then eliminating is a good strategy. Numbers that have 1 left over when counting by threes are 28, 31, 34 and 37. Now when counting by fives we have to have 2 left over, so only 37 is correct.) — 4

15 $6, $18 — 4

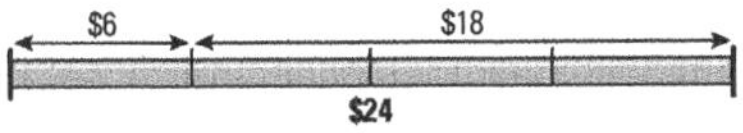

16 72 km — 3

17 42 — 3

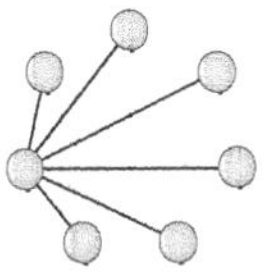

Each person sent 6 cards.

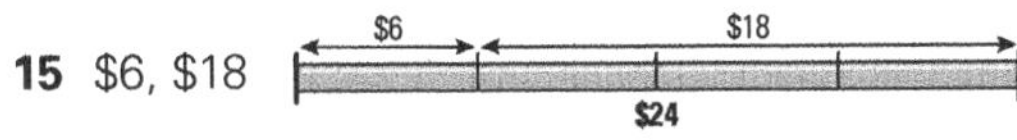

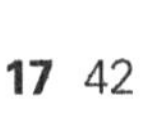

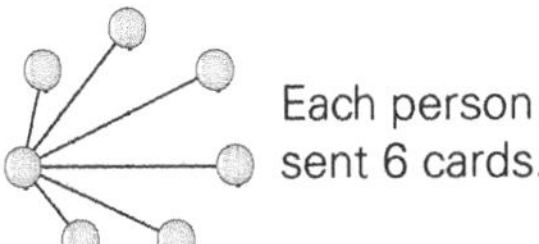

Possible marks

18 **a** 18 — 3
b 12 — 3

19 **a** — 2

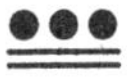

b 9 — 2

20 — 4

Rectangle	Number of matches
4	18
5	22

21 — 3

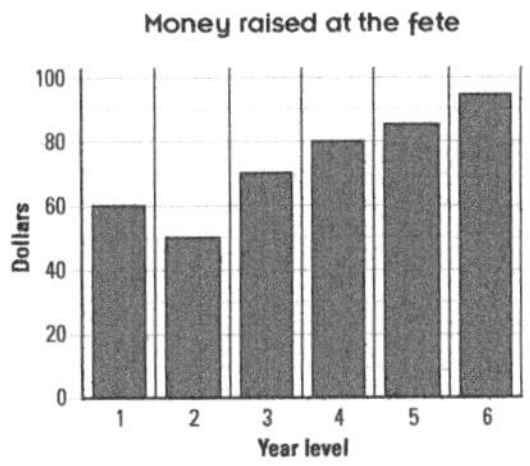

22 **a** 4 (80c each) — 3
b $5.60 — 3

23 **a** 1 chance in 4 — 3
b brown hair — 3

24 **a** 22 kg — 2
b 7 kg — 4

25 **a** 15 — 4
(3, 13, 23, 33, 43, 30, 31, 32, 34, 35, 36, 37, 38, 39)

b 81 — 5
(9 for 1 to 9
20 for 10 to 19
20 for 20 to 29
20 for 30 to 39
12 for 40 to 45)

Total **100**

TEST Solutions

YEAR 4 TEST 3 — Screening Test 3 (page 62)

Question	Answer	Possible marks
1	6000	2
2 a	$13	2
b	$91	2
3	2:45 p.m.	2
4	65 cm	3
5	$9	3
6	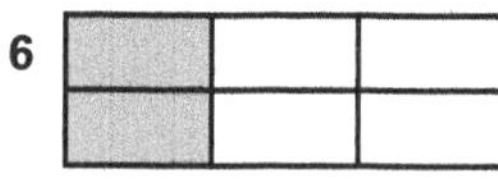 Any 2 rectangles can be shaded.	2
7	7 (working backwards)	3
8	2·09, 2·9, 2·92, 2·99	3
9	South 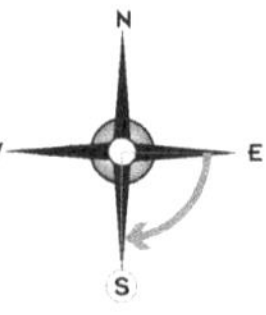	3
10	79 kg (He lost 5 kg.)	3
11 a	18	3
b	90	3
12 a	Leon	2
b	Jason	2
c	3·1 secs	2
13 a	18 litres	3
b	16 days	3
14	[5] 5 [2] + 1 [7] 3 = 7 2 5	3
15 a	37 (Note 10th term = 1 + 9 × 4)	3
b	21st term = 81 = 1 + 20 × 4	3
16	10	2
17 a	15	3
b	12	3
18 a	8 people	2
b	5 tables	3
19 a		2
b	time 10:40	3
c	 time 12:55	3
20	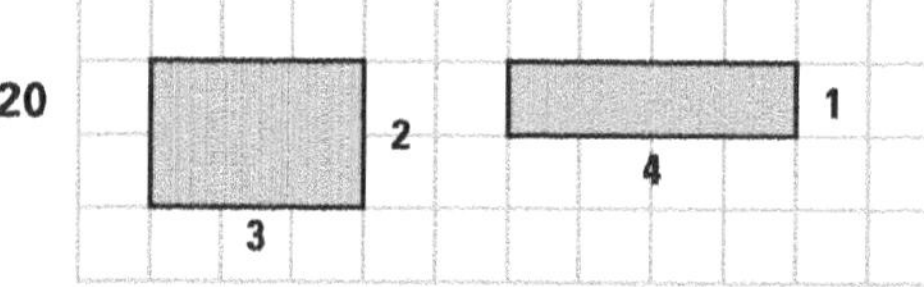	4
21	= 5 kg	2
	= 6, 7, 8 or 9 kg	2
	= 1, 2, 3, 4 kg (less than 5 kg)	2
	= 11 kg or 12 kg, 13 kg etc. (more than 10 kg)	2
22	20, 21, 22	3
23 a	133 cm	4
b	188 cm	2
24	2 bicycles, 5 tricycles	3
	Total	**100**

TEST Solutions

YEAR 4 TEST 4 Screening Test 4 (page 66)

	Possible marks
1 A	3
2 D	3
3 D	3
4 B	3
5 A	3
6 C	3
7 C	3
8 D	3
9 C	3
10 A	3
11 A	3
12 B	3
13 D	4
14 B	4
15 B	4
16 A	4
17 C	4
18 D	4
19 C	4
20 B	4
21 C	4
22 C	4
23 D	4
24 A	4
25 C	4
26 C	4
27 A	4
28 D	4
Total	**100**

TEST Solutions

YEAR 5 TEST 1 — Screening Test 1 (page 70)

Question	Answer	Possible marks
1 a	980	**2**
b	310	**2**
2 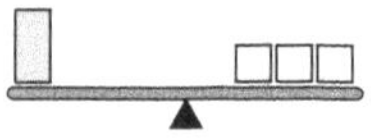		**2**
3	(see table below)	**4**

3	5	7
0	5	0
2	1	2

Question	Answer	Possible marks
4 a	8 m	**2**
b	15 ribbons	**2**
5 a	\$12.50	**2**
b	20 stamps	**2**
c	B For A 1 pen costs 69c. For B 1 pen costs 60c.	**2**
6	32 (2 × 10 + 2 × 4 + 4)	**3**
7 a	Writing equivalent fractions for $\frac{1}{4}$ and $\frac{1}{2}$ and then writing the fractions between them, it is possible to continue this process for a long time. Only a limited number of examples have been given.	**3**

			and				
$\frac{1}{4}$			and				$\frac{1}{2}$
$\frac{2}{8}$			$\frac{3}{8}$				$\frac{4}{8}$
$\frac{3}{12}$	$\frac{4}{12}=\frac{1}{3}$			$\frac{5}{12}$			$\frac{6}{12}$
$\frac{4}{16}$	$\frac{5}{16}$		$\frac{6}{16}=\frac{3}{8}$	$\frac{7}{16}$			$\frac{8}{16}$
$\frac{6}{24}$	$\frac{7}{24}$	$\frac{8}{24}=\frac{1}{3}$	$\frac{9}{24}=\frac{3}{8}$	$\frac{10}{24}=\frac{5}{12}$	$\frac{11}{12}$		$\frac{12}{24}$
$\frac{25}{100}$	$\frac{27}{100}$	$\frac{29}{100}$	$\frac{31}{100}$	$\frac{33}{100}$	$\frac{37}{100}$	$\frac{39}{100}$	$\frac{50}{100}$

Question	Answer	Possible marks
b	$\frac{3}{8}$ (From line two we can see that $\frac{3}{8}$ is exactly halfway between these fractions.)	**2**
8	7 cm^2	**3**
9	37	**3**
10 a	(see table below)	**3**

A	B	C	D	E	F	G
	19		20		21	
22		23		24		25
	26		27		28	

Question	Answer	Possible marks
b i	F (all multiples of 7 are in this column)	**2**
ii	A	**2**
11 a	faces = 6, edges = 12, vertices = 8	**3**
b	faces = 7, edges = 15, vertices = 10	**3**
12 a	4 hours	**2**
b	5 hrs 20 mins	**1**
13 a	2 times	**3**
b	28 times	**3**
14	9	**2**
15	85	**3**
16	$\frac{1}{4}$ (Note: △ PAS + △ QRB = $\frac{1}{2}$)	**3**
17	4 Phillip 6 George 5 Elizabeth 8 Sylvia 7 Mary	**3**
18	10 km	**3**
19	18 triangles	**3**
20 a 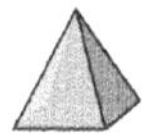		**2**
b 		**2**
c 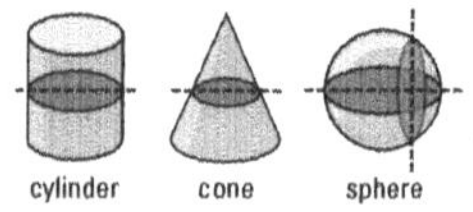	cylinder cone sphere (sphere can be cut vertically or horizontally)	**3**
21 a	\$6.00	**2**
b	6 hours	**2**
22 a	5 2 9 5 2 9 5 2 9 5 2 A = 2, B = 9	**3**
B	7 4 5 7 4 5 7 4 5 7 4 A = 5, B = 7	**3**
23	50 m 2 lengths + 1 width = 42 m; 1 length + 2 widths = 33 m; adding these two, 3 lengths + 3 widths = 75 m; 1 length + 1 width = 25 m; 2 lengths + 2 widths = 50 m	**4**
24 a	13 people (1 + 3 + 9 = 13)	**3**
b	3 hours (1 + 3 + 9 + 27 = 40)	**3**
	Total	**100**

TEST Solutions

YEAR 5 TEST 2 — Screening Test 2 (page 74)

		Possible marks
1	12 and 4	**3**
2	17	**3**
3	\$4.20	**3**
4	11 girls	**3**
5	32 cm	**3**
6	**a** 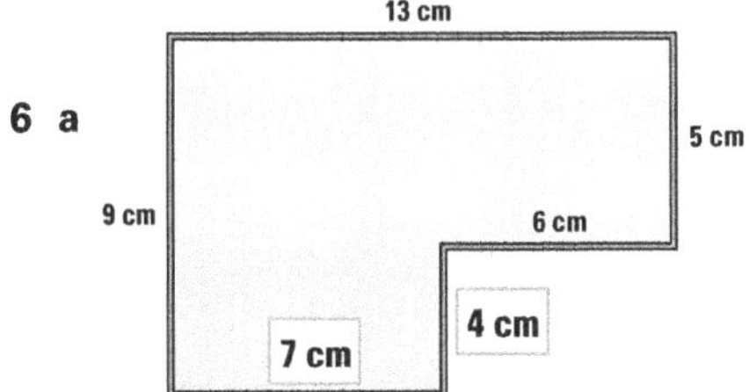	**2**
	b 44 cm	**2**
	c 93 cm^2	**3**
7	617 280	**3**
8	**a** $5^2 - 1 = 24 = 4 \times 6$ $6^2 - 1 = 35 = 5 \times 7$	**2**
	b $99^2 - 1 = 98 \times 100 = 9800$	**3**
9	24 times	**2**
10	\$5.50	**3**
11	$\frac{1}{3} = \frac{2}{6} = \frac{3}{9} = \frac{4}{12} = \frac{5}{15} = \frac{6}{18} = \frac{7}{21} = \frac{8}{24} = \frac{9}{27} = \frac{10}{30}$ $= \frac{100}{300}$ and so on.	**2**
12	$2\frac{11}{12}$	**2**
13	**a** 82 km	**2**
	b 984 km	**2**
14	**a** 27 cm^3	**2**
	b 189 cm^3 (7 cubes)	**2**
15	**a** younger = \$160, elder = \$320	**2**
	b younger = \$120, elder = \$360	**2**
16	**a** 30 days	**2**
	b 5 days	**2**
	c 90 men	**2**

		Possible marks
17	\$60	**4**
18	$\frac{3}{8}$	**3**
19	**a** $\frac{1}{12}$	**2**
	b 1 chance in 4	**2**
	c 4 chances in 12 or 1 chance in 3	**2**
	d 1 chance in 3 $(\frac{1}{4} + \frac{1}{12} = \frac{1}{3})$	**2**
20	30 plants 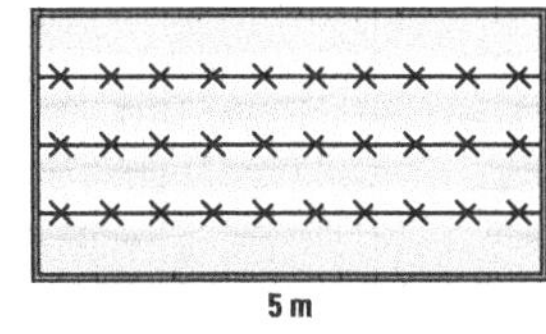	**5**
21	area of each white square = 9 cm^2 Side X = 3 cm 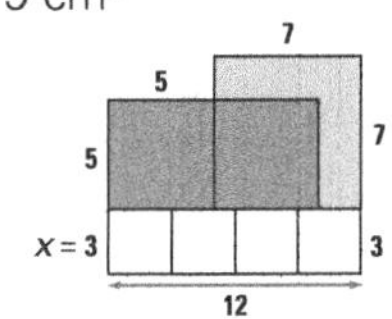	**3**
22	360 gm (Note: mass must be divisible by 30.)	**3**
23	7, 6, 3, 9, 2	**3**
24	90 minutes (LCM of 6, 10, 9)	**3**
25	A – 1 B = 8 or 7 C = 0 or 9 D = 2 or 4 $\begin{array}{r} 1802 \\ 180 \\ 10 \\ + \quad 1 \\ \hline 1993 \end{array}$ *or* $\begin{array}{r} 1794 \\ 179 \\ 19 \\ + \quad 1 \\ \hline 1993 \end{array}$	**8**
26	**a** 8	**1**
	b 24	**1**
	c 8	**1**
	Total	**100**

TEST Solutions

YEAR 5 TEST 3

Screening Test 3 (page 78)

		Possible marks
1	5 hours 35 minutes	**2**
2	$\frac{3}{8}$	**2**
3	4	**2**
4	36 m^2 (side 6 m)	**3**
5	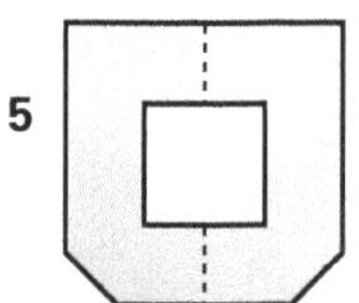	**3**
6	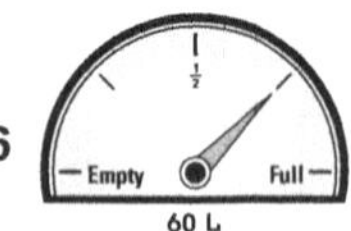	**2**
7	**a** \$1.50	**2**
	b 75c	**2**
8		**2**
9	**a** \$259	**2**
	b \$120	**2**
10	**a** 3	**2**
	b 9	**2**
11	**a** 15	**2**
	b 24	**2**
12	28°C	**2**
13	**a** (table below)	**2**
	b B	**3**
	c D	**3**
14	11, 21, 31, 41 (any two of these)	**2**
15	6 cm^2	**3**
16	4 days	**3**

13 a

A	B	C	D	E	F
30	29	28	27	26	
	31	32	33	34	35

		Possible marks
17	24 minutes	**3**
18	**a** 120 km	**2**
	b 30 km	**2**
	c 3 hours 30 minutes or $3\frac{1}{2}$ hours	**2**
19	**a** 144 cm^2	**3**
	b 1·5 cm	**3**
20	**a** 90c	**3**
	b 120	**3**
21	**a** 3 out of 5	**2**
	b 1 out of 5	**2**
	c 3 out of 5	**2**
	d 2 out of 5	**2**
22	**a** Row 5 1	
	Row 4 3	
	Row 3 6	
	Row 2 10	
	Row 1 15	**5**
	b 35	**2**
23	7	**4**
24	\$8 \$11	**6**
25	8 and 125	**4**
	Total	**100**

TEST Solutions

YEAR 5 TEST 4 Screening Test 4 (page 82)

	Possible marks
1 D	2
2 A	2
3 D	2
4 C	2
5 C	2
6 B	3
7 A	3
8 D	3
9 B	3
10 A	3
11 A	3
12 D	3
13 A	3
14 C	3
15 D	3
16 C	3
17 A	3
18 B	3
19 D	3
20 B	3
21 A	3
22 A	3
23 B	3
24 C	3
25 B	3
26 D	3
27 C	3
28 A	3
29 C	3
30 B	3
31 A	3

	Possible marks
32 C	3
33 B	3
34 A	3
35 B	3
Total	100

TEST Solutions

YEAR 6 TEST 1 Screening Test 1 (page 86)

Question	Answer	Possible marks
1 a	$1\frac{1}{4}$	2
b	$\frac{3}{8}$	2
c	$\frac{1}{4}$	2
d	$\frac{3}{2}$	2
2	8462	3
3 a	27, 20	2
b	2, $\frac{2}{3}$	2
c	31, 63	2
4 a	0·6	2
b	0·69	2
c	0·71	2
d	0·75	2
5	15 metres	3
6 a	300 metres	3
b	5000 steps	3
7	$14 \div 1\frac{3}{4} = 8$	2
8 a	volume = 64, so each side = 4 cm	2
b	1 cm	2
9	10 times	3
10	$\frac{1}{3}$ of pole = 15 m; $\frac{2}{3}$ of pole = 30 m; entire pole = 45 m	2
11	14 cm^2	3
12	240 cards (To get maximum, cards must be cut out as shown.) $20 \times 12 = 240$ cards	3
13	There are 7 equal parts of money, Sandra has 5 parts and Joy has 2. 7 parts = \$56, 1 part = \$8. Sandra has \$24 more than Joy.	3
14 a	One way is by vertical slicing $(4 \times 3 \times 3) + (3 \times 3) + (2 \times 3) + 3 = 54$	3
b	There are 13 **x**s, where **x** marks a cube with 3 red faces; the 13th is in the far corner on the bottom row.	3
15	12 different meals SFI, SFC, SCI, SCC, SVI, SVC PFI, PFC, PCI, PCC, PVI, PVC	3
16 a	There are 9. 11, 22, ……., 88, 99	2
b	30. Listing possibilities: 101, 111, 121, … , 191 202, 212, … , 292 303, 313, … , 393	3
17 a	1 hour 40 minutes, then 20 minutes	2
b	average speed $= \frac{\text{total distance}}{\text{total time}} = \frac{115}{2} = 57{\cdot}5$ km/h (It is incorrect to average the two speeds given!)	2
18 a	F	1
b	A	1
c	A and B	1
d	C and F	1
e	B	2
19	60 boys (12 days for 70 boys; 1 day for 12×70 boys; 14 days for $\frac{12 \times 70}{14}$ boys = 60 boys)	3
20 a	216 cm^2	2
b	$36 \times 6 = 216$ cm^2.	2
c	$36 \times 6 + 4 \times 2 = 224$ cm^2	2
21	14 green and 7 blue tickets	2
22	33	2
23	$\frac{5}{6}$ $(1 - \frac{1}{2} + \frac{1}{2} - \frac{1}{3} + \frac{1}{3} - \frac{1}{4} + \frac{1}{4} - \frac{1}{5} + \frac{1}{5} - \frac{1}{6} = 1 - \frac{1}{6})$	3
24 a	$\frac{5}{12}$ $\frac{1}{6} + \frac{1}{4} = \frac{5}{12}$	3
b	2 h 24 min $\frac{12}{5}$ hours	3
	Total	**100**

TEST Solutions

YEAR 6 TEST 2 — Screening Test 2 (page 90)

		Possible marks
1	5c	2
2	**a** 2·25	2
	b 0·15	2
3	**a** 74%	2
	b 7%	2
4	$\frac{1}{4}$	2
5	**a** $\frac{11}{12}$	2
	b $7\frac{3}{4}$	2
	c 8	2
	d 26	2
	e $12\frac{3}{4}$	2
6	3·061, 3·099, 3·48, 3·6	2
7	8	2
8	\$8 (child's ticket = \$4)	2
9	72 (18 are divisible; 90 – 18 = 72)	2
10	13, 15, 17	2
11	☐ = 6	2
12	**a** 8 kg	2
	b 10 weeks	2
13	$\frac{6}{15} = \frac{2}{5} = 40\%$	3
14	1 exercise book costs \$2; 1 novel costs = \$8 3 novels cost 12 × excercise books Total 16 × excercise books = \$32 1 × excercise books = \$2	3
15	112 cubes top 2 layers = 8 next 2 layers = 32 bottom 2 layers = 72	3
16	7:58 a.m. (in 24 hours gained 480 seconds = 8 minutes)	3

17 a

Number of octagons	1	2	3	4	5	6
Number of sides	8	15	22	29	36	43

Possible marks: 3

		Possible marks
17	**b** 78 (11 × 7 + 1)	3
18	Friday (every 12 days does both)	3
19	64 cm² (*X* = 8 cm)	3
20	8 do neither	3

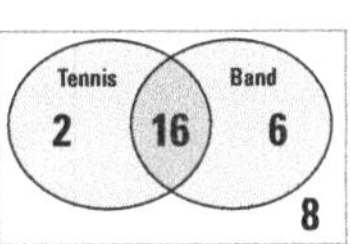

21 6 x 2 + 5 x 2 + 4
= 26 blocks — 3

15 m
18 m

		Possible marks
22	**a** 12 × 4 × 1·5 = 72 m³ = 72 000 L	2
	b 4 litres per 10 secs = 24 litres per minute time taken to fill pool with 72 000 L $= \frac{72\,000}{24}$ minutes = 3000 minutes = 50 hours	3
23	**a** 10 cm	2
	b 60 cm²	2
	c $50 + 50 + 60 + 2 \times \frac{1}{2} \times 6 \times 4 = 184$ cm	3
24	**a** 7 × 5 × 4	2
	b 7 × 5 × 4 – 5 × 3 × 2 = 140 – 30 = 110	2
25	**a i** 3	2
	ii 72	2
	iii 9	2
	iv 4	2
	v 60	2

b (3 marks)

	yards	feet	inches
		1	9
+		1	10
	1	0	7

c (3 marks)

	yards	feet	inches
	1	2	8
×			4
	7	1	8

Note: 32 inches = 2 ft 8 inches.

Total 100

TEST Solutions

YEAR 6 TEST 3 Screening Test 3 (page 94)

		Possible marks
1	10 (10 pairs each equal to 1)	**3**
2	**a** \$3.70 (4 apples cost 74c)	**3**
	b \$26 565 (increase of \$1265)	**3**
3	**a** \$42.50	**3**
	b 28 kg (8 kg excess)	**3**
4	**a** 3 seconds	**2**
	b 15 seconds	**2**
5	4 mins	**3**
	If the train is moving at 60 km/h, it will take 1 minute to travel 1 km so 4 minutes to travel 4 km. Note that the front of the train takes 3 minutes to pass through the tunnel, but since the train is 1 km long it will take 4 minutes.	
6	Since 36 = 24 + 12, you will need $\frac{1}{2}$ of each ingredient as well.	
	6 + 3 = 9 cups of cornflakes	
	4 + 2 = 6 whites of eggs	
	2 + 1 = 3 tablespoons of butter	
	40 + 20 = 60 grams of cocoa	
	60 + 30 = 90 grams of nuts	
	$1\frac{1}{8}$ cup of sugar	**4**

7

A	B	C	D	E
11	12	13	14	15
20	19	18	17	16
21	22	23	24	25
30	29	28	27	26

		Possible marks
	a B	**2**
	b E	**2**
	c B	**2**
8	● = 4, △ = 3, so ●●●△△ = 18	
9	7 × 10c (4 × 20c, 3 × 50c)	**3**
10	6 × 4 = 24 outfits (since with each blouse I can wear any 4 of the skirts)	**3**
11	27 possibilities You can draw a tree diagram or list all the possibilities.	**6**
	A77, A78, A79, A87, A88, A89, A97, A98, A99	
	(9 possibilities starting with B and 9 with C, altogether 27 possibilities *or* 3 ways of choosing a letter and 3 ways of choosing first number and 3 ways of choosing second number, a total of 3 x 3 x 3 = 27 possibilities)	
12	**a** A = 2 cm²	**2**
	B = $7\frac{1}{2}$ cm²	**2**
	C = 3 cm²	**2**
	b Area of figure = $35 - (\frac{1}{2} \times 2 \times 2 + \frac{1}{2} \times 2 \times 3 + \frac{1}{2} \times 3 \times 5) = 22\frac{1}{2}$ cm²	**3**
13	Since the cars are travelling towards each other	
	a 150 km	**3**
	b 150 km 60 min before they meet	
	15 km 6 min before they meet	
	$7\frac{1}{2}$ km 3 min before they meet	**3**
14	distance = 370 km	**3**
	average speed = $\frac{\text{total distance}}{\text{total time}}$	
	$= \frac{370}{5}$ km/h = 74 km/h	**3**
15	A suitable scale could be 2 m = 1 cm *or* 4 m = 1 cm (diagram drawn to scale)	**3**
16	B	**3**
17	42	**3**
18	Area of wall needed = 3 x 5 – 2 = 13 m²	
	number of bricks needed = 13 x 40 = 520	**3**
19	8 cm	**3**
	24 cm = 6 equal parts	
	4 cm = 1 part	
	original side of square = 8 cm	
20	6	**3**
21	90 = 10 × 9	**3**
22	36 = 6 × 3 × 2	**3**
23	**a** \$48 (US)	**3**
	b \$75 (Aust.)	**4**
	Note: \$5 (Aust.) = \$4 (US)	
24	22 100 = 4 × 5525	**4**
	Total	**100**

TEST Solutions

YEAR 6 TEST 4 Screening Test 4 (page 98)

Question	Answer	Possible marks
1	A	2
2	A	2
3	D	2
4	A	2
5	D	2
6	C	2
7	A	2
8	D	2
9	B	2
10	B	2
11	A	3
12	D	3
13	C	3
14	A	3
15	B	3
16	D	3
17	C	3
18	B	3
19	B	3
20	B	3
21	C	3
22	A	3
23	B	3
24	C	3
25	D	3
26	A	3
27	A	4
28	C	4
29	B	4
30	D	4
31	D	4
32	C	4
33	C	4
34	D	4
	Total	**100**

Ideas for teaching

mathematically able students

The mathematically able student

All too often, the exceptionally gifted and talented child is neglected and discouraged within our schools. The goals of excellence and equity incorporate a responsibility to ensure that these children … are nurtured and challenged to the limit of their ability. (*Excellence and Equity NSW. Curriculum Reform 1989*)

It is important to understand that the population of able students forms a normal distribution, with the exceptionally gifted at the very end. In many schools, the proportion of able students could be up to 15 per cent of all students.

It is a tremendous challenge to meet the needs of the talented maths student. There is no clear-cut path which can be followed. Teachers need to adapt a variety of approaches and adjust them depending on the ability and motivation of the student.

The spread of ability in mathematics is more marked than in many other subjects. The Cockcroft Report (a large study into the teaching of mathematics in schools in England) suggests that there is likely to be a seven-year spread of ability in mathematics for the new entry in a secondary school.

The mathematically able student:

- is capable of work that is academically two to three years ahead of his/her peers
- is capable of processing double the volume of work
- is original in his/her thinking
- is able to work and concentrate for long periods
- requires less reinforcement, revision and concrete data
- has an ability to quickly understand and grasp the essence of a problem
- demonstrates swiftness in reasoning, logical thought and argument
- demonstrates the ability to formulate generalisations
- is able to relate one problem to another
- is prepared to use initiative in tackling something different
- is self-confident in a new mathematical situation.

The identification process

How to identify mathematically talented students remains an unsolved problem. At present there are no effective means of correctly identifying these students. Teachers should use a diverse combination of methods for judging student performance and potential. Talents may emerge gradually, so students' abilities should be constantly reviewed. For talents to emerge, it is also essential that students be given rich, challenging material that encourages mathematical thinking. Without this, mathematically promising students may not even exhibit any of the expected characteristics. The *Enrich-e-matics 3rd Edition* student books are one source of such material for students between the ages of 6 and 13.

- It is necessary to use *multiple identification procedures*, both objective and subjective.
- The identification process must be a *constant, ongoing process*. There needs to be continuous evaluation and review. A maths profile should be built for each student and it should be continually updated throughout their primary school years.
- Identification needs to be *organised* and *justifiable*.

Methods for judging student performance and potential could include:

- teacher nomination based on observation of the student, anecdotal records, an original remark, an unexpected question, a leap to the abstract
- parent nomination
- student performance in tests of problem solving, higher order thinking skills and maths ability
- results of class tests, maths competitions and aptitude tests.

The identification process may be inhibited if:

- students disguise or conceal their ability to seek peer acceptance or to avoid appearing different
- students lack motivation to achieve
- there is a lack of opportunities and incentives to discover and pursue students' interests and strengths.

Peer interaction of students of similar ability is important to help to avoid these problems.

For many people, a catalyst is needed for their talents to emerge—for example:

- an *environmental catalyst*, such as opportunity in a gifted and talented program at school
- an *intrapersonal catalyst*, such as self-confidence, self-esteem, autonomy, persistence, interest, motivation or initiative
- an *interpersonal catalyst*, such as a significant person's encouragement.

What can schools do to cater for the mathematically able student?

The needs of mathematically able students can be met by:

- ability groupings
- enrichment classes held regularly
- acceleration, only for the exceptionally gifted
- participating in maths competitions. Preparation, training and follow-up after each competition is recommended. Maths competitions are a forum to display talent
- special abilities camps and problem-solving days between schools
- mentors, and resource and learning centres.

For these students, the most productive and powerful learning environment is one where there are children who share their own commitment to and enjoyment in learning. Offering enrichment to the highly able student once a week, or occasionally, is of little value if he/she is often bored in class.

Students should have the opportunity:

- to do challenging problems on all topics studied, at a level far beyond syllabus requirements
- to work on 'competition-style' problems
- to reach into more abstract, complex topics and treat those topics in depth
- to study topics not in the syllabus, including creative and recreational topics
- to skip already mastered material from the existing curriculum
- to do only some of a set of exercises, and not be made to do all that the rest of the class have to do.

We need to differentiate the maths curriculum with respect to:

- *content*—facts, knowledge, concepts students must learn, principles and generalisations they must develop. Flexibility is very important with respect to depth of treatment and time spent on topics
- *process*—appropriate teaching methods, enthusiasm and flexibility of the teacher, who is ready to modify conditions to meet the needs of these students. Teaching methods should encourage creative thinking and inquiry processes. The teacher's challenge is to help students reach their full potential. It is most important that teachers emphasise the *process* of solving problems, not the product; that is, that they stress how a problem is solved, not just what the correct answer is. The teacher should try to create an atmosphere in which there is tolerance of everyone's ideas
- *product*—the 'ends' of the instruction, tangible or intangible
- *environment*—the setting, both physical and psychological, in which learning occurs.

Differentiating the maths curriculum requires a total reconsideration of the basic course. It is necessary to ask if there are sections of the basic course that can be omitted, speeded up, modified or replaced. Curriculum differentiation needs to commence with the basics rather than after the regular program has been completed.

Curriculum compacting is very important when teaching talented students. It is the process of identifying learning objectives, pre-testing students for prior mastery of these objectives, and eliminating needless teaching of those aspects of the course where mastery can be documented.

Group work, discussions, investigations and open-ended questions

To gain the maximum advantage from the *Enrich-e-matics 3rd Edition* books, encourage students to discuss their solutions in small groups, with their teacher or their parents at home. This discussion of ideas enhances learning and helps to develop appropriate language for communication of mathematical ideas and develop confidence in problem solving.

Research shows that children learn a great deal from each other during group work. Active participation is maximised. Many students feel more comfortable in small group settings, are more willing to take a risk in doing a challenging problem, to explain their reasoning, to question and to respond to each other's ideas.

Working with other students of similar ability helps the isolation and loneliness that some talented students experience to break down. However, having said all this, if a child wishes to work alone, then he/she should not be forced to work in a group.

Important aspects of mathematical investigations and open-ended questions are:

- There are no known outcomes at the beginning.
- Students formulate their own questions, explore different possibilities and ask 'What if …?'
- Each student or group will explore a problem to the depth of their ability. Most problems can be investigated with varying degrees of sophistication. The same problem can thus be a challenge to students of a wide range of both age and ability.

A guide to problem solving

In 1982, the Cockcroft Report stated: 'The ability to solve problems is at the heart of mathematics'. Problem solving is the process of applying previously learnt knowledge to new and unfamiliar situations. Teachers should discuss the various strategies that can be used in solving a problem. The problems in this book often have several different methods shown. Useful strategies are as follows:

Understand the problem

- Read the problem carefully, possibly two or three times.
- Find the important relevant information.
- Decide what you are looking for.

Explore and decide on a plan

- Think about all the different problem-solving strategies that you can use:
  - Draw up a table.
  - Look for a pattern.
  - Make an organised list.
  - Look for a simpler problem of the same type.
  - Work backwards.
  - Make a model or act it out.
  - Use 'guess and check'.
  - Work systematically.
  - Eliminate possibilities.
  - Use easier numbers.
  - Draw a picture or diagram.
  - Use concrete materials.

Stuck?

- Ask yourself the following questions:
  - Is there enough information?
  - Have I checked what I am looking for?
  - What relevant knowledge can I use?
  - Have I used all the relevant information?
  - What strategies might be helpful?
  - Would it help to divide the problem into several parts?
  - Have I assumed (unconsciously) something that is incorrect?
- When really stuck, do something else and come back to the problem later with a fresh mind.

Finished?

- Ask yourself these questions:
  - Does the answer make sense? Is the solution realistic?
  - Have I answered the question that was asked?
  - Should there be one solution or several solutions?
  - Is it possible to do the problem in another, possibly more elegant, way?
- Compare your method with the methods used by other students, and be ready to accept and appreciate different methods. By explaining and teaching, your own insight is intensified.
- Reflect on the problem done. Think about the method you have used. Have you learnt something new? How similar or different was this problem to others that you have done? If you were stuck, what was your stumbling block?

Summary

For a quick guide, students can use the following:

- What am I given?
- What do I know, and what can I use?
- Have I checked my solution and answer?
- What am I looking for, what do I want?
- What should I do to solve the problem?

How and when should the teacher intervene?

What should the teacher do when a student says he or she can't do a problem? Find out if the student has understood the question. Ask them to restate it in their own words and check through the above guide for problem solving.

- Does the student know the necessary theory or have the necessary knowledge to do the problem?
- Restate the problem using smaller numbers or ask the student to do a simpler problem.
- Would concrete material or diagram help?
- Help students by asking questions that stimulate their thinking.

Intervention should enable the student to progress. It should be used only to avoid undue frustration. It should never be used to the extent that it robs the student of the joy associated with success.

Qualities that should be rewarded and encouraged

Perseverance—This is very important; it means not giving up too easily but 'sticking it out' until a solution or a method is apparent. Problem solving cannot be rushed; it takes time and it may be frustrating.

Willingness to try and to take risks—Making mistakes and getting stuck is a necessary and expected part of learning. Able students are quite happy to venture into the unknown and are willing to try.

Flexibility—Solving questions in a number of ways should be encouraged.

Active involvement physically and mentally—Students need to draw diagrams and make extensive use of concrete materials such as interlocking cubes, geoboards, balances and various grid and dot papers.

Skill in presentation—Students' work should be recorded and reported with as much explanation as possible.

Ability to make predictions—Having discovered a pattern by using a number of cases, the able student can extend the pattern and make a prediction or generalisation.

Reflection—It is important to encourage students to reflect on a problem after having solved it, and to think about their own thinking. They should be ready to explain why they did what they did.

Commitment—To become great mathematicians, talented students should be challenged continually and given regular opportunities to practise thinking mathematically.

Creativity and originality—Students frequently come up with unexpected results and it is most satisfying for a teacher to watch the way in which students' confidence and sophistication grow.

Spinners

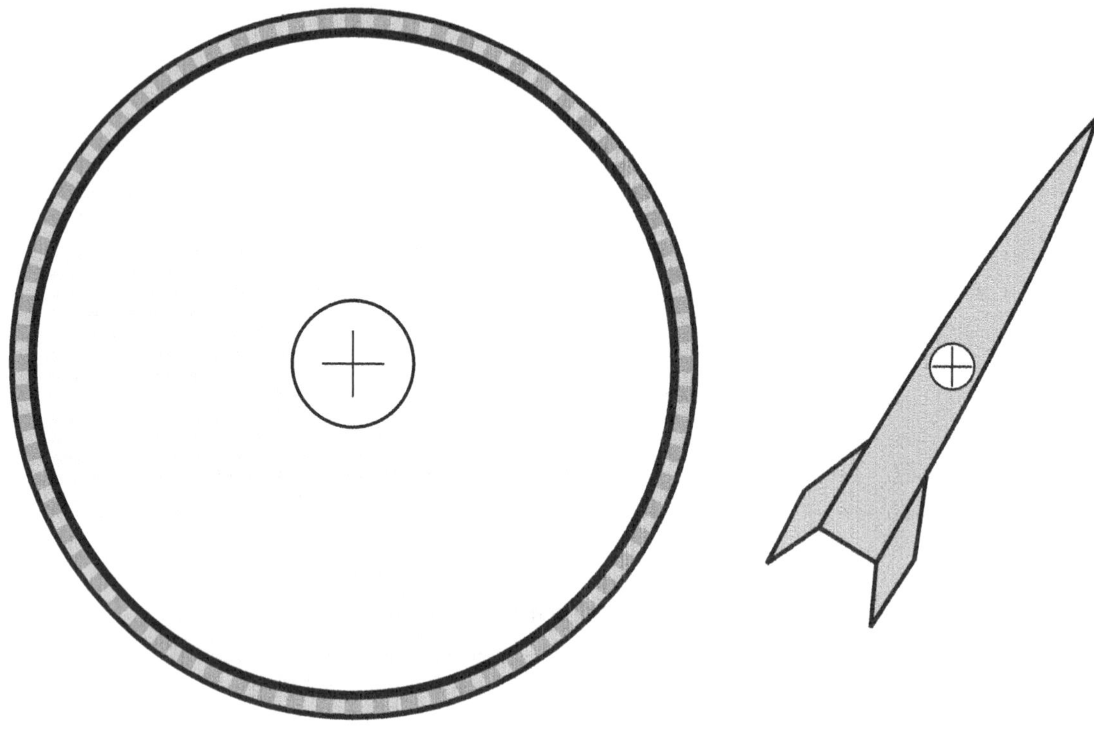

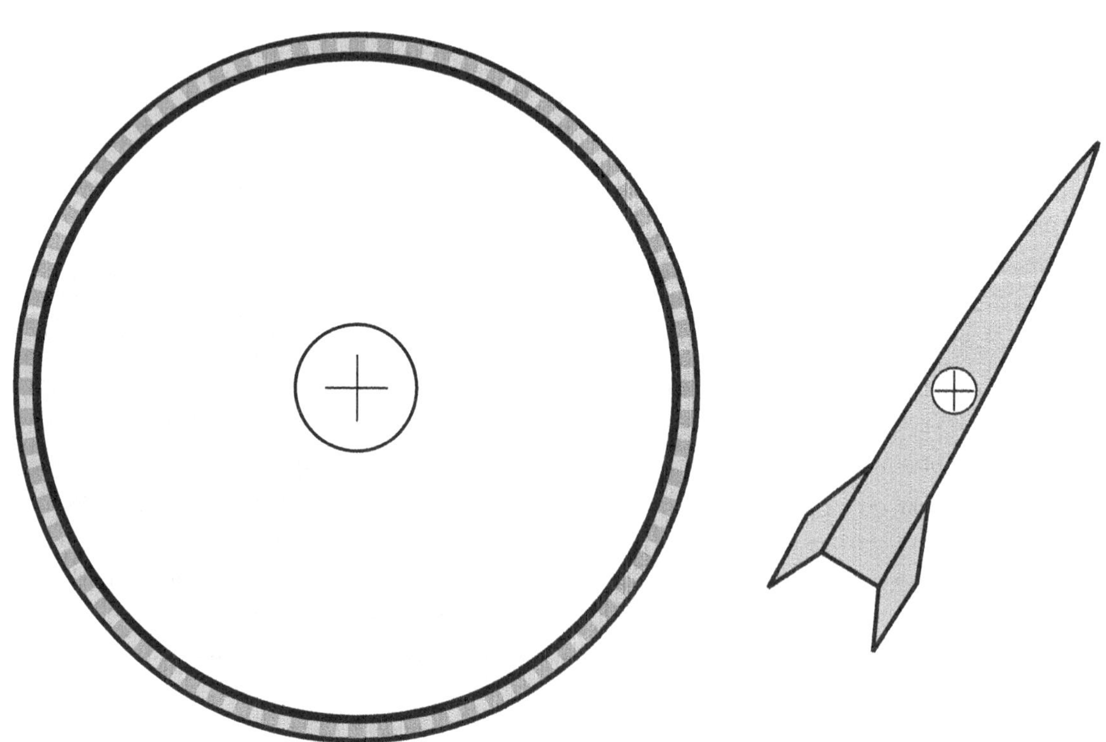

Symmetry

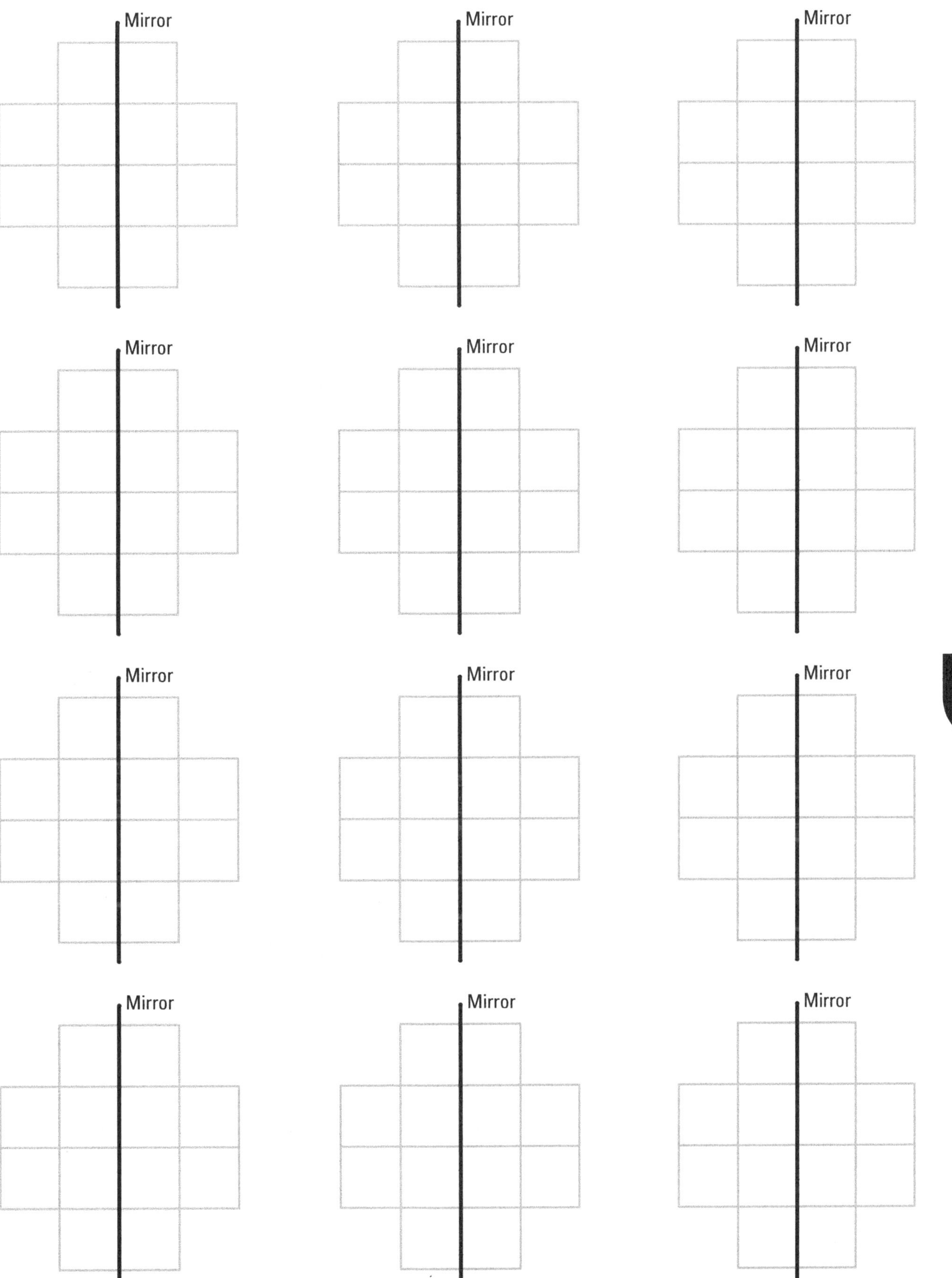

How many triangles?

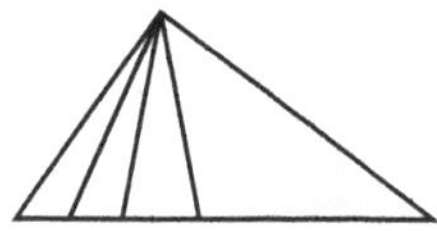 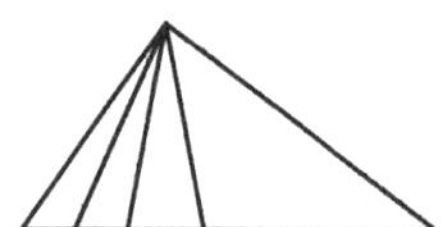  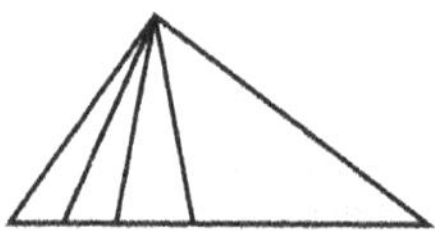 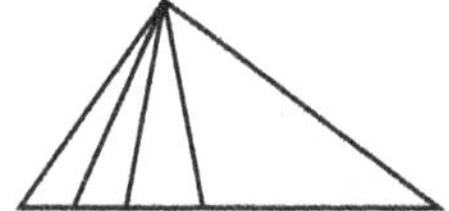

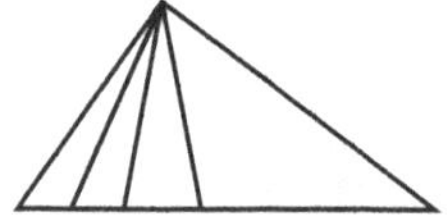   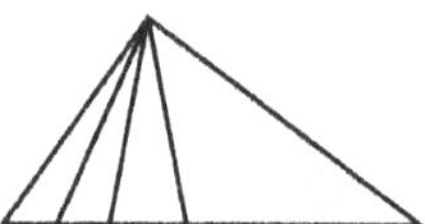 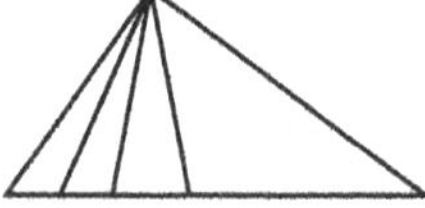

 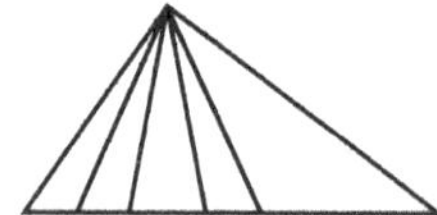 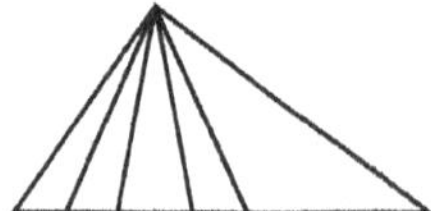 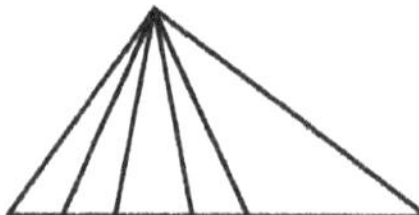 

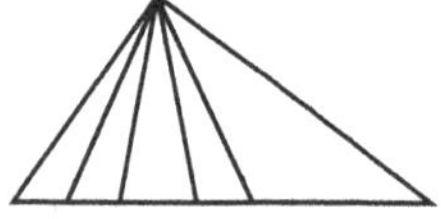 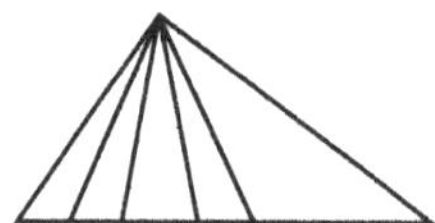  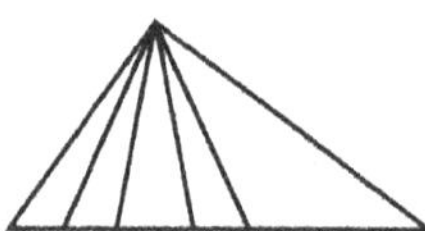 

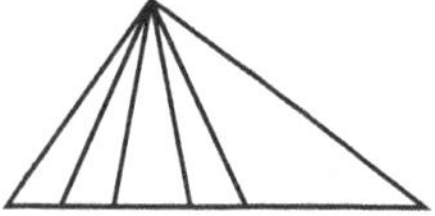    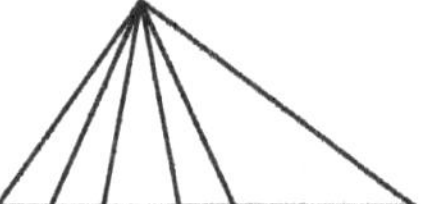

1 cm square paper

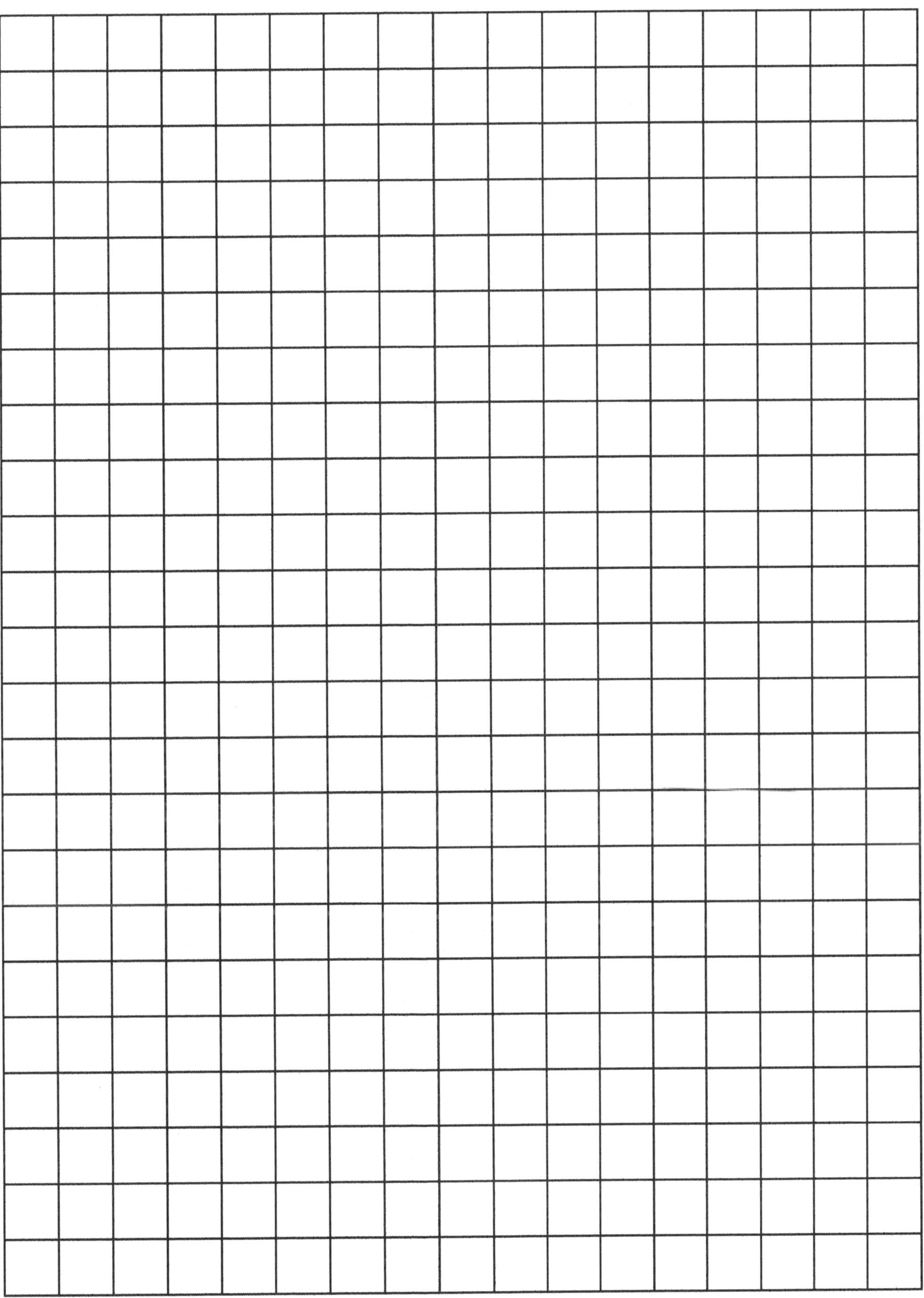

How many squares?

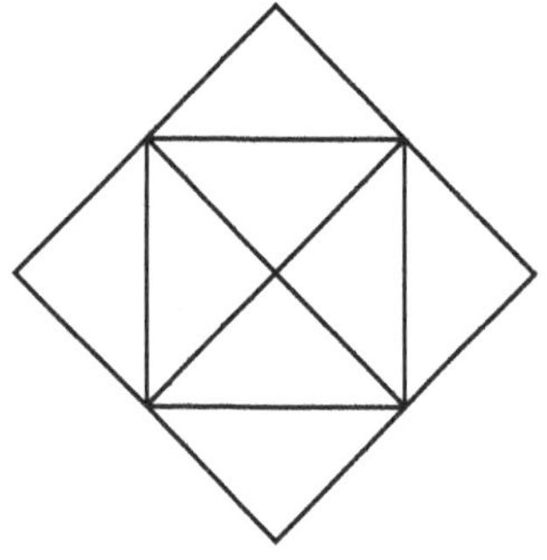
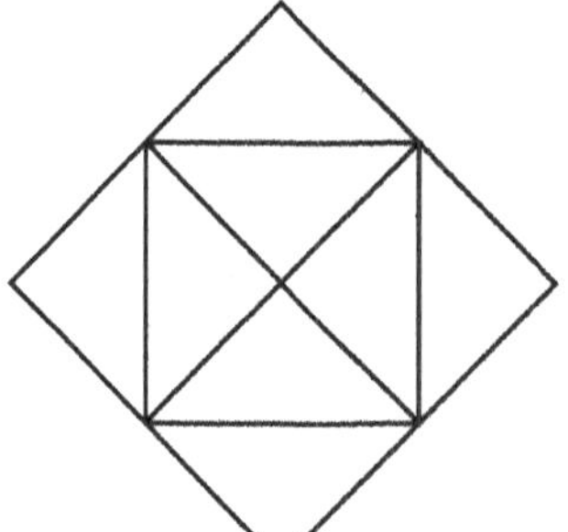
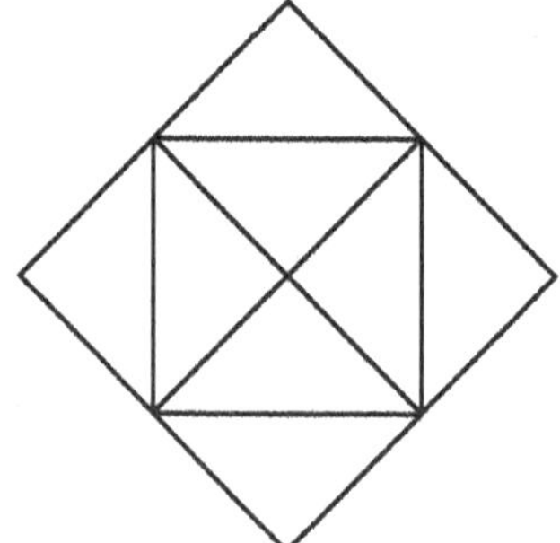

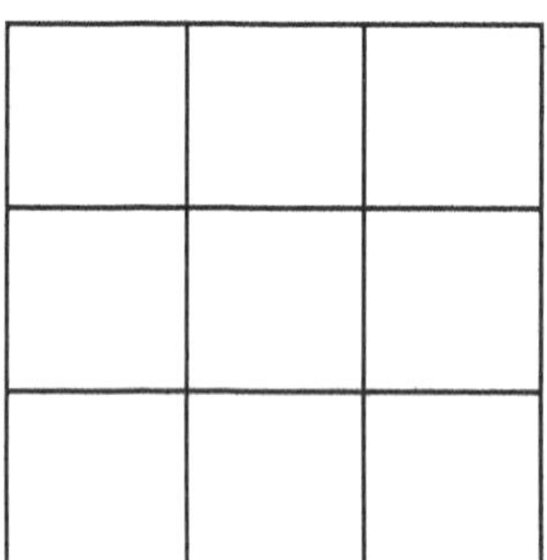
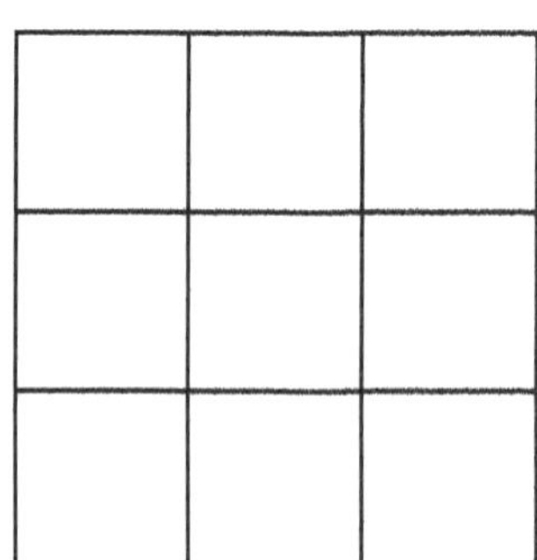
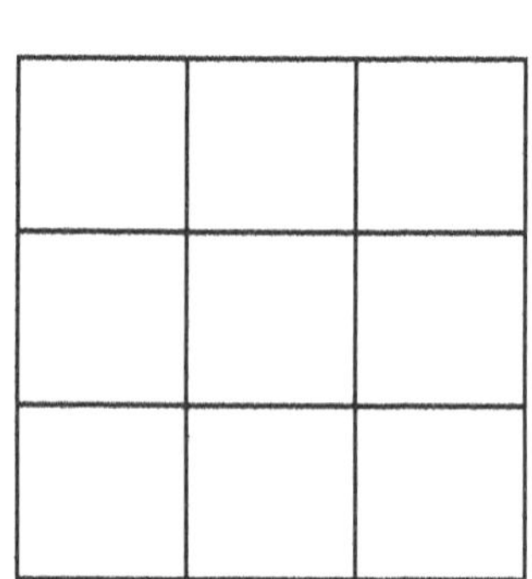

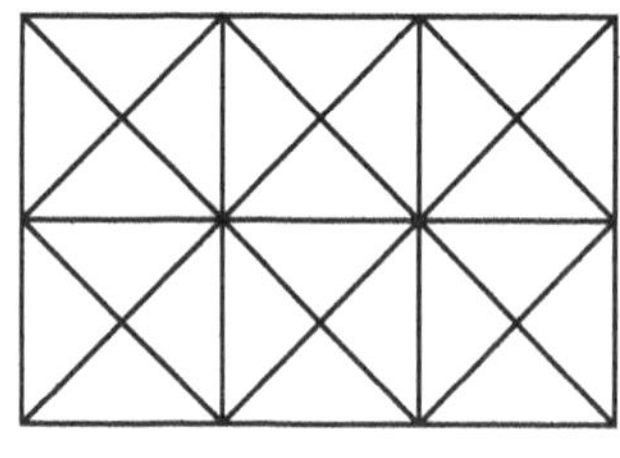
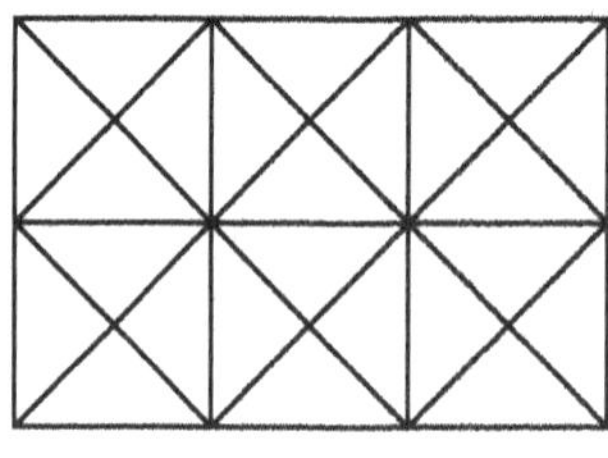
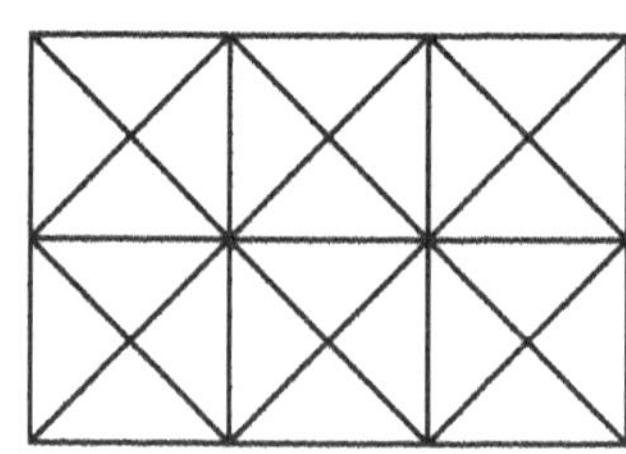

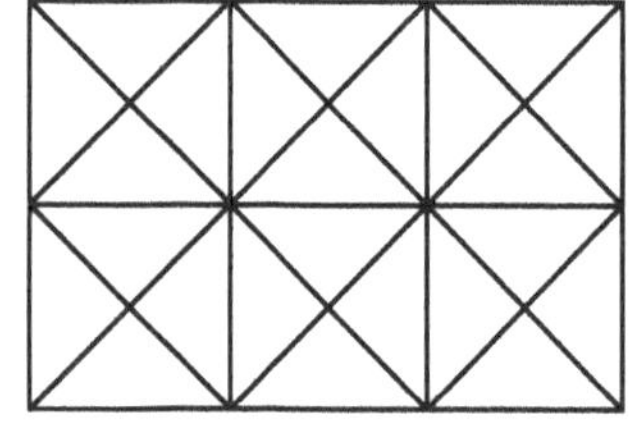
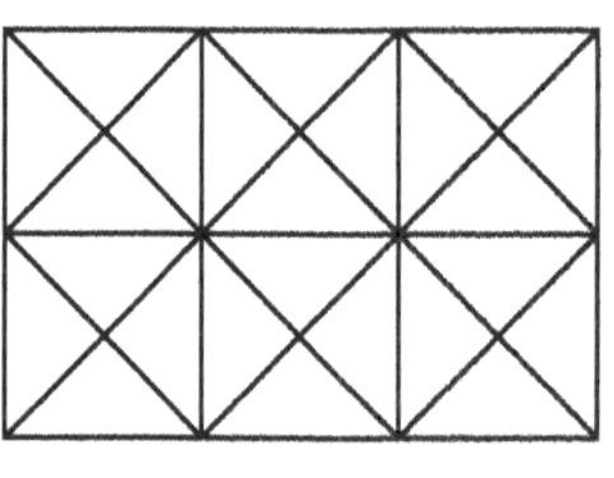
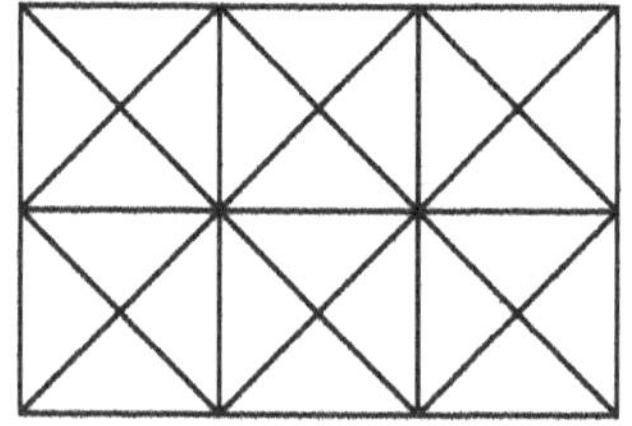

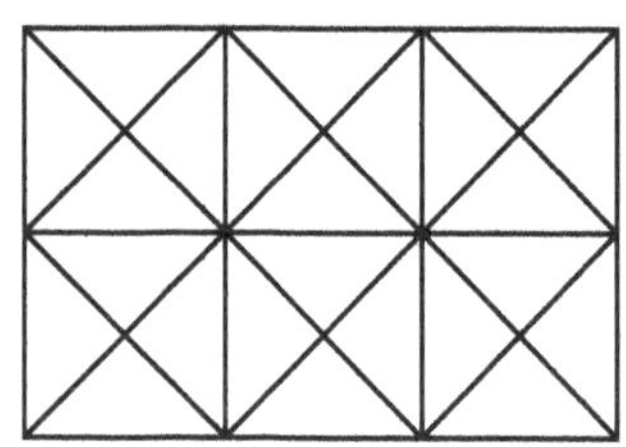
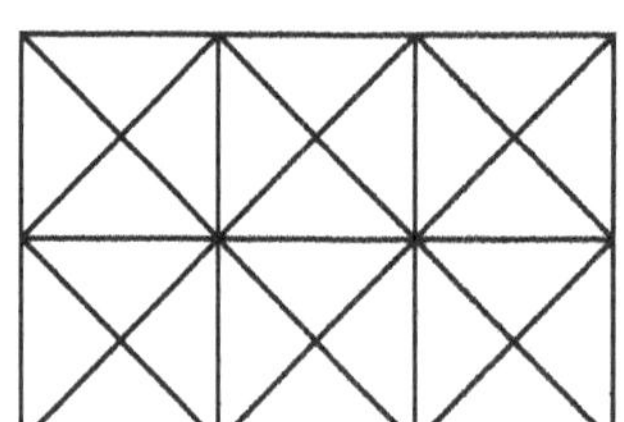
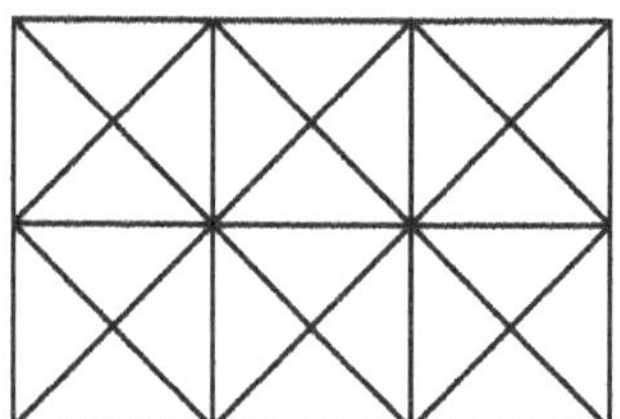

Two copies of this page will be required.

How many triangles?

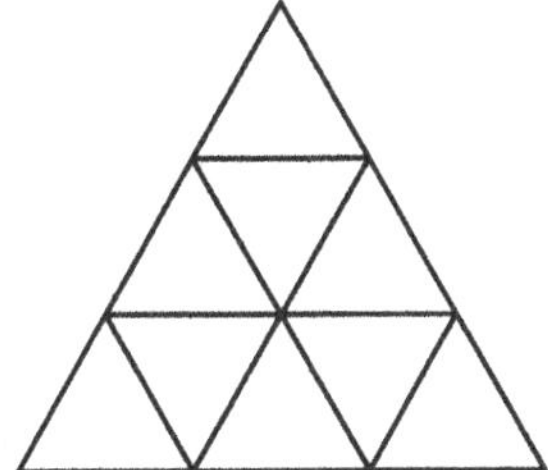
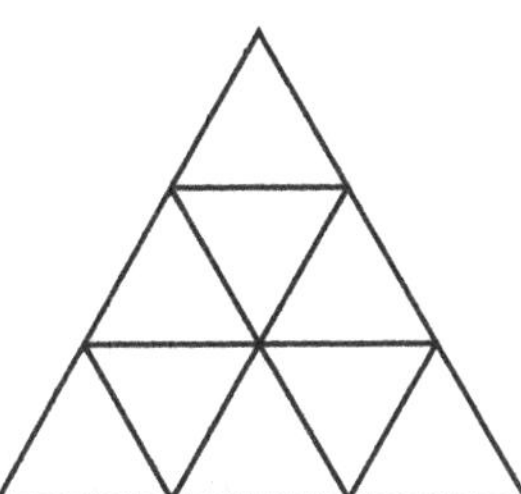
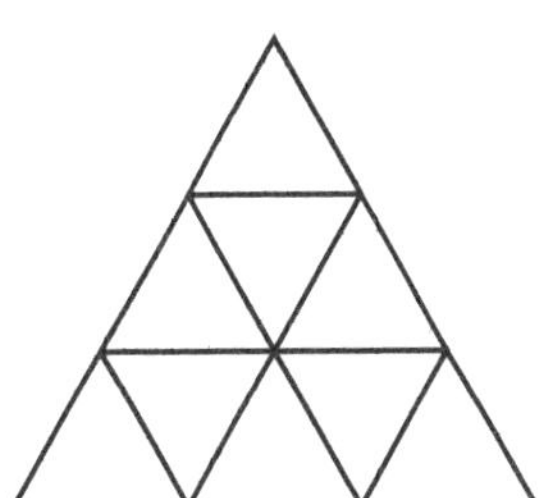
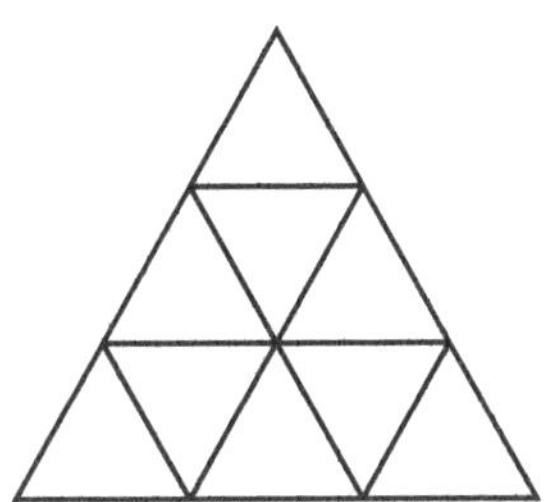

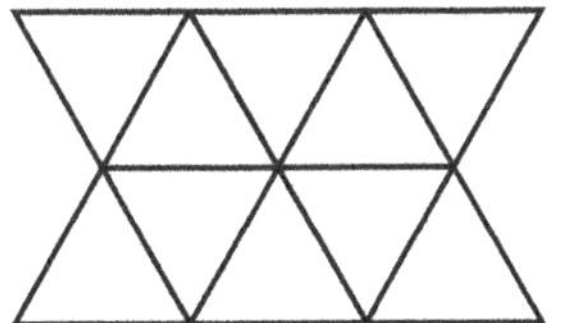
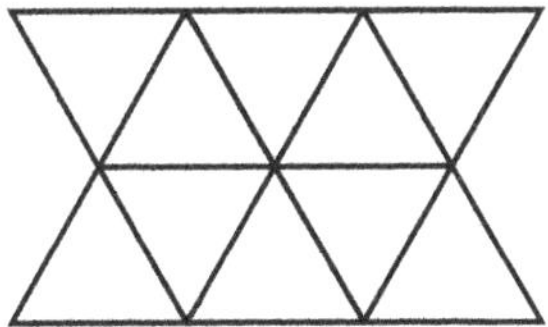
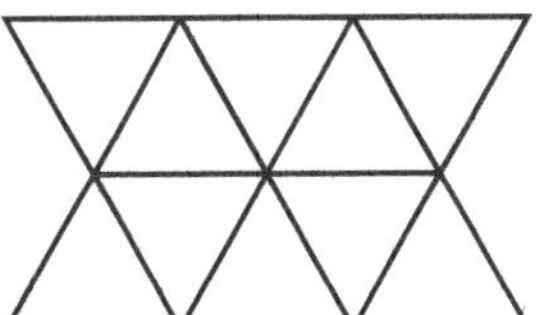
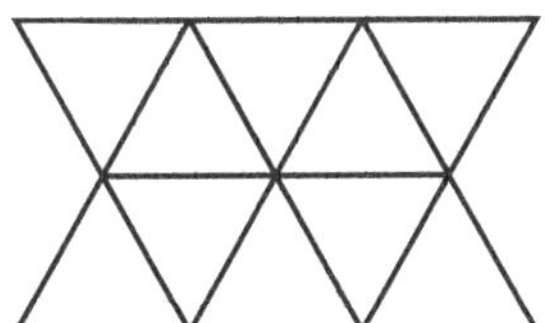

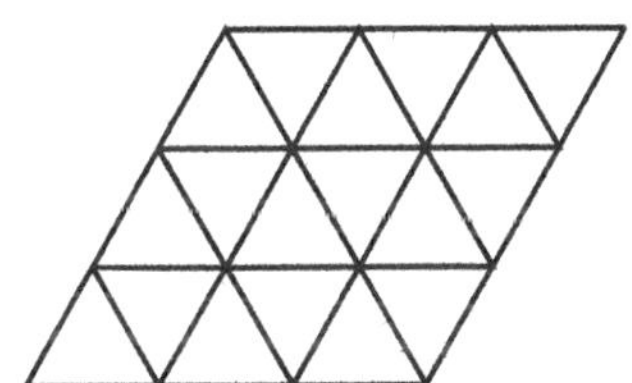
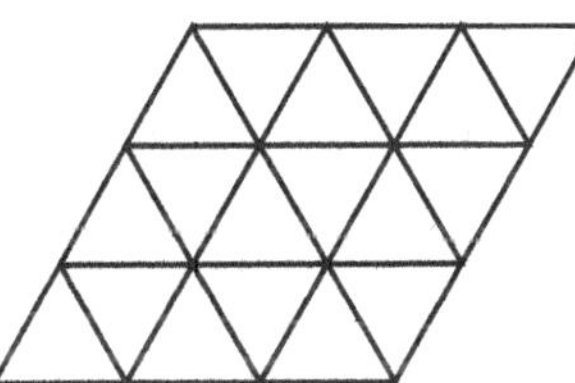
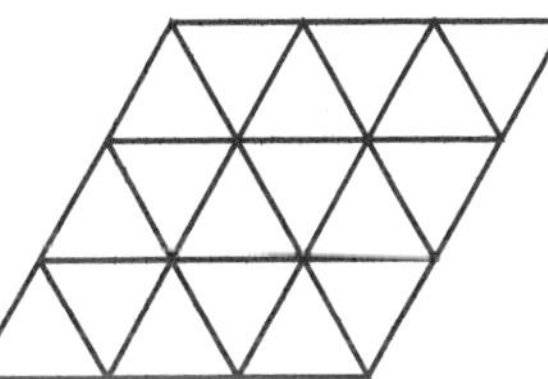
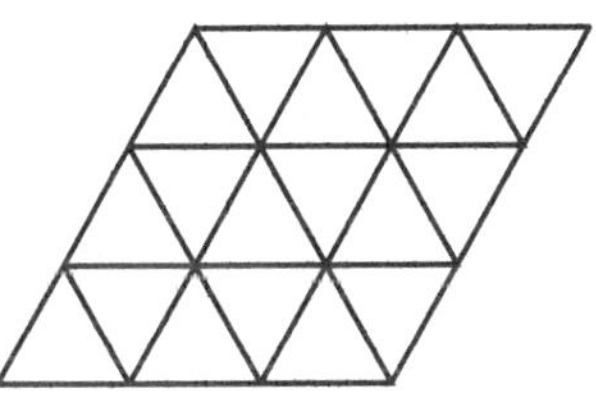

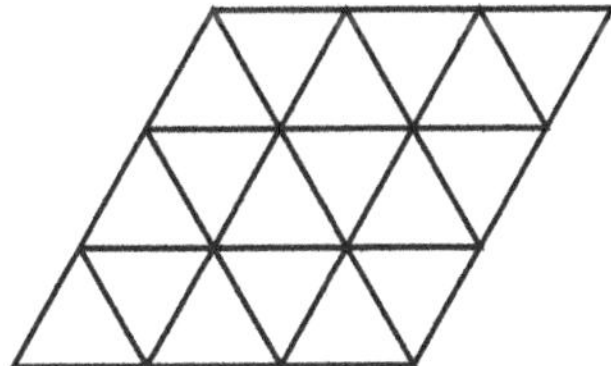
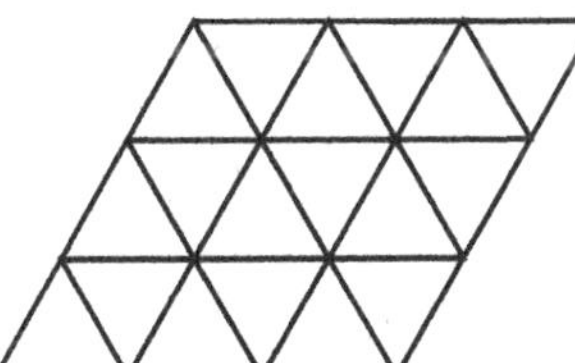
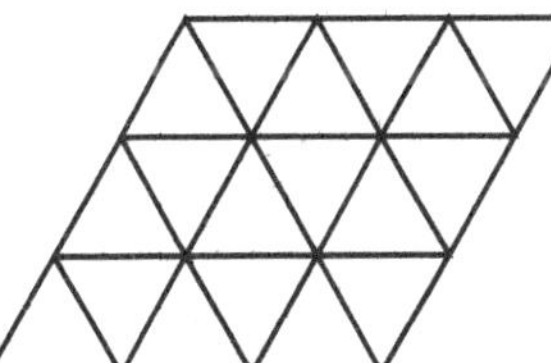
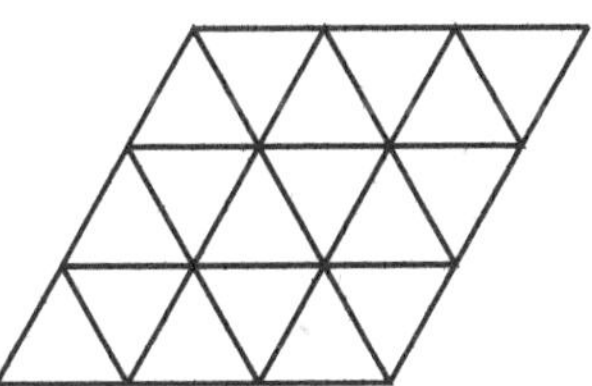

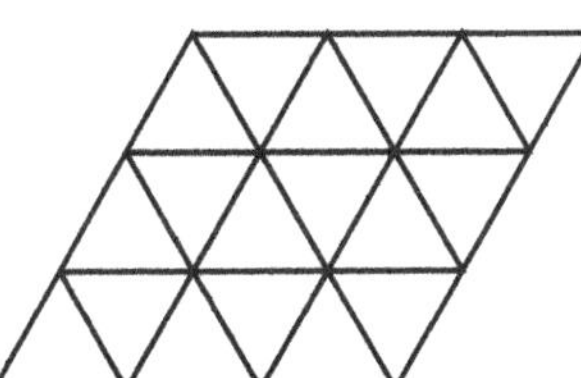
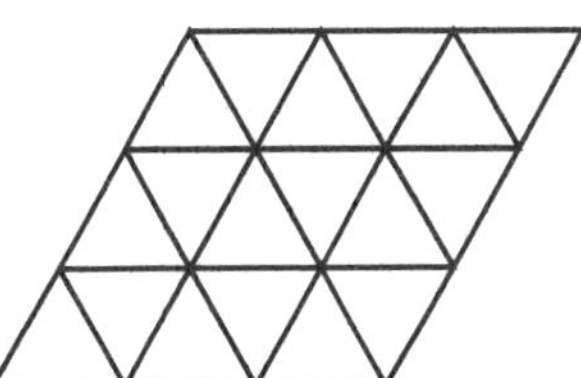

1 cm square dot paper

2 cm square paper

Prisoners in cells and Magic squares

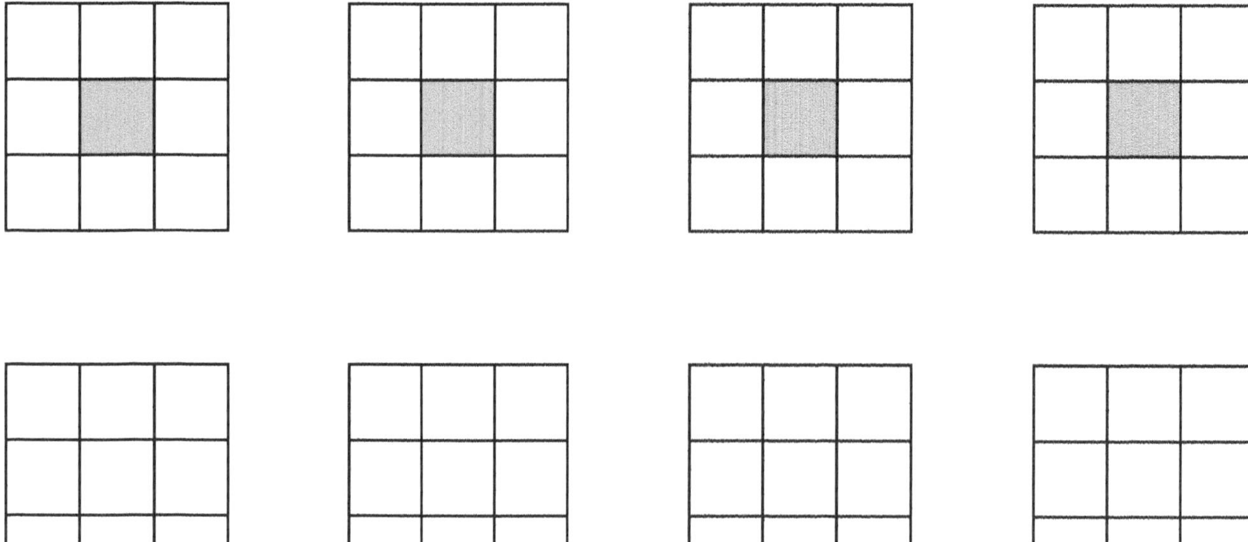

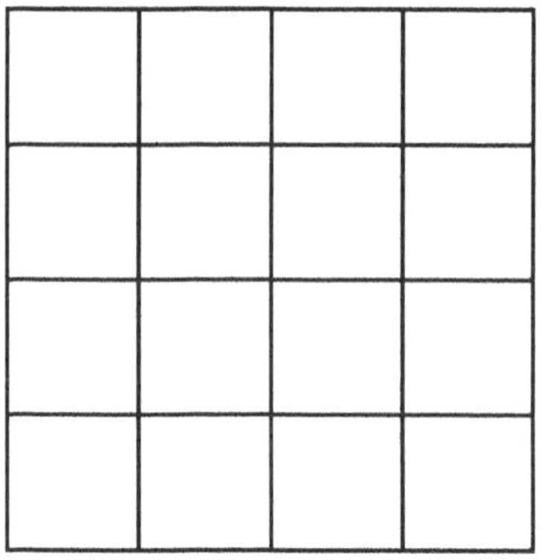

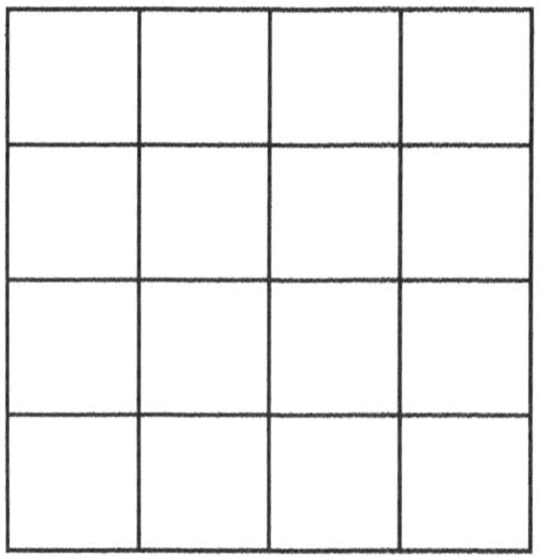

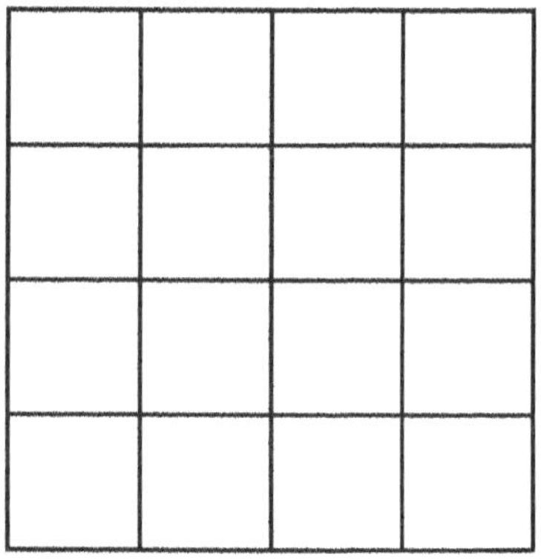

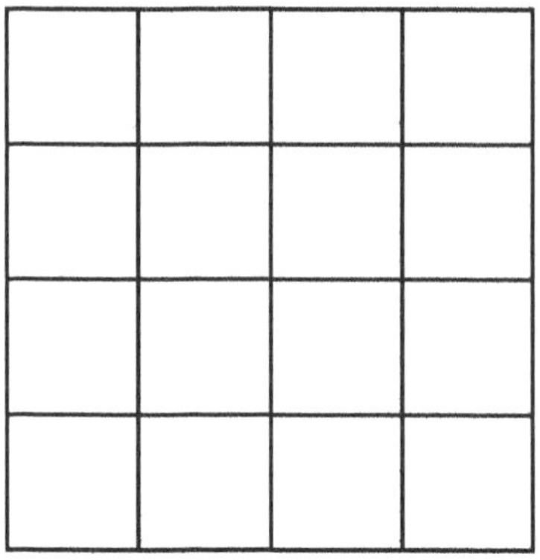

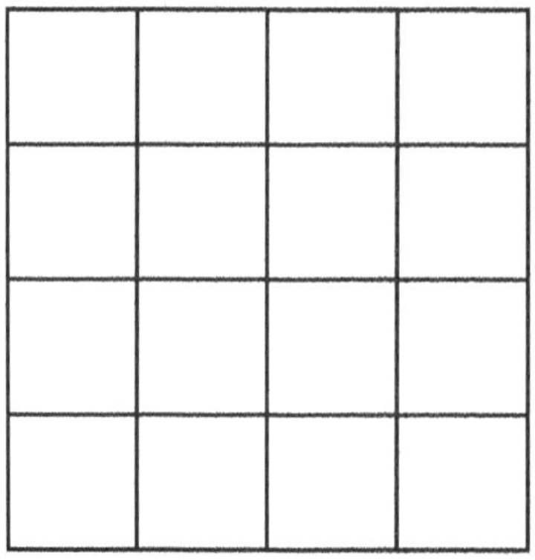

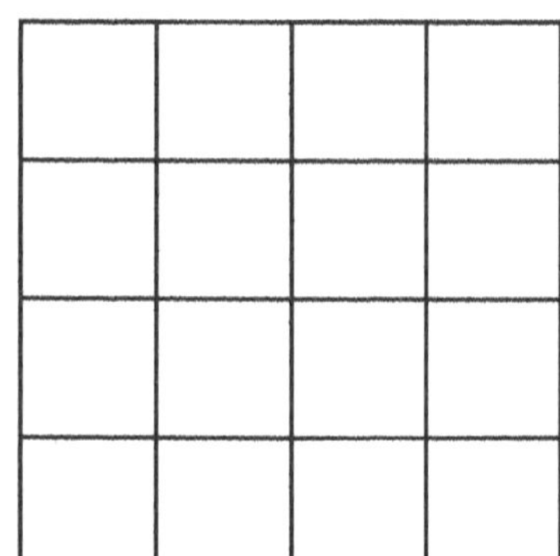

How many?

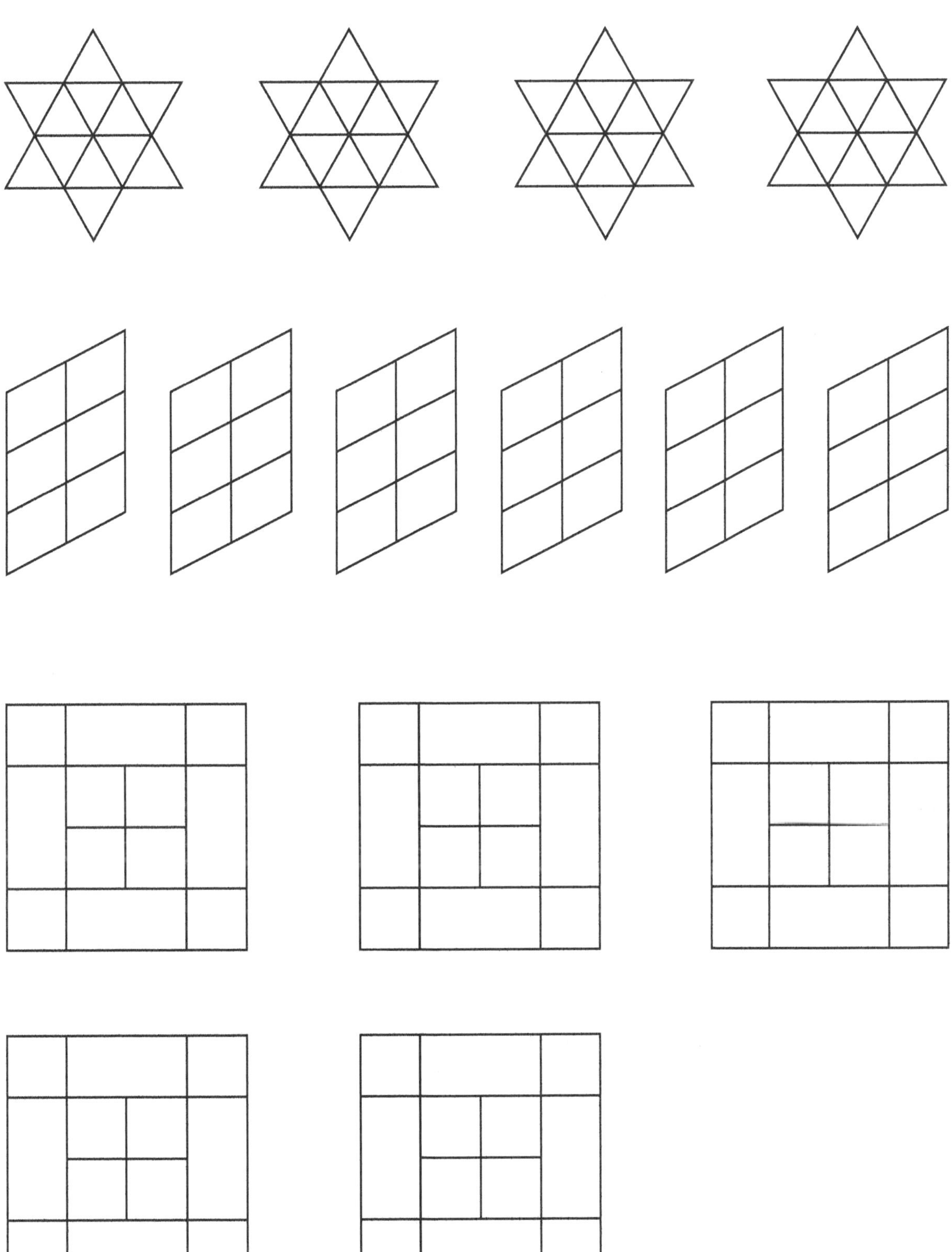

Two copies of this page will be required.

Curves

36/72
1/37
2/38
3/39
4/40
5/41
6/42
7/43
8/44
9/45
10/46
11/47
12/48
13/49
14/50
15/51
16/52
17/53
18/54
55/19
56/20
57/21
58/22
59/23
60/24
61/25
62/26
63/27
64/28
65/29
66/30
67/31
68/32
69/33
70/34
71/35

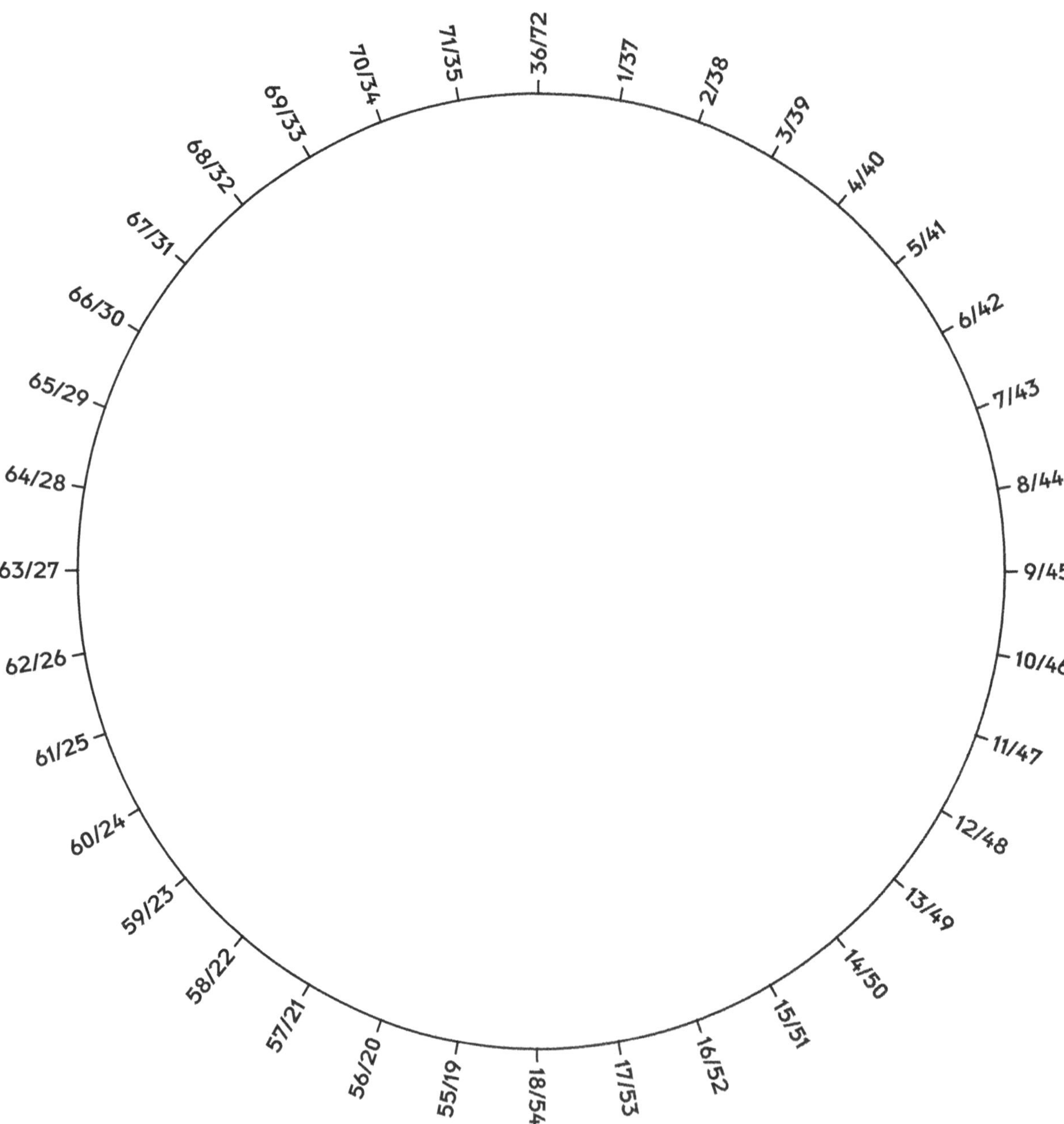

STUDENT BOOK Solutions

BOOK 1

Total of ten (page 1)

1 2 3 4 5 6 7 8 9 10

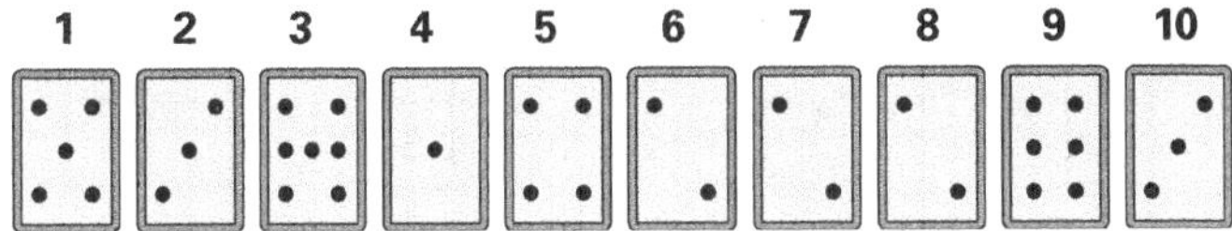

Total of twelve (page 2)

1 2 3 4 5 6 7 8 9 10

Number sentences (page 3)

1 1	**2** 2	**3** 5	**4** 4	**5** 5
6 1	**7** 4	**8** 3	**9** 4	**10** 5
11 2	**12** 3	**13** 4	**14** 5	

Add or subtract (page 4)

1 +	**2** –	**3** –	**4** +
5 –	**6** +	**7** –	**8** –
9 +	**10** –	**11** –	**12** +
13 +	**14** –	**15** +	**16** –

Number patterns (page 5)

Students should be encouraged to write down the differences between the numbers in each pattern as shown in the example. Saying the differences aloud ('add 3' etc.) may be helpful.

1 13, 15	**2** 13, 16	**3** 7, 5	**4** 12, 14
5 25, 30	**6** 50, 60	**7** 50, 40	**8** 43, 53
9 55, 66	**10** 16, 14		

What comes next? (page 6)

Students should write down the differences between the numbers given and then continue the pattern.

1 24, 27	**2** 15, 18	**3** 14, 13	**4** 44, 33
5 13, 11	**6** 15, 10	**7** 51, 61	**8** 44, 54
9 11, 8	**10** 30, 32	**11** 20, 15	**12** 22, 20
13 47, 37	**14** 17, 21		

Continue the pattern (page 7)

1 8642 8642 8642	**2** 12345 12345
3 357 99 357 99	**4** 1223 1223 1223
5 9876 9876 9876	**6** 2468 2468 2468
7 30 220 30 220	**8** 7339 7339 7339
9 5445 5445 5445	**10** 97579 97579
11 5311 5311 5311	**12** 123 321 123 321

Bricks (page 8)

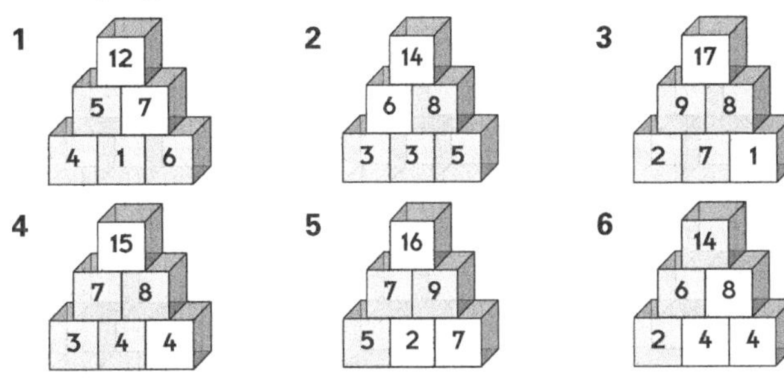

Counter puzzles (page 9)

In question 1, since the sum of the numbers in a line has to be 8, and since 2 + 1 = 3, the missing number in the row is 5.

1 is the middle number, so two numbers that add to 7 are needed and there are two possible solutions as shown.

Students should complete the row or column in which two numbers are given first.

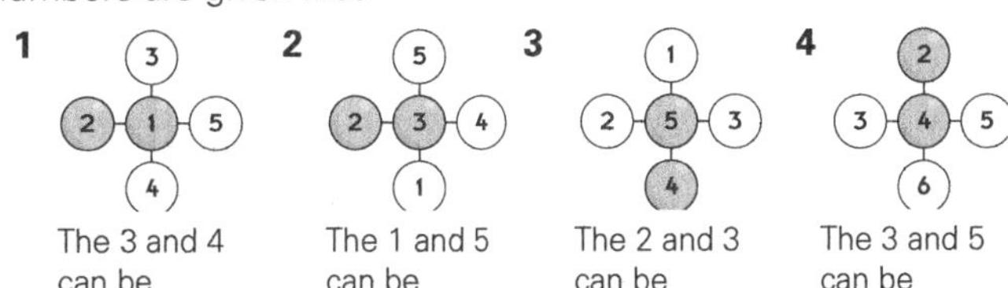

1 The 3 and 4 can be interchanged. **2** The 1 and 5 can be interchanged. **3** The 2 and 3 can be interchanged. **4** The 3 and 5 can be interchanged.

What number fits? (page 10)

Students should first complete the row or column in which two numbers are given.

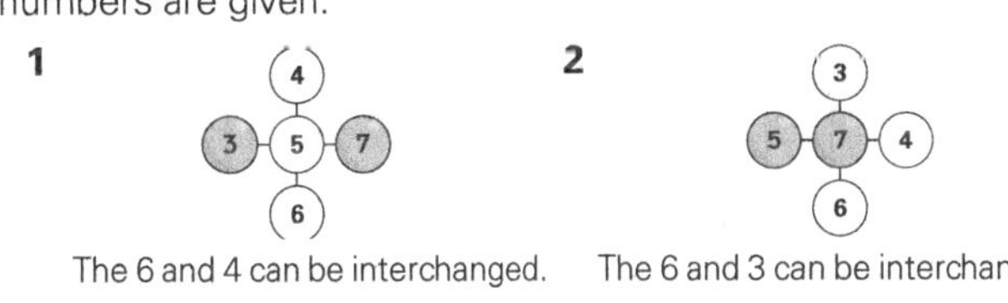

1 The 6 and 4 can be interchanged. **2** The 6 and 3 can be interchanged.

3

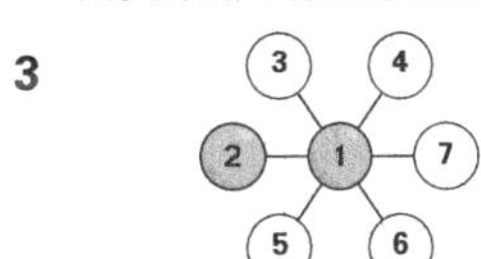

The 3 and 6 can be interchanged, or the 4 and 5 can be interchanged, and 3 and 6 can be swapped with 4 and 5.

4

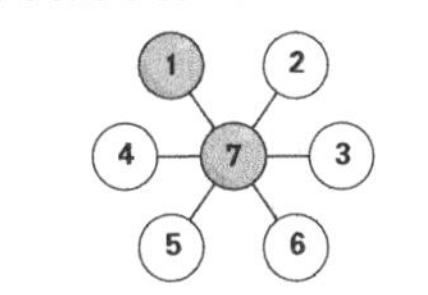

The 2 and 5 can be interchanged, or the 4 and 3 can be interchanged, and 2 and 5 can be swapped with 4 and 3.

Place the numbers (page 11)

Students will need to use the 'guess and check' problem-solving strategy to solve these problems. They should find counters very useful.

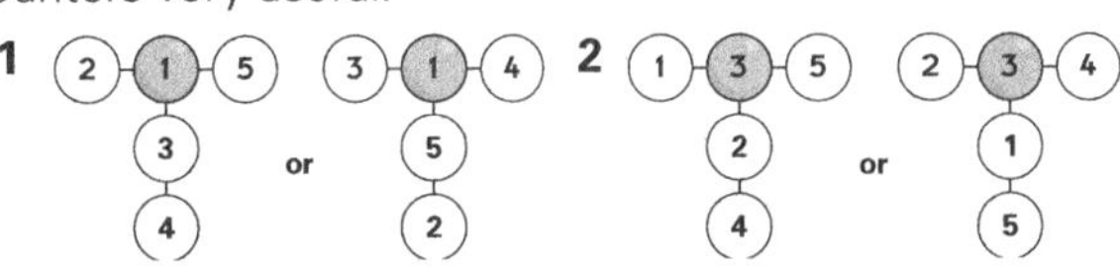

1 2 and 5 or 3 and 4 can be swapped. **2** 1 and 5 or 2 and 4 can be swapped.

Place the numbers (page 11) ***continued...***

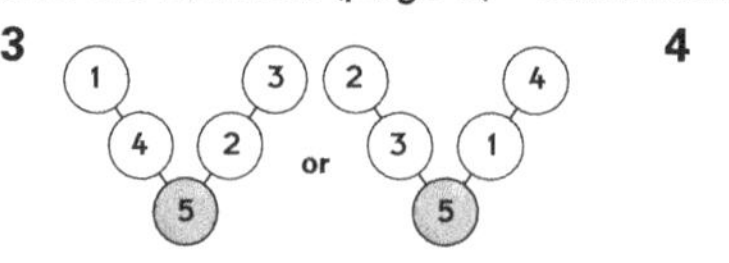

1 and 4 or 3 and 2 can be swapped.

Numbers in circles (page 12)

Students are only expected to do these questions after using the guess and check problem-solving technique a few times. Using their numbered counters should be helpful and estimation is also important. In questions 2 and 3, students may need to be reminded that each number can be used only once.

1
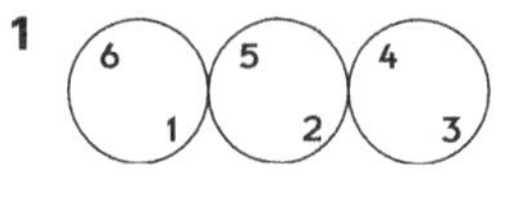

2
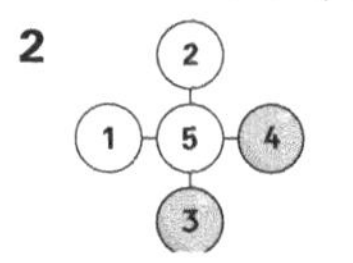

3
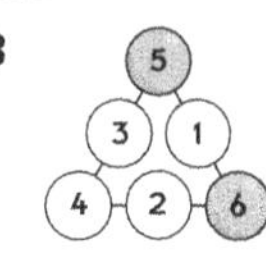

Numbers in triangles (page 13)

Students should be encouraged to persevere and use 'guess and check' to work out their solution. Counters should be very helpful.

1
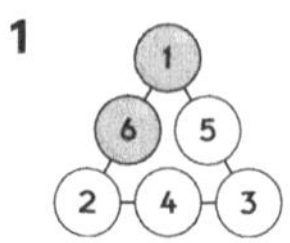

2
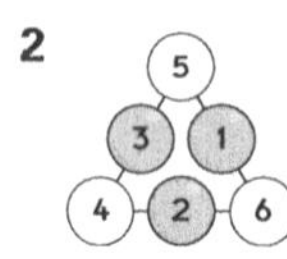

3
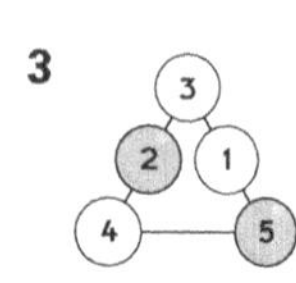

4
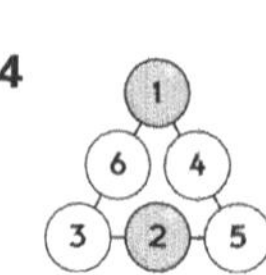

Abacus numbers (page 14)

1 13 **2** 4 **3** 11 **4** 22 **5** 15
6 20 **7** 24 **8** 34 **9** 31

Which abacus number? (page 15)

1 25 **2** 6 **3** 12 **4** 44 **5** 18
6 50 **7** 51 **8** 43 **9** 37

Abacus addition (page 16)

1 12, 13 or 22 **2** 4, 5 or 14 **3** 11, 12 or 21
4 22, 23 or 32 **5** 15, 16 or 25 **6** 20, 21 or 30

Abacus subtraction (page 17)

1 12, 11 or 2 **2** 15, 14 or 5 **3** 11, 10 or 1
4 22, 21 or 12 **5** 34, 33 or 24 **6** 24, 23 or 14

Abacus challenge (page 18)

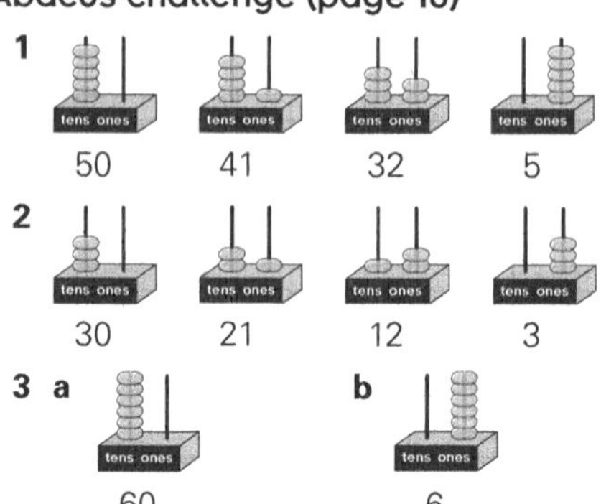

1 50 41 32 5

2 30 21 12 3

3 a 60 **b** 6

Grid sums (page 19)

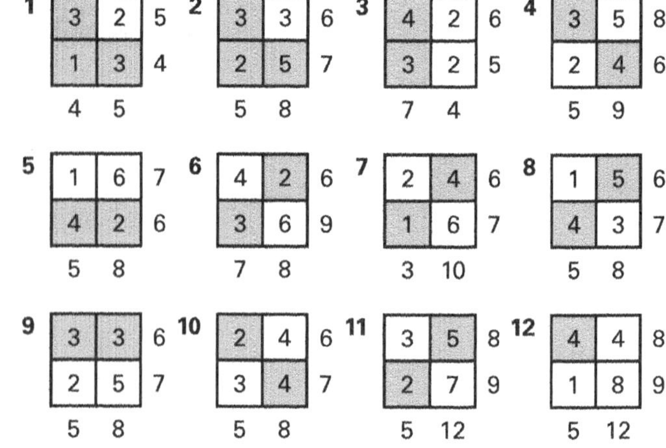

Find my start (page 20)

1 8 **2** 12 **3** 8 **4** 12
5 22 **6** 6 **7** 4

Parts of shapes (page 21)

1 2, 8 **2** 2, 6 **3** 4, 5 **4** 3, 6 **5** 4, 9
6 4, 8 **7** 2, 4 **8** 6, 12 **9** 4, 6

Fractions (page 22)

Tick 1, 2, 4, 5 and 6. Colour 3 and 7.

Find the value (page 23)

1 16c **2** 9c **3** 40c **4** 44c **5** 20c **6** 12c
7 20c **8** 60c **9** 60c **10** 28c **11** 60c **12** 24c

Halves and quarters (page 24)

Students should be able to work out these problems by completing the picture in each question.

1 4 **2** 10 **3** 14 **4** 8 **5** 12 **6** 20

How many apples? (page 25)

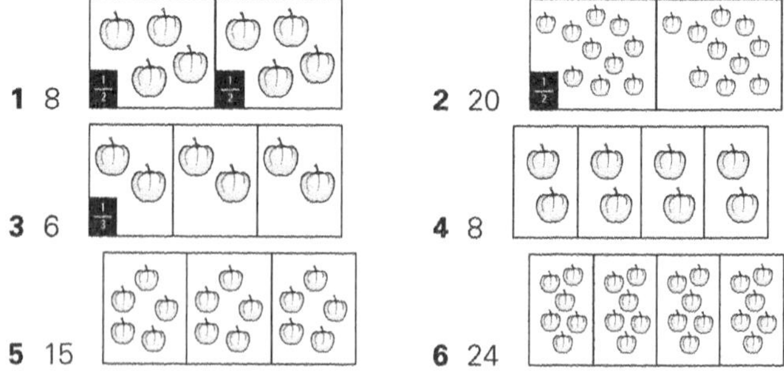

1 8 **2** 20
3 6 **4** 8
5 15 **6** 24

Number line (page 26)

1 3 **2** 4 **3** 4 **4** 6 **5** 4 **6** 12
7 20 **8** 12 **9** 10 **10** 6 **11** 2 **12** 20

Bowling pins (page 27)

1 a 2 and 7 **b** 3 and 9 *or* 4 and 8
2 a 2 and 3 and 7 **b** 2 and 4 and 8 *or* 3 and 4 and 7
3 a 6 and 5 *or* 3 and 8 **b** 6 and 7 *or* 8 and 5 *or* 9 and 4

4 **a** 4 and 5 and 6 *or* 5 and 3 and 7
  b 3 and 6 and 8 *or* 4 and 5 and 8 *or* 4 and 6 and 7 *or* 3 and 5 and 9
5 4 and 6 and 3 and 7 *or* 3 and 4 and 5 and 8

Missing numbers (page 28)

1 5, 8, 4, 3, 3, 6, 4, 3, 2, 4, 6 **2** 2, 2, 3, 7, 5

Dice puzzles (page 29)

1 The spots on opposite faces add to 7.
2 a 5 **b** 3
3 a 2 **b** 6
4 a 1 **b** 4
5 First die: 5 and 2 should be on opposite faces. Second die: 4 and 3 should be on opposite faces.
6 5 is the sum; 1 spot, 4 spots

Who am I? (page 30)

1 8 **2** 6 **3** 8 **4** 5 **5** 7
6 13 **7** 15 **8** 40 **9** 25 **10** 16

Space arithmetic (page 31)

1 = 1, = 3 **2** = 2, = 3 **3** = 5, = 1
4 = 1, = 3 **5** = 4, = 2 **6** = 2, = 4
7 = 5, = 7 **8** = 7, = 10 **9** = 4, = 2

Shape puzzles (page 32)

1 = 4, = 5 **2** = 5, = 3 **3** = 6, = 10 **4** = 3, = 5
5 = 2, = 9 **6** = 5, = 2 **7** = 5, = 4

Find my rule (page 33)

1

IN:	6	11	1	8	2	5	9	10	4	Rule
OUT:	8	13	3	10	**4**	**7**	**11**	**12**	**6**	**+ 2**

2

IN:	6	11	3	8	2	5	9	10	4	Rule
OUT:	7	12	4	9	**3**	**6**	**10**	**11**	**5**	**+ 1**

3

IN:	6	11	1	8	2	5	9	10	4	Rule
OUT:	5	10	0	7	**1**	**4**	**8**	**9**	**3**	**– 1**

4

IN:	4	5	2	7	6	11	1	8	3	Rule
OUT:	7	8	5	10	**9**	**14**	**4**	**11**	**6**	**+ 3**

5

IN:	2	5	9	10	4	7	6	11	13	Rule
OUT:	0	3	7	8	**2**	**5**	**4**	**9**	**11**	**– 2**

6

IN:	6	11	1	8	2	5	9	10	4	Rule
OUT:	10	15	5	12	**6**	**9**	**13**	**14**	**8**	**+ 4**

7

IN:	6	3	8	5	9	10	4	7	11	Rule
OUT:	3	0	5	2	**6**	**7**	**1**	**4**	**8**	**– 3**

8

IN:	5	9	10	4	7	6	3	1	8	Rule
OUT:	10	14	15	9	**12**	**11**	**8**	**6**	**13**	**+ 5**

Sum of digits (page 34)

1 $12 + 3 = 15$ $13 + 2 = 15$ $21 + 3 = 24$ $23 + 1 = 24$ $31 + 2 = 33$ $32 + 1 = 33$

The largest sum is 33. The smallest sum is 15.

2 a $43 + 2 = 45$ or $42 + 3 = 45$ **b** $23 + 4 = 27$ or $24 + 3 = 27$

The largest difference is 31. The smallest difference is 9.

Difference of digits (page 35)

1 $12 - 3 = 9$ $13 - 2 = 11$ $21 - 3 = 18$ $23 - 1 = 22$ $31 - 2 = 29$ $32 - 1 = 31$

2 a $43 - 2 = 41$ **b** $23 - 4 = 19$

Money (page 36)

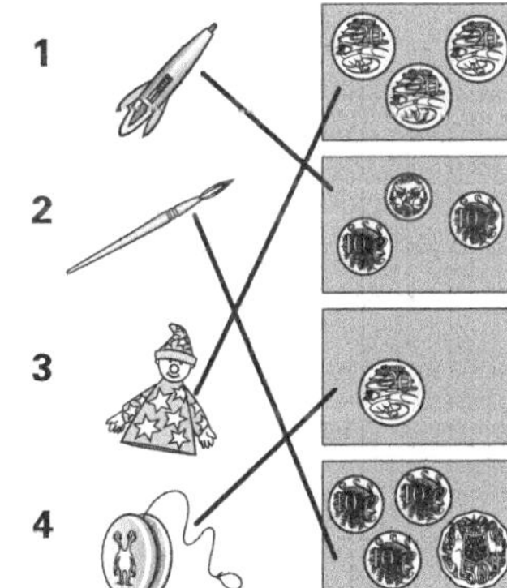

Shopping (page 37)

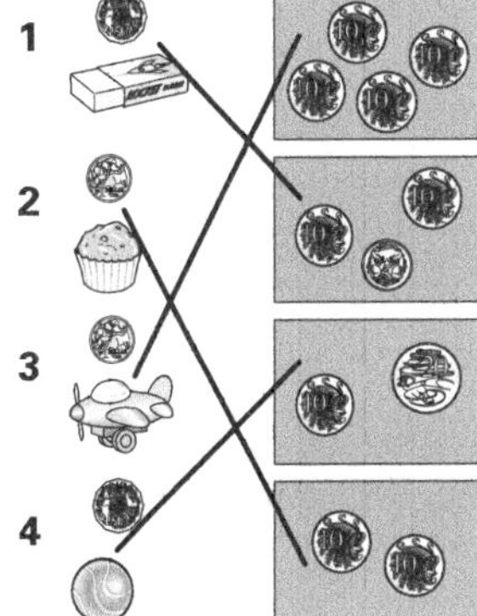

Patterns (page 38)

It may be helpful to say the shapes aloud in the groups that the patterns occur. For example, in the first question say: 'circle, circle, line, circle, circle, line, circle, circle, line, …'

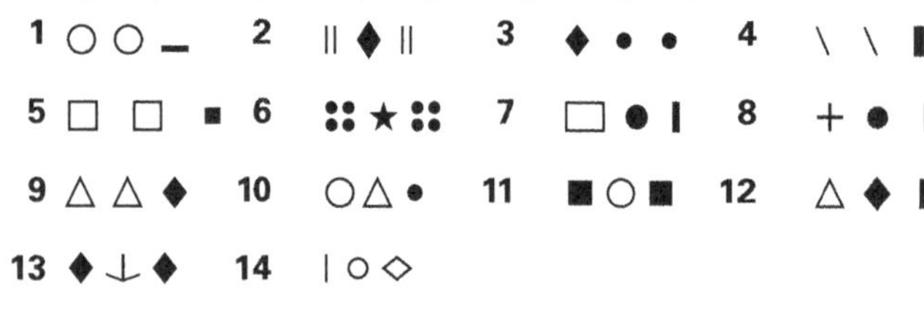

Tile patterns (page 39)

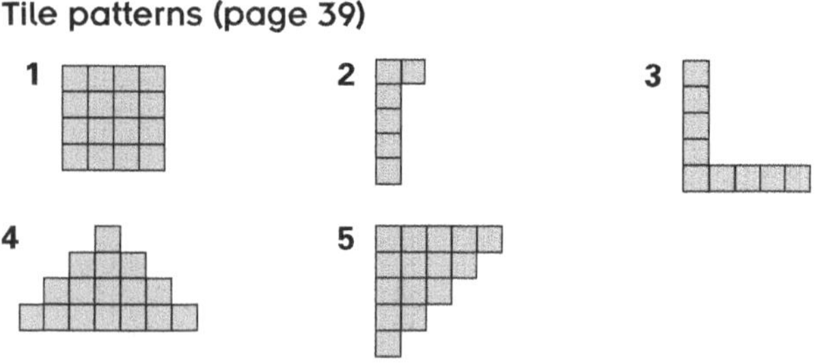

Square tiles (page 40)

1 6 **2** 10 **3** 5

4

Number of steps in staircase	Number of squares
2	3
3	6
4	10
5	15
6	21
7	28

What's the chance? (page 41)

Answers will vary and need careful discussion to ensure that children understand the concept of 'impossible', 'certain' and 'uncertain'.

Shading blocks (page 42)

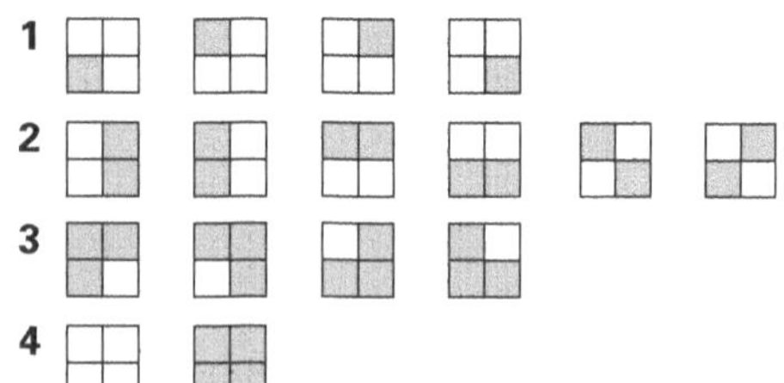

Different shading (page 43)

Flag patterns (page 44)

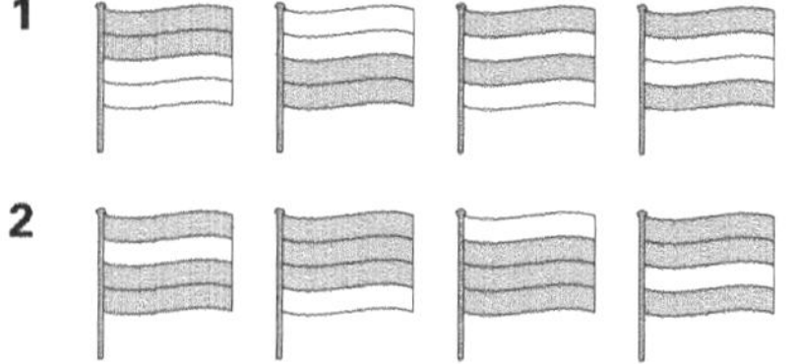

Shading circles (page 45)

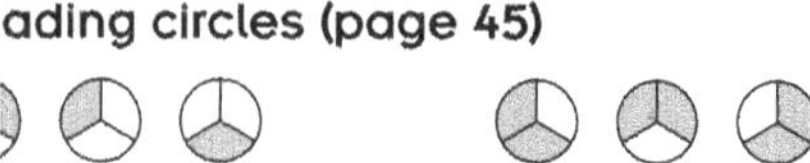

Picture graphs (page 46)

1

month	picture	rainfall
May	💧💧💧	6 mm
June	💧💧💧💧	8 mm
July	💧💧	4 mm
August	💧💧💧💧💧💧	12 mm

2

month	picture	rainfall
September	💧	2 mm
October	💧💧💧💧💧	10 mm
November	💧💧💧💧	8 mm
December	💧💧	4 mm

Birthday graphs (page 47)

1

month	picture	No. of children
January	웃웃웃	9
February	웃웃	6
March	웃웃웃웃	12
April	웃웃웃웃웃	15
May	웃웃웃웃웃웃웃	21

2 May **3** February **4** 63

Favourite subjects (page 48)

1 6 **2** 4 **3** Computing
4 Sport **5** Maths and Music **6** 2

Weather (page 49)

1 True **2** False **3** True
4 False **5** True **6** True

Test results (page 50)

1 6 **2** 7 **3** 2 **4** 2

5

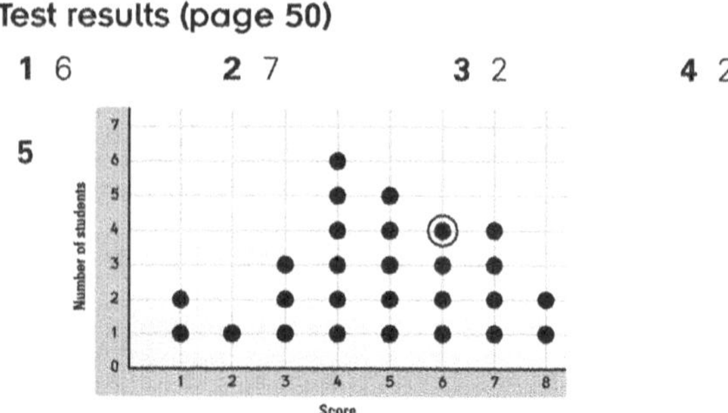

6 27 including students who did it late **7** 6 **8** 6

Class captain (page 51)

1 **a** Tom **b** Lee **c** Josh **d** Ann **e** Emma

2 35 **3**

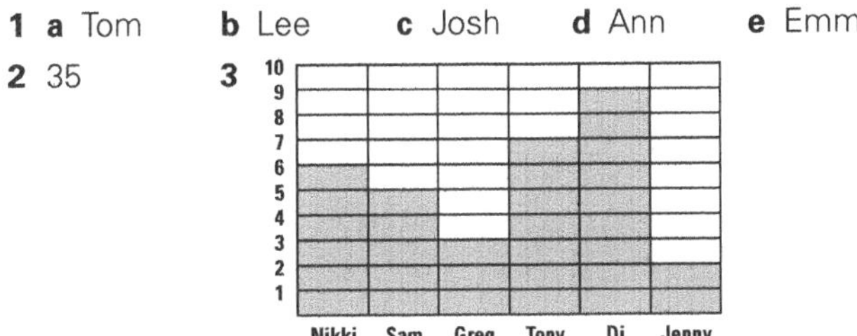

How long? (page 52)

1 4 **2** 2 **3** 6 **4** 3
5 8 **6** 6 **7** 12 **8** 10

Measure the rocket (page 53)

1 3 **2** 4 **3** 6 **4** 4
5 40 **6** 50 **7** 8 **8** 10

Long or short? (page 54)

1 8 **2** 20 **3** 12 **4** 12
5 15 **6** 8 **7** 15 **8** 6

Find the temperature (page 55)

1 10°C **2** 20°C **3** 40°C **4** 8°C **5** 35°C
6 9°C **7** 25°C **8** 15°C **9** 7°C **10** 30°C

Time (page 56)

Students may require a clock with hands that they can move.

1 4 p.m. **2** 7 p.m. **3** 5 p.m. **4** 1 p.m.
5 2 p.m. **6** 10 p.m. **7** 9 p.m. **8** 4 p.m.
9 5 p.m. **10** 9 p.m. **11** 2 p.m. **12** 12 noon
13 11 a.m. **14** 8 a.m.

What day? (page 57)

Students need to be made aware of the seven-day week cycle. They should be able to generalise this result by the end of this exercise. Encourage them to make up similar questions of their own and quiz other classmates.

1 **a** Thursday **b** Tuesday **c** Saturday **d** Wednesday
2 **a** Sunday **b** Friday **c** Tuesday **d** Saturday
3 **a** Wednesday **b** Saturday **c** Monday **d** Monday

May days (page 58)

Encourage discussion about the information in the calendar, for example: 'What is special about the numbers written under one another in each column?', 'What would be the date one day after 31 May?'

1 Saturday **2** 4 **3** 4 **4** Thursday
5 Wednesday **6** Thursday **7** 31 **8** No

Check the calendar (page 59)

1 31 **2** 4 **3** 5 **4** Monday
5 Monday **6** Thursday **7** Thursday **8** 18 days
9 16 days **10** 4

Scales (pages 60 and 61)

1 1 kg **2** 5 kg **3** 10 kg **4** 3 kg **5** 8 kg
6 9 kg **7** 5 kg **8** 10 kg **9** 14 kg **10** 4 kg
11 7 kg **12** 3 kg **13** 13 kg **14** 7 kg

Perimeter (page 62)

1 10 cm **2** 12 cm **3** 16 cm **4** 16 cm **5** 16 cm
6 14 cm **7** 18 cm **8** 14 cm **9** 16 cm

Temperature facts? (page 63)

1 False **2** True **3** False **4** False
5 True **6** False **7** True **8** False
9 True **10** True **11** False **12** False

Measurement facts? (page 64)

1 False **2** False **3** False **4** True **5** False
6 True **7** True **8** True **9** False **10** False
11 True **12** True **13** True **14** False **15** True

Building blocks (page 65)

1 4, 5, 8 **2** 6, 9, 10 **3** 7, 11, 12

Moving blocks (page 66)

1 5, 6, 8 **2** 4, 9, 10 **3** 7, 11, 12

Enlarging figures (page 67)

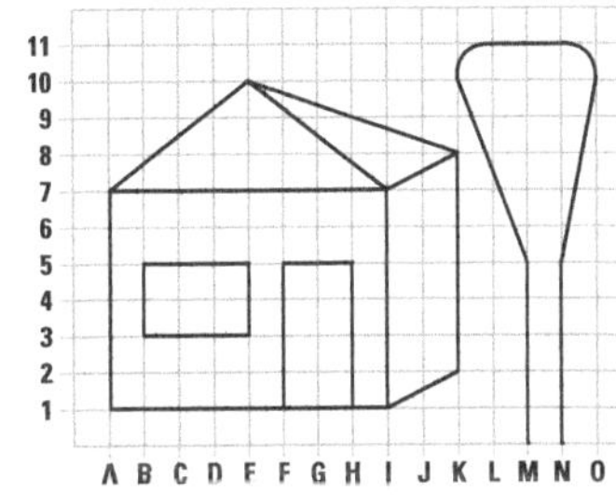

Same shapes (page 68)

1, 14 2, 19 3, 11 4, 20 5, 25
6, 15 7, 26 8, 21 9, 22 10, 23
12, 18 13, 16 17, 24

Combining shapes (page 69)

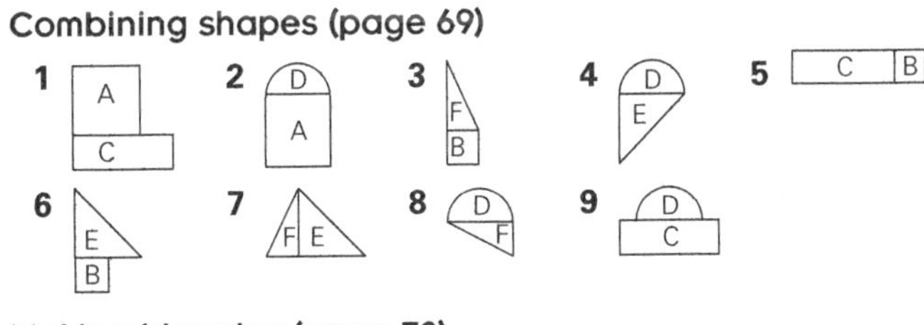

Making triangles (page 70)

1 **2** 12

3 **4** 15

5 18

Making squares (page 71)

Students should be encouraged to look for a pattern in each question. For example:

1

2 13

3

4 16

5 19

No. of squares	No. of sticks
1	4
2	7
3	10
4	13
5	16
6	19

In a line (page 72)

1 a 2 b 5 c 11

2 a 5 b 6 c 7

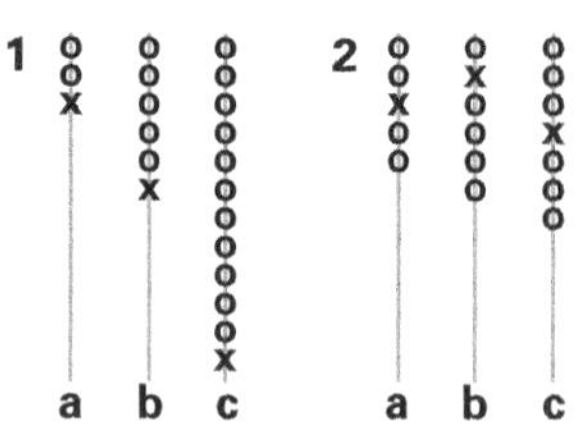

Match the objects (page 73)

1 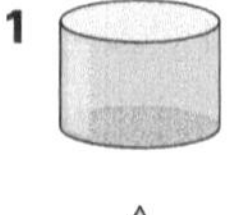

2 

3 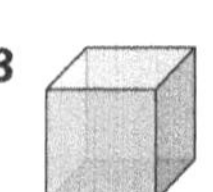

4 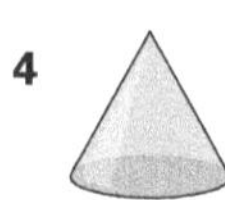

5 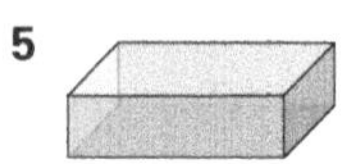

Tessellations (page 74)

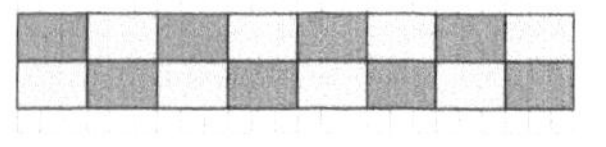

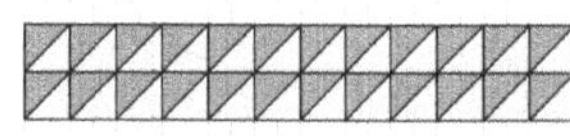

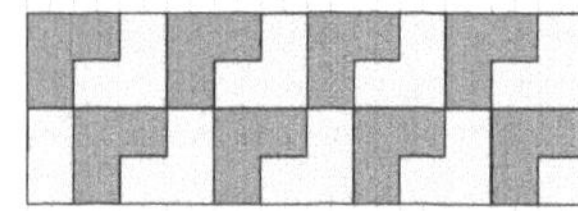

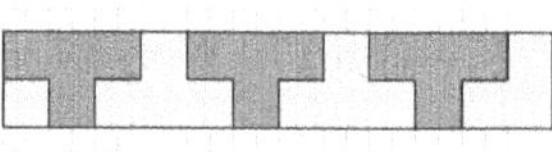

Spatial reasoning (page 75)

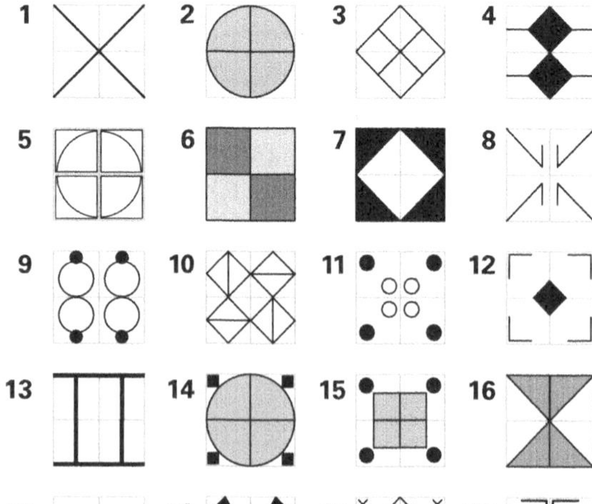

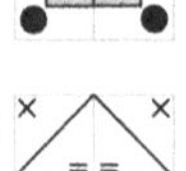

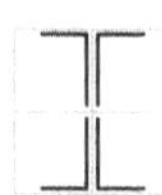

Shapes and values (page 76)

1 = 2

2 = 2, = 3

3 = 4, = 1

4 = 2, = 5

5 = 7, = 4

6 = 5, = 3

7 = 10, = 5

8 = 6, = 1

9 = 8, = 9

Odd one out (page 77)

1 C 2 C 3 B 4 B 5 C 6 A

Complete the picture (page 78)

1 2 3 4 5

How many squares? (page 79)

1 9 2 10 3 12 4 15 5 11

6 15 7 12 8 13 9 12

Shading halves (page 80)

There are many possible solutions. These are some of them.

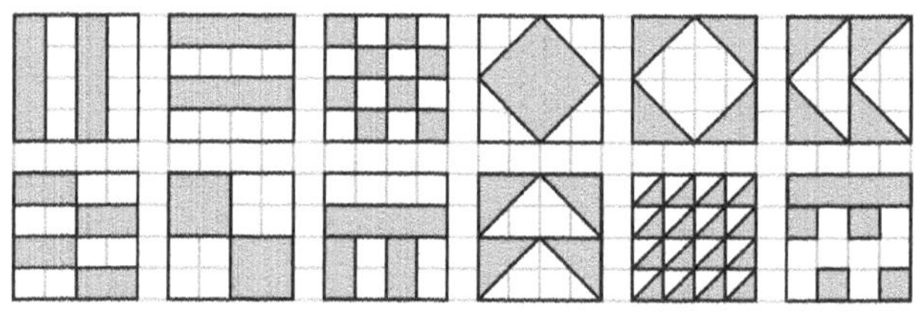

Drawing lines (page 81)

There are many possible solutions. These are some of them.

1 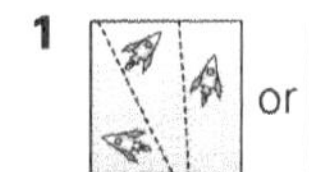 or 

2 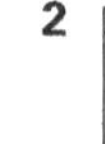 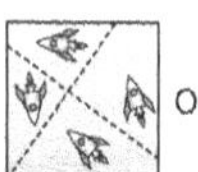 or 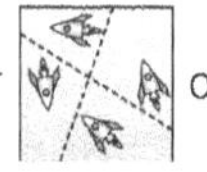 or 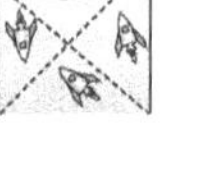

3 a  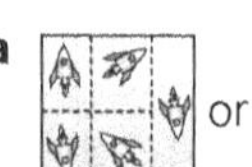 or  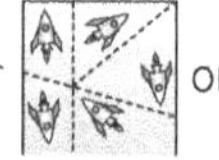 or 

b 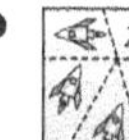 or 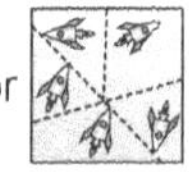

No threes (page 82)

Not all possibilities are listed and students are not expected to find more than two solutions.

Students should experiment by placing counters in the empty squares.

1 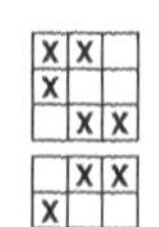 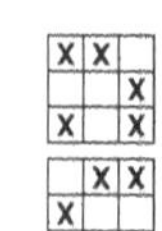 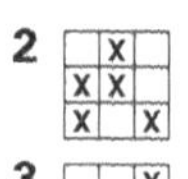

2 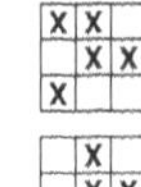 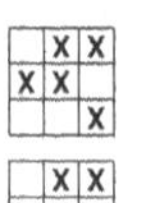

3 

Balancing act (page 83)

Students could find it helpful if the balances are expressed in words as shown in the first two questions below.

1 

If one square balances two circles, how many circles will two squares balance?

2 

If two stars balance four squares, then four stars will balance ... squares.

3 

4 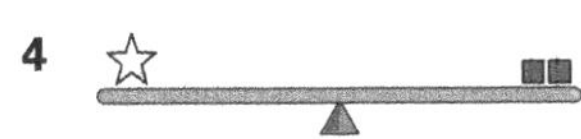

5 

6 

Balance (page 84)

Again, expressing the balances in words is very helpful for students. If one square balances three circles and one oval balances two circles, then a square and an oval will balance ... circles.

1 

2 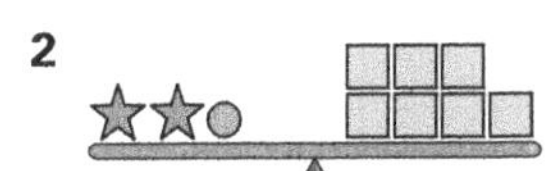

3 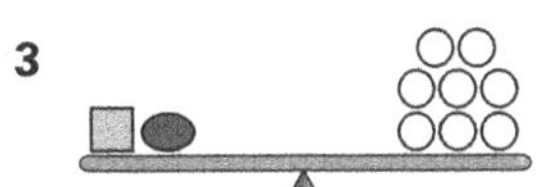

4 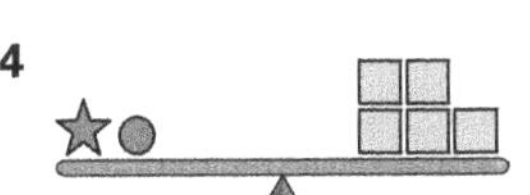

5 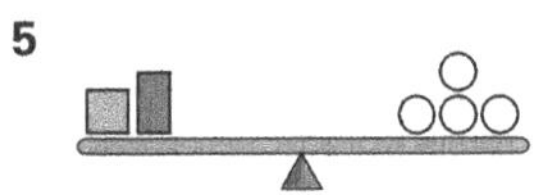

6 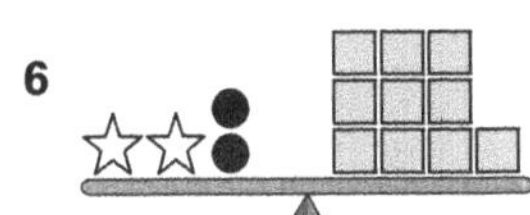

Find your balance (page 85)

1 8 **2** 6 **3** 11 **4** 7

5 5 **6** 4 **7** 1

Number of pins? (page 86)

1 a 8 **b** 10

2 a 7 **b** 9

3 a 6 **b** 7

An interesting discussion can follow about the patterns in each question. An extension question can be asked after each of the above questions, for example: 'How many pins are required for seven pieces of paper?'

Grid puzzles (page 87)

1 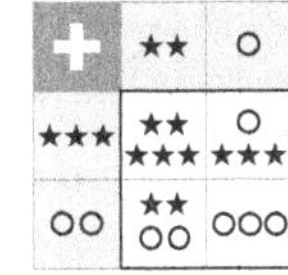

2 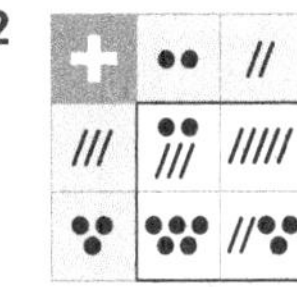

3 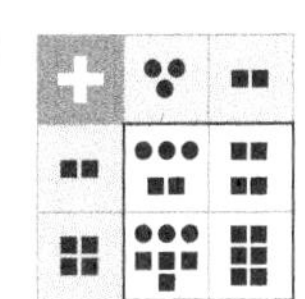

4

+	II	+++
+	II+	++ ++
I	III	+++ I

5 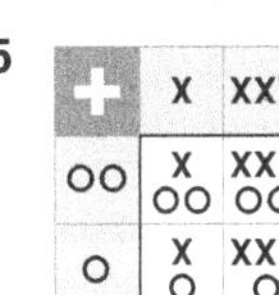

6 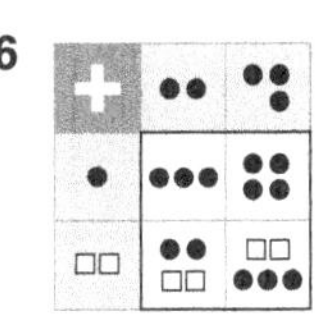

Gridlock (page 88)

1 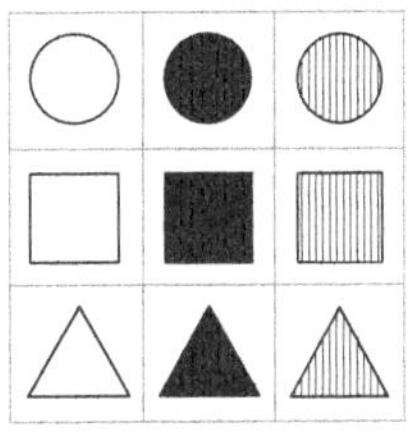

2 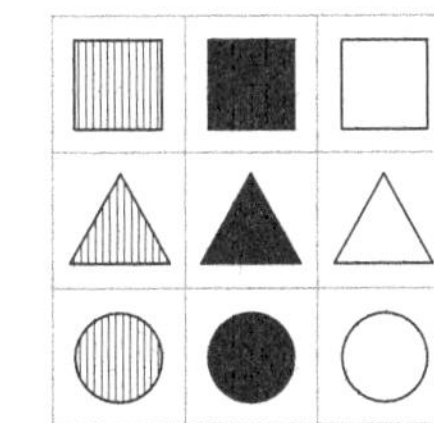

3 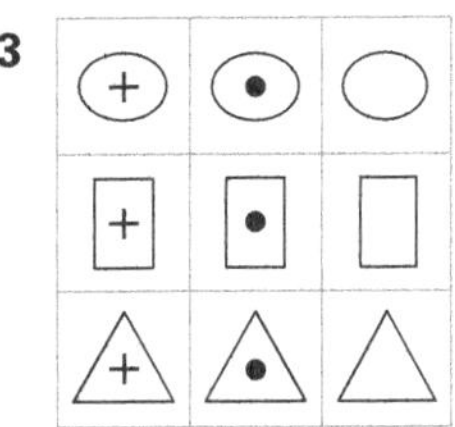

4 

5 

6 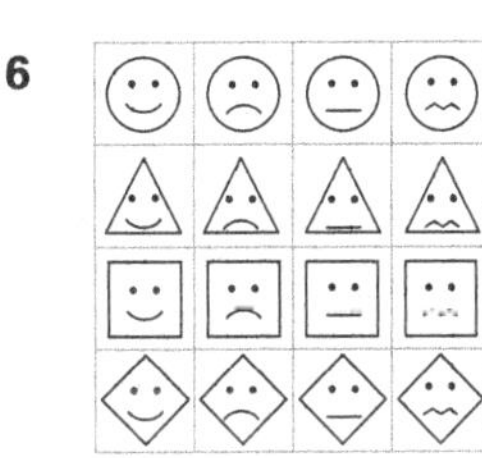

STUDENT BOOK Solutions

BOOK 2

Table squares (page 1)

1

+	4	1
2	6	3
1	5	2

2

+	5	3
1	6	4
3	8	6

3

+	6	2
4	10	6
3	9	5

4

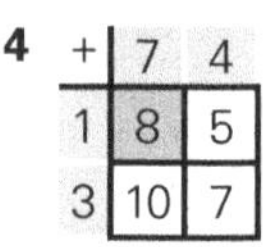

+	7	4
1	8	5
3	10	7

5

+	6	5
1	7	6
5	11	10

6

+	2	3
6	8	9
7	9	10

7

+	8	2
2	10	4
5	13	7

8

+	6	5
3	9	8
4	10	9

9

+	8	7
1	9	8
4	12	11

Addition squares (page 2)

1

+	3	2	5
1	4	3	6
4	7	6	9
3	6	5	8

2

+	5	3	9
4	9	7	13
2	7	5	11
7	12	10	16

3

+	8	2	7
3	11	5	10
5	13	7	12
6	14	8	13

4

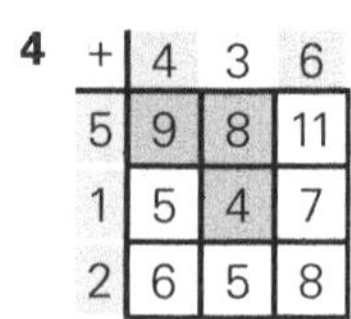

+	4	3	6
5	9	8	11
1	5	4	7
2	6	5	8

5

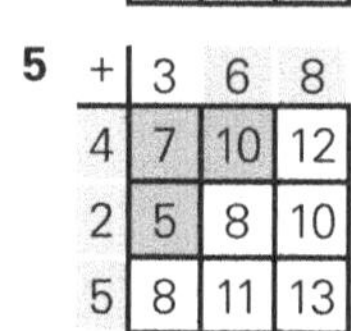

+	3	6	8
4	7	10	12
2	5	8	10
5	8	11	13

6

+	2	5	4
8	10	13	12
6	8	11	10
3	5	8	7

7

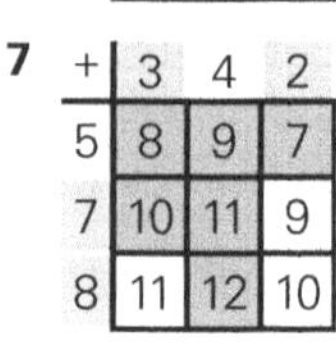

+	3	4	2
5	8	9	7
7	10	11	9
8	11	12	10

8

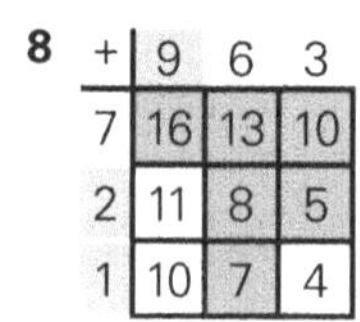

+	9	6	3
7	16	13	10
2	11	8	5
1	10	7	4

9

+	8	4	3
7	15	11	10
5	13	9	8
2	10	6	5

Bricks (page 3)

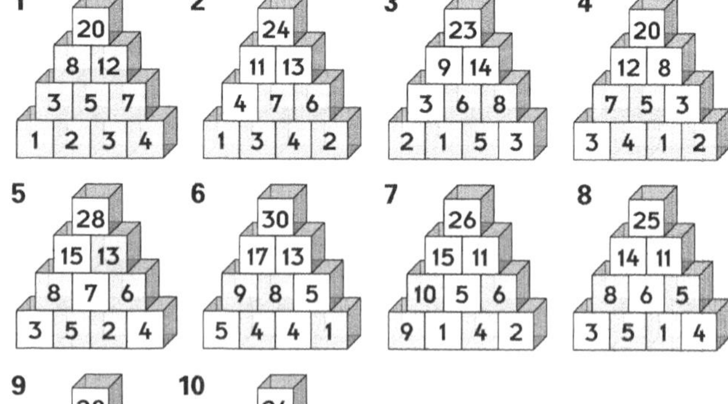

9
28
15 13
9 6 7
7 2 4 3

10
24
15 9
10 5 4
6 4 1 3

Number sentences (page 4)

Students can use 'guess and check' to work out these problems. (Not all possibilities are listed.)

1 5 + 4 + 1 = 10 *or* 4 + 4 + 2 = 10 *or* 5 + 3 + 2 = 10 *or* 3 + 3 + 4 = 10

2 5 − 4 + 1 = 2 *or* 4 − 3 + 1 = 2 *or* 2 − 1 + 1 = 2 *or* 3 − 2 + 1 = 2

3 5 − 3 + 2 = 4 *or* 4 − 2 + 2 = 4 *or* 3 − 1 + 2 = 4 *or* 5 − 4 + 3 = 4 *or* 4 − 3 + 3 = 4 *or* 3 − 2 + 3 = 4 *or* 2 − 1 + 3 = 4

4 3 + 4 + 1 = 8 *or* 5 + 2 + 1 = 8 *or* 3 + 3 + 2 = 8 *or* 2 + 2 + 4 = 8

5 5 − 4 − 1 = 0 *or* 3 − 2 − 1 = 0 *or* 4 − 3 − 1 = 0 *or* 5 − 3 − 2 = 0 *or* 5 − 2 − 3 = 0 *or* 4 − 2 − 2 = 0 *or* 4 − 1 − 3 = 0 *or* 3 − 1 − 2 = 0

6 3 + 3 + 3 = 9 *or* 4 + 2 + 3 = 9 *or* 4 + 4 + 1 = 9 *or* 5 + 1 + 3 = 9

7 5 − 2 + 4 = 7 *or* 5 − 1 + 3 = 7 *or* 5 − 3 + 5 = 7 *or* 4 − 2 + 5 = 7 *or* 3 − 1 + 5 = 7 *or* 4 − 1 + 4 = 7

8 5 + 4 + 3 = 12 *or* 4 + 4 + 4 = 12 *or* 5 + 5 + 2 = 12

9 4 − 4 + 1 = 1 *or* 3 − 3 + 1 = 1 *or* 2 − 2 + 1 = 1 *or* 5 − 5 + 1 = 1

10 5 + 4 − 1 = 8 *or* 5 + 5 − 2 = 8

11 4 + 3 − 2 = 5 *or* 5 + 4 − 4 = 5 *or* 5 + 3 − 3 = 5 *or* 4 + 2 − 1 = 5 *or* 5 + 2 − 2 = 5 *or* 5 + 1 − 1 = 5 *or* 4 + 4 − 3 = 5 *or* 5 + 5 − 5 = 5

12 2 + 2 + 1 = 5 *or* 3 + 1 + 1 = 5

13 3 + 3 + 1 = 7 *or* 3 + 2 + 2 = 7 *or* 5 + 1 + 1 = 7 *or* 4 + 2 + 1 = 7

14 4 − 2 + 1 = 3 *or* 3 − 1 + 1 = 3 *or* 5 − 3 + 1 = 3 *or* 3 − 2 + 2 = 3 *or* 5 − 4 + 2 = 3 *or* 4 − 3 + 2 = 3

15 5 + 3 + 3 = 11 *or* 5 + 5 + 1 = 11 *or* 4 + 4 + 3 = 11 *or* 5 + 4 + 2 = 11

Asking for more than one solution is an excellent extension for the more able student.

Addition and subtraction (page 5)

1 +, + **2** +, − **3** −, − **4** +, − **5** −, −
6 −, + **7** −, + **8** +, − **9** −, + **10** +, +
11 −, − **12** −, + **13** −, + **14** −, +

Number patterns (page 6)

Students should write down the difference between the numbers as shown in the example. They should be encouraged to say these differences aloud. For example, in question 1: 'Add 2, add 1, add 2, add 1, add 2'. Now students continue the pattern and say 'add 1' and write down 12. Next they say 'add 2' and write down 14.

1 12, 14 **2** 22, 27 **3** 16, 14 **4** 11, 15
5 20, 15 **6** 50, 60 **7** 18, 26 **8** 26, 31
9 34, 24 **10** 30, 21 **11** 22, 29 **12** 28, 36

Find my rule (page 7)

1

IN:	6	11	1	8	2	5	9	10	4	Rule
OUT:	9	14	4	11	**5**	**8**	**12**	**13**	**7**	**+ 3**

2

IN:	6	3	11	8	7	5	9	10	4	Rule
OUT:	3	0	8	5	**4**	**2**	**6**	**7**	**1**	**– 3**

3

IN:	6	11	1	8	2	5	9	10	4	Rule
OUT:	13	18	8	15	**9**	**12**	**16**	**17**	**11**	**+ 7**

4

IN:	6	11	8	12	7	5	9	10	4	Rule
OUT:	2	7	4	8	**3**	**1**	**5**	**6**	**0**	**– 4**

5

IN:	6	11	13	8	2	5	9	10	4	Rule
OUT:	4	9	11	6	**0**	**3**	**7**	**8**	**2**	**– 2**

6

IN:	6	11	1	8	2	5	9	10	4	Rule
OUT:	15	20	10	17	**11**	**14**	**18**	**19**	**13**	**+ 9**

7

IN:	6	11	13	8	12	5	9	10	7	Rule
OUT:	1	6	8	3	**7**	**0**	**4**	**5**	**2**	**– 5**

8

IN:	6	11	1	8	2	5	9	10	4	Rule
OUT:	17	22	12	19	**13**	**16**	**20**	**21**	**15**	**+ 11**

9

IN:	6	11	1	8	2	5	9	10	4	Rule
OUT:	12	22	2	16	**4**	**10**	**18**	**20**	**8**	**× 2**

10

IN:	6	3	1	8	2	5	9	10	4	Rule
OUT:	4	7	9	2	**8**	**5**	**1**	**0**	**6**	**– 10**

or sum of IN and OUT is 10

In this last question the teacher can give the hint: 'Try adding'.

Follow the rules (page 8)

Teachers can make a game of this activity by asking what is happening to my IN number to get my OUT number? Ask students to express the rule in words as well as symbols as shown below.

1

IN:	6	11	1	8	12	15	7	20	14	Rule
OUT:	11	16	6	13	**17**	**20**	**12**	**25**	**19**	**+ 5**

2

IN:	16	11	13	8	12	15	9	10	6	Rule
OUT:	10	5	7	2	**6**	**9**	**3**	**4**	**0**	**– 6**

3

IN:	6	11	1	8	2	5	9	10	4	Rule
OUT:	30	55	5	40	**10**	**25**	**45**	**50**	**20**	**× 5**

4

IN:	6	11	1	8	2	5	9	10	4	Rule
OUT:	16	21	11	18	**12**	**15**	**19**	**20**	**14**	**+ 10**

5

IN:	6	11	1	8	2	5	9	10	4	Rule
OUT:	26	31	21	28	**22**	**25**	**29**	**30**	**24**	**+ 20**

6

IN:	6	3	1	8	2	5	9	10	4	Rule
OUT:	66	33	11	88	**22**	**55**	**99**	**110**	**44**	**× 11**

7

IN:	6	3	1	8	2	5	9	10	4	Rule
OUT:	60	30	10	80	**20**	**50**	**90**	**100**	**40**	**× 10**

8

IN:	6	3	1	8	2	5	9	10	4	Rule
OUT:	62	32	12	82	**22**	**52**	**92**	**102**	**42**	**× 10 + 2**

9

IN:	6	11	1	8	2	5	9	10	4	Rule
OUT:	18	33	3	24	**6**	**15**	**27**	**30**	**12**	**× 3**

10

IN:	15	10	55	30	25	60	5	20	35	Rule
OUT:	3	2	11	6	**5**	**12**	**1**	**4**	**7**	**÷ 5**

Related rows (page 9)

1

13	5	9	6	8	7
16	8	12	9	11	10

rule: + 3

2

12	20	4	7	2	6
17	25	9	12	7	11

rule: + 5

3

14	7	16	5	9	6
12	5	14	3	7	4

rule: – 2

4

15	4	11	7	5	20
25	14	21	17	15	30

rule: + 10

5

5	8	1	5	2	4
50	80	10	50	20	40

rule: x 10

6

4	1	12	3	9	8
11	8	19	10	16	15

rule: + 7

7

6	2	8	3	9	4
12	4	16	6	18	8

rule: x 2

8

4	10	7	1	5	11
13	19	16	10	14	20

rule: + 9

9

14	26	18	32	12	16
3	15	7	21	1	5

rule: – 11

Numbers in circles (page 10)

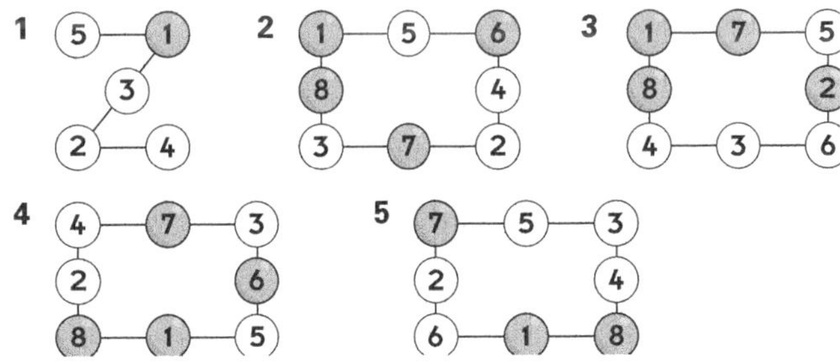

Number puzzles (page 11)

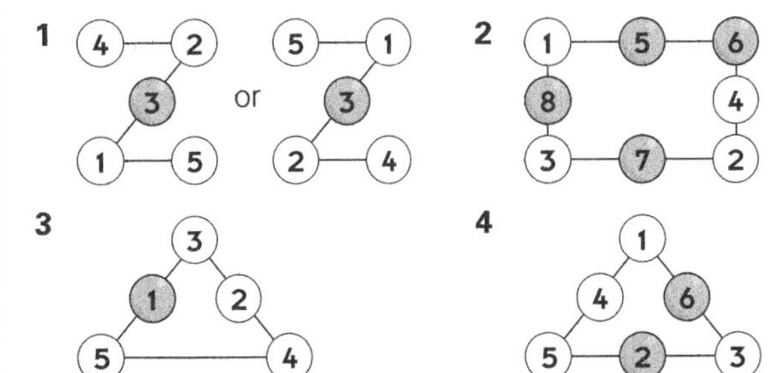

Number lines (page 12)

1 6 **2** 2 **3** 9 **4** 20 **5** 4 **6** 40
7 10 **8** 12 **9** 25 **10** 9 **11** 20 **12** 25

Star values (page 13)

1 7 **2** 12 **3** 8 **4** 10 **5** 30 **6** 25
7 15 **8** 11 **9** 15 **10** 12 **11** 13 **12** 14

Equal space (page 14)

1 6 **2** 6 **3** 30 **4** 10 **5** 8 **6** 8
7 10 **8** 19 **9** 16 **10** 12 **11** 18 **12** 20
13 9 **14** 21

Which sign? (page 15)

1 +, − **2** −, − **3** +, − **4** +, + **5** −, +
6 −, − **7** +, − **8** −, + **9** +, − **10** +, −
11 +, − **12** −, + **13** +, − **14** −,− **15** +, −
16 +, − **17** −, − **18** −, + **19** −, − **20** +, −

Calculating numbers (page 16)

1 6 = 3 + 3 8 = 3 + 5
9 = 3 + 3 + 3 10 = 5 + 5
11 = 3 + 3 + 5 12 = 3 + 3 + 3 + 3
13 = 5 + 5 + 3 14 = 3 + 3 + 3 + 5
15 = 5 + 5 + 5 *or* 3 + 3 + 3 + 3 + 3
16 = 5 + 5 + 3 + 3 17 = 5 + 3 + 3 + 3 + 3
18 = 5 + 5 + 5 + 3 *or* 3 + 3 + 3 + 3 + 3 + 3
19 = 5 + 5 + 3 + 3 + 3 20 = 5 + 5 + 5 + 5
21 = 5 + 5 + 5 + 3 + 3

2 There are several possible solutions for some of the numbers. Not all possibilities are listed here.
6 = 3 + 3 *or* 5 + 4 − 3 *or* 5 + 5 − 4
7 = 4 + 3 *or* 5 + 5 − 3 *or* 4 + 4 + 4 − 5
8 = 4 + 4 *or* 5 + 3
9 = 4 + 5 *or* 3 + 3 + 3
10 = 5 + 5 *or* 3 + 3 + 4
11 = 3 + 3 + 5 *or* 4 + 4 + 3 *or* 5 + 5 + 5 − 4
12 = 4 + 4 + 4 *or* 3 + 3 + 3 + 3 *or* 5 + 5 + 5 − 3
13 = 5 + 5 + 3 *or* 4 + 4 + 4 + 4 − 3
14 = 5 + 5 + 4 *or* 3 + 3 + 4 + 4
15 = 5 + 5 + 5 *or* 3 + 3 + 3 + 3 + 3 *or* 4 + 4 + 4 + 3
16 = 5 + 5 + 3 + 3 *or* 4 + 4 + 4 + 4 *or* 5 + 5 + 5 + 5 − 4
17 = 4 + 4 + 4 + 5 *or* 5 + 5 + 4 + 3 *or* 5 + 5 + 5 + 5 − 3
18 = 5 + 5 + 5 + 3 *or* 3 + 3 + 3 + 3 + 3 + 3 *or* 5 + 4 + 5 + 4
19 = 5 + 5 + 5 + 4 *or* 5 + 5 + 3 + 3 + 3
20 = 5 + 5 + 5 + 5 *or* 4 + 4 + 4 + 4 + 4

Calculator fun (page 17)

1 11 = 7 + 4 8 = 4 + 4
12 = 4 + 4 + 4 14 = 7 + 7
16 = 4 + 4 + 4 + 4 20 = 4 + 4 + 4 + 4 + 4
15 = 7 + 4 + 4 18 = 7 + 7 + 4
19 = 4 + 4 + 4 + 7 21 = 7 + 7 + 7
22 = 7 + 7 + 4 + 4 25 = 7 + 7 + 7 + 4

2 5 = 4 + 4 + 4 − 7 6 = 7 − 4 + 7 − 4
10 = 7 + 7 − 4 17 = 7 + 7 + 7 − 4
9 = 7 − 4 + 7 − 4 + 7 − 4 *or* 4 + 4 + 4 + 4 − 7
13 = 7 + 7 + 7 − 4 − 4 *or* 4 + 4 + 4 + 4 + 4 − 7

Find my start (page 18)

1 12 **2** 10 **3** 9 **4** 10
5 5 **6** 5 **7** 6

Missing numbers (page 19)

1 3, 2, 7, 7, 9 **2** 18, 22, 4, 3, 6 **3** 5, 13, 1, 6, 10

House numbers (page 20)

A discussion of even and odd numbers would be helpful. Encourage students to use 'guess and check' to work out their solution, always remembering that they must be adding two consecutive odd, or two consecutive even numbers to find the sum of the required number.

sum of two house numbers	one house number	next door house number
8	3	5
4	1	3
20	9	11
14	6	8
24	11	13
16	7	9
26	12	14
18	8	10

Possible house numbers (page 21)

1 **a** 47 or 74 **b** 59 or 95 **c** 16 or 61
d 259 or 295 or 529 or 592 or 925 or 952
e 147 or 174 or 417 or 471 or 714 or 741
f 338 or 383 or 833 **g** 449 or 494 or 944
h 105 or 150 or 501 or 510 (a three-digit number can not start with 0). It is important to discuss with students the value of working systematically.

2 To reach this generalisation, students need to discuss question 1 very carefully.
a 2 **b** 6 **c** 3

Odd one out (page 22)

1 24 the other numbers are multiples of 5 or end in a 5, *or*
5 the other numbers are two-digit numbers
2 37 the other numbers are units or one-digit numbers
3 23 the other numbers are multiples of 10 or end in a 0
4 16 the other numbers are odd
5 12 the other numbers are multiples of 5 or end in a 5
6 97 the other numbers end in a 9
7 17 the other numbers are even *or*
6 the other numbers are two-digit numbers
8 20 the other numbers are between 50 and 59
9 25 the other numbers are even or all end in 2
10 36 the other numbers are multiples of 11
11 100 the other numbers are between 90 and 99
12 13 the other numbers are less than 10 or one-digit

Who am I? (page 23)

Students should list all possible solutions after one or two clues and then eliminate the possibilities as further clues are given.

So, in question 1, a student would write after the first two lines that the possibilities are: 12, 14, 16, 18. Then, since the sum of the digits is 9, we know that 12, 14 and 16 do not satisfy this condition and we are left with 18 as the solution.

1 18 **2** 11 **3** 15 **4** 6
5 35 **6** 40 or 22 **7** 24

Sharing (page 24)

1 2 sweets, 4 sweets
2 2 sweets, 4 sweets, 8 sweets
3 2 sweets, 4 sweets, 5 sweets, 10 sweets

Multiple choice (page 25)

1 C **2** D **3** C **4** D **5** A **6** A

Make your choice (page 26)

1 C and D **2** B **3** A, B and D
4 B and D **5** B and C **6** A

How many cars? (page 27)

Since only four children can fit into a car, an extra car is needed if there are 1, 2 or 3 children left over. It does not matter if each car is not full.

1 a 2 **b** 3 **c** 6 **d** 2
e 3 **f** 5 **g** 7
2 **a** 2 **b** 5 **c** 5 **d** 7

Money (page 28)

1 5 **2** 4 **3** 5 **4** 10
5 2 **6** 4 **7** 25 **8** 20

Find the value (page 29)

1 16c **2** 20c **3** 24c **4** 16c **5** 15c **6** 24c
7 80c **8** 50c **9** $10 **10** $27 **11** $12 **12** $8

Shape values (page 30)

Students should first work out the fraction shaded.

1 4c **2** 5c **3** 4c **4** 3c
5 5c **6** 4c **7** 5c **8** 2c
9 $10 **10** $4 **11** $30 **12** $15
13 $6 **14** $4 **15** $12 **16** $12

Money with fractions (page 31)

1 a 40c **b** 80c
2 a 80c **b** $1.60
3 a $10 **b** $30
4 a $5 **b** $10

Money problems (page 32)

1 a 60c **b** 50c **c** 5 **d** 4 **e** no **f** 4
2 a 70c **b** 80c **c** 3 **d** no **e** yes **f** 2

Making money (page 33)

1 a

5c	10c	20c
3	1	0
1	2	0
1	0	1
5	0	0

b

5c	10c	20c
0	1	1
2	0	1
0	3	0
2	2	0
4	1	0
6	0	0

2

5c	10c	20c
2	0	2
0	3	1

3

5c	10c	20c	50c
2	2	1	1
0	5	0	1
0	2	4	0

Shopping problems (page 34)

1 a 2 comic books or 1 balloon and 1 ice block
b $1.40, 60c **c** $3 **d** no **e** $3
2 a $1.50 **b** no **c** 50c

Squared-paper patterns (page 35)

1 **2**

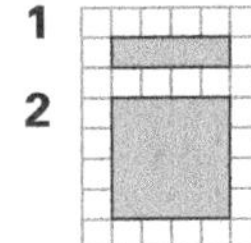

3 **4**

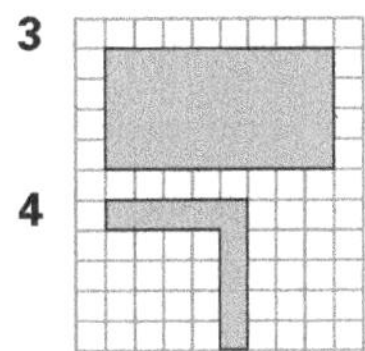

5 **6**

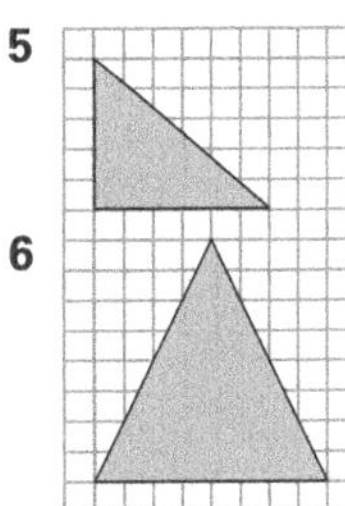

Square tiles (page 36)

1 9 **2** 16 **3** 5

4

Number of steps in staircase	Number of squares
2	4
3	9
4	16
5	25
6	36
7	49

As an extension, the teacher could discuss how the number of squares forms the pattern of square numbers.

Spinner experiments (page 37)

Teachers will need to explain and discuss the meaning of 'predict' and the results of experiments and how results of a large number of experiments will closely resemble results of prediction.

1 a 8 **b** 20
2 a 8 **b** 4 **c** 20
3 a 5 **b** 30

Spinner experiments continued (page 38)

Students may need some help with recording their results. Using a table to tally results is recommended. Discuss with students how predictions compare to experiment results.

For question 1, students should discuss why the numbers for A and B were the same for the prediction.

Spinner experiments continued (page 38) ***continued...***

For question 2, students should discuss why the numbers for Y and Z were the same, and why X was double for the prediction.

For question 3, students should discuss why the numbers for P, Q and R were the same for the prediction.

1. For 16 spins the arrow will land in A approximately 8 times and in B approximately 8 times.
   For 20 spins the arrow will land in A approximately 10 times and in B approximately 10 times.
2. For 16 spins the arrow will land in X approximately 8 times and in Z and Y approximately 4 times.
   For 20 spins the arrow will land in X approximately 10 times and in Z and Y approximately 5 times.
3. For 15 spins the arrow will land in P, Q and R approximately 5 times.
   For 30 spins the arrow will land in P, Q and R approximately 10 times.

Lights on/off (page 39)

1

2

Dice graphs (page 40)

1. Ideally all the columns should be the same height. In fact, if the experiment is carried out enough times, the columns will eventually equalise as each number on an unbiased die is equally likely to occur. If a die is thrown 60 times, the likelihood of each number occurring is 10.
2. 7 should be the winning total for the majority of the students, as that is the most likely sum on two dice. However, since it is an experiment and it is not repeated a large number of times, answers could vary and are acceptable. 6, 8 or even 5 or 9 may also be winning totals, though these are not as likely. Winning totals of 2 or 10 are most unlikely.

   Probability (total 2) = $\frac{1}{36}$ Probability (total 3) = $\frac{2}{36}$
   Probability (total 4) = $\frac{3}{36}$ Probability (total 5) = $\frac{4}{36}$
   Probability (total 6) = $\frac{5}{36}$ Probability (total 7) = $\frac{6}{36}$
   Probability (total 8) = $\frac{5}{36}$ Probability (total 9) = $\frac{4}{36}$
   Probability (total 10) = $\frac{3}{36}$ Probability (total 11) = $\frac{2}{36}$
   Probability (total 12) = $\frac{1}{36}$

   These probabilities are for the teacher only and demonstrate that if two dice were thrown 36 times we should get the following graph.

Most students should get a graph similar to this shape.

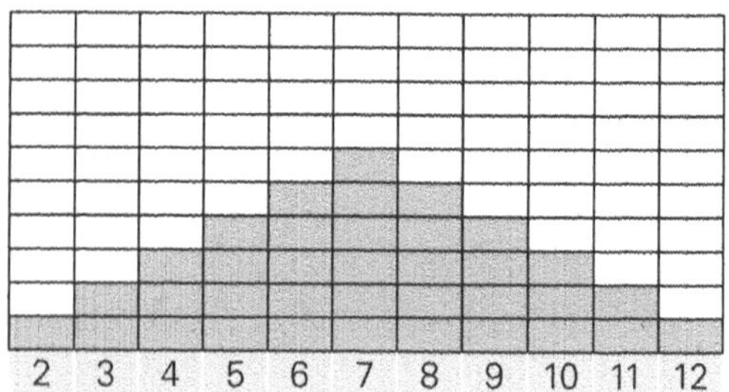

Gym-time graphs (page 41)

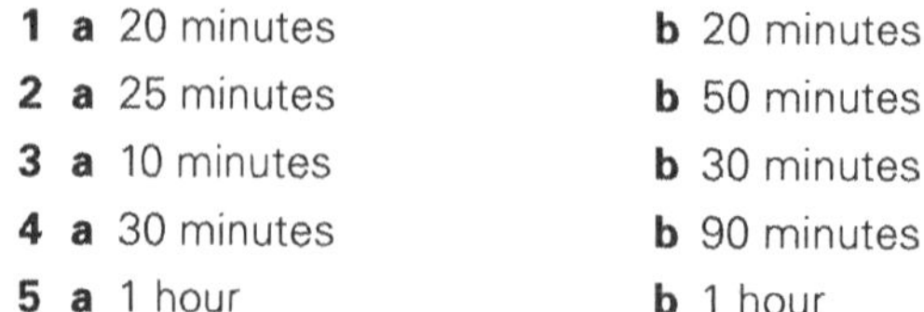

1. **a** 20 minutes **b** 20 minutes
2. **a** 25 minutes **b** 50 minutes
3. **a** 10 minutes **b** 30 minutes
4. **a** 30 minutes **b** 90 minutes
5. **a** 1 hour **b** 1 hour

Working out (page 42)

1. **a** 20 minutes **b** 20 minutes **c** 60 minutes
2. **a** 2 hours **b** 2 hours
3. **a** 90 minutes **b** 120 minutes or 2 hours
4. **a** 2 hours **b** 6 hours
5. **a** 1 hour **b** 4 hours

Find the temperature (page 43)

1 9°C **2** 11°C **3** 30°C **4** 10°C **5** 55°C
6 8°C **7** 15°C **8** 25°C **9** 6°C **10** 8°C

What's the temperature? (page 44)

1 8°C **2** 13°C **3** 6°C **4** 30°C **5** 33°C
6 18°C **7** 30°C **8** 25°C **9** 30°C **10** 26°C

Temperature facts? (page 45)

1 True **2** True **3** True **4** False
5 False **6** False **7** True **8** False
9 True **10** False **11** True **12** True
13 True **14** False **15** False **16** False

Week days (page 46)

Students should be encouraged to discover that the days in a week form a seven-day cycle.

1. **a** Saturday **b** Thursday **c** Friday **d** Friday **e** Friday
2. **a** Sunday **b** Monday **c** Sunday **d** Sunday
3. **a** Friday **b** Wednesday **c** Friday **d** Friday

Time (page 47)

1 7 p.m. **2** 9 p.m. **3** 8 p.m. **4** 2 p.m.
5 4 p.m. **6** 12 noon **7** 10 a.m. **8** 10 a.m.
9 11 a.m. **10** 11 a.m. **11** 10 a.m. **12** 10 a.m.

Time facts? (page 48)

1 False **2** True **3** False **4** False **5** False
6 True **7** False **8** False **9** False **10** True
11 False **12** False **13** False **14** True

BOOK 2 Solutions

Calendar (page 49)

1 31 days **2** 5 **3** 9 **4** 8 December
5 24 November **6** Wednesday **7** Sunday
8 Monday **9** 11 days **10** 7

Scales (page 50)

1 3.5 kg **2** 5 kg **3** 8 kg **4** 8 kg **5** 17 kg
6 1.5 kg **7** 6.5 kg **8** 11 kg **9** 5 kg

Find the mass (page 51)

1 3 cubes, 9 grams **2** 3 cubes, 9 grams
3 4 cubes, 12 grams **4** 6 cubes, 18 grams
5 5 cubes, 15 grams **6** 6 cubes, 18 grams
7 7 cubes, 21 grams **8** 6 cubes, 18 grams
9 12 cubes, 36 grams

Mass facts? (page 52)

1 False **2** True **3** True **4** True **5** False
6 False **7** False **8** True **9** True **10** True
11 True **12** False **13** False **14** True **15** False

Measurement facts? (page 53)

These questions require discussion and checking by measuring. Students can be encouraged to make up questions of their own.

1 False **2** False **3** True **4** False **5** True
6 False **7** False **8** True **9** True **10** False
11 False **12** False **13** True **14** False **15** True
16 False

True or false? (page 54)

1 False **2** False **3** True **4** True **5** False
6 False **7** True **8** True **9** True **10** False
11 True **12** True **13** True **14** True **15** False

Measuring (page 55)

1 a 10 cups
b 15 cups
c 35 cups

Number of litres	Number of cups
1	5
4	20
2	10
3	15
7	35

2 a 3 metres
b 20 ribbons

Number of metres	Number of ribbons
1	4
3	12
5	20

3 a 5 pizzas
b 12 children

Number of medium pizzas	Number of equal slices
1	3
5	15
4	12

Measuring containers (page 56)

1 a 4 = 2 + 2 **b** 12 = 5 + 5 + 2
c 11 = 5 + 2 + 2 + 2 **d** 1 = 5 − 2 − 2

2 a 10 = 5 + 5 **b** 2 = 5 − 3
c 1 = 3 + 3 − 5 or 1 = 5 + 5 − 3 − 3 − 3
d 11 = 5 + 3 + 3

Perimeter (page 57)

1 12 cm **2** 16 cm **3** 14 cm
4 18 cm **5** 16 cm **6** 22 cm

S can be placed anywhere and is only needed to help students to know when they have completed walking around the figure.

Multiple choice—mass (page 58)

1 B **2** A **3** B **4** D **5** C **6** B

Symmetry patterns (page 59)

There are 20 possible answers altogether.

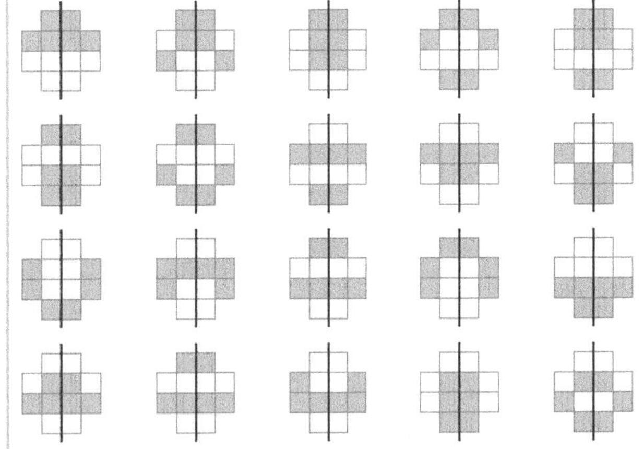

Symmetry (page 60)

1 A V X K H Y U C O

2

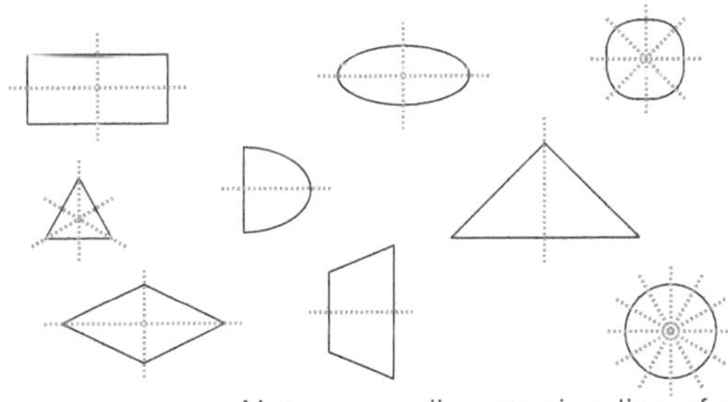

Note: every diameter is a line of symmetry.

Four squares (page 61)

There are five different solutions. Each shape can be rotated and/or reflected.

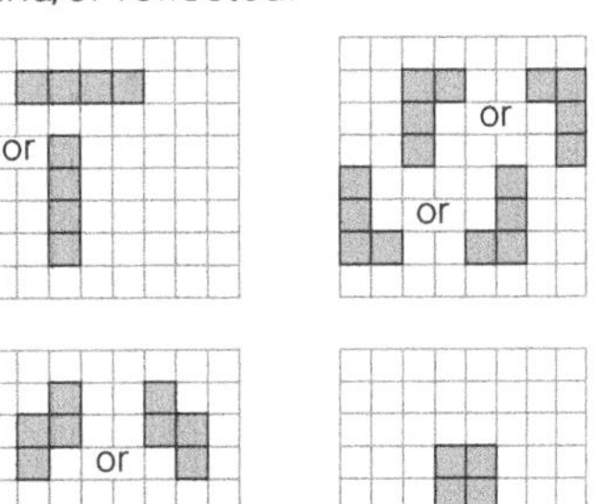

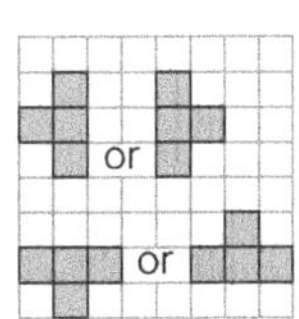

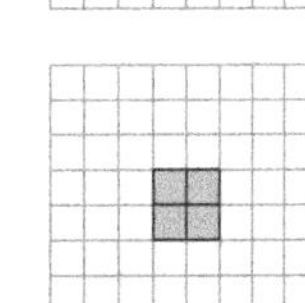

Tessellations (page 62)

Tiling patterns (page 63)

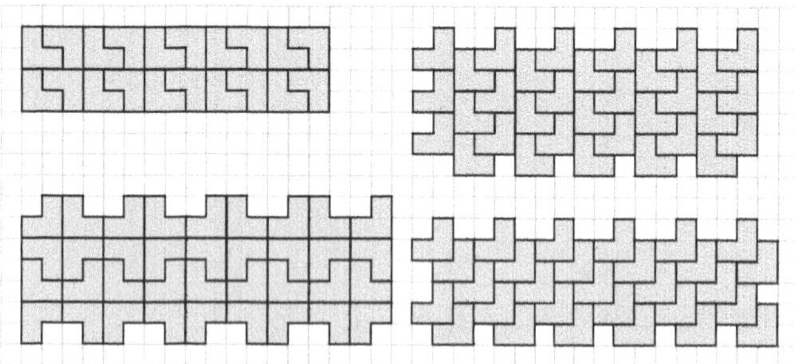

Grid position (page 64)

1  E7  D1

2 3 units, square

3 5 units and 3 units, rectangle

4 F6, 3 units

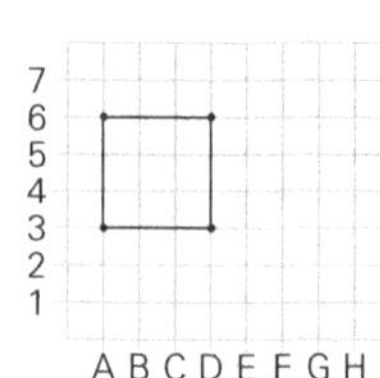

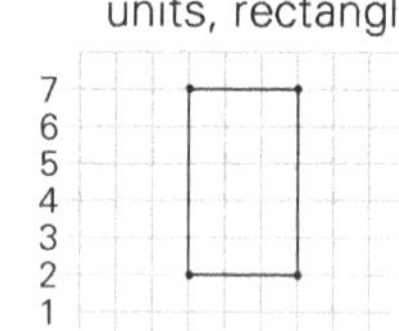

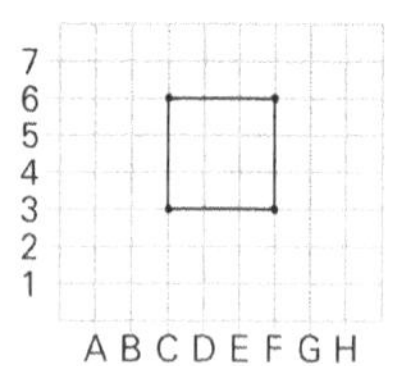

Which shape? (page 65)

R E C T A N G L E S

Combining shapes (page 66)

1 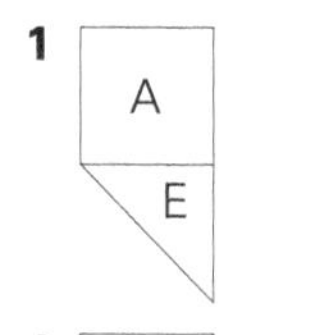

2 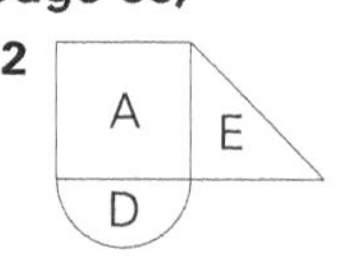

3 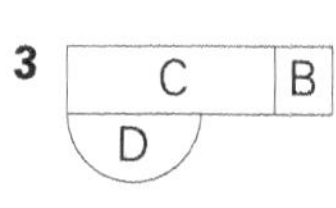

4 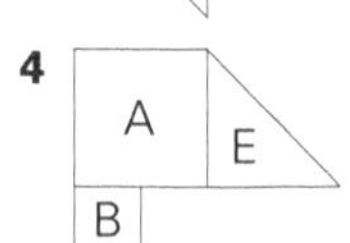

5 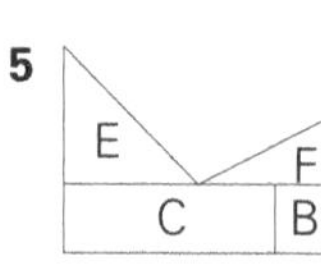

6 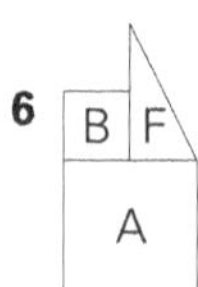

How many cubes? (page 67)

1 **a** 8 **b** 7 **c** 1 **d** 27 **e** 19

2 **a** 9 **b** this solid

Solids (page 68)

1. 6 faces, 12 edges, 8 corners
2. 5 faces, 9 edges, 6 corners
3. 5 faces, 8 edges, 5 corners
4. 4 faces, 6 edges, 4 corners
5. 7 faces, 15 edges, 10 corners

Toothpick geometry (page 69)

1 **a** 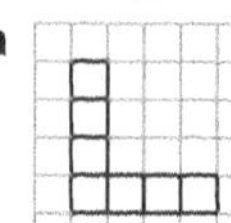 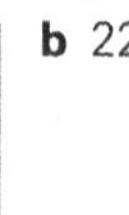 **b** 22

2 **a** 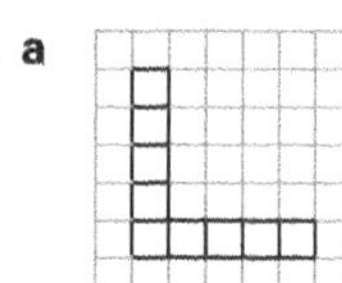 **b** 28

3 34

4

first	second	third	fourth	fifth	sixth
4	10	16	22	28	34

5 40

Fruit drinks (page 70)

1 **a** 2 **b** 3, 3, 3 **c** 12

2 **a** 4 **b** 3

Sharing (page 71)

1 5 litres **2** 8 kg

The following are quite difficult exercises and students may need to be shown how to use a table and 'guess and check'. It is not expected that students will find the correct answer immediately without using 'guess and check'.

3 Encourage students to make a guess. Let us say it is 4 litres for Daniel. Then, since Lisa's share is 2 litres less, hers is 2 litres. These two amounts add to 6 litres, not 10 litres.

Daniel's share	Lisa's share	Total
4 litres	2 litres	6—too low so increase Daniel's share
5 litres	3 litres	8—still too low

Increase Daniel's share again and we reach the correct answer.

Daniel's share	Lisa's share	Total
6 litres	4 litres	10 litres

4 Maria's share: 8 kg; Sylvia's share: 6 kg

Maria's share	Sylvia's share
8 kg	6 kg

5 Jonathan's share: 7kg; Kevin's share: 3kg

Jonathan's share	Kevin's share
7 kg	3 kg

Shading shapes (page 72)

1 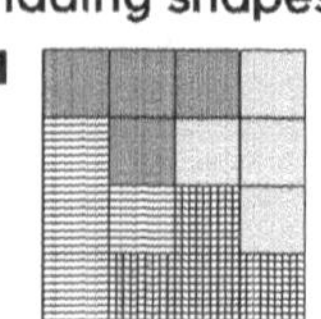

2 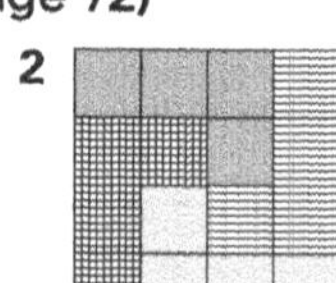

3 

Shading squares (page 73)

Children need to be shown what is meant by working systematically: that they must be organised in the way they shade. If they shade in a 'haphazard' manner, they will not know if they have covered all the possibilities.

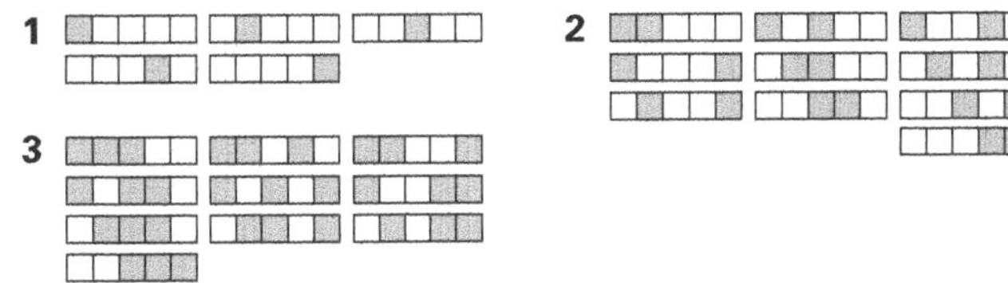

How many squares? (page 74)

1 9

2 4 14

How many rectangles? (page 75)

It is important that students colour in systematically. First, one rectangle at a time should be coloured in, then two at a time, then three and so on. Unless students work systematically, they are not likely to find all the possibilities.

1 a 6

b 10

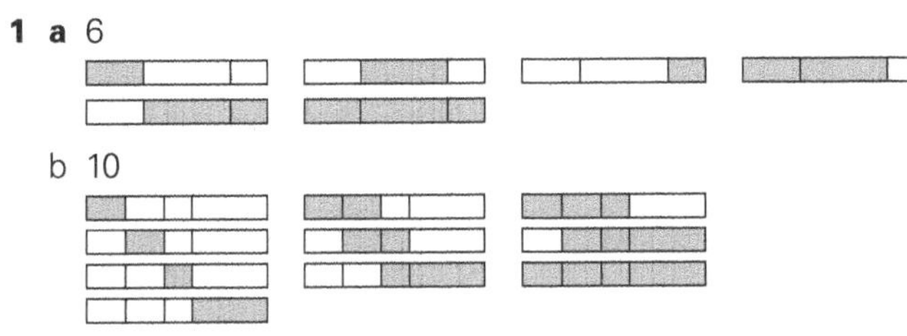

Balancing act (page 76)

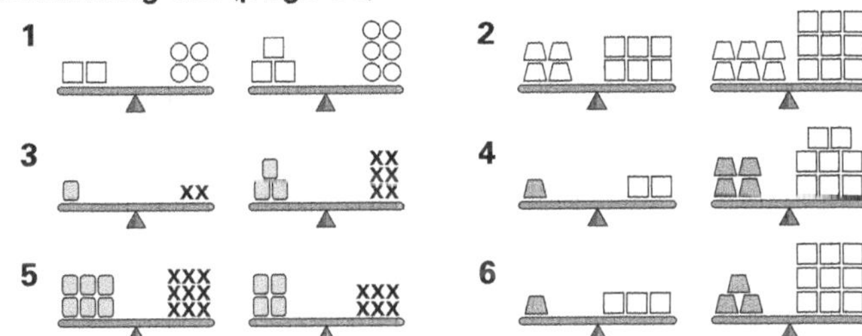

Balances (page 77)

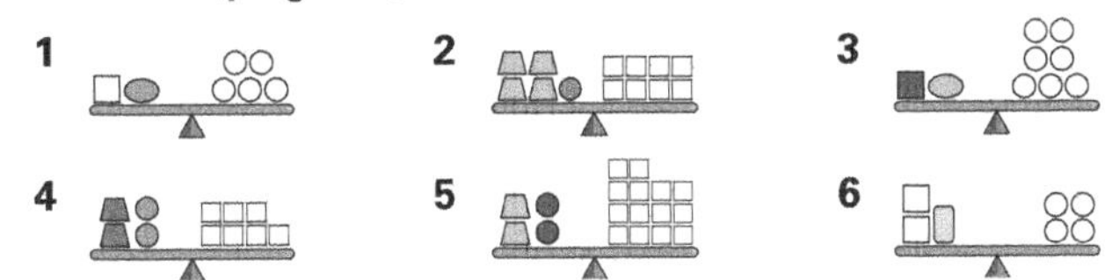

Balance each side (page 78)

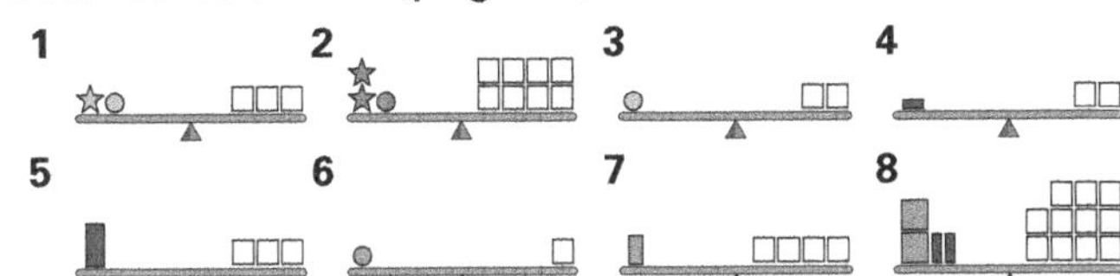

Saying the balances in words or drawing extra diagrams may be helpful to students.

For question 7, we could draw:

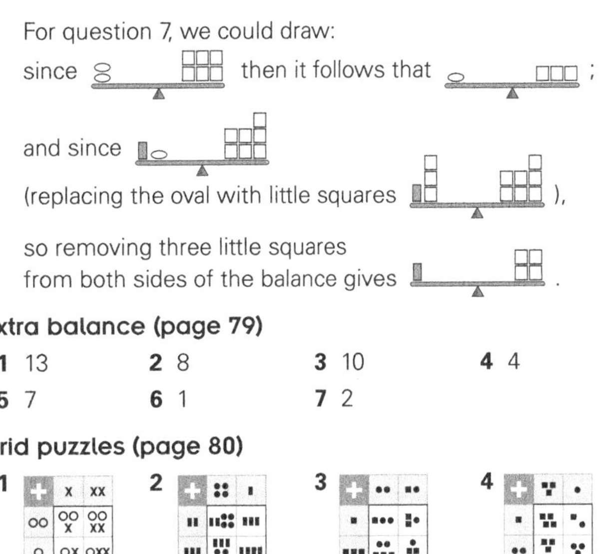

since … then it follows that …;

and since …

(replacing the oval with little squares …),

so removing three little squares from both sides of the balance gives … .

Extra balance (page 79)

1 13 **2** 8 **3** 10 **4** 4

5 7 **6** 1 **7** 2

Grid puzzles (page 80)

1

+	x	xx
oo	oo x	oo xx
o	ox	oxx

2

+	●●●●	▮
▮▮	▮▮●●●●	▮▮▮
▮▮▮	▮▮▮●●●●	▮▮▮▮

3

+	●●	■●
■	■●●	■■●
■■■	●●■■■	●■■■■

4

+	■■■	●
■	■■■■	■●
●●	■■■●●	●●●

5

+	●●●	+
●	●●●●	●+
++	●●● ++	+++

6

+	▲▲	●
▲▲	▲▲ ▲▲	▲▲ ●
●●●	▲▲ ●●●	●●●●

7

+	♥	▮
●	♥●	▮●
♥♥	♥♥♥	▮♥♥

8

+	O	=
••	•• O	= ••
—	— O	≡

Square puzzles (page 81)

1

+	★★★	■
★	★★★ ★	★■
■■	★★★ ■■	■■■

2

+	>	>>
00	00>	>> 00
000	000 >	000 >>

3

+	●●●●	▮●●
▮▮●	▮▮● ●●●●	▮▮▮●●●
▮▮▮	▮▮▮● ●●●	▮▮▮▮ ●●

4

+	++ ++	+ +
o+	o++ +++	++ o+
oo	o++ o++	oo ++

5

+	★★/	★//
//	★★ ///	★ ////
★ ★★	★★★★ ★/	★★★ //★

6

+	▲▲ ●▲	●●
●● ▲	▲▲ ▲▲ ●●●	●● ●● ▲
▲	▲▲● ▲▲	●● ▲

7

+	■ ///	■■
●●	///■ ●●	■■ ●●
■	■■ ///	■■ ■

8

+	x	xx oo
cc	cc x	cc oo xx
c o o	c o x o	xx ooo c o

Drawing lines (page 82)

Many solutions are possible, these are some examples only.

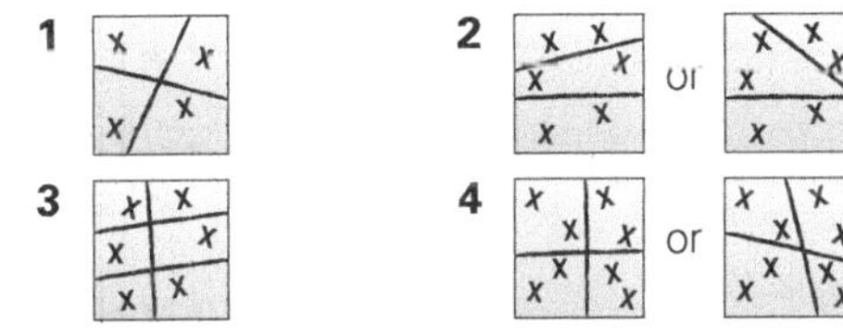

Students should experiment with toothpicks or matches to work out where to draw the lines.

How many lines? (page 83)

1 a 6 lines b 10 lines c 15 lines

2

Number of dots	Number of lines
3	3
4	6
5	10
6	15
7	21

Making shapes (page 84)

There are many possible solutions. Below are some possibilities.

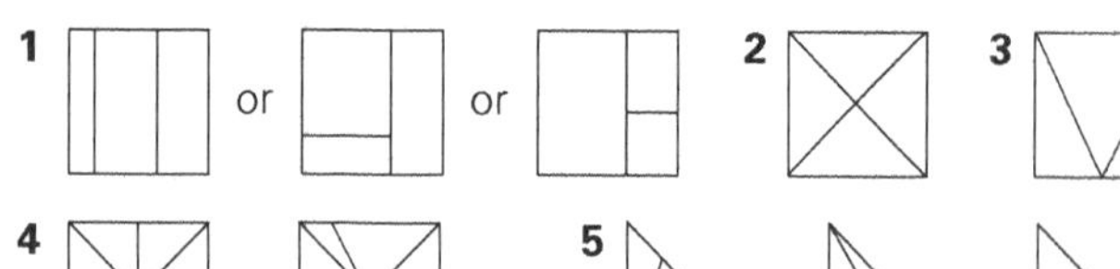

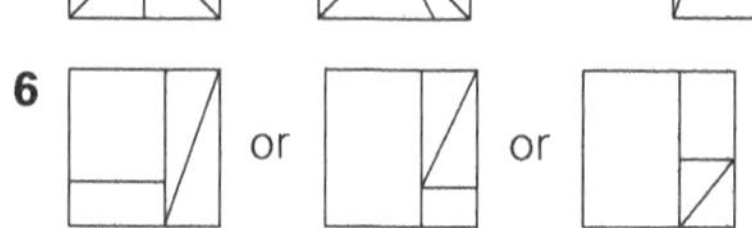

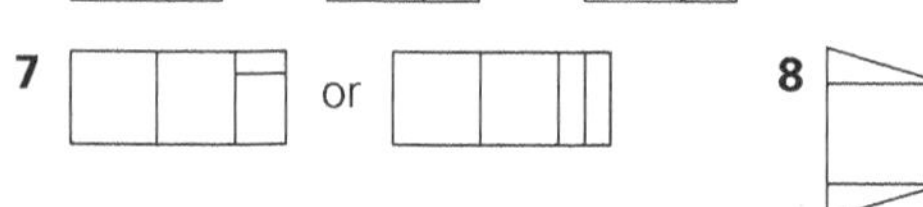

Puzzles with matches (page 85)

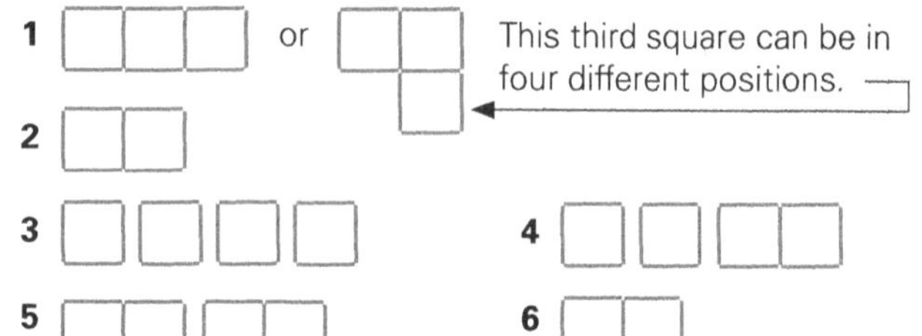

Match puzzles (page 86)

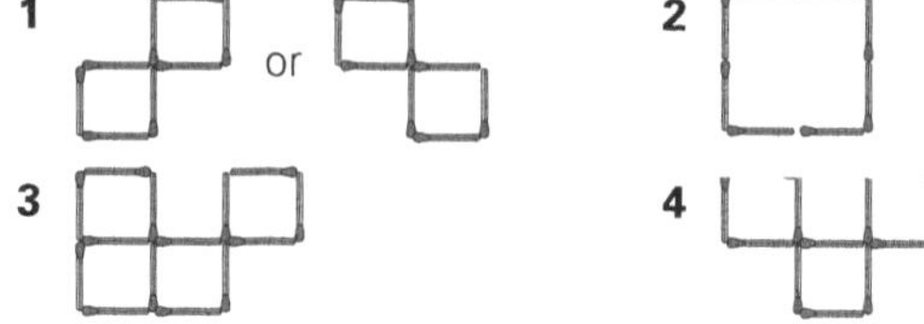

Cyclist routes (page 87)

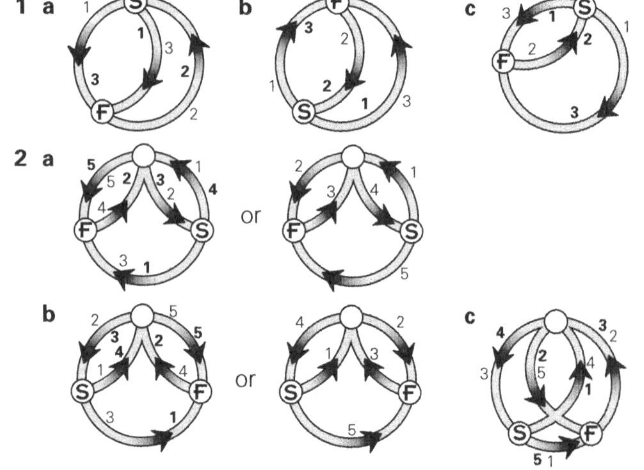

Dots on cubes (page 88)

1 14 **2** 14 **3** 18 **4** 16 **5** 18

6 18 **7** 18 **8** 18 **9** 24

STUDENT BOOK Solutions

BOOK 3

Number grids (page 1)

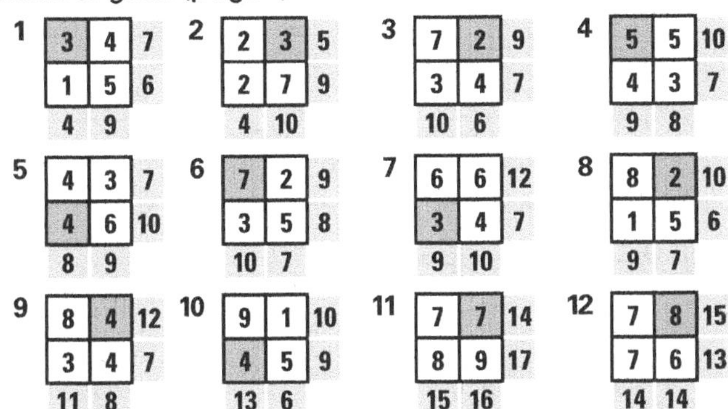

Students should begin by subtracting the numbers given in the grids from the relevant horizontal or vertical totals. In question 1, row 1: 7 – 3 = 4; and column 1: 4 – 3 = 1.

Grid puzzles (page 2)

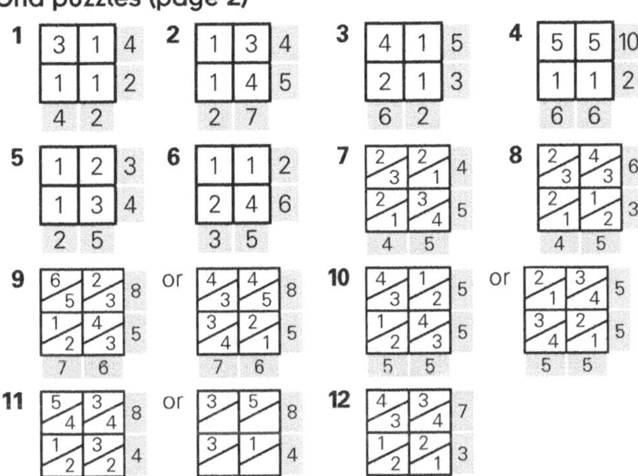

This is an excellent exercise in trial and error. Students could start question 1 by asking, 'Which two numbers add up to 4?' The possibilities are 3 and 1 or 2 and 2, but if 2 and 2 is used, the lower right-hand number would be 0, which is not a counting number. Therefore, 3 and 1 is the correct combination.

Missing numbers (page 3)

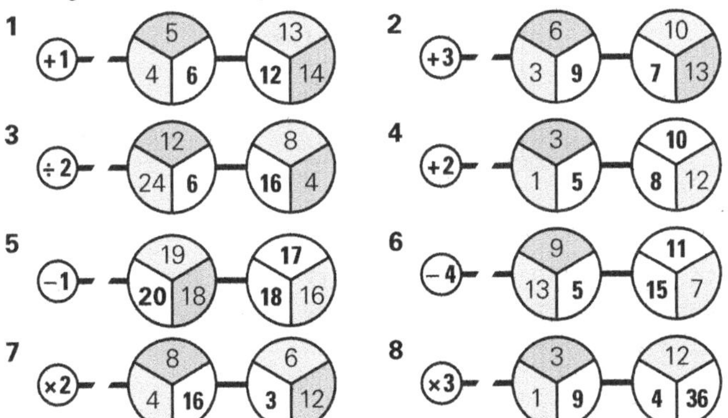

Number sentences (page 4)

1 2 + 3 + 4 = 9 **2** 6 + 5 – 2 = 9
3 4 + 3 – 5 = 2 or 5 + 3 – 6 = 2
4 6 + 3 – 5 = 4 or 6 + 2 – 4 = 4 or 5 + 3 – 4 = 4 or 5 + 2 – 3 = 4
5 5 + 3 + 2 = 10 **6** 5 + 4 + 2 = 11 or 6 + 3 + 2 = 11
7 6 – 3 – 2 = 1 **8** 6 – 4 – 2 = 0 or 5 – 3 – 2 = 0
9 6 – 5 + 2 = 3 or 5 – 4 + 2 = 3 or 4 – 3 + 2 = 3
10 2 × 3 + 4 = 10
11 2 × 5 – 3 = 7 or 2 × 6 – 5 = 7 or 3 × 4 – 5 = 7
12 2 × 4 – 6 = 2 or 2 × 3 – 4 = 2 **13** 2 × 3 – 6 = 0
14 2 × 4 – 5 = 3 **15** 2 × 3 + 5 = 11 or 2 × 4 + 3 = 11

This exercise assumes only knowledge of the 2 times table, but it will require a certain amount of insight and will involve trial and error. Students must be reminded that, in each case, all the numbers used must be different.

Number ideas (page 5)

1 +, – **2** ×, – **3** +, – **4** +, –
5 ×, – or 2 + 3 × 1 = 5 or (2 + 3) × 1 = 5
6 ×, × or 1 + 2 + 3 = 6 **7** ×, + **8** +, –
9 ×, – **10** ×, – **11** ×, – **12** ×, – **13** ×, –
14 –, + or 6 × 2 – 4 = 8 **15** +, – **16** ×, –
17 ×, + **18** ×, + **19** ×, – **20** ×, –

Numbers and symbols (page 6)

The following are some of the possible answers. Students can mark one another's work and then list all the possibilities they found.

Students will write number sentences to the level of their ability. They should be encouraged to use brackets, but it is not expected that many of them will have all these possibilities.

1 2 + 2 – (2 × 2) = 0
2 + 2 – (2 + 2) = 0
2 × 2 – 2 × 2 = 0
(2 × 2) × (2 – 2) = 0
(2 × 2 – 2) ÷ 2 = 1
2 × 2 ÷ 2 ÷ 2 = 1
2 ÷ 2 + 2 ÷ 2 = 2
(2 + 2 + 2) ÷ 2 = 3
2 + 2 – (2 – 2) = 4
2 + 2 + 2 – 2 = 4
2 × 2 × 2 ÷ 2 = 4
2 + 2 + 2 ÷ 2 = 5
2 × 2 × 2 – 2 = 6
2 × (2 + 2 ÷ 2) = 6
2 + 2 + 2 × 2 = 8
2 + 2 × 2 × 2 = 10
2 × (2 + 2 + 2) = 12

2 + 2 – 2 – 2 = 0
2 × 2 – 2 – 2 = 0
(2 + 2) × (2 – 2) = 0
(2 + 2 – 2) ÷ 2 = 1
2 ÷ 2 + 2 – 2 = 1
2 + 2 × (2 – 2) = 2
2 + 2 – 2 ÷ 2 = 3
(2 × 2 + 2) ÷ 2 = 3
2 × 2 – (2 – 2) = 4
2 × 2 + 2 – 2 = 4
2 × 2 × (2 ÷ 2) = 4
2 × 2 + 2 ÷ 2 = 5
(2 + 2) × 2 – 2 = 6
2 + 2 + 2 + 2 = 8
2 × 2 + 2 × 2 = 8
2 × (2 + 2 × 2) = 12
2 × 2 × 2 × 2 = 16

2 3 × (3 − 3) = 0
3 ÷ (3 ÷ 3) = 3
3 + 3 ÷ 3 = 4
3 + 3 + 3 = 9
3 × 3 + 3 = 12
3 + 3 − 3 = 3
3 × 3 ÷ 3 = 3
3 × 3 − 3 = 6
3 × (3 + 3) = 18
3 × 3 × 3 = 27

3 3 + 3 − 3 − 3 = 0
(3 + 3) × (3 − 3) = 0
3 × 3 ÷ (3 × 3) = 1
3 × 3 ÷ 3 ÷ 3 = 1
(3 × 3 − 3) ÷ 3 = 2
3 × 3 − 3 − 3 = 3
(3 × 3 + 3) ÷ 3 = 4
3 + 3 + 3 − 3 = 6
3 + 3 + (3 ÷ 3) = 7
3 × 3 + 3 − 3 = 9
3 × 3 × 3 ÷ 3 = 9
3 × 3 + 3 + 3 = 15
3 × 3 + 3 × 3 = 18
3 × 3 × 3 − 3 = 24
3 × 3 × 3 + 3 = 30
3 × 3 × 3 × 3 = 81
3 × 3 − 3 × 3 = 0
3 × 3 × (3 − 3) = 0
3 ÷ 3 × 3 ÷ 3 = 1
(3 + 3 − 3) ÷ 3 = 1
3 ÷ 3 + 3 ÷ 3 = 2
(3 + 3 + 3) ÷ 3 = 3
3 + 3 − (3 ÷ 3) = 5
(3 + 3) × 3 ÷ 3 = 6
3 × 3 − (3 ÷ 3) = 8
3 × 3 × (3 ÷ 3) = 9
3 × 3 + 3 ÷ 3 = 10
(3 × 3 − 3) × 3 = 18
3 × (3 + 3) + 3 = 21
3 × (3 + 3 + 3) = 27
(3 × 3 + 3) × 3 = 36

Square puzzles (page 7)

1

2	3	5
2	4	6
4	7	

A = 2
B = 3
C = 4

2

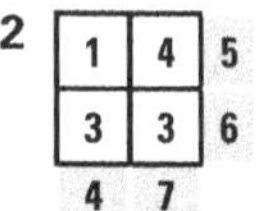

1	4	5
3	3	6
4	7	

A = 1
B = 4
C = 3

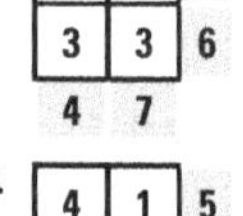

3

2	1	3
3	3	6
5	4	

A = 2
B = 3
C = 1

4

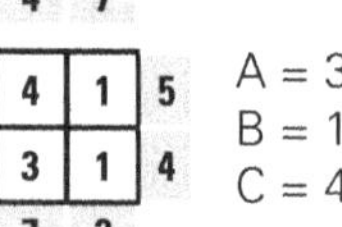

4	1	5
3	1	4
7	2	

A = 3
B = 1
C = 4

5

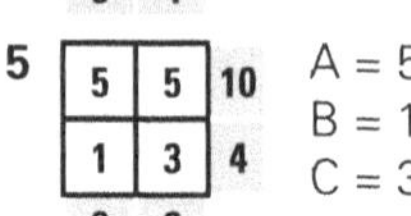

5	5	10
1	3	4
6	8	

A = 5
B = 1
C = 3

6

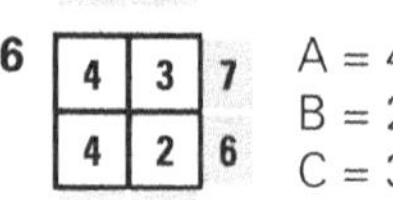

4	3	7
4	2	6
8	5	

A = 4
B = 2
C = 3

7

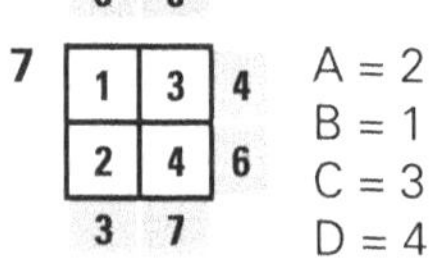

1	3	4
2	4	6
3	7	

A = 2
B = 1
C = 3
D = 4

8

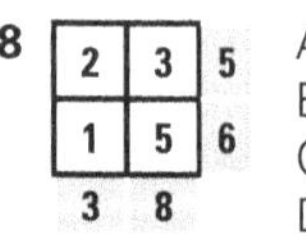

2	3	5
1	5	6
3	8	

A = 2
B = 3
C = 1
D = 5

9

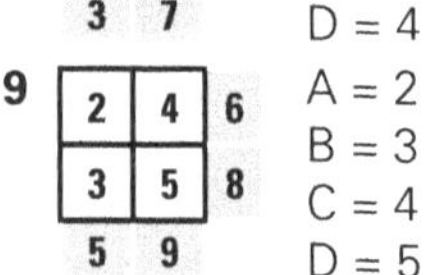

2	4	6
3	5	8
5	9	

A = 2
B = 3
C = 4
D = 5

10

5	1	6
3	4	7
8	5	

A = 5
B = 1
C = 4
D = 3

In the first six questions, students should start with the row or column that uses a letter twice.

Find my rule (page 8)

	RULE		Last box		
1	−3	×2	5	2	4
2	+4	×2	2	6	12
3	×2	−7	10	20	13
4	+9	−4	5	14	10
5	−6	+11	12	6	17
6	×3	−5	5	15	10
7	−4	×3	6	2	6
8	×3	×2	1	3	6
9	+2	×3	2	4	12
10	−9	×2	11	2	4

Students will need to know 2 and 3 times tables for these exercises.

What's the rule? (page 9)

	RULE		Last box		
1	×3	−4	8	24	20
2	×5	+2	4	20	22
3	−5	×3	6	1	3
4	+2	×5	6	8	40
5	+3	×10	7	10	100
6	×10	−7	6	60	53
7	×3	×2	4	12	24
8	×5	−11	4	20	9
9	×2	×5	4	8	40
10	−9	×3	12	3	9

Students will need to know 2, 3, 5 and 10 times tables for these exercises.

Which two numbers? (page 10)

1 a 1 + 4, 2 + 3
b 1 + 5, 2 + 4, 3 + 3
c 1 + 9, 2 + 8, 3 + 7, 4 + 6, 5 + 5
d 1 + 11, 2 + 10, 3 + 9, 4 + 8, 5 + 7, 6 + 6
e 1 + 14, 2 + 13, 3 + 12, 4 + 11, 5 + 10, 6 + 9, 7 + 8

2 a 1 × 6, 2 × 3
b 1 × 12, 2 × 6, 3 × 4
c 1 × 16, 2 × 8, 4 × 4
d 1 × 20, 2 × 10, 4 × 5
e 1 × 24, 2 × 12, 3 × 8, 4 × 6

3 a 4 − 1, 5 − 2, 6 − 3, 7 − 4, 8 − 5 and so on
b 6 − 1, 7 − 2, 8 − 3, 9 − 4, 10 − 5 and so on
c 7 − 1, 8 − 2, 9 − 3, 10 − 4, 11 − 5 and so on
d 11 − 1, 12 − 2, 13 − 3, 14 − 4, 15 − 5 and so on
e 18 − 1, 19 − 2, 20 − 3, 21 − 4, 22 − 5 and so on

Listing the possible solutions in a systematic way will ensure that all answers are given.

Find two numbers (page 11)

Students can do these questions by drawing up a table and listing all possibilities as suggested, or by using 'guess and check'. However, if they use 'guess and check', they should also be encouraged to record their trials in a table. They can also start with the fact that the product of the two numbers (in question 1) is 10. Therefore, the numbers can be 10 and 1 or 5 and 2.

1 2, 5 **2** 2, 8 **3** 3, 7 **4** 1, 9 **5** 2, 7
6 1, 8 **7** 2, 5 **8** 1, 7 **9** 3, 5 **10** 5, 6
11 4, 8 **12** 2,10 **13** 2, 3

	First number	Second number (1 more than first)	Sum	
1st guess	1	2	3	*Too low*
2nd guess	2	3	5	*Correct*

If students draw up a table and use 'guess and check', these relatively difficult questions should become easy to solve.

14 3, 5 **15** 2, 6 **16** 5, 7 **17** 5, 6 **18** 7, 8 **19** 9, 6
20 3, 6 **21** 6, 12 **22** 10, 20 **23** 7, 14 **24** 9, 18

Number puzzles (page 12)

Using the numbers 1 to 5, we are looking for two pairs that give the same sum; we can place the remaining number in the middle. There are three possibilities.

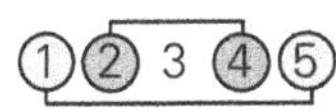

The three possibilities result in these solutions:

1 a 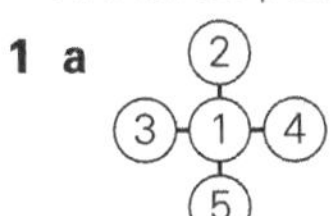 **b** 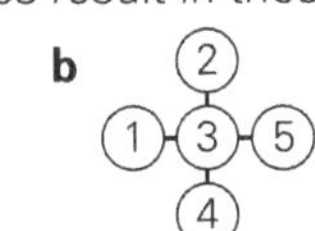 **c** 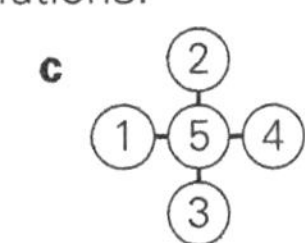

Naturally, we can swap the position of the number pairs, so that other possible solutions for **a** are:

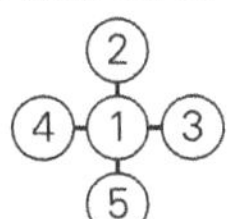 or 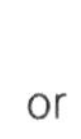 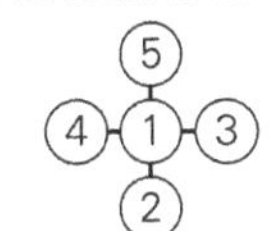 or 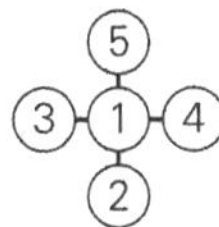

2 Since in **a** the sum of the numbers along each side is 12, students should place first 3 then 2 in the right-hand column. The remaining numbers to be arranged are 4, 5 and 6.

Since ◯ + ◯ + ② = 12, ◯ + ◯ = 10 and 4 and 6 are the only possibilities. Students can then use trial and error to find the way in which they will fit.

a 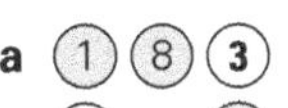 **b**  **c**   **d** 
 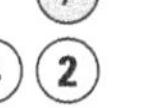   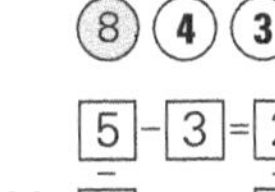

a			b			c			d		
1	8	3	6	3	4	5	6	3	6	2	7
5		7	2		8	1		7	1		5
6	4	2	5	7	1	8	2	4	8	4	3

3 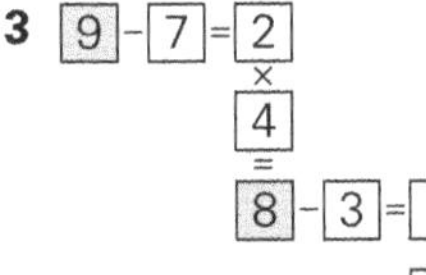 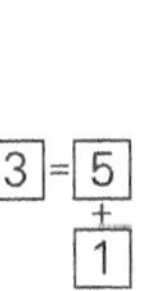

9 − 7 = 2; 2 × 4 = 8; 8 − 3 = 5; 5 + 1 = 6

4 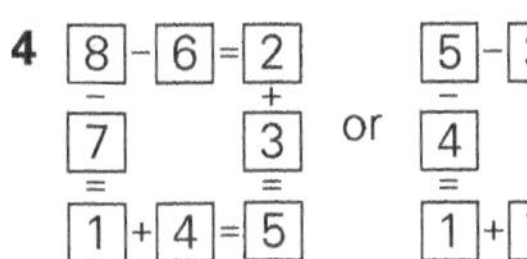

8 − 6 = 2; 8 − 7 = 1; 2 + 3 = 5; 1 + 4 = 5 or 5 − 3 = 2; 5 − 4 = 1; 2 + 6 = 8; 1 + 7 = 8

Magic squares (page 13)

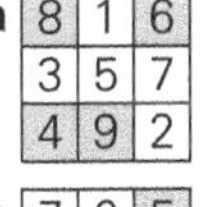
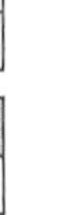
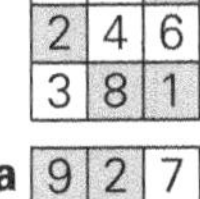
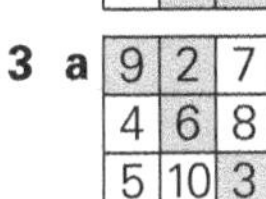
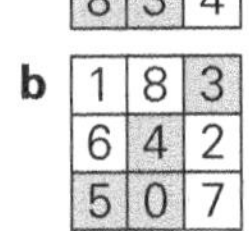
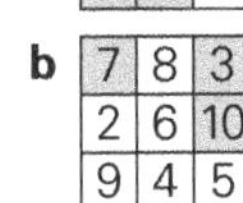
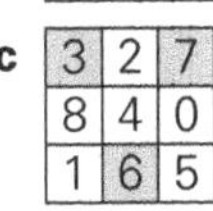

1 a

8	1	6
3	5	7
4	9	2

b

6	7	2
1	5	9
8	3	4

c

4	3	8
9	5	1
2	7	6

2 a

7	0	5
2	4	6
3	8	1

b

1	8	3
6	4	2
5	0	7

c

3	2	7
8	4	0
1	6	5

3 a

9	2	7
4	6	8
5	10	3

b

7	8	3
2	6	10
9	4	5

c

5	4	9
10	6	2
3	8	7

Patterns and sequences (page 14)

		Pattern			Pattern
1	10, 12, 14	+2	**2**	11, 13, 15	+ 2
3	23, 27, 31	+ 4	**4**	30, 37, 44	+ 7
5	50, 60, 70	+ 10	**6**	25, 30, 35	+ 5
7	23, 26, 29	+ 3	**8**	45, 55, 65	+ 10
9	33, 41, 49	+ 8	**10**	29, 31, 33	+ 2
11	10, 5, 0	− 5	**12**	42, 40, 38	− 2
13	50, 40, 30	− 10	**14**	11, 8, 5	− 3
15	26, 22, 18	− 4	**16**	28, 23, 18	− 5
17	20, 24, 28	+ 4	**18**	35, 42, 49	+ 7
19	10, 8, 6	− 2	**20**	56, 67, 78	+ 11
21	15, 18, 21	+ 3	**22**	30, 36, 42	+ 6
23	37, 46, 55	+ 9	**24**	53, 44, 35	− 9
25	35, 28, 21	− 7	**26**	29, 23, 17	− 6
27	86, 98, 110	+ 12	**28**	60, 45, 30	− 15
29	59, 72, 85	+ 13	**30**	56, 45, 34	− 11

Missing numbers (page 15)

		Pattern			Pattern			Pattern
1	9	+ 3	**2**	18	− 3	**3**	50	− 5
4	89	− 4	**5**	33	+ 4	**6**	67	+ 11

7 16 + 1, + 2, + 3, + 4, + 5 + 6, + 7
8 15 − 1, − 2, − 3, − 4, − 5, − 6, − 7
9 74 + 2, + 3, + 4, + 5, + 6, + 7
10 91 − 1, − 2, − 3, − 4, − 5, − 6
11 15 + 2, + 3, + 4, + 5, + 6, + 7, + 8
12 12 + 3, − 1, + 3, − 1, + 3, − 1 + 3, − 1
or 2 patterns 6, 8, 10, 12, 14 and 9, 11, 13, 15
13 30 2 patterns 2, 4, 6, 8, 10 and 10, 20, 30, 40
14 7 2 patterns 1, 3, 5, 7, 9 and 5, 10, 15, 20
15 13 2 patterns 3, 8, 13, 18, 23 and 10, 12, 14, 16
16 21 + 6, − 2, − 6, − 2, + 6, − 2, + 6, − 2
or 2 patterns 3, 7, 11, 15, 19 and 9, 13, 17, 21
17 77 2 patterns 97, 87, 77, 67, 57 and 95, 90, 85, 80
18 36 2 patterns 28, 29, 30, 31, 32 and 36, 35, 34, 33
19 29 2 patterns 39, 34, 29, 24, 19 and 7, 10, 13, 16
20 46 2 patterns 73, 64, 55, 46, 37 and 1, 5, 9, 13

Find three numbers (page 16)

1
△3 + △3 = 6 **Do 1st**
△3 + ② = [5]
[5] + ② = 7
[5] + [5] = 10 **Do 2nd**

2
△4 + △4 = 8 **Do 1st**
△4 + △4 + △4 = ⑥ + ⑥
[3] + [3] = ⑥ **Do 2nd** ↑

3
△7 + △7 = 14 **Do 1st**
△7 − ⑤ = 2 **Do 2nd**
△7 − [3] = 4
⑤ + △7 + [3] = 15

4
[6] + [6] = 12 **Do 1st**
△8 − [6] = ② **Do 2nd**
② + ② + ② = [6] ↓
△8 + ② = 10 **Do 3rd**

5
△4 + △4 + △4 = 12 **Do 1st**
△4 − [3] = 1 **Do 2nd**
[3] + [3] + [3] = ⑨
⑨ + △4 = 13

6
△2 + [6] = 8 **Do 2nd**
[6] + ⑩ = 16
△2 + ⑩ = 12 ← **or Do 2nd**
△2 + △2 + △2 + △2 = 8 **Do 1st** ↓

7
△6 + △6 = 12 **Do 1st**
△6 + ⑤ = [11]
△6 − ⑤ = 1
⑤ + ⑤ + ⑤ = 15 **Do 2nd** ↓

8
△9 + △9 = 18 **Do 1st**
△9 − [4] = 5 **Do 2nd**
⑫ − △9 = 3
⑫ − [4] = 8

Find three numbers (page 16) *continued...*

9 3 + 3 + 3 = 9 **Do 1st**
7 + 3 = 10 **Do 2nd**
13 − 7 = 6
13 − 3 = 10

10 6 + 6 + 6 = 18 **Do 1st**
11 + 11 = 22 **Do 2nd**
11 − 6 = 5
8 − 6 = 2

Equations with three unknowns are usually handled algebraically by 15-year-old students. However, by simple deductions and the 'guess and check' method they can be handled easily by these young children.

Which three numbers? (page 17)

1 6 + 6 = 12 **Do 1st**
6 + 4 = 10 **Do 2nd**
4 + 7 = 11

2 5 + 5 + 5 = 15 **Do 1st**
9 − 5 = 4 **Do 2nd**
10 − 9 = 1

3 3 + 3 + 3 = 9 **Do 1st**
12 − 3 = 9
21 − 12 = 9

4 4 + 2 = 6
4 − 2 = 2
2 + 2 = 4
1 + 2 = 3

Find two numbers whose sum is 6 and difference is 2.

5 9 + 3 = 12
9 − 3 = 6
3 + 3 + 3 = 9
1 + 9 = 10

6 4 + 4 = 8
4 + 8 = 12
2 + 2 = 4
2 + 8 = 14

7 2 + 2 + 2 = 3 + 3
2 + 3 = 5
8 − 3 = 5

8 5 + 3 = 8
5 − 3 = 2
3 + 3 = 6
6 − 5 = 1

9 4 + 3 = 7
4 + 4 + 4 = 3 + 3 + 3 + 3
4 + 4 = 8
8 − 3 = 5

10 5 + 5 = 9 + 1
9 − 5 = 4
9 + 5 = 14
3 + 3 + 3 = 9
9 − 3 = 6

Arrange the numbers (page 18)

1 a

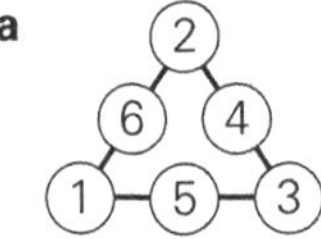

b

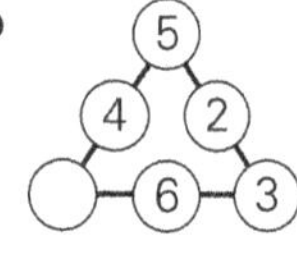

c

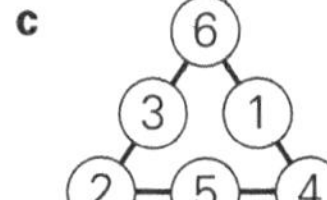

d

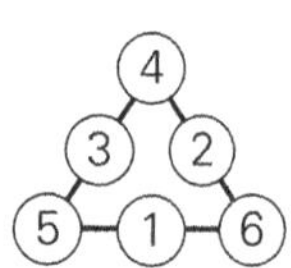

Note that six variations of this one basic solution can be found by rotating and reflecting the triangle.

2

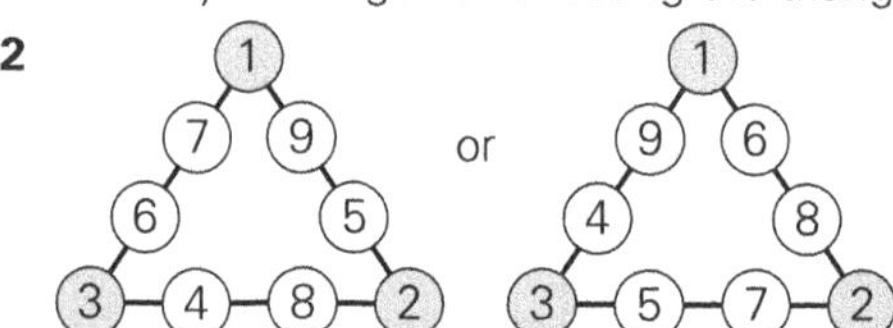

3 a 2, 3, 4, 7, 6, 5 around centre 1

b

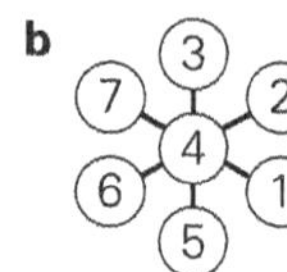

4 a 2, 3, 4, 5, 9, 8, 7, 6 around centre 1

b

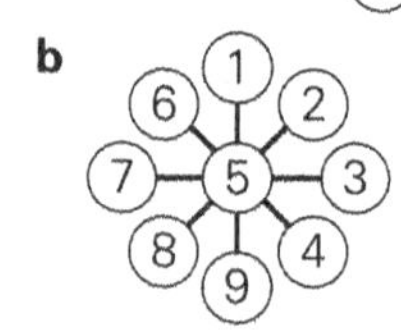

c 1, 2, 3, 4, 8, 7, 6, 5 around centre 9

There are many other possible solutions for questions 3 and 4, as the order of the lines can vary. Only the centre numbers must remain the same.

What's the value? (page 19)

1 × 1 = 1
1 + 1 = 2
2 × 2 = 2 + 2 = 4
2 + 0 = 2
1 + 1 + 1 = 3
3 × 2 = 6
2 + 3 = 5
5 + 1 = 6
5 + 2 = 7
7 + 2 = 9
3 + 3 + 3 = 9
9 − 1 = 8
8 ÷ 2 = 4

What's the question? (page 20)

The numbers given here are only some of the possible solutions—there are many, many more.

1 a 7 + 1, 6 + 2, 5 + 3, 4 + 4

b 3 + 2 + 2 + 1, 4 + 2 + 1 + 1, 3 + 3 + 1 + 1, 5 + 1 + 1 + 1

c 9 − 1, 10 − 2, 11 − 3, 12 − 4, 20 − 12

d 2 × 3 + 7 − 1 − 4, 10 × 2 + 5 − 9 − 8, 5 + 4 + 3 + 2 − 6

e For example, share 32 sweets among 4 children; *or* 16 ÷ 2, 24 ÷ 3, 40 ÷ 5

f $\frac{1}{2} \times 16$, $\frac{1}{3} \times 24$, $\frac{1}{5} \times 40$, $\frac{1}{10} \times 80$, $\frac{2}{5} \times 20$, $\frac{1}{4} \times 32$, $\frac{1}{6} \times 48$

2 a

10	11	12	13	14	15	16	17
+ 8	+ 7	+ 6	+ 5	+ 4	+ 3	+ 2	+ 1
18	18	18	18	18	18	18	18

b

11	12	13	14	15	16	17	18	19
+ 9	+ 8	+ 7	+ 6	+ 5	+ 4	+ 3	+ 2	+ 1
20	20	20	20	20	20	20	20	20

c

23	22	21	20	19	18	17	16	15
+ 1	+ 2	+ 3	+ 4	+ 5	+ 6	+ 7	+ 8	+ 9
24	24	24	24	24	24	24	24	24

3 $34 + 21 = 55$ $31 + 24 = 55$ $43 + 12 = 55$ $13 + 42 = 55$ **and** reversing rows: $21 + 34 = 55$ $24 + 31 = 55$ $12 + 43 = 55$ $42 + 13 = 55$

4 $65 + 34 = 99$ $64 + 35 = 99$ $56 + 43 = 99$ $46 + 53 = 99$ $34 + 65 = 99$ $35 + 64 = 99$ $43 + 56 = 99$ $53 + 46 = 99$

Past numbers (page 21)

1 a i 3 **ii** 20 **iii** 11 **iv** 33 **v** 15

b i YYYY

ii <YYY

iii <<YYYYYY

iv <<<

v <<<<YYYYY

2 a i 4 **ii** 6 **iii** 12 **iv** 15 **v** 19

b i $\bullet\bullet$ **ii** $\underline{\bullet\bullet}$ **iii** $\underline{\bullet\bullet\bullet\bullet}$ **iv** $\underline{\underline{\bullet}}$ **v** $\underline{\underline{\underline{\bullet}}}$

Continue the pattern (page 22)

1 $5 \times 9 = 50 - 5$
$6 \times 9 = 60 - 6$

2 $5 \times 8 = 50 - 10$
$6 \times 8 = 60 - 12$

3 $1 + 2 + 3 + 4 + 5 + 6 + 5 + 4 + 3 + 2 + 1 = 36 = 6 \times 6$
$1 + 2 + 3 + 4 + 5 + 6 + 7 + 6 + 5 + 4 + 3 + 2 + 1 = 49 = 7 \times 7$

4 $1 + 2 + 3 + 4 + 5 = 6 + 7 + 8 + 9 + 10 - 5 \times 5$
$1 + 2 + 3 + 4 + 5 + 6 = 7 + 8 + 9 + 10 + 11 + 12 - 6 \times 6$

5 $16 + 17 + 18 + 19 + 20 = 21 + 22 + 23 + 24$
$25 + 26 + 27 + 28 + 29 + 30 = 31 + 32 + 33 + 34 + 35$

6 $(5 - 4) \times (5 + 4) = 9 = 5 + 4$
$(6 - 5) \times (6 + 5) = 11 = 6 + 5$

7 $(5 - 1) \times (5 + 1) = 24$
$(6 - 1) \times (6 + 1) = 35$

8 $(5 - 3) \times (5 + 3) = 2 \times 8$
$(6 - 4) \times (6 + 4) = 2 \times 10$

Shapes and values (page 23)

1

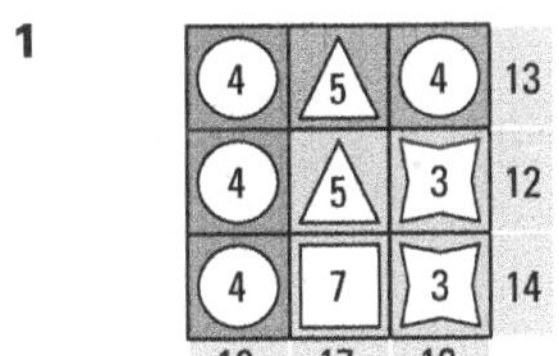

2

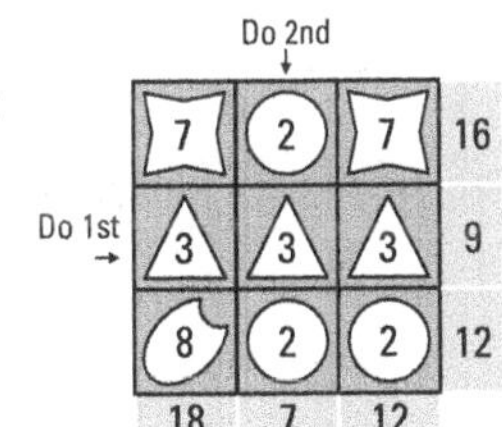

3

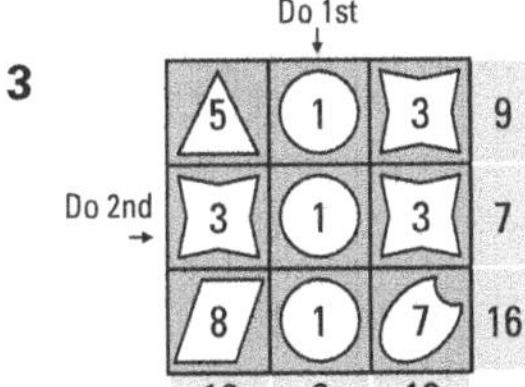

4

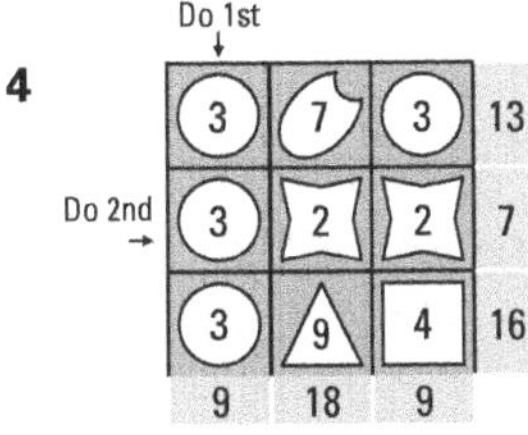

5

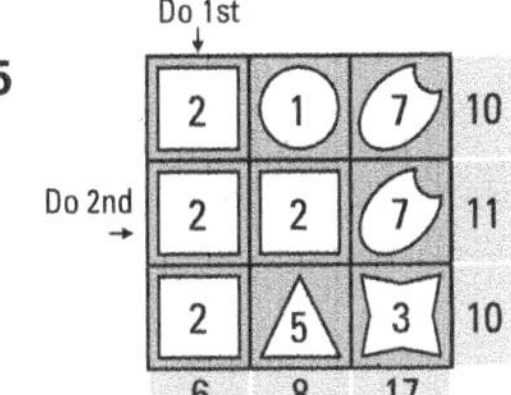

6

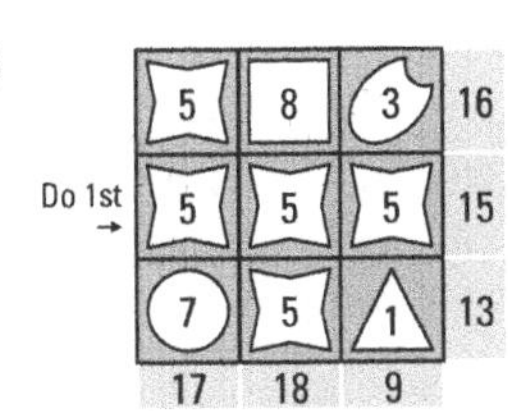

7

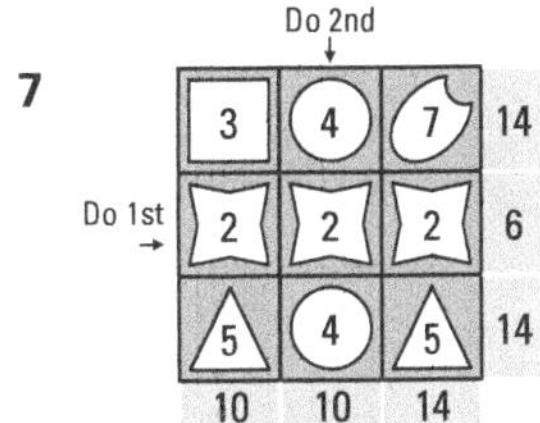

8

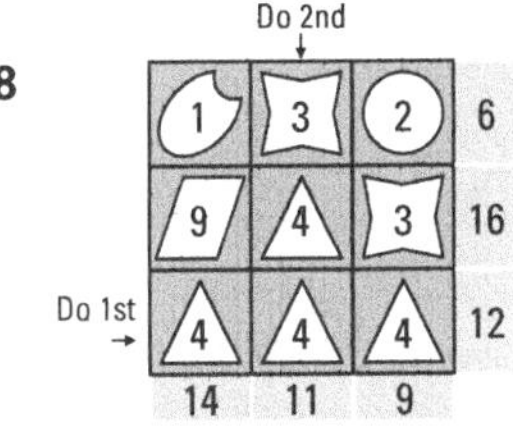

9

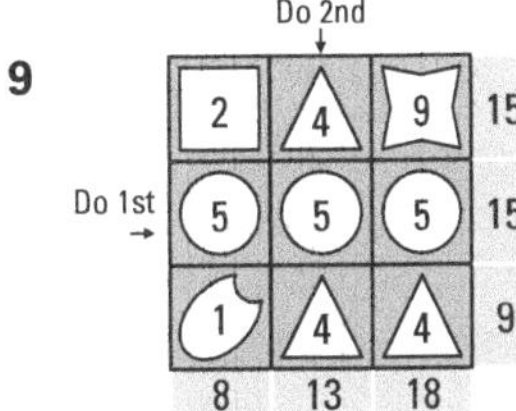

10

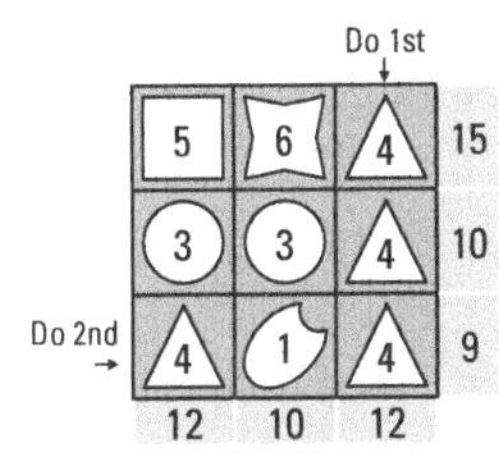

These questions develop logical thinking. Students should first work out the value of the shape that occurs three times in a row or column, then they will have to decide through careful observation which shape they will find the value of next.

Related rows (page 24)

Students will need to know 2, 3, 4, 5 and 10 times tables in order to do these exercises. If they are not familiar with square numbers, leave out sets 17–20.

		Rule			Rule			Rule
1	11	+ 5	**2**	4	× 2	**3**	4	− 8
4	50	× 10	**5**	8	× 4	**6**	15	+ 6
7	5	× 5	**8**	12	× 2	**9**	13	× 2 + 1
10	3	× 2 − 1	**11**	49	× 10 − 1	**12**	9	× 5 − 1
13	24	× 4	**14**	25	× 4 + 1	**15**	7	× 3 + 1
16	16	× 5 + 1	**17**	16	square	**18**	15	square − 1
19	3	− 5	**20**	26	square + 1			

Differences (page 25)

With this puzzle, students can be happily subtracting for hours. They should notice that the square before the final one will have four identical numbers at the vertices.

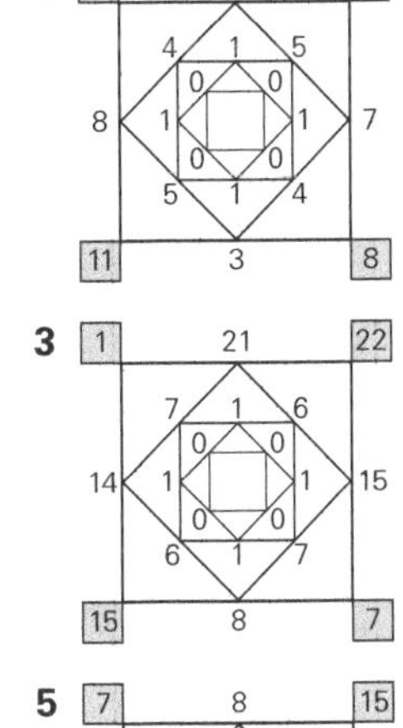

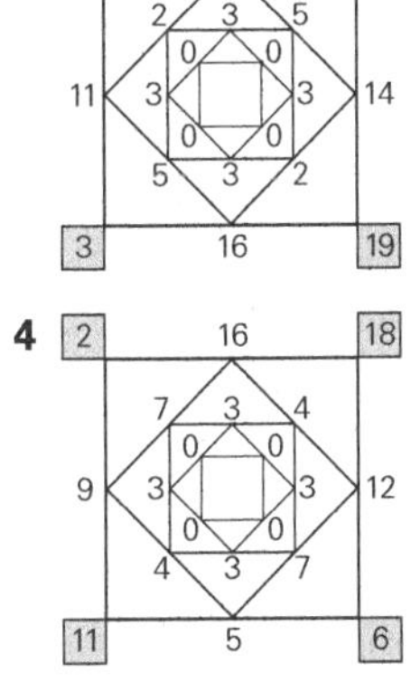

5

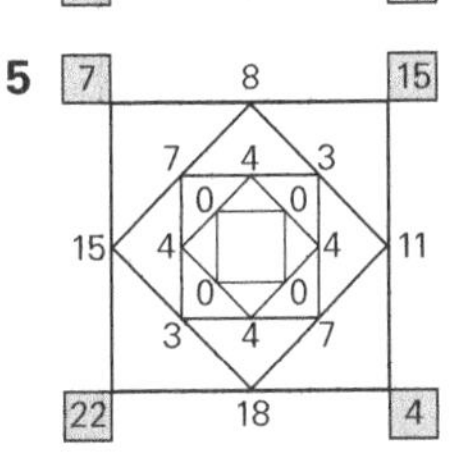

6

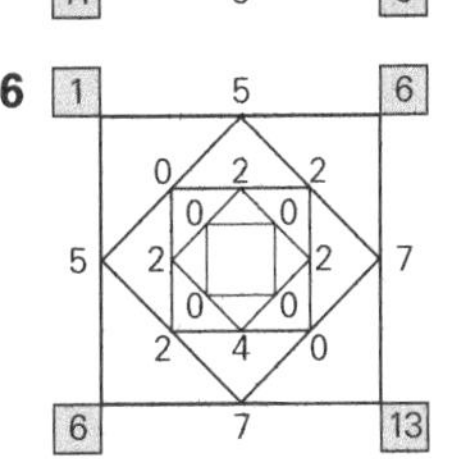

Differences (page 25) ***continued...***

7 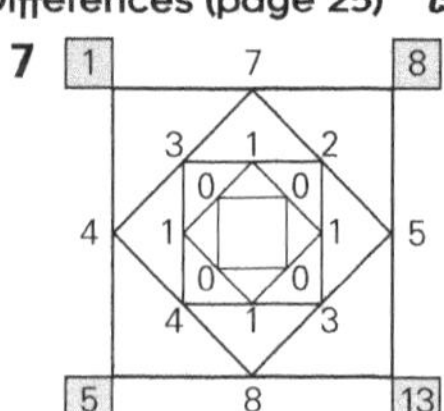

8 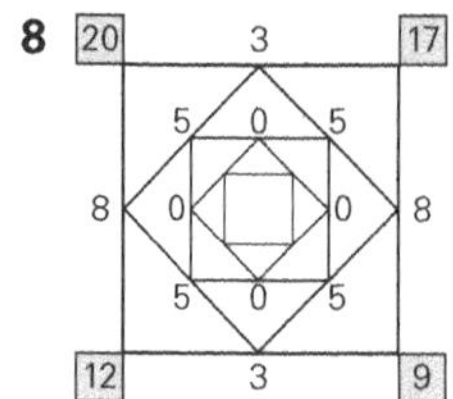

9 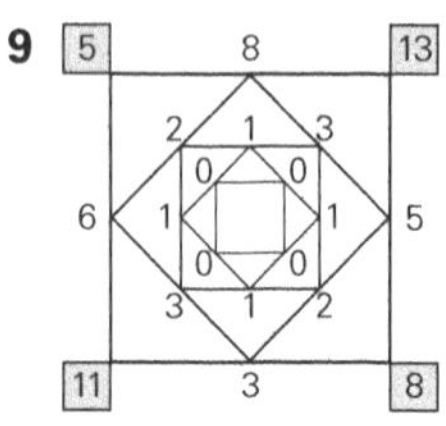

Students can make up their own questions. They will have to choose their own corner numbers.
Note that some groups of corner numbers will give sequences that are longer than 5 squares.

Number line (page 26)

1 17 **2** 170 **3** 1700 **4** 147
5 227 **6** 1170 **7** 464 **8** 31

Students must note the scale very carefully in each question.

Find the value (page 27)

1 (10 – 9) + (8 – 7) + (6 – 5) + (4 – 3) + (2 – 1) = 5 × 1 = 5
2 (100 – 90) + (80 – 70) + (60 – 50) + (40 – 30) + (20 – 10) = 5 × 10 = 50
3 (120 – 100) + (80 – 60) + (40 – 20) = 3 × 20 = 60
4 (40 – 35) + (30 – 25) + (20 – 15) + (10 – 5) = 4 × 5 = 20
5 (18 – 15) + (12 – 9) + (6 – 3) = 3 × 3 = 9
6 (24 – 20) + (16 – 12) + (8 – 4) = 3 × 4 = 12
7 (100 – 95) + (90 – 85) + (80 – 75) + (70 – 65) + (60 – 55) = 5 × 5 = 25
8 (600 – 500) + (400 – 300) + (200 – 100) = 3 × 100 = 300
9 (150 – 125) + (100 – 75) + (50 – 25) = 3 × 25 = 75
10 (300 – 270) + (240 – 210) + (180 – 150) + (120 – 90) + (60 – 30) = 5 × 30 = 150

162

Counting-frame arithmetic (page 28)

1 a 2 **b** 200 **c** 200 **d** 101
2 a 12 **b** 12 **c** 3 **d** 300
e 30 **f** 300 **g** 201
3 a 4 **b** 13 **c** 22 **d** 31
e 40 **f** 103 **g** 112 **h** 121
i 130 **j** 202 **k** 211 **l** 220
m 301 **n** 310 **o** 400

Shading parts (page 29)

Some possible solutions are drawn below. All the shadings can be done in reverse for question 1, so that the white parts can be shaded and the shaded parts left white. Also, all the shapes can be turned. Students can turn this book 90°, so that ◫ is the same as ⊟ in the first question.

1 There are a very large number of solutions.
Please remember that this is only a sample.

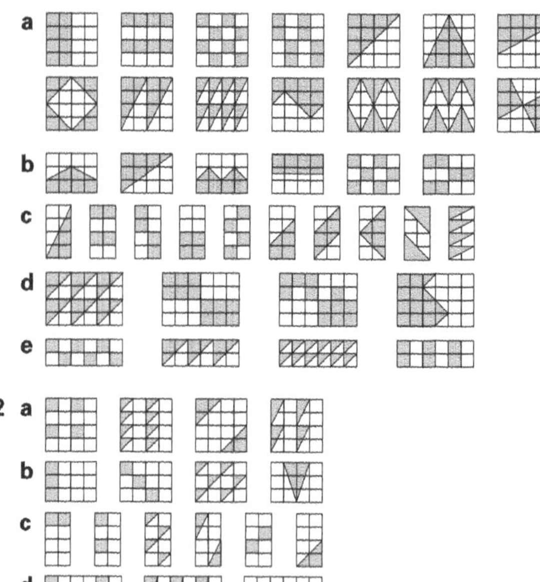

2 a
b
c
d
e

What fraction's shaded? (page 30)

A solution is given for each question. However, there are many possible solutions.

1 $\frac{1}{2}$ **2** $\frac{1}{2}$ **3** $\frac{1}{2}$ **4** $\frac{1}{4}$ **5** $\frac{1}{2}$
6 $\frac{3}{8}$ **7** $\frac{1}{2}$ **8** $\frac{3}{8}$ **9** $\frac{1}{2}$ **10** $\frac{5}{8}$

How old? (page 31)

1 Douglas is 5 years old now. In 3 years' time he will be 8.
2 Nadia is 19 now. In 5 years' time she will be 24.
3

Andrew	Tina	Jeremy	Sum of ages
11	7	16	34

Andrew is 11.

4

Brenda	Jacob	Tammy	Sum of ages
8	4	10	22

5 The twins are 6 years old.
6 Each triplet is 5 years old.
7

Lydia	Yvonne	Greg	Barbara
7	8	14	4

Coins (page 32)

Students must work systematically to ensure they list all possible amounts.

1 20c, 40c, 60c, 5c, 25c, 45c, 65c
2 10c, 20c, 30c, 40c, 50c, 5c, 15c, 25c, 35c, 45c, 55c
3 20c, 40c, 50c, 70c, 90c
4 10c, 50c, 60c, \$1, \$1.10, \$1.50, \$1.60, \$2, \$2.10
5 10c, 20c, 50c, 60c, 70c, \$1, \$1.10, \$1.20, \$1.50, \$1.60, \$1.70

Money problems (page 33)

1 5 **2** 8

3 Expressing the answer as a table, and working systematically, will help to make sure that all possible ways are found.

20c	10c	5c
2	–	–
1	2	–
1	1	2
1	0	4
–	4	–
–	3	2
–	2	4
–	1	6
–	–	8

4 a \$3.00 **b** 75c **c** \$2.25

5 a \$2.40 **b** \$3.60 **c** 40c **d** 80c

6 Sheila spent \$3 on lunch. She had \$3 left. She then spent \$1.50 on a magazine. She now has \$1.50.

7 The strategy to use in this problem is to work backwards. Melanie had \$4 before she put half of it in the bank. She had \$8 before she spent half of it. Melanie's pocket money is \$8.

8 a 6 lollies, 10c change

b 4 biscuits, 12c change

How much money? (page 34)

1 a 18 **b** 7 **c** 27

2 a 40c **b** 50c

3 \$5.90

	Week 1	Week 2	Week 3	Week 4	Total
Spent	\$1.80	70c	\$2.10	\$1.50	
Saved	\$1.20	\$2.30	\$0.90	\$1.50	\$5.90

4 a 50c + 20c + 20c, 50c + 20c + 10c + 10c

b \$1 + 50c + 10c, \$1 + 20c + 20c + 10c + 10c, 50c + 50c + 20c + 20c + 10c + 10c

5 \$1.60 Total spent = \$2 + \$1.40 = \$3.40
Change = \$5 – \$3.40 = \$1.60

Missing shapes (page 35)

It will be helpful in these exercises if students repeat the patterns orally, as suggested in the first two solutions below.

1 The pattern is 'square, circle, square, two circles', and then this is repeated.

2 The pattern is 'oval, square, square', and this is repeated.

3 **4** **5** **6** 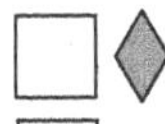

7 **8** **9** **10**

Patterns with squares (page 36)

1 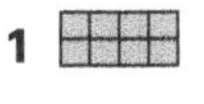 **2** 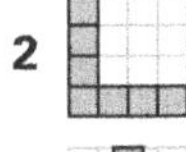 **3** 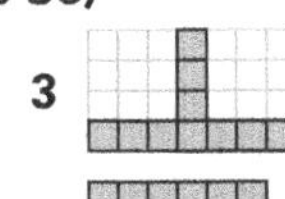 **4** 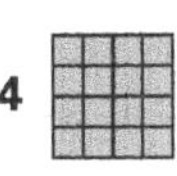

5 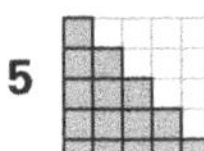 **6** 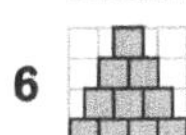 **7** 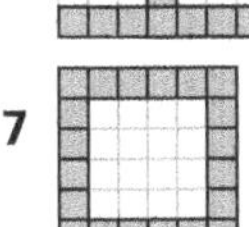

8

	1st shape	2nd shape	3rd shape	4th shape	5th shape
1	2	4	6	8	10
2	1	3	5	7	9
3	1	4	7	10	13
4	1	4	9	16	25
5	3	6	10	15	21
6	1	3	6	10	15
7	8	12	16	20	24

Faces (page 37)

In these exercises, it is necessary to work systematically.

1 It is worthwhile to discuss with students how they intend to draw the 12 faces. If they have not previously come across problems that involve working systematically, the procedure should be explained.
In this case, it would be best to draw first all the possible faces with, say, curly hair and eyes closed, then those with curly hair and eyes open, and then all possible combinations with straight hair.
Of course, all the possible combinations with a happy mouth could equally have been drawn first. The important thing is to work to a plan—to establish an order in which the features will be used, not to select them haphazardly. Without a systematic approach, it is extremely difficult to work out what is missing and to ensure that all possibilities have been accounted for.

2 

Building blocks (page 38)

1 Y/Y/Y, Y/R/R, R/R/R, R/R/Y, R/Y/R, R/Y/Y

2 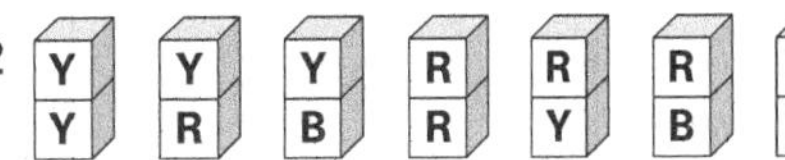

It is important to work systematically.

Shading circles (page 39)

Working systematically is essential in this exercise. It is important to discuss shading combinations with students, explaining which are different and which are not.

1

2 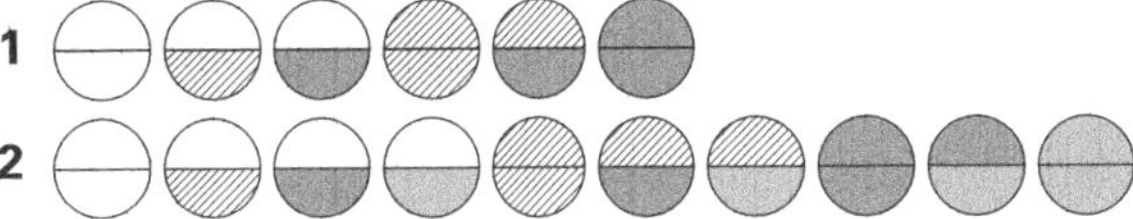

Shading sectors (page 40)

1 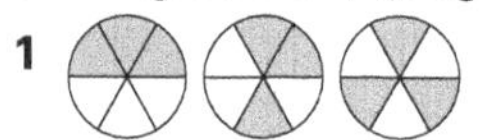

2 6 patterns:

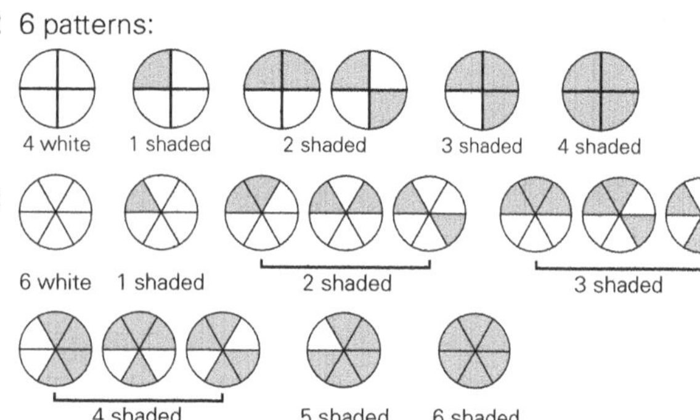

The difficulty students have with this exercise is in recognising that what appears to be a different pattern is the same as an existing one seen from a different angle. It might be helpful to cut out circles for them to work with. They will then have a chance to turn the shaded circles around and check that all are different.

Colour these flags (page 41)

Students will need guidance to work systematically in these questions.

1 

2 Altogether there are 27 possible colour combinations. Given here are the remaining 24.

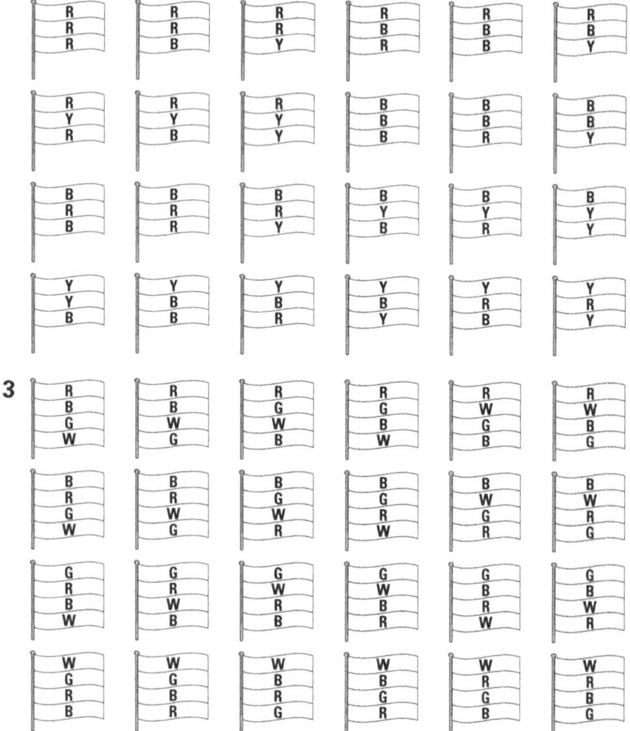

Chance (page 42)

1 a 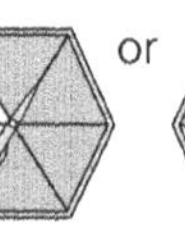 b 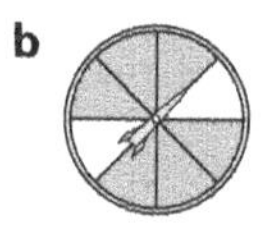 c 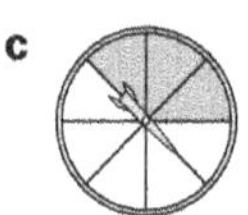

2 a 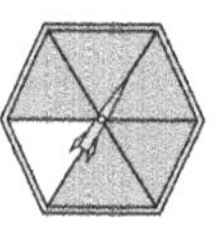 or 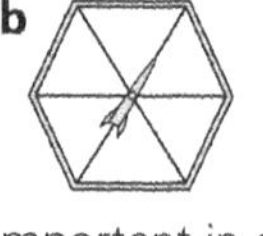 b c 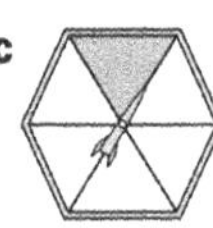

Note: the number of sectors is important in questions 1 and 2, not their positions.

3 **a** one chance in five or $\frac{1}{5}$ **b** two in five chance or $\frac{2}{5}$
c three in five chance or $\frac{3}{5}$ **d** zero

4 yellow, green, red, blue

How many routes? (page 43)

It is essential that students work systematically when doing these exercises. They need to adopt a system such as 'First mark all routes that move up (or go north)', although that may not necessarily be the choice.

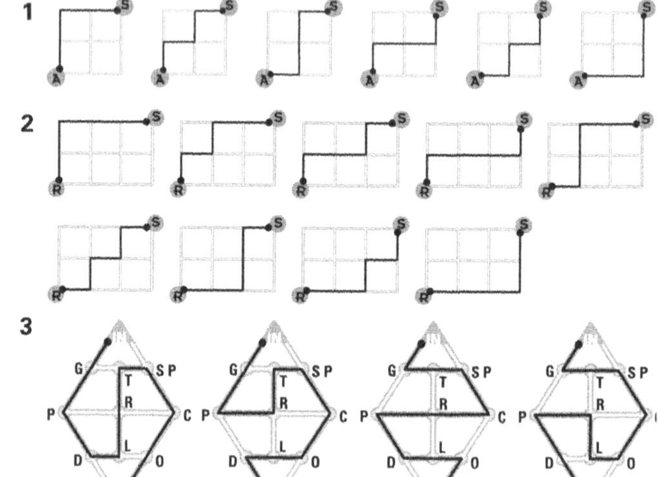

Vegetable patch (page 44)

1

Front	
P	C
Z	T
O	L
M	S
Back	

2 There are many different ways to describe the position of the flowers in this garden. Your teacher can check your work.

Graphs (page 45)

1 **a** between 2 p.m. and 3 p.m.
b between noon and 1 p.m.
c 40 **d** between 11 a.m. and noon
e 10 a.m.–11 a.m. and 1 p.m.–2 p.m.
f 70 **g** 320 **h** $640

2 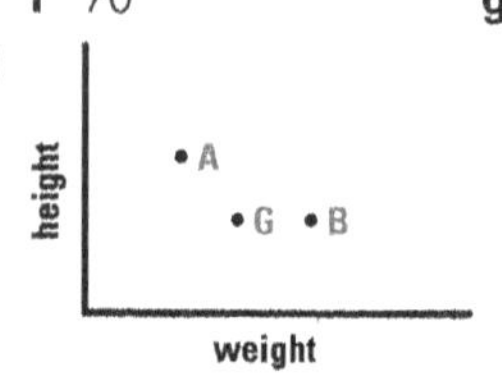

Working with graphs (page 46)

1 a 5 cm **b** 20 cm **c** 30 cm
2 a 30 **b** 15
3 a 24 **b** 48

Temperature readings (page 47)

1 a 25 **b** 26 **c** 24 **d** 29
2 a 31 **b** 33 **c** 30 **d** 30
3 a 32 **b** 30 **c** 33 **d** 32

Temperature challenges (page 48)

1 a 10 a.m. to 11 a.m. **b** 11 a.m. to noon **c** 12°C
d 14°C **e** 8°C **f** 9 a.m.
2 a 11 a.m. **b** highest 26°C and lowest 18°C
c 23°C **d** between 10 a.m. and 11 a.m.
e 9 a.m. and 1 p.m.
f Between 8 a.m. and 9 a.m. the temperature rose 2°C.
Between 9 a.m. and 10 a.m. the temperature rose 2°C.
Between 10 a.m. and 11 a.m. the temperature rose 4°C.
Between 11 a.m. and 1 p.m. the temperature dropped 6°C.
Between 1 p.m. and 3 p.m. the temperature rose 2°C.
Between 3 p.m. and 4 p.m. the temperature didn't change.

Find the length (page 49)

1 a 6 cm **b** 9 cm **c** 4 cm **d** 5 cm
2 a 7 cm, 2 cm **b** 13 cm, 4 cm **c** 7 cm, 9 cm

Measuring lengths (page 50)

1

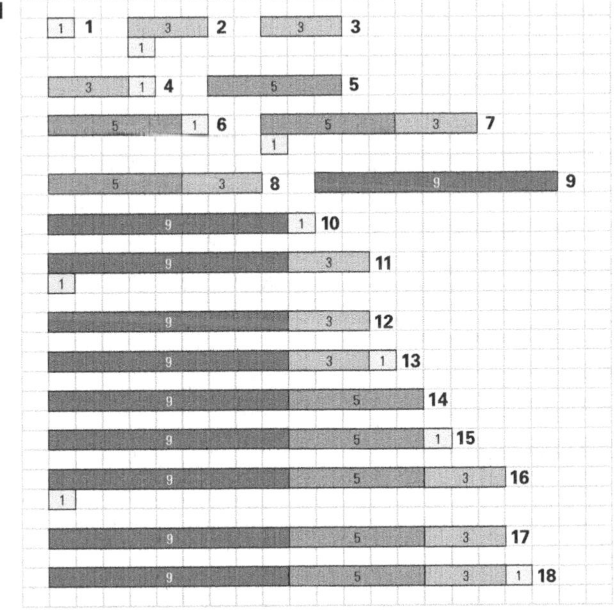

Note that we can write these solutions in the following way as well.

2 = 3 − 1 4 = 3 + 1 6 = 5 + 1
7 = 3 + 5 − 1 8 = 3 + 5 10 = 9 + 1
11 = 9 + 3 − 1 12 = 9 + 3 13 = 9 + 3 + 1
14 = 9 + 5 15 = 9 + 5 + 1
16 = 9 + 5 + 3 − 1 17 = 9 + 5 + 3
18 = 9 + 5 + 3 + 1

2 5 = 7 + 1 − 3 6 = 7 − 1 8 = 7 + 1
9 = 7 + 3 − 1 10 = 7 + 3 11 = 7 + 3 + 1
12 = 5 − 3 13 = 15 + 1 − 3 14 = 15 − 1
16 = 15 + 1 17 = 15 + 3 − 1 18 = 15 + 3
19 = 15 + 3 + 1 20 = 15 + 7 + 1 − 3
21 = 15 + 7 − 1 22 = 15 + 7 23 = 15 + 7 + 1
24 = 15 + 7 + 3 − 1 25 = 15 + 7 + 3
26 = 15 + 7 + 3 + 1

How many tiles? (page 51)

1 12 + 12 + 4 + 4 = 32
or 10 × 2 + 4 x 2 + 4 = 32
corners
or 10 + 10 + 6 + 6 = 32

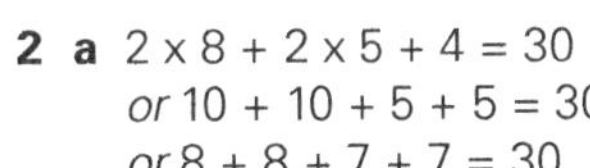

2 a 2 x 8 + 2 x 5 + 4 = 30
or 10 + 10 + 5 + 5 = 30
or 8 + 8 + 7 + 7 = 30

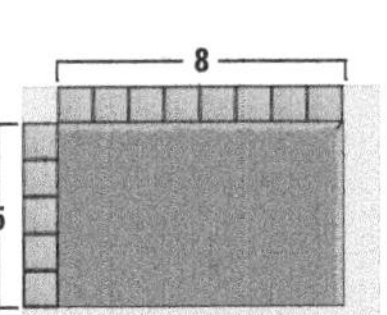

b 2 x 5 + 2 x 4 + 4 = 22
or 7 + 7 + 4 + 4 = 22
or 5 + 5 + 6 + 6 = 22

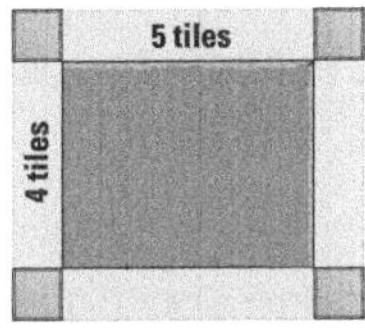

c 2 x 6 + 2 x 5 + 4 = 26
or 8 + 8 + 5 + 5 = 26
or 6 + 6 + 7 + 7 = 26

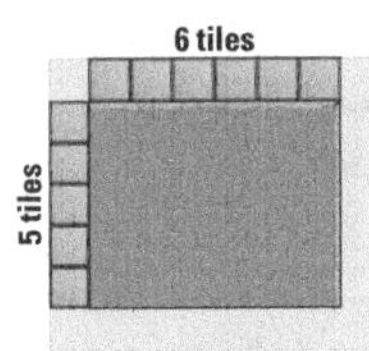

Time puzzles (page 52)

1 a Monday **b** Monday **c** Wednesday
2 If two days ago it was Wednesday, today is Friday.
a Friday **b** Friday **c** Friday
3 There are 60 minutes in 1 hour, 30 minutes in $\frac{1}{2}$ hour, 15 minutes in $\frac{1}{4}$ hour, 45 minutes in $\frac{3}{4}$ hour.
a 90 minutes **b** 195 minutes
4 a 10:20 a.m. **b** 7:45 a.m. **c** 1 hour 40 minutes
5 a dressed: 7:17 **b** breakfast eaten: 7:26
c room cleaned: 7:32 **d** clarinet practised: 7:57
e school reached: 8:17 **f** 13 minutes
6 a 1 October **b** 5 November

Time challenges (page 53)

1 a February, March, July, September
b Saturday, Sunday, Tuesday, Wednesday
c summer, autumn, winter, spring
2 a 11:59 **b** 12:03 **c** 11:48
3 a 145 seconds **b** 1 min 50 s
4 a 270 m (90 m in 1 min) **b** 60 litres (12 L in 1 min)
5 23 min (late train arrives at 2:10 p.m., early train at 2:33 p.m.)

Time challenges (page 53) *continued...*

6 10:45 a.m. (planes leave at 10:00 a.m., 10:15 a.m., 10:30 a.m., 10:45 a.m., 11:00 a.m.)

7 a 15 s + 45 s
15 s + 15 s + 30 s
15 s + 15 s + 15 s + 15 s

b 45 s + 45 s
45 s + 15 s + 30 s
30 s + 30 s + 15 s + 15 s
45 s + 15 s + 15 s + 15 s
30 s + 15 s + 15 s + 15 s + 15 s
15 s + 15 s + 15 s + 15 s + 15 s + 15 s

Train timetable (page 54)

1 21 minutes 2 11 minutes 3 25 minutes
4 a 7 minutes b 7:47
5 7:15 6 13 minutes

Read the dial (page 55)

1 a 20 L b 10 L c 30 L
2 a 30 L b 15 L c 20 L
3 a 18 L b 60 L c 27 L
4 a 25 L b 25 L c 75 L

Weighing parcels (page 56)

1 14 kg, 2 kg, 8 kg 2 13 kg, 7 kg, 3 kg
3 18 kg, 3 kg, 5 kg 4 11 kg, 3 kg, 5 kg

Mass problem solving (page 57)

1

Larri's mass	Rhonda's mass	The sum
21 kg	17 kg	38 kg

2

Veronica's mass	Damien's mass	The sum
24 kg	27 kg	51 kg

3

Mass of parcel	Mass of book	Total mass
8 kg	2 kg	10 kg

4

Daniel's fruit	Peter's fruit	Total mass of fruit
7 kg	10 kg	17 kg

5

Mass of banana	Mass of drink	Total mass
200 g	300 g	500 g

6

Mass of apple	Mass of pear	Total mass
350 g	250 g	600 g

7

Mass of pear	Mass of pineapple	Total mass
0.2 kg = 200 g	1.8 kg = 1800 g	2 kg = 2000 g

Find the mass (page 58)

1 650 g 2 450 g
3 3.7 kg or 3 kg 700 g
4 850 g The right-hand scale shows 950 g.
Box + 100 g = 950 g. Therefore box = 850 g.
5 1.65 kg or 1 kg 650 g Jug + 50 g = 1 kg 700 g
Jug = 1 kg 650 g
6 2.45 kg or 2 kg 450 g sugar + 200 g = 2 kg 650 g
sugar = 2 kg 450 g

Travelling (page 59)

1 a 35 km b 18 km
c Charring Cross d
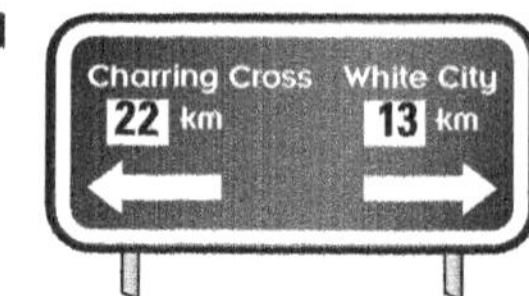

2 a 7
b level 1
c level 2

Count the shapes (page 60)

1 a 2 b 4 c 6 d 6 e 9 f 4
2 a 4 b 5 c 6 d 8 e 3
3 a 3 b 4 c 2 d 2 e 6 f 4
g 5 h 6

Counting cubes (page 61)

It is essential that students have access to little cubes—they may need them in order to answer question 1, and should definitely use them to answer question 2.

1 a 6 b 9 c 6 d 9 e 8 f 18
g 24 h 18 i 6 j 12 k 9 l 18
2 a 9 b 6 c 7

Packing boxes (page 62)

1 a l = 16 cm w = 9 cm h = 4 cm
b l = 12 cm w = 8 cm h = 9 cm
2 a 48 (6 cubes along length, 4 cubes along width, 2 cubes along height)
b 6 (3 cubes along length, 2 cubes along width, 1 cube high)

Rolling boxes (page 63)

This is an excellent exercise in visual perception. If students have the slightest difficulty doing it, they should definitely experiment with cut-out shapes.

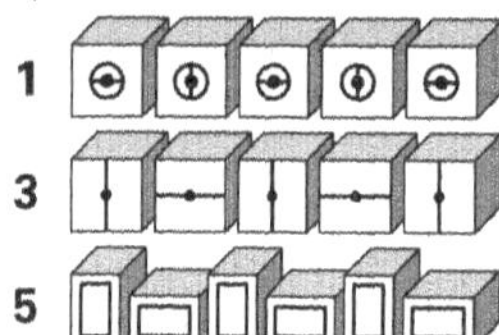
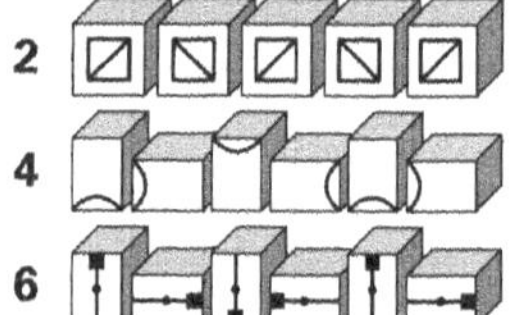

Torn rectangles (page 64)

1 8 2 12 3 14 4 15
5 18 6 9 7 20 8 18

Position (page 65)

1 a G8, G9, G10, G11, G12 or E6, E7, E8, E9, E10
b G13 c F10, F11, F12, F13

2   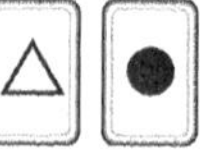

3 11
4 Niva, Joel, Gabriel, Asher and Emma

Tessellations (page 66)

These tessellations can be coloured in two different ways.

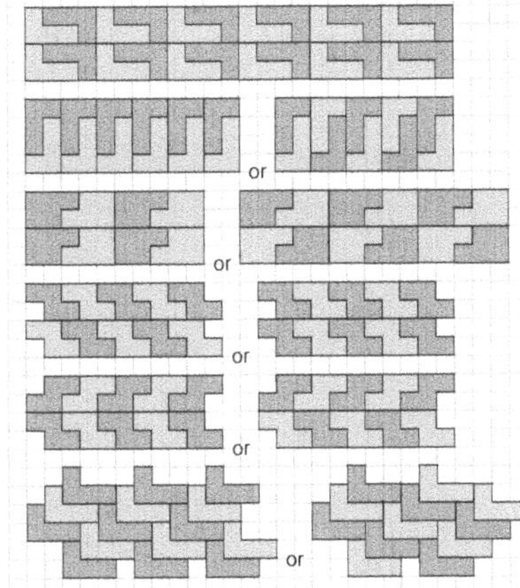

Solid shapes (page 67)

1 **A** rectangular prism **B** square pyramid
C hexagonal prism **D** cylinder
E cube **F** triangular prism
G pentagonal pyramid **H** cone

2 **a** D **b** C **c** H **d** G **e** F **f** G

3 6 **4** 12 **5** 5

Bearings (page 68)

1 **a** east **b** south **c** north

2 **a** 3 blocks south
b 5 blocks east
c 3 blocks south then 3 blocks west, or 3 blocks west then 3 blocks south
AND 1 block south then 5 blocks east, or 5 blocks east then 1 block south
d 4 blocks south then 3 blocks east, or 3 blocks east then 4 blocks south
e 4 blocks south then 2 blocks west, or 2 blocks west then 4 blocks south

Challenging cubes (page 69)

Students should make this net of a cube and construct the cube to help them visualise the different views.

1 **a**  **b**  **c** 
d    **e** 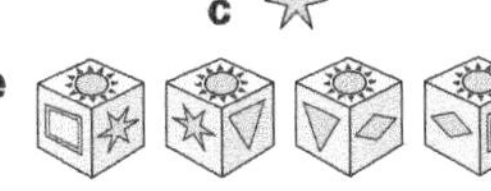

2 **a**  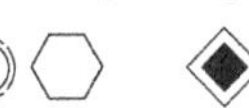 **b**  
c

Painting solids (page 70)

It would be helpful for students to build these solids and then work out the number of black faces.

1 4, 2, 6 **2** 4, 4, 1, 9
3 4, 6, 2, 12 **4** 8, 12, 4, 24

How many blocks? (page 71)

1 **a** 18 **b** 12 **c** 60 (Bottom layer has 30 blocks)
2 **a** 24 **b** 8 **c** 32 (Bottom layer has 16 blocks)

Amazing mathematics (page 72)

The best way to treat these puzzles is to call out the instructions and let the whole class follow them. The students will be fascinated by the fact that in each case the answer is always the same, regardless of the number thought of.

The exercises are set out below for the teacher's interest only—students are not expected to do this. However, if you call the number 'blob', very bright children can follow this.

1 The result is always 4. To work through the problem, let the original number be ▲.

▲
▲••••••
▲▲••••••••••••
▲▲••••••••
▲••••
••••

2 The result is always the number you originally thought of. Let the original number be ■.

■
■•••••
■■■•••••••••••••••
■■■••••••
■••
■

3 The result is always 5. Let the original number be ●.

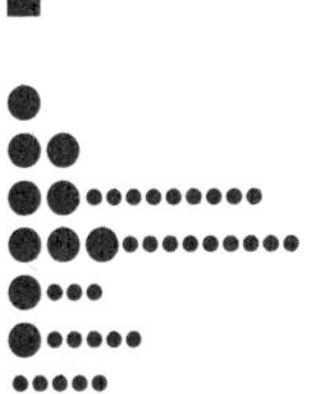

Logic using scales (page 73)

It would be helpful if students could do some experiments with a set of scales before tackling these exercises. Also, a discussion of the examples might help some of them.

1 ★★ Halve each side.
2 ★★★★ Halve each side.
3 ★★ Find one-third of each side.
4 ★★★ Find one-third of each side.
5 ★★ Remove one □ from each side.
6 ★★★★★ Remove one □ from each side.
7 ★ Remove one □ from each side, then halve.
8 ★★ Remove one □ from each side, then halve.
9 ★★★ Remove one ★ from each side.
10 ★★★★ Remove one □ and ★ from each side.
11 ★★★★ Remove one ★ from each side, then halve.
12 ★★★ Remove two ★s from each side, then halve.

Balanced scales (page 74)

1 ★★★★★★ Use 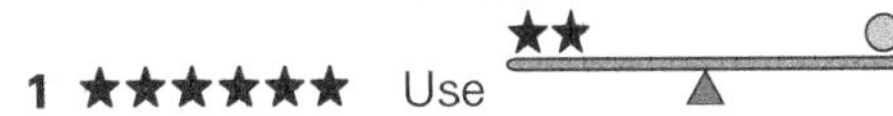
2 ★★★★★★★★★ Triple first scale.
3 ★★★★ Replace ○ with ★★★ on first scale.
4 ★★★★★ Replace □ with ★★★ on first scale.
5 ★★★★ Replace each ○ with ★★ on first scale.
6 ★★ Replace □ with ★★★ on first scale:

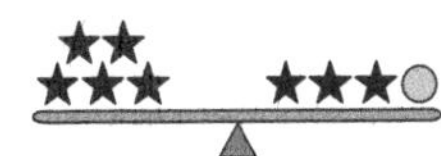
Now remove ★★★ from both sides.

Balanced scales (page 74) ***continued...***

7 ★★ Replace each ○ with ★ on first scale:

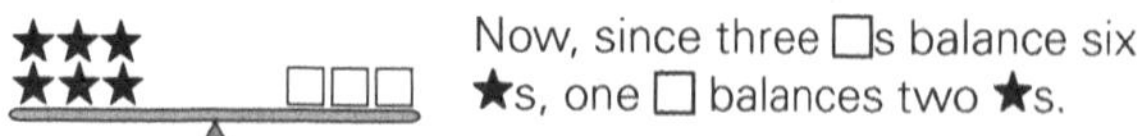

Now remove ★ from both sides.

8 ★★★ Replace each ○ with ★★ on first scale:

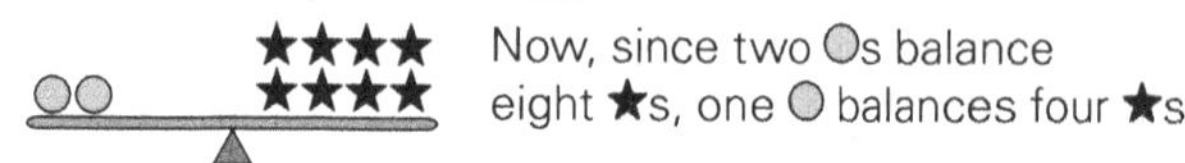

Now, since two □s balance six ★s, one □ balances three ★s.

9 ★★ Replace each ○ with ★★★:

Now, since three □s balance six ★s, one □ balances two ★s.

10 ★★★★ Replace each □ with ★★★:

Now, since two ○s balance eight ★s, one ○ balances four ★s.

Shapes and symbols (page 75)

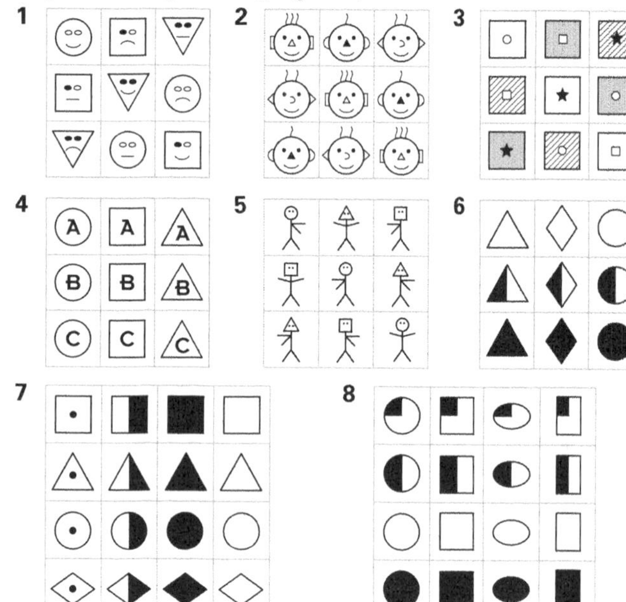

How many? (page 76)

Students could count the number of **X**s row by row or column by column. However, in each case it is easier if they look for a pattern, as suggested in the solutions, and they should be encouraged to do this.

1 If square is filled in: 49; missing in triangle shape: 10; therefore, number of **X**s = 39

2 If rectangle is filled in: 56; missing in square shapes: 8; therefore, number of **X**s = 48

3 In top triangle: 16; in lower triangle: 16; however, one is repeated; therefore, number of **X**s = 31

4 Five rows of four **X**s each = 20; number remaining = 6; therefore, number of **X**s = 26

5 If rectangle is filled in: 72; missing in triangle shapes: 17; therefore, number of **X**s = 55

6 Figure is symmetrical; number in top half = 20; therefore, number of **X**s = 40

Shaded figures (page 77)

The values of the large squares are written below them.

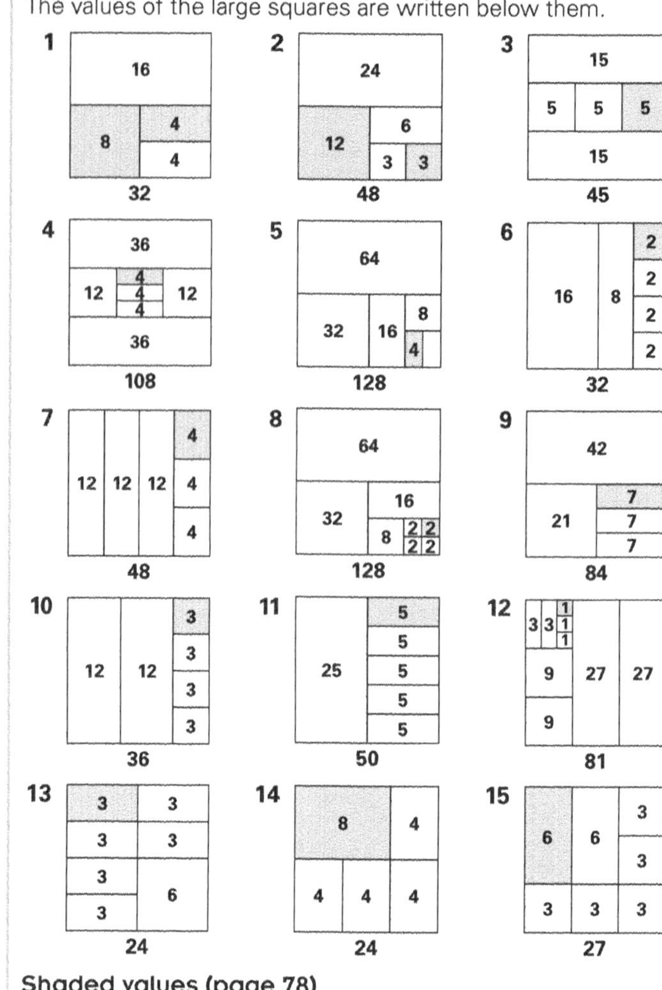

Shaded values (page 78)

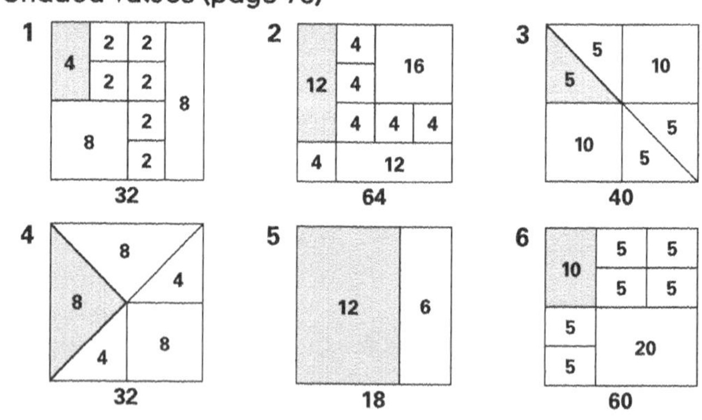

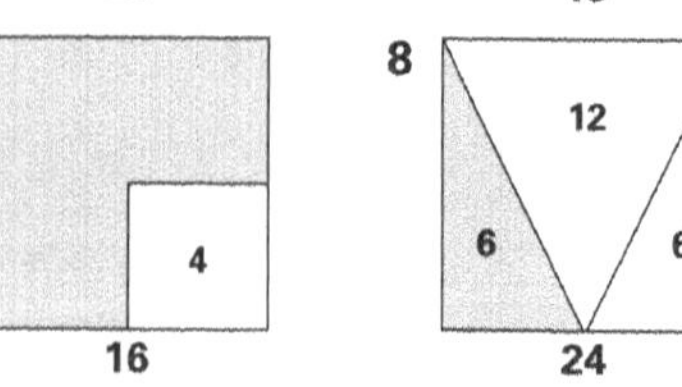

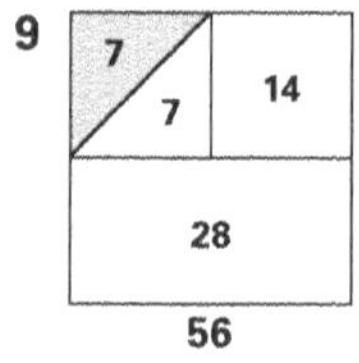

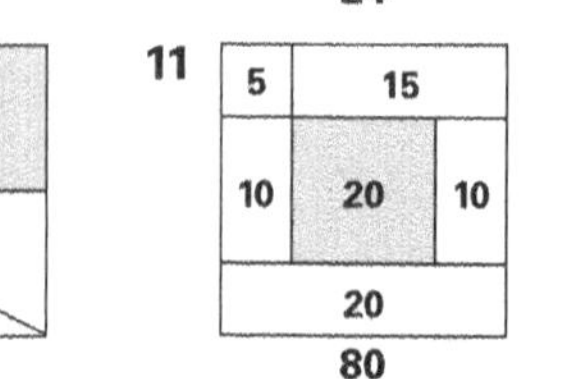

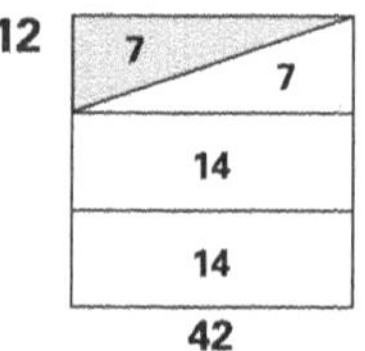

13 4, 4, 8, 8, 8, 16, 12, 12 — 72

14 9, 3, 12, 24 — 48

15

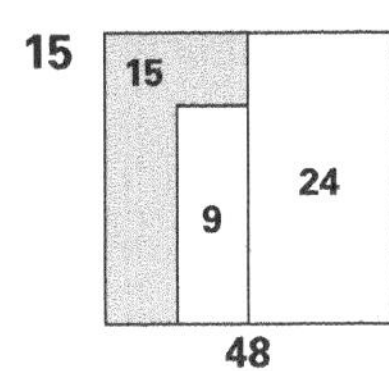

Triangular numbers (page 79)

1

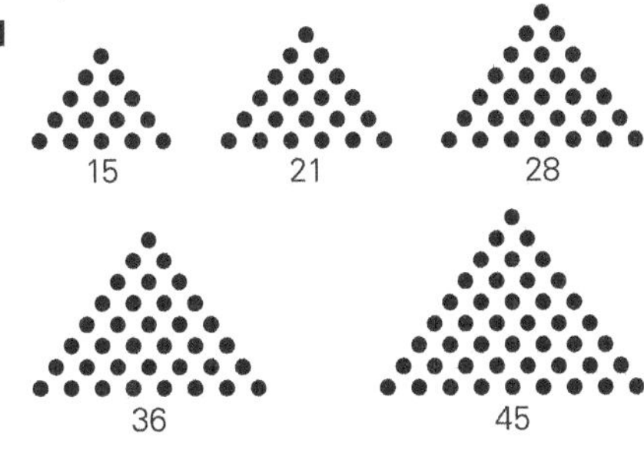

Students could also build these towers using centicubes. Discuss with them the fact that as the towers grow, the base layers increase by one extra number each time. This is the reason for the pattern of differences shown in question 2.

2

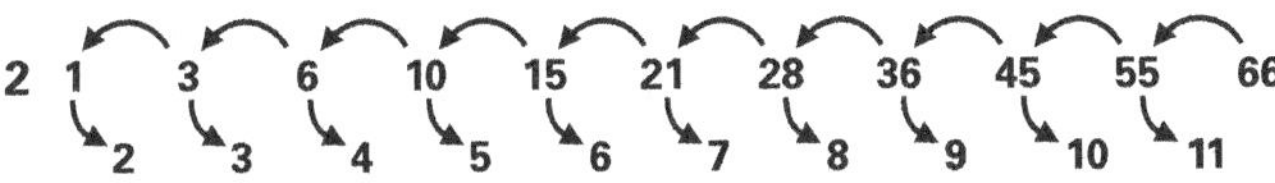

3 An experiment should be done with students shaking hands in the classroom. They really enjoy it, and get quite a thrill when they discover that the resulting numbers follow the pattern (above) of triangular numbers.

a

Members present	Handshakes
2	1
3	3
4	6
5	10
6	15
7	21
8	28
9	36
10	45
11	55

b 7 members **c** 11 members

Joining the dots (page 80)

1

Number of points	Number of lines
3	3
4	6
5	10
6	15
7	21
8	28
9	36

2 a 5 **b** 15

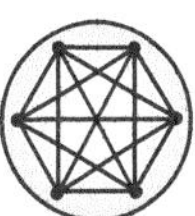

Discuss the similarity of this question to the one above. As an extension, students could be asked to do similar questions; for example: 'A family of six all buy presents for each other. How many presents are bought altogether?' In this case, discuss families of three, four and five first. The results will equal twice the triangular numbers—that is, 6, 12, 20 and 30.

How many triangles? (page 81)

1

1 fold

Single triangles	2
Double triangles	1
Total	**3**

The shading in the figure below covers a double triangle.

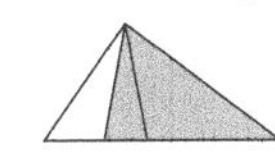

2 folds

Single triangles	3
Double triangles	2
Triple triangles	1
Total	**6**

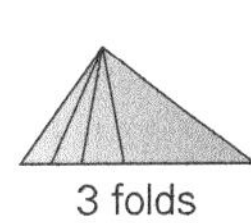

3 folds

Single triangles	4
Double triangles	3
Triple triangles	2
Large triangles	1
Total	**10**

Note the pattern here.

When counting the total number of triangles, students should set out their work as indicated above. They should be asked to shade the triangles, as shown here, using the diagram masters in this book.

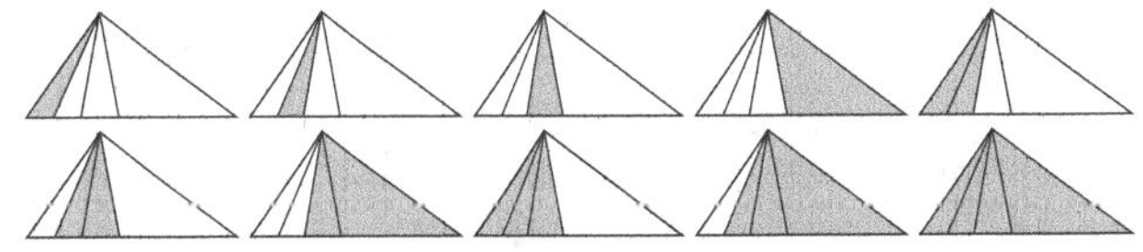

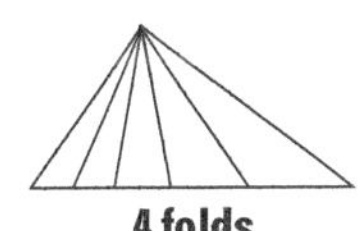

4 folds

Single triangles	5
Double triangles	4
Triple triangles	3
Quadruple triangles	2
Large triangles	1
Total	15

Folds	0	1	2	3	4	5	6	7	8
Triangles	1	3	6	10	15	21	28	36	45

2 It is important to note that the total number of triangles follows the pattern of the triangular numbers. Students should find this result fascinating. Completing the pattern (as above), we can easily show that the number of triangles for 8 folds is 45.

BOOK 3 Solutions

Count the triangles (page 82)

1

Single	6
Double	3
Triple	6
Large	1
Total	**16**

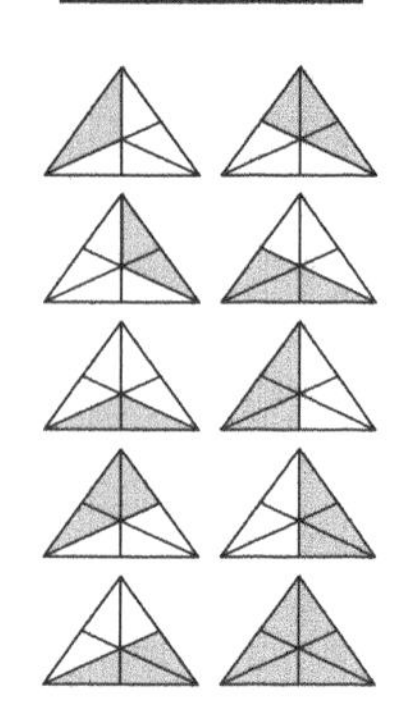

2

Single	8
Double	8
Large	2
Total	**18**

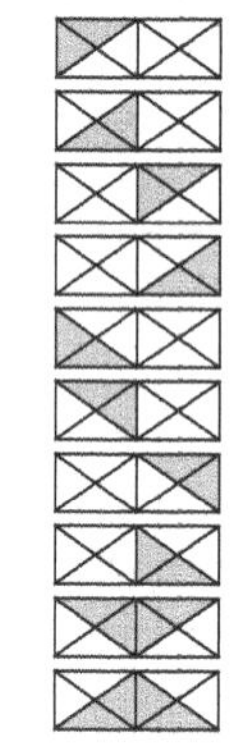

3

Single	6
Double	3
Triple	6
Large	1
Total	**16**

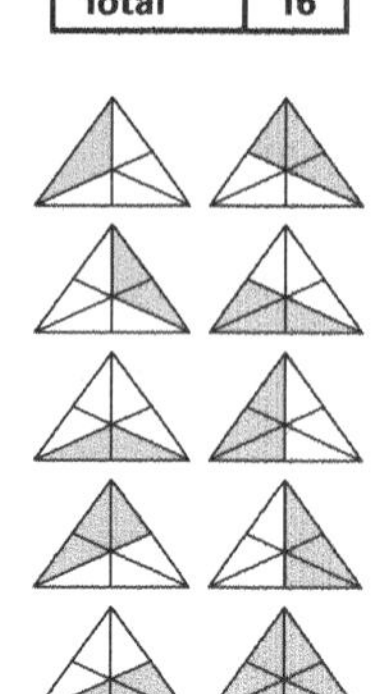

Logic with figures (page 83)

1  **2** 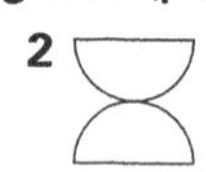 **3** 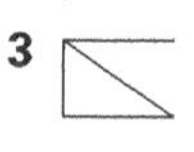 **4**  **5** 

6  **7** 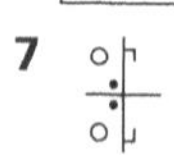 **8** 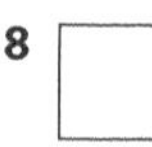 **9** 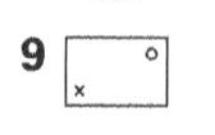 **10** 

Matchstick puzzles (page 84)

								Rule
1	Number of diamonds	1	2	3	4	5	→ 10	4 × ◇
	Number of matches	4	8	12	16	20	→ 40	
2	Number of triangles	1	2	3	4	5	→ 10	3 × △
	Number of matches	3	6	9	12	15	→ 30	
3	Number of squares	1	2	3	4	5	→ 10	4 × □
	Number of matches	4	8	12	16	20	→ 40	
4	Length of rectangles (matches)	1	2	3	4	5	→ 10	2 × ▭ + 2
	Number of matches	4	6	8	10	12	→ 22	
5	Number of double triangles	1	2	3	4	5	→ 10	5 × ◊
	Number of matches	5	10	15	20	25	→ 50	
6	Size of squares	1	2	3	4	5	→ 10	4 × sides of □
	Number of matches	4	8	12	16	20	→ 40	

Hexagon puzzles (page 85)

1 a 16 **b** 6, 11, 16, 21, 26, 31, 36

c 26 = 6 + 4 × 5 **d** 4 (6 + 3 × 5 = 21)

e 6 (Note for the teacher: for *N* hexagons we need 6 + (*N* – 1) × 5 matches.)

2 a 14 **b** 6, 10, 14, 18, 22, 26, 30

c 26 (6 + 5 × 4 = 26) **d** 5 (6 + 4 × 4 = 22)

(Note for the teacher: for *N* hexagon tables = 6 + (*N* – 1) × 4 people can be seated.)

Puzzles with matches (page 86)

Students should be provided with matches so that this can be a practical exercise.

1 a **b** **c** or

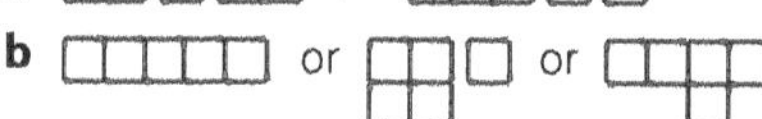

2 a or
b or or

3 a **b**

4 a or **b** or or

5 a or or or

b or or other shapes

6 **7 a** **b** or

c **d**

This is a difficult puzzle, as students are not expected to make squares of different sizes. They may need a hint.

Folding paper (page 87)

It is worthwhile having a discussion with students before starting this experiment. They should first be asked how many times they think a sheet of paper can be folded in half, and then experiment with paper of different sizes and thickness. They usually find the results amazing.

1 a

Folds	1	2	3	4	5	6	7	8
Rectangles	2	4	8	16	32	64	128	256

Continue the pattern to predict the results of 7 and 8 folds.

b The maximum number of times that a sheet of paper can be folded is 8; however, depending on the size and thickness of the paper, it may be possible to fold it only 6 or 7 times.

c 256

2 a

Folds	1	2	3	4	5	6
Creases	1	3	7	15	31	63
Rectangles	2	4	8	16	32	64

Continue the pattern to predict the results of 5 and 6 folds. Again, depending on the size and thickness of the paper, it may be possible to fold it only 4 times.

b 64

Secret codes (page 88)

1 a HOLIDAYS ARE FUN

b

2 THIS IS VERY HARD

3 A – 25 B – 26 C – 1 D – 2 E – 3 F – 4
G – 5 H – 6 I – 7 J – 8 K – 9 L – 10
M – 11 N – 12 O – 13 P – 14 Q – 15 R – 16
S – 17 T – 18 U – 19 V – 20 W – 21 X – 22
Y – 23 Z – 24

a i 1 25 18 **ii** 6 19 12 5 16 23 **iii** 11 25 18 6 17

b HOPE YOU ARE ENJOYING THIS

STUDENT BOOK Solutions

BOOK 4

Patterns and sequences (page 1)

1 21, 25, 29	**2** 51, 58, 65	**3** 24, 29, 34
4 11, 7, 3	**5** 37, 31, 25	**6** 88, 85, 82
7 20, 24, 28	**8** 25, 31, 37	**9** 56, 67, 78
10 66, 77, 88	**11** 3, $3\frac{1}{2}$, 4	**12** 4, $3\frac{1}{2}$, 3
13 13, 14, 17	**14** 10, 11, 13	**15** 28, 30, 37
16 49, 47, 46	**17** 29, 25, 23	**18** 32, 25, 21

Find my rule (page 2)

Students should be encouraged to express the rule in words. If they have difficulty with the exercises, they can be given a hint: in number 11, for example, you could say, 'Try multiplying by 2'.

That would give this result:

6	3	7	8	2
12	6	14	16	

and it is easy to see that you must subtract 1 each time to match the numbers in the second row:

11	5	13	15	

Therefore, the rule is 'multiply by 2, then subtract 1'.

						Rule
1	3	7	4	8	2	+ 3
	6	10	7	11	5	
2	7	12	8	9	15	− 4
	3	8	4	5	11	
3	4	2	7	5	3	× 2
	8	4	14	10	6	
4	6	1	8	10	4	× 2 +1
	13	3	17	21	9	
5	1	2	3	4	5	10 −
	9	8	7	6	5	
6	3	4	5	6	7	20 −
	17	16	15	14	13	
7	9	7	12	6	10	+ 6
	15	13	18	12	16	
8	6	3	9	11	4	+ 10
	16	13	19	21	14	
9	6	3	9	7	2	× 10
	60	30	90	70	20	
10	1	3	5	7	9	× 2 + 1
	3	7	11	15	19	
11	6	3	7	8	2	× 2 − 1
	11	5	13	15	3	
12	1	2	3	4	5	× 10 − 1
	9	19	29	39	49	
13	6	9	3	7	5	× 10 + 1
	61	91	31	71	51	
14	6	9	13	17	5	− 5
	1	4	8	12	0	
15	14	8	10	18	6	÷ 2
	7	4	5	9	3	
16	3	6	4	7	1	× 5
	15	30	20	35	5	
17	3	6	4	7	1	+ 11
	14	17	15	18	12	
18	23	16	14	17	21	− 13
	10	3	1	4	8	

Missing numbers (page 3)

						Rule
1	4	7	3	8	2	× 6
	24	42	18	48	12	
2	44	12	45	20	13	− 8
	36	4	37	12	5	
3	5	1	3	4	7	× 4
	20	4	12	16	28	
4	5	2	1	3	6	× 11
	55	22	11	33	66	
5	4	7	3	8	9	× 2
	8	14	6	16	18	
6	4	7	3	8	9	× 2 − 1
	7	13	5	15	17	
7	4	7	3	8	9	× 2 + 1
	9	15	7	17	19	
8	7	3	5	2	6	× 10 − 1
	69	29	49	19	59	
9	8	4	3	7	1	× 10 + 4
	84	44	34	74	14	
10	8	4	14	18	20	÷ 2 − 1
	3	1	6	8	9	
11	8	1	3	7	5	× 2 + 2
	18	4	8	16	12	
12	8	1	3	7	5	× 3
	24	3	9	21	15	
13	8	1	3	7	5	× 3 − 1
	23	2	8	20	14	
14	2	4	1	7	8	× 3 + 1
	7	13	4	22	25	
15	5	7	4	8	12	× 5
	25	35	20	40	60	
16	5	7	4	8	12	× 5 − 1
	24	34	19	39	59	
17	5	7	4	8	1	× 3 + 2
	17	23	14	26	5	
18	5	7	4	8	2	× 2 − 2
	8	12	6	14	2	
19	5	7	4	3	1	square
	25	49	16	9	1	
20	6	2	3	8	1	square − 1
	35	3	8	63	0	

Find two numbers (page 4)

1 4, 1	**2** 3, 2	**3** 5, 2	**4** 6, 1	**5** 3, 4
6 7, 1	**7** 5, 3	**8** 6, 2	**9** 5, 5	**10** 8, 2
11 6, 4	**12** 9, 2	**13** 6, 5	**14** 7, 4	**15** 10, 1
16 6, 6	**17** 11, 1	**18** 7, 5	**19** 8, 4	**20** 9, 3
21 6, 5	**22** 8, 3	**23** 12, 1	**24** 7, 6	**25** 10, 3
26 10, 8	**27** 11, 7	**28** 17, 1	**29** 19, 1	**30** 15, 5

Exploring numbers (page 5)

1 99	**2** 10	**3** 999	**4** 100	**5** 98
6 103	**7** 987	**8** 101	**9** 102	**10** 998
11 109	**12** 972	**13** 9 (1, 2, 3, ..., 9)		

14 90 (10, 11, 12, ..., 99)

15 18 (50, 51, 52, ..., 59, 15, 25, 35, 45, 65, 75, 85, 95) — (50, 51, 52, ..., 59): 10; (15, 25, 35, 45, 65, 75, 85, 95): 8

16 72 (90 − 18 as 55 has already been counted)

17 101 **18** 995 **19** 100 **20** 999

Shape values (page 6)

1

5	5	5	15
4	3	4	11
4	1	7	12
13	9	16	

2

4	4	4	12
3	3	5	11
4	2	1	7
11	9	10	

3

5	1	8	14
5	3	4	12
5	1	5	11
15	5	17	

4

2	1	8	11
2	2	2	6
7	7	5	19
11	10	15	

5

5	4	1	10
4	4	4	12
2	8	8	18
11	16	13	

6

6	6	6	18
6	9	1	16
3	5	5	13
15	20	12	

7

8	1	1	10
8	1	7	16
8	5	3	16
24	7	11	

8

7	7	5	19
8	7	4	19
2	7	7	16
17	21	16	

9

7	7	4	18
6	6	6	18
3	8	4	15
16	21	14	

10

9	9	9	27
1	7	5	13
4	4	5	13
14	20	19	

The four operations (page 7)

1

■	5	8	6	6	9	4	7	6	9	9	8	7	8
△	2	3	2	3	7	2	3	5	2	3	7	6	5
■ + △	7	11	8	9	16	6	10	11	11	12	15	13	13
■ − △	3	5	4	3	2	2	4	1	7	6	1	1	3
■ × △	10	24	12	18	63	8	21	30	18	27	56	42	40

2

■	4	5	7	5	3	5	4	6	7	5	5	6	8
△	3	2	2	1	2	4	2	2	1	3	2	5	7
△ × △	9	4	4	1	4	16	4	4	1	9	4	25	49
■ + △	7	7	9	6	5	9	6	8	8	8	7	11	15
■ × △	12	10	14	5	6	20	8	12	7	15	10	30	56

3

■	4	7	2	5	8	10	7	6	8	5	9	6	10
△	3	5	1	3	1	8	2	4	4	4	3	3	3
2 × ■	8	14	4	10	16	20	14	12	16	10	18	12	20
■ − △	1	2	1	2	7	2	5	2	4	1	6	3	7
■ × △	12	35	2	15	8	80	14	24	32	20	27	18	30

4

■	9	10	12	12	24	20	24	8	36	28	18	18	21
△	3	2	3	6	3	5	6	4	9	4	2	6	3
3 × △	9	6	9	18	9	15	18	12	27	12	6	18	9
■ + △	12	12	15	18	27	25	30	12	45	32	20	24	24
■ ÷ △	3	5	4	2	8	4	4	2	4	7	9	3	7

These questions are a challenging way to test students' knowledge and understanding of the four operations.

Table squares (page 8)

1

+	3	5	4	8
2	5	7	6	10
7	10	12	11	15
9	12	14	13	17
6	9	11	10	14

2

+	5	3	6	4
2	7	5	8	6
7	12	10	13	11
1	6	4	7	5
8	13	11	14	12

3

+	4	1	7	8
7	11	8	14	15
12	16	13	19	20
8	12	9	15	16
1	5	2	8	9

4

+	11	5	8	6
14	25	19	22	20
10	21	15	18	16
7	18	12	15	13
3	14	8	11	9

5

+	2	9	1	8
7	9	16	8	15
5	7	14	6	13
11	13	20	12	19
3	5	12	4	11

6

+	12	3	7	9
4	16	7	11	13
8	20	11	15	17
2	14	5	9	11
7	19	10	14	16

7

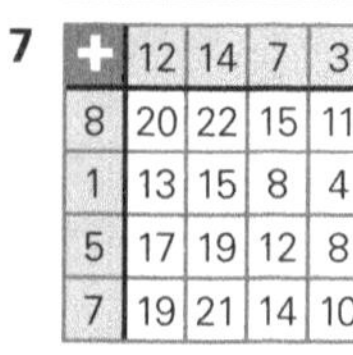

+	12	14	7	3
8	20	22	15	11
1	13	15	8	4
5	17	19	12	8
7	19	21	14	10

8

+	6	10	3	8
6	12	16	9	14
5	11	15	8	13
7	13	17	10	15
10	16	20	13	18

9

+	11	5	8	12
2	13	7	10	14
5	16	10	13	17
8	19	13	16	20
6	17	11	14	18

10

+	4	5	6	9
8	12	13	14	17
2	6	7	8	11
4	8	9	10	13
7	11	12	13	16

Letters and symbols (page 9)

1 1 **2** 2 **3** 4 **4** 2 **5** 1 **6** 0
7 6 **8** 3 **9** 5 **10** 6 **11** 5 **12** 3
13 4 **14** 4 **15** 5 **16** 5
17 2, 3 or 4, 6 or 6, 9 **18** 3, 4 or 6, 8 **19** 5, 3
20 9 **21** 3 **22** 4 **23** 7 **24** 8 **25** 4

What's my message? (page 10)

1 HAVE A GOOD WEEKEND

Code	K	D	N	E	W	H	V	A	O	G
	0	1	2	3	4	5	6	7	8	9

2 YOU ARE CLEVER

Code	R	V	A	U	O	L	C	E	Y
	1	2	3	4	5	6	7	8	9

Square puzzles (page 11)

1 A = 3, B = 1, C = 2, D = 4, E = 5

5	2	2	9
5	1	3	9
1	4	3	8
11	7	8	

2 A = 3, B = 5, C = 4, D = 2, E = 1

3	5	4	12
3	4	1	8
2	2	1	5
8	11	6	

3 A = 2, B = 5, C = 1, D = 3, E = 4

5	2	1	8
3	4	4	11
3	2	1	6
11	8	6	

4 A = 4, B = 1, C = 3, D = 2, E = 5

4	5	4	13
1	3	2	6
1	5	3	9
6	13	9	

5 A = 4, B = 3, C = 5, D = 2, E = 1

2	1	5	8
3	4	5	12
5	4	2	11
10	9	12	

Find the values (page 12)

In puzzle 1, E = 5 is easy to find. To work out the value of the other letters, we need to use trial and error.

In the last row, A + D = 3, since A + D + 3 = 6. Try A = 2 and D = 1; the result would be:

3	3	5	11
1		5	11
2	1	3	6
6	9	13	

and the value of C would be 5, but this cannot be, as E = 5. So try A = 1 and D = 2, which leads to the correct solution.

1 A = 1, B = 3, C = 4, D = 2, E = 5

3	3	5	11
2	4	5	11
1	2	3	6
6	9	13	

2 A = 3, B = 1, C = 4, D = 5, E = 2

4	1	2	7
1	5	3	9
2	2	4	8
7	8	9	

Start with the third row and work out C.

3 A = 1, B = 2, C = 3, D = 4, E = 5

3	2	4	9
1	5	3	9
4	3	2	9
8	10	9	

From row 1 or column 3 B + C = 5 and then from the second column work out E, using the fact that B + C = 5. Then use 'guess and check'.

4 A = 2, B = 5, C = 4, D = 1, E = 3

1	3	5	9
1	4	3	8
2	5	4	11
4	12	12	

Start with the first column and work out A. Then use the fact that B + C = 9 in the third row to work out E.

5 A = 3
B = 4
C = 2
D = 5
E = 1

5	4	2	11
4	1	3	8
3	2	5	10
12	7	10	

You can either use 'guess and check' or the fact that A + D = 8 to work out B in the first column.

Odd and even (page 13)

1 Reinforce the results of completing the table squares. In **1a** for instance, emphasise that 'even number plus even number gives even number'.

a

+	2	4	6
2	4	6	8
4	6	8	10
6	8	10	12

b

×	2	4	6
2	4	8	12
4	8	16	24
6	12	24	36

c

+	1	3	5
1	2	4	6
3	4	6	8
5	6	8	10

d

×	1	3	5
1	1	3	5
3	3	9	15
5	5	15	25

e

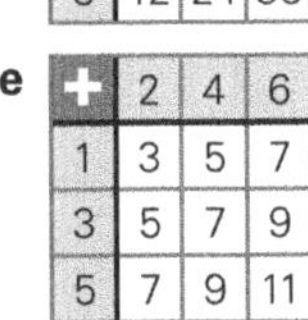

+	2	4	6
1	3	5	7
3	5	7	9
5	7	9	11

f

×	2	4	6
1	2	4	6
3	6	12	18
5	10	20	30

2 These tables require students to make generalisations.

a

+	O	E
O	E	O
E	O	E

b

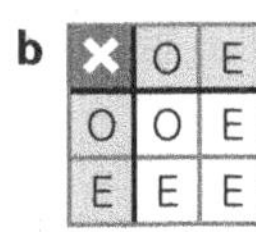

×	O	E
O	O	E
E	E	E

3 Though examples do not prove a result, they can clarify the meaning of a statement. Therefore, students should be encouraged to try several examples before completing a sentence.

a even **b** even **c** even **d** even
e even **f** odd **g** even **h** odd
i odd **j** odd **k** even **l** even
m either even or odd

Find three numbers (page 14)

1 □5, △7, ○4

2 □6, △4, ○8

3 □8, △4, ○11 Think of two numbers that add to 15 and whose difference is 7.

4 □8, △3, ○9 Find two numbers that add to 12 and whose difference is 6.

5 □2, △10, ○7

6 □12, △5, ○7 Find two numbers whose sum is 17 such that one number is 7 more than the other.

7 □7, △8, ○9

8 □3, △6, ○12

Number sentences (page 15)

1 Students should first play around with the numbers 8, 4, 2 and 1, putting symbols between them. They will most probably work out many of the solutions in this way, but may have difficulty with the order of operations. Some of the possible answers are listed below.

a $1 = 8 - 4 - 2 - 1$
$= 8 \div (4 \times 2 \times 1)$
$= 8 - (4 + 2) - 1$

b $2 = 8 - 4 - (2 \times 1)$
$= 8 \div 4 \div 2 + 1$
$= 8 - 4 - (2 \div 1)$
$= (8 - 4 - 2) \div 1$
$= (8 - 4) \div 2 \div 1$
$= 8 - (4 + 2) \times 1$
$= 8 - (4 + 2) \div 1$

c $3 = 8 - 4 - 2 + 1$
$= 8 \div 4 + 2 - 1$
$= (8 - 4) \div 2 + 1$

d $4 = (8 \div 4) + 2 \times 1$
$= 8 \div 4 + 2 \div 1$

e $4 = (8 + 4) \div (2 + 1)$
$= (8 - 4) \div (2 - 1)$
$= (8 - 4) \times (2 - 1)$

f $5 = 8 - 4 + 2 - 1$
$= 8 \div 4 + 2 + 1$
$= 8 \div 4 \times 2 + 1$

g $6 = 8 - 4 + 2 \times 1$
$= 8 - 4 \div 2 \times 1$
$= 8 - 4 + 2 \div 1$

h $6 = (8 + 4) \div (2 \times 1)$
$= (8 - 4) + (2 \div 1)$
$= (8 \div 4) \times (2 + 1)$

i $7 = 8 - 4 + 2 + 1$
$= 8 - 4 \div 2 + 1$
$= (8 + 4) \div 2 + 1$

j $8 = 8 \times (4 - 2 - 1)$
$= 8 \div (4 - 2 - 1)$

k $9 = 8 + 4 - 2 - 1$
$= 8 + 4 \div 2 - 1$
$= (8 - 4) \times 2 + 1$

l $10 = 8 + 4 - 2 \times 1$
$= 8 + 4 \div 2 \times 1$
$= 8 + 4 - 2 \div 1$

m $11 = 8 + 4 \div 2 + 1$
$= 8 + 4 - 2 + 1$

n $12 = (8 + 4) \times (2 - 1)$
$= (8 + 4) \div (2 - 1)$
$= (8 - 4) \times (2 + 1)$

o $13 = 8 + 4 + 2 - 1$

p $14 = 8 + 4 + 2 \times 1$
$= 8 + 4 + 2 \div 1$

q $15 = 8 + 4 + 2 + 1$
$= 8 \times 4 \div 2 - 1$
$= 8 + 4 \times 2 - 1$

r $15 = 8 \times (4 - 2) - 1$

s $16 = 8 \times (4 \div 2) \div 1$
$= 8 \times (4 \div 2) \times 1$
$= 8 \times (4 - 2) \times 1$
$= 8 \times (4 - 2) \div 1$
$= 8 + (4 \times 2) \times 1$
$= 8 + (4 \times 2) \div 1$

t $24 = (8 + 4) \times 2 \times 1$

2 **a** $5 = 2 + 2 + 2 \div 2$
$5 = 2 \times 2 + 2 \div 2$
$5 = 2^2 + 2 \div 2$

b $16 = 2 \times 2 \times 2 \times 2$

c $8 = 2 \times 2 + 2 \times 2$
$8 = 2 + 2 + 2 + 2$
$8 = 2 \times 2 + 2 + 2$

d $44 = 22 + 22$

e $26 = 22 + 2 + 2$

Magic squares (page 16)

First complete the lines in which three numbers are given. In square **1 a** they are marked with stars.

1 **a**

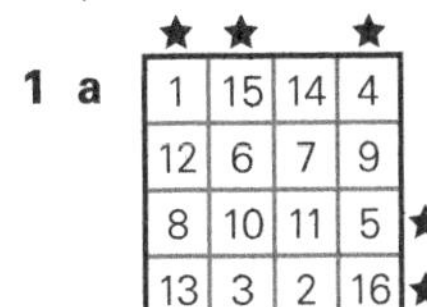

★	★		★	
1	15	14	4	
12	6	7	9	
8	10	11	5	★
13	3	2	16	★

b

13	8	12	1
3	10	6	15
2	11	7	14
16	5	9	4

You will have to look at the sum on the diagonal here.

2 The shaded squares show the rows in which the magic numbers can be found.

a

10	7	4	19
5	18	11	6
17	2	9	12
8	13	16	3

Magic number = 40

b

1	14	7	12
15	4	9	6
10	5	16	3
8	11	2	13

Magic number = 34

c

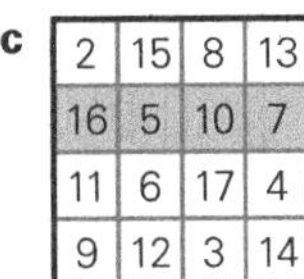

2	15	8	13
16	5	10	7
11	6	17	4
9	12	3	14

Magic number = 38

Square differences (page 17)

These square differences are based on the algebraic fact that
$\square^2 - \triangle^2 = (\square + \triangle) \times (\square - \triangle)$;

if $\square$ and $\triangle$ are consecutive numbers,

then $\square - \triangle = 1$ and $\square^2 - \triangle^2 = \square + \triangle$;

if $\square - \triangle = 2$, then $\square^2 - \triangle^2 = 2 \times (\square + \triangle)$;

if $\square - \triangle = 3$, then $\square^2 - \triangle^2 = 3 \times (\square + \triangle)$ and so on.

1 $5^2 - 3^2 = 2 \times 8 = 16$
$6^2 - 4^2 = 2 \times 10 = 20$

2 $6^2 - 3^2 = 3 \times 9 = 27$
$7^2 - 4^2 = 3 \times 11 = 33$

3 $7^2 - 3^2 = 4 \times 10 = 40$
$8^2 - 4^2 = 4 \times 12 = 48$

4 $8^2 - 3^2 = 5 \times 11 = 55$
$9^2 - 4^2 = 5 \times 13 = 65$

5 $9^2 - 7^2 = 2 \times 16 = 4 \times 8 = 32$
$11^2 - 9^2 = 2 \times 20 = 4 \times 10 = 40$

6 $8^2 - 6^2 = 2 \times 14 = 4 \times 7 = 28$
$10^2 - 8^2 = 2 \times 18 = 4 \times 9 = 36$

7 $4^2 - 3^2 = 4 + 3 = 7$
$5^2 - 4^2 = 5 + 4 = 9$

8 $5^2 - 1^2 = 4 \times 6 = 24$
$6^2 - 1^2 = 5 \times 7 = 35$

9 $6^2 - 2^2 = 4 \times 8 = 32$
$7^2 - 2^2 = 5 \times 9 = 45$

10 $7^2 - 6^2 + 5^2 - 4^2 = 7 + 6 + 5 + 4$
$8^2 - 7^2 + 6^2 - 5^2 = 8 + 7 + 6 + 5$

Primes (page 18)

1 a

Number	Factors	No. of factors
8	1, 2, 4, 8	4
9	1, 3, 9	3
10	1, 2, 5, 10	4
11	1, 11	2
12	1, 2, 3, 4, 6, 12	6
13	1, 13	2
14	1, 2, 7, 14	4
15	1, 3, 5, 15	4
16	1, 2, 4, 8, 16	5
17	1, 17	2
18	1, 2, 3, 6, 9, 18	6
19	1, 19	2
20	1, 2, 4, 5, 10, 20	6
21	1, 3, 7, 21	4
22	1, 2, 11, 22	4
23	1, 23	2
24	1, 2, 3, 4, 6, 8, 12, 24	8
25	1, 5, 25	3
26	1, 2, 13, 26	4

Number	Factors	No. of factors
27	1, 3, 9, 27	4
28	1, 2, 4, 7, 14, 28	6
29	1, 29	2
30	1, 2, 3, 5, 6, 10, 15, 30	8
31	1, 31	2
32	1, 2, 4, 8, 16, 32	6
33	1, 3, 11, 33	4
34	1, 2, 17, 34	4
35	1, 5, 7, 35	4
36	1, 2, 3, 4, 6, 9, 12, 18, 36	9
37	1, 37	2
38	1, 2, 19, 38	4
39	1, 3, 13, 39	4
40	1, 2, 4, 5, 8, 10, 20, 40	8

b The prime numbers have exactly two factors.

c Only square numbers have exactly three factors. In fact only the squares of the prime numbers 4, 9, 25, 49 … have three factors only.

2 ☒ (2) (3) ~~4~~ (5) ~~6~~ (7) ~~8~~ ~~9~~ ~~10~~
(11) ~~12~~ (13) ~~14~~ ~~15~~ ~~16~~ (17) ~~18~~ (19) ~~20~~
~~21~~ ~~22~~ (23) ~~24~~ ~~25~~ ~~26~~ ~~27~~ ~~28~~ (29) ~~30~~
(31) ~~32~~ ~~33~~ ~~34~~ ~~35~~ ~~36~~ (37) ~~38~~ ~~39~~ ~~40~~
(41) ~~42~~ (43) ~~44~~ ~~45~~ ~~46~~ (47) ~~48~~ ~~49~~ ~~50~~
~~51~~ ~~52~~ (53) ~~54~~ ~~55~~ ~~56~~ ~~57~~ ~~58~~ (59) ~~60~~
(61) ~~62~~ ~~63~~ ~~64~~ ~~65~~ ~~66~~ (67) ~~68~~ ~~69~~ ~~70~~
(71) ~~72~~ (73) ~~74~~ ~~75~~ ~~76~~ ~~77~~ ~~78~~ (79) ~~80~~
~~81~~ ~~82~~ (83) ~~84~~ ~~85~~ ~~86~~ ~~87~~ ~~88~~ (89) ~~90~~
~~91~~ ~~92~~ ~~93~~ ~~94~~ ~~95~~ ~~96~~ (97) ~~98~~ ~~99~~ ~~100~~

Prime patterns (page 19)

1 a

A	B	C	D	E	F
☒	(2)	(3)	~~4~~	(5)	~~6~~
(7)	~~8~~	~~9~~	~~10~~	(11)	~~12~~
(13)	~~14~~	~~15~~	~~16~~	(17)	~~18~~
(19)	~~20~~	~~21~~	~~22~~	(23)	~~24~~
~~25~~	~~26~~	~~27~~	~~28~~	(29)	~~30~~
(31)	~~32~~	~~33~~	~~34~~	~~35~~	~~36~~
(37)	~~38~~	~~39~~	~~40~~	(41)	~~42~~
(43)	~~44~~	~~45~~	~~46~~	(47)	~~48~~
~~49~~	~~50~~	~~51~~	~~52~~	(53)	~~54~~
~~55~~	~~56~~	~~57~~	~~58~~	(59)	~~60~~
(61)	~~62~~	~~63~~	~~64~~	~~65~~	~~66~~
(67)	~~68~~	~~69~~	~~70~~	(71)	~~72~~
(73)	~~74~~	~~75~~	~~76~~	~~77~~	~~78~~
(79)	~~80~~	~~81~~	~~82~~	(83)	~~84~~
~~85~~	~~86~~	~~87~~	~~88~~	(89)	~~90~~
~~91~~	~~92~~	~~93~~	~~94~~	~~95~~	~~96~~
(97)	~~98~~	~~99~~	~~100~~		

1 b i F

ii A

iii A and E

iv 6

2 a

☒	(2)	(3)	~~4~~
(5)	~~6~~	(7)	~~8~~
~~9~~	~~10~~	(11)	~~12~~
(13)	~~14~~	~~15~~	~~16~~
(17)	~~18~~	(19)	~~20~~
~~21~~	~~22~~	(23)	~~24~~
~~25~~	~~26~~	~~27~~	~~28~~
(29)	~~30~~	(31)	~~32~~
~~33~~	~~34~~	~~35~~	~~36~~
(37)	~~38~~	~~39~~	~~40~~
(41)	~~42~~	(43)	~~44~~
~~45~~	~~46~~	(47)	~~48~~
~~49~~	~~50~~	~~51~~	~~52~~
(53)	~~54~~	~~55~~	~~56~~
~~57~~	~~58~~	(59)	~~60~~
(61)	~~62~~	~~63~~	~~64~~
~~65~~	~~66~~	(67)	~~68~~
~~69~~	~~70~~	(71)	~~72~~
(73)	~~74~~	~~75~~	~~76~~
~~77~~	~~78~~	(79)	~~80~~
~~81~~	~~82~~	(83)	~~84~~
~~85~~	~~86~~	~~87~~	~~88~~
(89)	~~90~~	~~91~~	~~92~~
~~93~~	~~94~~	~~95~~	~~96~~
(97)	~~98~~	~~99~~	~~100~~

b

☒	(2)	(3)	~~4~~	(5)
~~6~~	(7)	~~8~~	~~9~~	~~10~~
(11)	~~12~~	(13)	~~14~~	~~15~~
~~16~~	(17)	~~18~~	(19)	~~20~~
~~21~~	~~22~~	(23)	~~24~~	~~25~~
~~26~~	~~27~~	~~28~~	(29)	~~30~~
(31)	~~32~~	~~33~~	~~34~~	~~35~~
~~36~~	(37)	~~38~~	~~39~~	~~40~~
(41)	~~42~~	(43)	~~44~~	~~45~~
~~46~~	(47)	~~48~~	~~49~~	~~50~~
~~51~~	~~52~~	(53)	~~54~~	~~55~~
~~56~~	~~57~~	~~58~~	(59)	~~60~~
(61)	~~62~~	~~63~~	~~64~~	~~65~~
~~66~~	(67)	~~68~~	~~69~~	~~70~~
(71)	~~72~~	(73)	~~74~~	~~75~~
~~76~~	~~77~~	~~78~~	(79)	~~80~~
~~81~~	~~82~~	(83)	~~84~~	~~85~~
~~86~~	~~87~~	~~88~~	(89)	~~90~~
~~91~~	~~92~~	~~93~~	~~94~~	~~95~~
~~96~~	(97)	~~98~~	~~99~~	~~100~~

2 c When the numbers are in four columns, all the primes except 2 occur in columns 1 and 3.

When the numbers are in five columns, all primes: ending in 1 are in column 1; ending in 7 are in column 2; ending in 3 are in column 3; ending in 9 are in column 4.

5 is the only prime number in column 5. The rest of the numbers in this column are the multiples of 5.

3

100	99	98	(97)	96	95	94	93	92	91
65	**64**	63	62	(61)	60	(59)	58	57	90
66	(37)	**36**	35	34	33	32	(31)	56	(89)
(67)	38	(17)	**16**	15	14	(13)	30	55	88
68	39	18	(5)	**4**	(3)	12	(29)	54	87
69	40	(19)	6	☒	(2)	(11)	28	(53)	86
70	(41)	20	(7)	8	**9**	10	27	52	85
(71)	42	21	22	(23)	24	**25**	26	51	84
72	(43)	44	45	46	(47)	48	**49**	50	(83)
(73)	74	75	76	77	78	(79)	80	**81**	82

The square numbers (shaded) are all along one main diagonal and another diagonal.

Goldbach's conjecture (page 20)

Although all possible solutions are listed here, students are not expected to find every one.

1 6 = 3 + 3
2 8 = 3 + 5
3 12 = 5 + 7
4 14 = 7 + 7 = 3 + 11
5 16 = 5 + 11 = 3 + 13
6 18 = 7 + 11 = 5 + 13
7 20 = 7 + 13 = 3 + 17
8 22 = 11 + 11 = 5 + 17 = 3 + 19
9 24 = 11 + 13 = 7 + 17 = 5 + 19
10 26 = 13 + 13 = 7 + 19 = 3 + 23
11 28 = 11 + 17 = 5 + 23
12 30 = 13 + 17 = 11 + 19 = 7 + 23
13 32 = 13 + 19 = 3 + 29
14 34 = 17 + 17 = 11 + 23 = 5 + 29 = 3 + 31
15 36 = 17 + 19 = 13 + 23 = 7 + 29 = 5 + 31
16 38 = 19 + 19 = 7 + 31
17 40 = 17 + 23 = 11 + 29 = 3 + 37
18 44 = 13 + 31 = 7 + 37 = 3 + 41
19 62 = 31 + 31 = 19 + 43 = 3 + 59
20 100 = 47 + 53 = 41 + 59 = 29 + 71 = 17 + 83 = 11 + 89 = 3 + 97

Which number remains? (page 21)

1 8 **2** 16 **3** 15 **4** 6 **5** 4 **6** 15
7 6 **8** 16 **9** 45 **10** 9 **11** 12 **12** 25

Arrange the numbers (page 22)

1 There are two very difficult tasks to be done in this exercise:
- **a** to find the common sum for each row;
- b to find the eleven numbers in the circles.

There are many possibilities for placing the numbers in the circles, so 'guess and check' is not a good strategy. There are three possibilities:

6 as the centre number

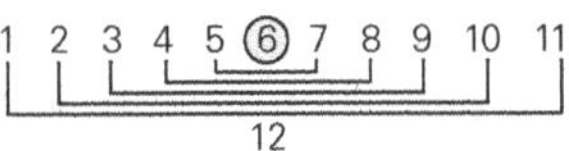

There are 5 pairs of numbers. Each pair adds to 12.

11 as the centre number

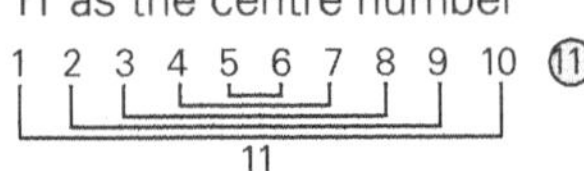

There are 5 pairs of numbers. Each pair adds to 11.

1 as the centre number

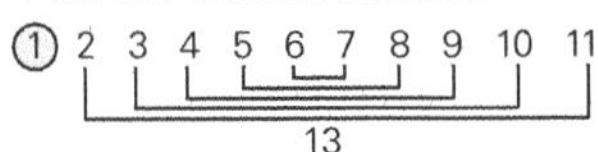

There are 5 pairs of numbers. Each pair adds to 13.

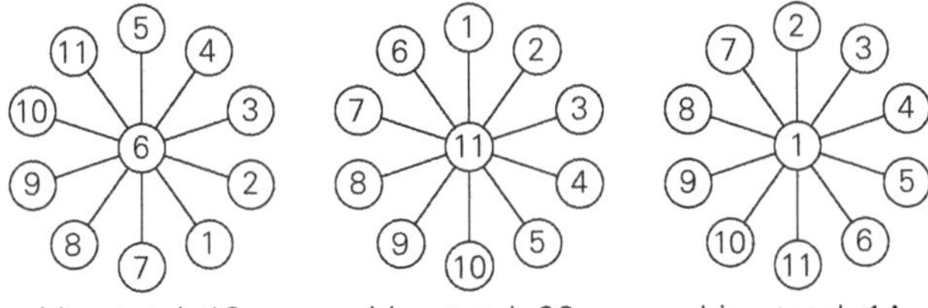

Note: There are many other possible answers, as the position of the pairs of numbers can be changed. We need to find pairs of numbers that have the same sum.

2 7 − 4 = 3
× 2 =
6 − 5 = 1
+ 8 = 9

3 a Do first

b

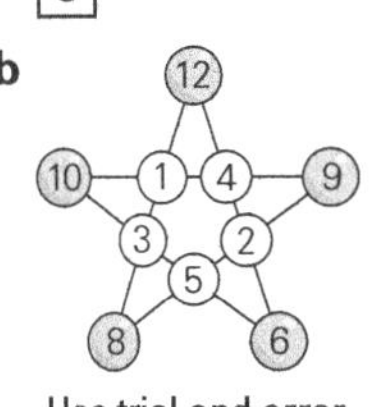

c

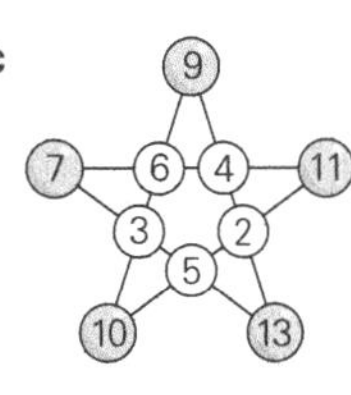

Use trial and error.

4 This is an excellent exercise in trial and error and working systematically.
Students should first work out the possible unit digits:

1 5 2 3 4
0 6 9 8 7

Now, remembering that all four digits must be different, the possibilities are:

91	21	81	31	71	41	51	61
+20	+90	+30	+80	+40	+70	+60	+50

Remember that in these solutions you are carrying:

95	15	85	25	75	35	72	32	62
+16	+96	+26	+86	+36	+76	+39	+79	+49
42	93	13	63	43	94	14	84	24
+69	+18	+89	+48	+68	+17	+97	+27	+87

Triangle patterns (page 23)

This is an excellent exercise in trial and error.

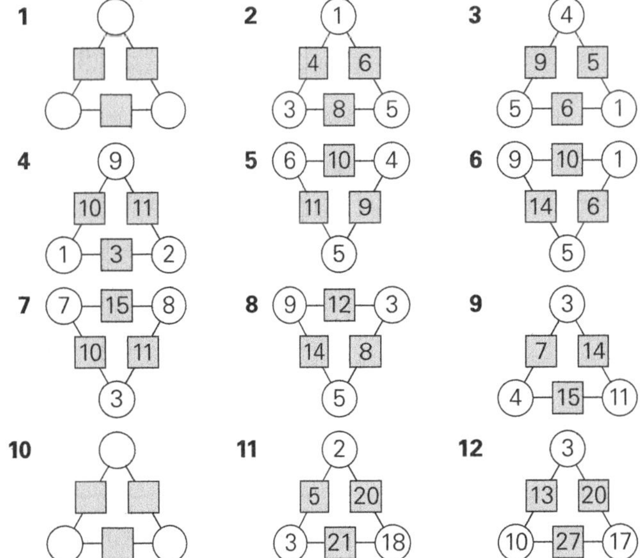

The quickest way of doing these is as follows:
1 Add the three numbers in the squares.
2 Halve this answer.
3 The result is the sum of each number in a square and the circle opposite it.

Triangle patterns (page 23) ***continued...***

In exercise 9, the sum of the numbers in the square is 36, so half this result is 18.

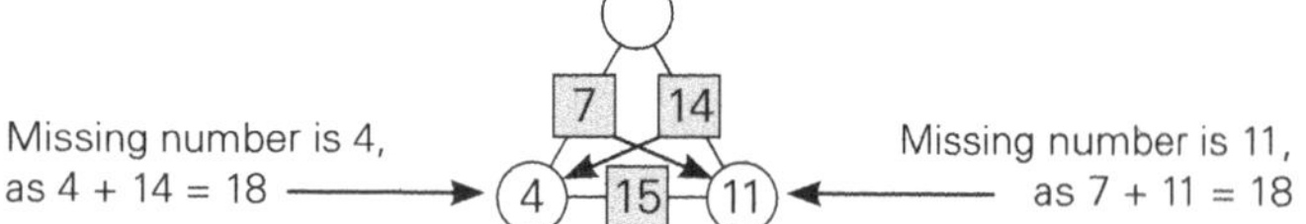

What's my question? (page 24)

All the possible solutions are listed.

1 $\begin{array}{r}50\\\times 2\\\hline 100\end{array}$ $\begin{array}{r}25\\\times 4\\\hline 100\end{array}$ $\begin{array}{r}20\\\times 5\\\hline 100\end{array}$

2 $\begin{array}{r}12\\\times 3\\\hline 36\end{array}$ $\begin{array}{r}18\\\times 2\\\hline 36\end{array}$

3 $\begin{array}{r}32\\\times 2\\\hline 64\end{array}$ $\begin{array}{r}16\\\times 4\\\hline 64\end{array}$

4 $\begin{array}{r}27\\\times 2\\\hline 54\end{array}$ $\begin{array}{r}18\\\times 3\\\hline 54\end{array}$

5 $\begin{array}{r}24\\\times 2\\\hline 48\end{array}$ $\begin{array}{r}12\\\times 4\\\hline 48\end{array}$ $\begin{array}{r}16\\\times 3\\\hline 48\end{array}$

6 $\begin{array}{r}12\\\times 7\\\hline 84\end{array}$ $\begin{array}{r}21\\\times 4\\\hline 84\end{array}$ $\begin{array}{r}28\\\times 3\\\hline 84\end{array}$ $\begin{array}{r}14\\\times 6\\\hline 84\end{array}$

7 $\begin{array}{r}36\\\times 2\\\hline 72\end{array}$ $\begin{array}{r}18\\\times 4\\\hline 72\end{array}$ $\begin{array}{r}24\\\times 3\\\hline 72\end{array}$ $\begin{array}{r}12\\\times 6\\\hline 72\end{array}$

8 $\begin{array}{r}48\\\times 2\\\hline 96\end{array}$ $\begin{array}{r}24\\\times 4\\\hline 96\end{array}$ $\begin{array}{r}12\\\times 8\\\hline 96\end{array}$ $\begin{array}{r}32\\\times 3\\\hline 96\end{array}$ $\begin{array}{r}16\\\times 6\\\hline 96\end{array}$

Dartboard problems (page 25)

1 a 3, 5, 6, 8, 9, 10, 11, 12 and all further numbers are possible.

b 1,2, 4 and 7 are the only impossible scores.

3 + 3 = 6
3 + 5 = 8
3 + 3 + 3 = 9
5 + 5 = 10
3 + 3 + 5 = 11
3 + 3 + 3 + 3 = 12
3 + 5 + 5 = 13
3 + 3 + 3 + 5 = 14
5 + 5 + 5 = 15

Most students will use trial and error. Once three consecutive possible scores are found—8, 9 and 10— there is no need to search further; you can add 3 to any of these, so that all remaining numbers are possible.

2 a 2, 4, 6, 7, 8, 9 and all further numbers are possible.

b 1, 3, and 5 are the only impossible scores. This time, once we have two consecutive possible scores the search is over.

3 22, 18 , 24 or 16

9 + 7 + 3 + 3 = 22
9 + 9 + 3 + 1 = 22
7 + 7 + 3 + 1 = 18
9 + 9 + 3 + 3 = 24
7 + 3 + 3 + 3 = 16

Since the scores are all odd numbers and since four odd numbers give an even number, no odd number is a possible total.

4 List all possible combinations that would achieve a score of 34.

9	6	2	No. of throws
0	0	17	17
0	1	14	15
0	2	11	13
0	3	8	11
2	0	8	10
0	4	5	9
2	2	2	6

a Smallest number of throws is 6

b Greatest number of throws is 17

c 9 + 9 = 18
6 + 6 + 6 = 18
6 + 6 + 2 + 2 + 2 = 18
6 + 2 + 2 + 2 + 2 + 2 + 2 = 18
2 + 2 + 2 + 2 + 2 + 2 + 2 + 2 + 2 = 18

Missing digits (page 26)

1 $\begin{array}{r}43\\+32\\\hline 75\end{array}$ **2** $\begin{array}{r}134\\+255\\\hline 389\end{array}$ **3** $\begin{array}{r}482\\+265\\\hline 747\end{array}$ **4** $\begin{array}{r}381\\+595\\\hline 976\end{array}$

5 $\begin{array}{r}274\\+389\\\hline 663\end{array}$ **6** $\begin{array}{r}732\\+378\\\hline 1110\end{array}$ **7** $\begin{array}{r}78\\-24\\\hline 54\end{array}$ **8** $\begin{array}{r}379\\-227\\\hline 152\end{array}$

9 $\begin{array}{r}87\\-34\\\hline 53\end{array}$ **10** $\begin{array}{r}42\\-38\\\hline 4\end{array}$ **11** $\begin{array}{r}324\\-286\\\hline 38\end{array}$ **12** $\begin{array}{r}432\\-179\\\hline 253\end{array}$

13 $\begin{array}{r}13\\\times 5\\\hline 65\end{array}$ **14** $\begin{array}{r}69\\\times 3\\\hline 207\end{array}$ **15** $\begin{array}{r}54\\\times 7\\\hline 378\end{array}$

16 $\begin{array}{r}223\\\times 4\\\hline 892\end{array}$ $\begin{array}{r}248\\\times 4\\\hline 992\end{array}$ **17** $\begin{array}{r}273\\\times 4\\\hline 1092\end{array}$ $\begin{array}{r}298\\\times 4\\\hline 1192\end{array}$

18 $\begin{array}{r}246\\\times 3\\\hline 738\end{array}$ **19** $\begin{array}{r}365\\\times 7\\\hline 2555\end{array}$ **20** $\begin{array}{r}177\\\times 9\\\hline 1593\end{array}$

Open-ended questions (page 27)

1 a 6, 12, 18, 24, … any multiple of 6

b 10, 20, 30, 40, … any multiple of 10

c 2, 3, 5, 7, 11, 13, 17, 19, 23, 29 these are all prime numbers less than 30 with exactly two factors

d 4, 9, 169 these are the squares of prime numbers and have exactly three factors

2 a 6 or 12 or 18, or any multiple of 6

b 12 or 24 or 36, or any multiple of 12

c 12 or 24 or 36, or any multiple of 12

d 10 or 20 or 30, or any multiple of 10

3 19, 28 or 37

4 1 × 2 × 12, 1 × 3 × 8, 1 × 4 × 6, 2 × 2 × 6, 3 × 2 × 4

5 4, 16, 36, 64

6 11, 31, 41, 61, 71

7 There are an infinite number of answers to these questions. Some possible answers are:

a $\frac{1}{2}+\frac{1}{2}, \frac{1}{3}+\frac{2}{3}, \frac{1}{4}+\frac{3}{4}, \frac{1}{5}+\frac{4}{5}, \frac{2}{5}+\frac{3}{5}, \ldots$

b $2+\frac{1}{2}, 2\frac{1}{4}+\frac{1}{4}, 2\frac{1}{6}+\frac{2}{6}, 1\frac{1}{4}+1\frac{1}{4}, \ldots$

c $1-\frac{1}{4}, 2-1\frac{1}{4}, 3-2\frac{1}{4}, 1\frac{1}{2}-\frac{3}{4}, \frac{7}{8}-\frac{1}{8}, \ldots$

d $\frac{1}{2}\times\frac{2}{1}, \frac{1}{3}\times\frac{3}{1}, \frac{1}{4}\times\frac{4}{1}, \frac{3}{2}\times\frac{2}{3}, \frac{3}{4}\times\frac{4}{3}, \ldots$

What fraction's shaded? (page 28)

There are many possible solutions. Here are a few of them.

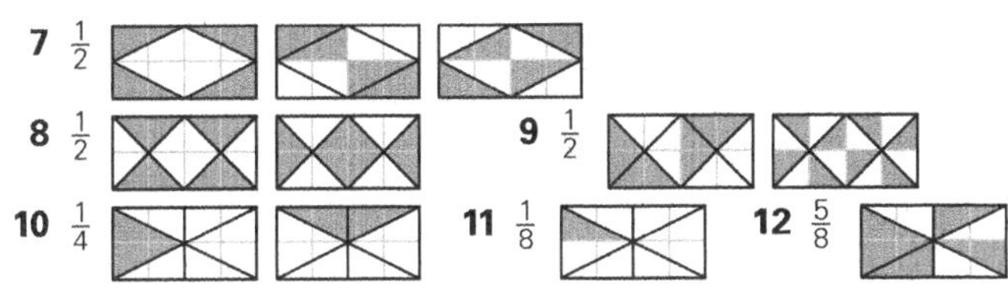

Hexagon fractions (page 29)

1 $\frac{1}{6}$ **2** $\frac{1}{3}$ **3** $\frac{1}{2}$ **4** $\frac{1}{3}$ **5** $\frac{1}{3}$ **6** $\frac{1}{2}$

7 $\frac{1}{2}$ **8** $\frac{2}{3}$ **9** $\frac{1}{2}$ **10** $\frac{1}{6}$ **11** $\frac{1}{12}$ **12** $\frac{5}{12}$

Pizza maths (page 30)

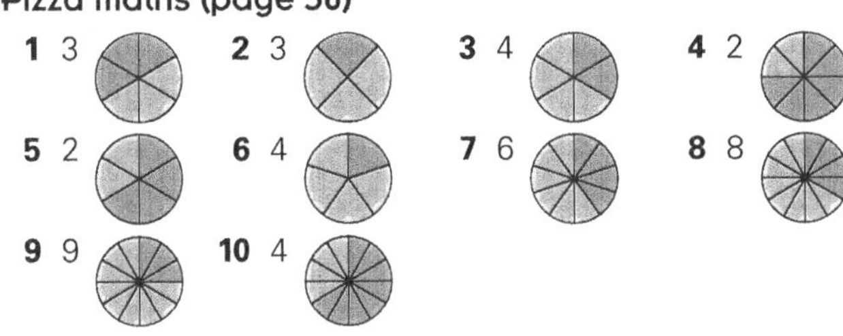

Problem-solving fractions (page 31)

1 Present for his mother = \$100, \$100 remains, $\frac{1}{2}$ of the remainder some books = \$50. So he has \$50 left.

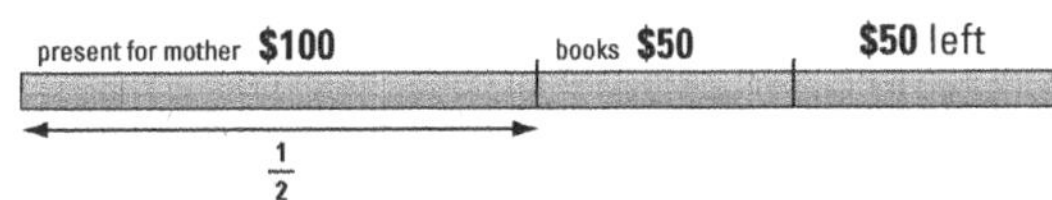

2 Eddie spent \$50 on food and had \$200 left. After spending $\frac{1}{4}$ of this rent, he had \$150 left.

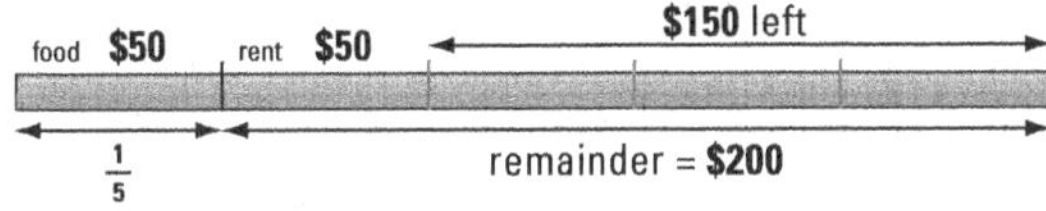

3 Half of the remainder is also \$12. So \$24 is half of the original amount. Dylan had \$48 at the start.

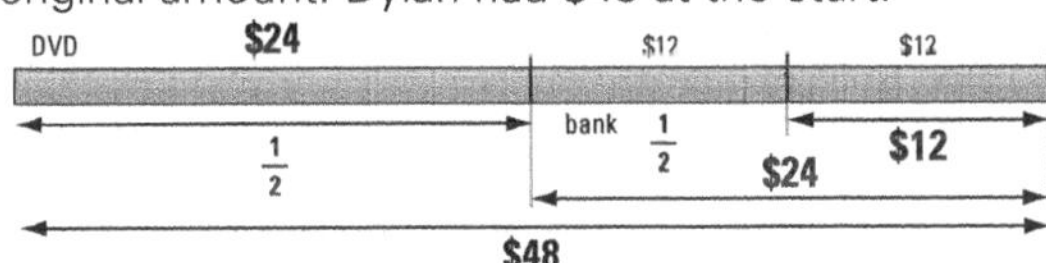

4 \$20 is half, she also spent \$20 at the fun park. So each third is \$20. Romy had \$60 at the start.

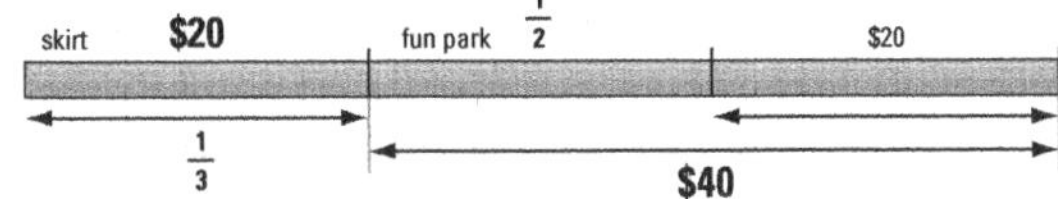

Problems with fractions (page 32)

1 Jordana is also given \$20 and \$40 is $\frac{4}{5}$ of the original amount, each $\frac{1}{5}$ is \$10. Daniel had \$50 at the start.

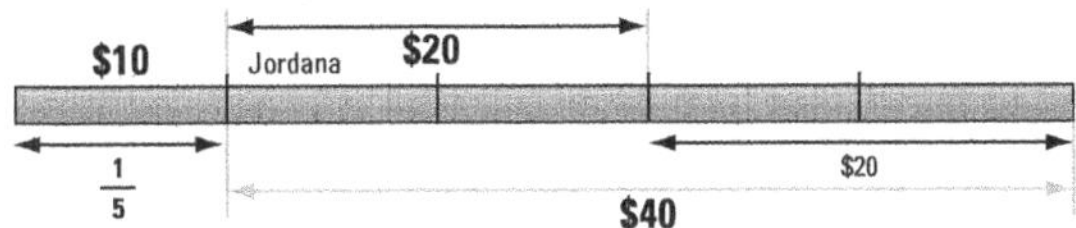

2 a Lauren spent \$6 on a CD, so half the amount is \$18.

b Lauren had \$36 at the start.

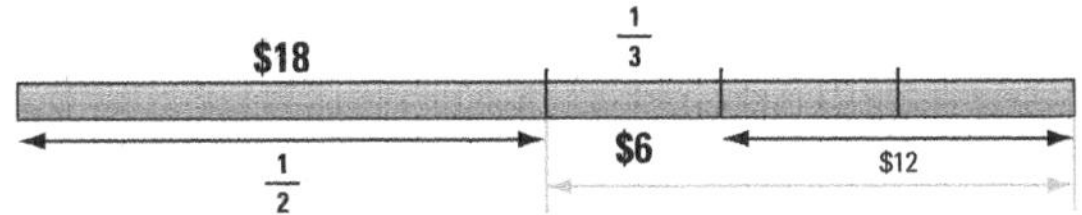

3 a Matthew spent \$20 at the concert.

b Matthew had \$80 at the start.

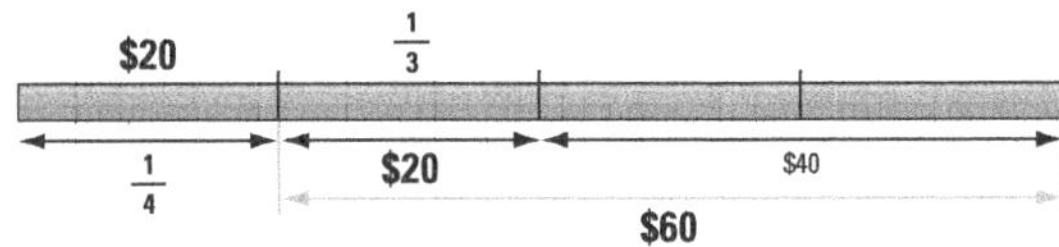

4 a Lee spent \$120 on the skirt. So after she spent a third of the money, Lee has \$480 left. Her jacket cost \$240.

b Lee had \$720 at the start.

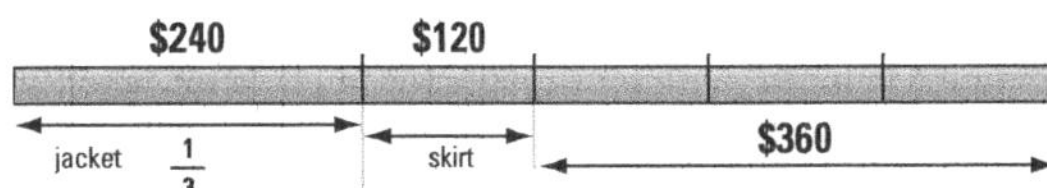

Number machine (page 33)

1 a 6 **b** 6 **c** 8 or 7 **d** 14 or 13

2 a i 19 → 10 → 5 → 3 → 2 → 1

ii 28 → 14 → 7 → 4 → 2 → 1

b i 6 or 5

ii

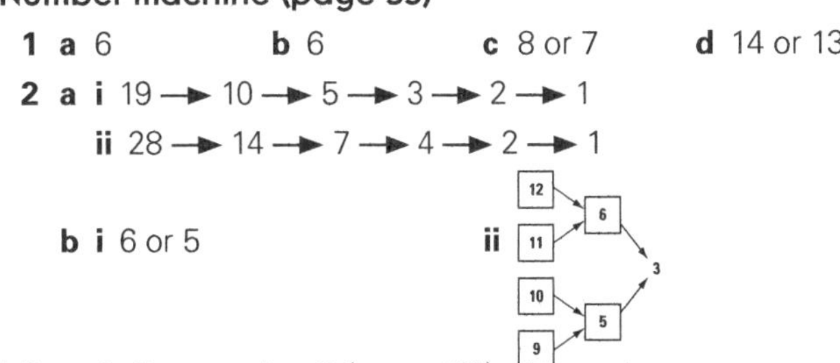

Where's the number? (page 34)

1 a

L	M	N	O	P	Q	R
4		3		2		1
	5		6		7	
11		10		9		8
	12		13		14	
18		17		16		15
	19		20		21	
25		24		23		22

b Q **c** R **d** O **e** Q **f** R

2 a

T	U	V	W	X	Y	Z
1	2	3	4	5	6	7
14	13	12	11	10	9	8
15	16	17	18	19	20	21
28	27	26	25	24	23	22
29	30	31	32	33	34	35
42	41	40	39	38	37	36
43	44	45	46	47	48	49

b T **c** Z **d** T **e** T **f** Z

Money problems (page 35)

Note that these questions can be done using 'guess and check'. However, another way, a possibly more efficient way of doing them, is to consider doing a table and putting the amount more in that column first. Then subtract this from the total.

Drink	Bottle
\$1	

And now split the remainder amount \$1 equally, that is 50c each. So Drink cost \$1.50 and Bottle 50c.

1 a 50c **b** 40 c **c** 25 c

2 a

Shirt	Tie
\$20	
+ \$40	\$40
= \$60	\$40

b

Shirt	Tie
\$50	
+ \$25	\$25
= \$75	\$25

c

Shirt	Tie
\$90	
+ \$5	\$5
= \$95	\$5

3 Book \$55, Pen \$5

4 Peter \$35, Daniel \$32

Peter	Daniel
\$3	
+ \$32	\$32
= \$35	\$32

Money problems (page 35) *continued...*

5

Postcards (30c)	Letters (40c)	Total spent	
2	3	$1.80	– *too high*
4	1	$1.60	

Note: since we know that $1.60 is even, Lydia bought an even number of postcards.

6 6 pencils, 3 sharpeners

7 **a** 2, 8 **b** 6, 4 **c** 2, 10 **d** 8, 4

At the market (page 36)

1 **a** 5 **b** 10 **c** 10 **d** 15

2 **a** 2 **b** 6

3 **a** 1 kg **b** 3 kg **c** 3 kg **d** 1·5 kg

4 **a** 25 **b** 30 **5** 50c

Shopping (page 37)

In the first four problems, it is best to calculate the cost first of one article and then the others.

1 $32 **2** $4.50 **3** 1 tea towel = $7
a $35 **b** $63

4 1 pencil = 40c, ∴ 2 pencils = 80c

5 **a i** 1 **ii** 10 **iii** 3

b Since 1 apple is 25c, the cost is:
i $1.25 **ii** $2.25 **iii** $7.50

6 **a i** 3 apples + 3 oranges = $6

Add the two amounts. At this stage do not try to work out the cost of 1 orange or 1 apple.

ii 1 apple + 1 orange = $2. Find $\frac{1}{3}$ of the above amount.

iii Since 1 apple + 2 oranges = $2.75
1 orange = 75c

iv 1 apple = $1.25. Use the result in **ii**.

b i 5 lemons + 5 pears = $6. Add the two amounts.

ii 1 lemon + 1 pear = $1.20. Find $\frac{1}{5}$ of the above amount.

iii 2 lemons + 2 pears = $2.40. Double the amount in **ii**.

iv Since 2 lemons + 3 pears = $3.20
1 pear = 80c. Use the result in **iii**.

7 If 1 drink + 3 sandwiches = $9.50
2 drinks + 4 sandwiches = $14.00
So 1 drink + 2 sandwiches = $7.00
1 sandwich = $2.50
∴ 1 drink = $2.00

Missing shapes (page 38)

1 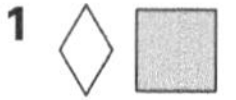 2    3  4 

5 6 **7** D D **8** M D

9 ★ ★ 10

Halves and quarters (page 39)

1 These are only some of the possible solutions. There are many more, including any of these with the shaded and the white sections reversed. This exercise reinforces the concept of one-half and one-quarter, while allowing students to be creative.

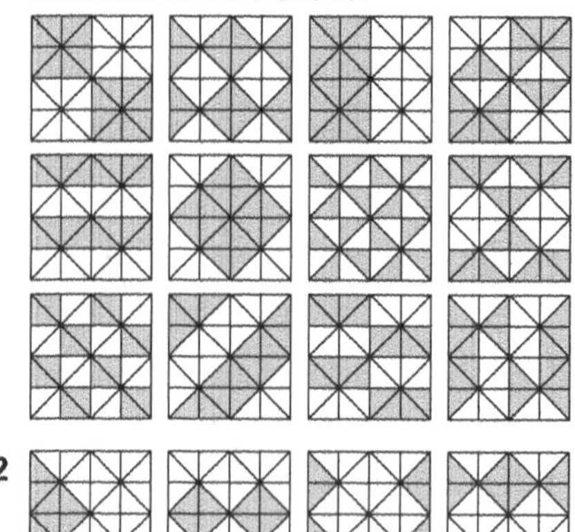

2

How many possibilities? (page 40)

1 **a i** 8 × 1, 4 × 2 **ii** 12 × 1, 4 × 3, 6 × 2
iii 14 × 1, 7 × 2 **iv** 18 × 1, 9 × 2, 6 × 3
v 21 × 1, 7 × 3 **vi** 20 × 1, 10 × 2, 5 × 4
vii 25 × 1, 5 × 5 **viii** 24 × 1, 12 × 2, 8 × 3, 6 × 4

b 24 The rectangles are 24 × 1, 12 × 2, 8 × 3, 6 × 4.

2 **a** 12 There are 4 ways of choosing the first digit and 3 ways of choosing the second digit. Students should list the possibilities working systematically.

15, 16, 18, 51, 56, 58, 61, 65, 68, 81, 85, 86

b 6 There are 2 ways of choosing the second digit to ensure it is even and there are 3 ways of choosing the first digit, namely we can have 16, 18, 56, 58, 68 or 86.

c 6 The number can only start with 6 or 8, then there are 3 different ways of choosing the last digit.
61, 65, 68 81, 85, 86

3 **a** 3 pairs **b** 10 pairs **c** 45 pairs

 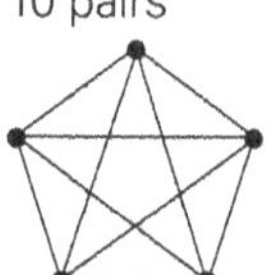 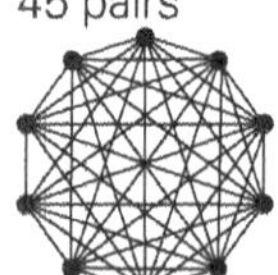

Note that every child can have 9 different partners giving 10 x 9 = 90. However, a pair consists of 2 children, hence 45 pairs. AB is the same pair as BA.

4 6 × 5 ÷ 2 = 15 matches

Counting techniques (page 41)

1 **a** 6 possibilities
56 57
65 67
75 76

b 9 possibilities
55 56 57
65 66 67
75 76 77

or with a tree diagram:

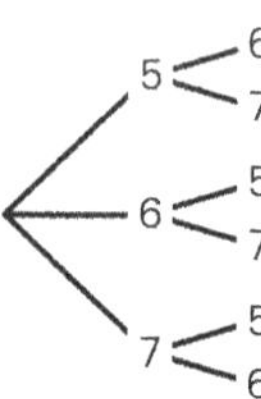

2 a 6 possibilities

567 576 657
675 756 765

or with a tree diagram:

5 — 6 — 7
5 — 7 — 6
6 — 5 — 7
6 — 7 — 5
7 — 5 — 6
7 — 6 — 5

b 27 possibilities

555 556 557
565 566 567
575 576 577
655 656 657
665 666 667
675 676 677
755 756 757
765 766 767
775 776 777

3 The first method given is difficult, and should be explained only to the extremely talented students.

1st digit can be chosen in 4 different ways
2nd digit can be chosen in 3 different ways
3rd digit can be chosen in 2 different ways
4th digit can be chosen in 1 way only.
Therefore total number of possibilities = 4 x 3 x 2 x 1
= 24

or

List the possibilities starting with 5:

5678 5768 5867
5687 5786 5876

If we start with 6, 7 or 8, in each case there are again 6 possibilities, and so there is a total of 24.

or

List all 24 possibilities, working systematically.

or

Use a tree diagram:

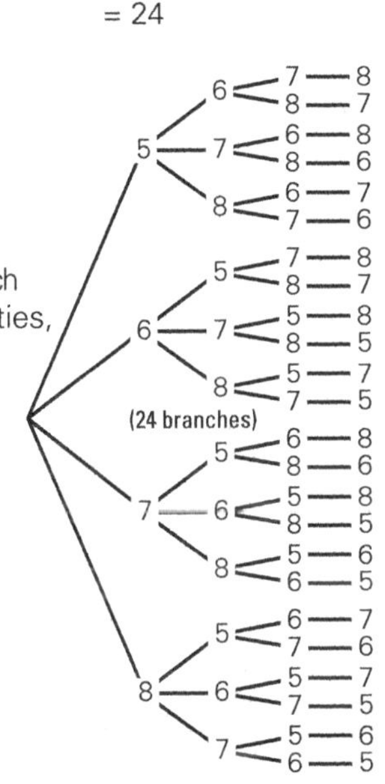

4 a i 3 (for 5, 15 and 25)
ii 11 (for 2, 12, 20, 21, 22, 23, 24, 25, 26, 27)
iii 2 (for 10 and 20)
b i 19 (for 5, 15, 25, 35, 45, 50, 51, 52, 53, 54, 55, 56, 57, 58, 59, 65, 75, 85)
ii 19 (for 2, 12, 20, 21, 22, 23, 24, 25, 26, 27, 28, 29, 32, 42, 52, 62, 72, 82)
iii 8 (for 10, 20, 30, 40, 50, 60, 70, 80)

What chance? (page 42)

1 a 2 chances out of 20 or 1 chance in 10
b The person brought a sandwich from home.

2 There are many different possible ways of shading the rectangles as long as the number of rectangles shaded is the same in each question.

a 1 in 5 (shade any 3 out of the 15 rectangles)

b no chance (no shading at all)

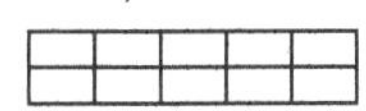

c certain (shade all rectangles)

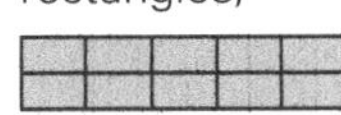

d 1 in 4 (shade any 5 out of the 20 rectangles)

3 a 120 **b** 3 chances out of 12 or 1 chance in 4
c orange **d** 5

e

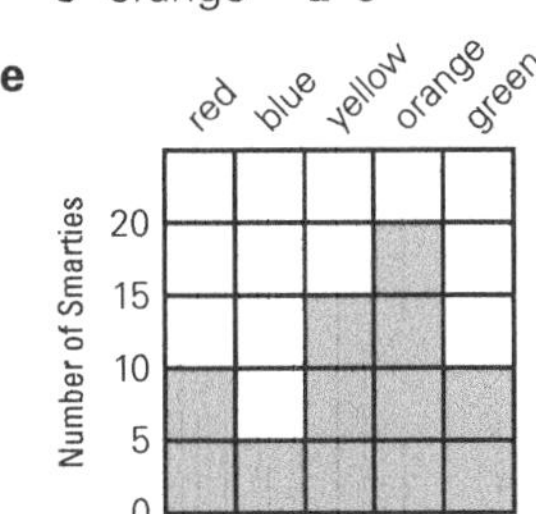

Spinner probabilities (page 43)

1 3 chances out of 25, $\frac{3}{25}$ **2** $\frac{2}{25}$
3 6 **4** $\frac{13}{25}$ **5** 2 and 10 or 3 and 9 or 4 and 8 or 5 and 7
6 0 **7** $\frac{6}{25}$ **8** $\frac{3}{25}$

Picture graphs (page 44)

1 a 25 × 5 = 125 **b** bus **c** taxi
d 25 **e** tram
2 a 15 **b** Monday and Thursday
c 21 **d** Tuesday **e** 35 × 3 = 105

Picture this (page 45)

1 a 18 **b** 15 **c** 12 **d** chips **e** pies, 3
f 28 × 3 = 84 **g** 9 left (6 sold)
2 a i 24 **ii** 8 **iii** 6 **iv** 18
b 72

Graphs (page 46)

1

George						
Emanuel						
Laura						
Phoebe						
Mimi						

2 a 2 km **b** 1 hour **c** 3 km
d Faster. She travelled 3 km in 1 hour to George and only 2 km in 1 hour to Helena.
3 a 12 **b** 36

Sector graph (page 47)

1 a 2 hours **b** 4 hours
2 a 12 hours **b** 6 hours **c** 36 hours
3 a 5 hours **b** 15 hours **c** 20 hours

Perimeter (page 48)

1

P (cm)	Possible rectangles	Dimensions
8	3	3×1, 1×3, 2×2
10	4	4×1, 3×2, 1×4, 2×3
12	5	5×1, 4×2, 3×3, 2×4, 1×5
14	6	6×1, 5×2, 4×3, 3×4, 2×5, 1×6
16	7	7×1, 6×2, 5×3, 4×4, 3×5, 2×6, 1×7

All measurements are in centimetres and the lengths of sides that need to be worked out are in bold type.

2 a 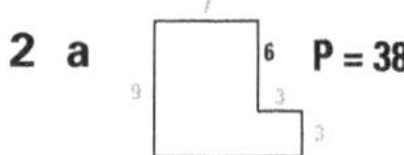

b 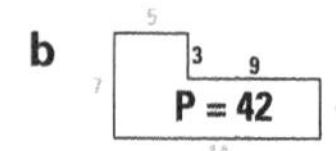

c 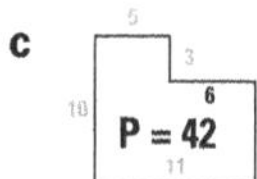

d 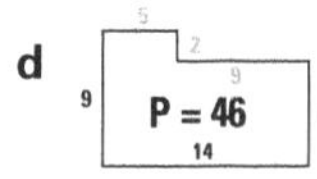

e 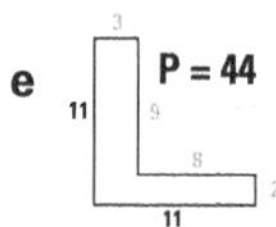

f 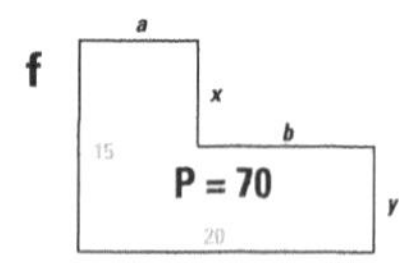

There are many possible solutions to figure **f**, but the perimeter will always be 70 cm regardless of length of the sides, since ***x*** and ***y*** will always add to 15 and ***a*** and ***b*** to 20.

Find the perimeter (page 49)

1 a All the figures have a perimeter of 12 cm.

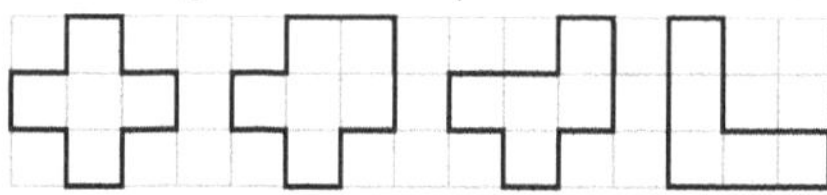

Some other shapes with a perimeter of 12 cm

b All the figures have a perimeter of 16 cm.

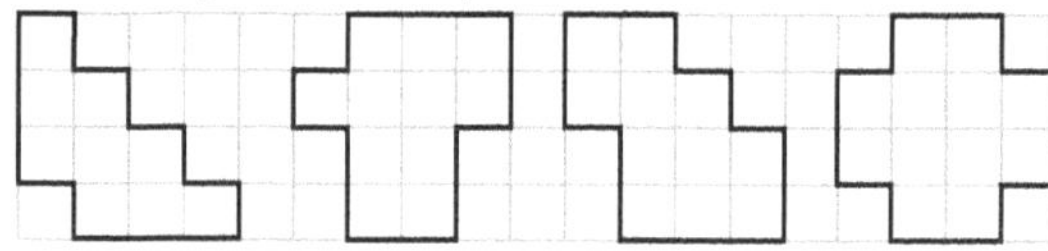

Some other shapes with a perimeter of 16 cm

This is an amazing result: all the figures have the same perimeter. They are some of the shapes that can be drawn when the corners of squares are pushed in.

2 Diane's shortcut and Alan's path cover exactly the same distance. So if the children leave home at the same time and walk at the same speed, they will arrive at school at exactly the same time. In fact, Diane's path is not really a shortcut.

Area investigation (page 50)

These exercises offer an excellent way to teach or reinforce the concept of area. A student who can draw at least six of the examples given below will have a good understanding of this concept. It would be a good idea if each student's work were to be marked by two others before you saw it. Discussions should also be encouraged.

1 a, c, d, f, g, h, i, j

2 These are only some of the possible shapes. For both **2a** and **2b** there are many other possible solutions.

a 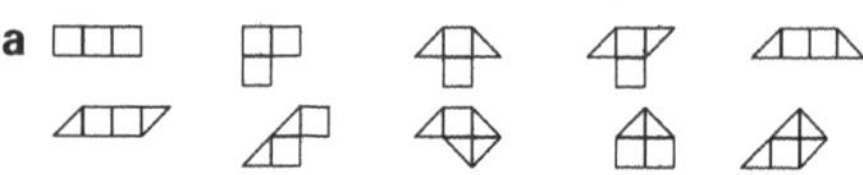

b 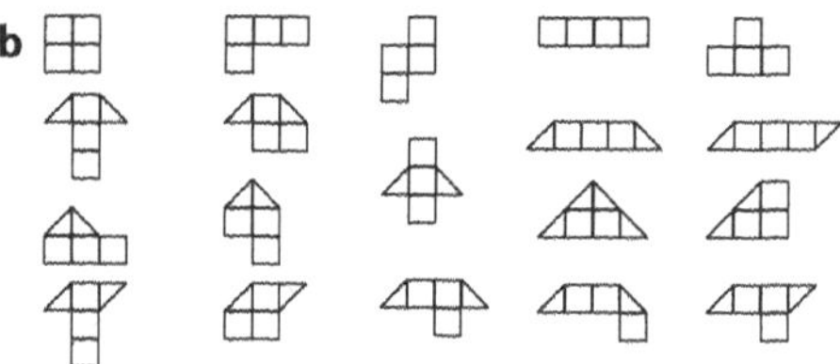

Perimeter and area (page 51)

Any of these shapes can be reflected or rotated for a correct solution.

1 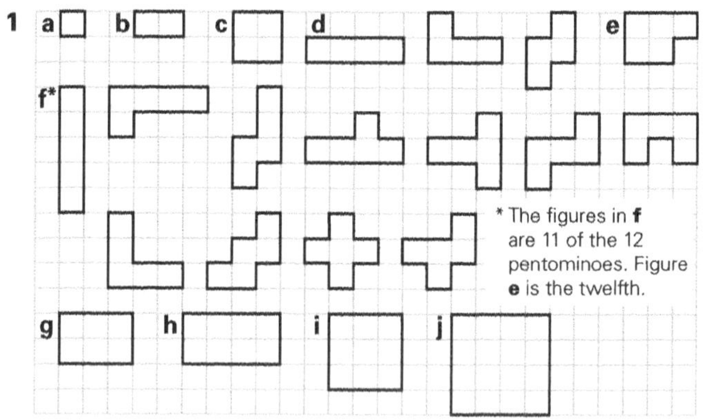

Squared or dot paper is essential for these exercises.

Taping boxes (page 52)

1 a 20 + 8 + 20 + 8 = 56 cm

b 18 + 7 + 18 + 7 = 50 cm

c 50 − 15 − 15 = 20 cm
2 × height = 20 cm
height = 10 cm

d 60 − 2 × 25 = 10 cm
2 × height = 10 cm
height = 5 cm

2 a Tape = 2 × length + 2 × width + 4 × height
= 2 × 20 + 2 × 6 + 4 × 8
= 84 cm

b 2 × 18 + 2 × 7 + 4 × 5 = 70 cm

c 2 × 15 + 2 × 5 = 40 cm
100 − 40 = 60 cm
4 × height = 60 cm
height = 15 cm

d 2 × 24 + 2 × 12 = 72 cm
100 − 72 = 28 cm
4 × height = 28 cm
height = 7 cm

Show the temperature (page 53)

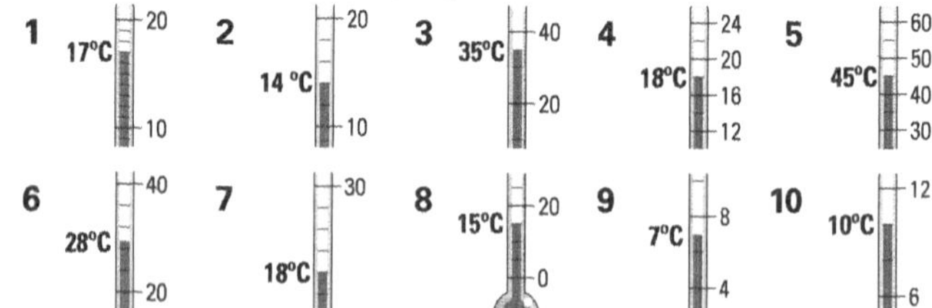

Match the answer (page 54)

1 C	**2** E	**3** D	**4** A	**5** B
6 D	**7** C	**8** A	**9** E	**10** B
11 D	**12** E	**13** B	**14** C	**15** A
16 C	**17** E	**18** B	**19** A	**20** D

Measurement using tables (page 55)

1 a 550 g, 450 g **b** 850 g, 350 g

2 Michelle $16, Mia $8

3 Imran 16 kg, Bruce 8 kg, Gordon 11 kg

4 4 packages, 8 chocolate mints, Total spent $1.20

5

	Bananas	Grapes	Cost
a	2 kg	3 kg	$18
b	4 kg	0	$12
	0	3 kg	$12
	2 kg	$1\frac{1}{2}$ kg	$12
	3 kg	$\frac{3}{4}$ kg	
	1kg	$2\frac{1}{4}$ kg	
	$1\frac{1}{3}$ kg	2 kg	
	$2\frac{2}{3}$ kg	1 kg	

There are many possible answers to **5 b** and only some are given here in the table. Students are unlikely to give more than the first two.

All the solutions can be represented on the graph; some are identified by the dots.

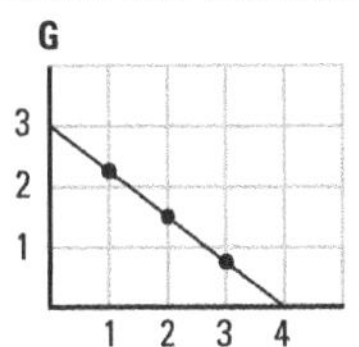

An informal discussion about all the possible solutions is worthwhile.

Fractions in measurement (page 56)

1 a $3 **b** $1\frac{3}{4}$ kg

2 200 mL

3 a 200 cm **b** 180 cm

4 a 36 L **b** $54 **c** 528 km

5 a 160 ml **b** 200 mL

6 a $2.50 **b** $3.75 **c** $25

7 $1\frac{1}{2}$ hours

Questioning time (page 57)

1 12 min **2** 28 min **3** 9 h 46 min

4 a 16 hours **b** 7 p.m.

5 11 hours 30 minutes **6** 41 hours 40 minutes

7 9:50 a.m. In 1 hour, clock loses 10 x 60 sec = 10 minutes

8 9:15 p.m. In 9 hours, clock loses 9 x 5 = 45 minutes

9 6:48 p.m. In 6 hours, clock loses 6 x 2 = 12 minutes

10 12:12 p.m. In 12 hours, clock loses 24 x 2 = 48 minutes

Time challenges (page 58)

1 Monday

				Yesterday	Today	Tomorrow
Tues	Wed	Thurs	Frid	Sat	Sun	Mon

2 a Wednesday **b** Monday **c** Friday

3 a 25 days

b i Saturday **ii** 9 days

c 80 months **d** 2 years 4 months

4 Saturday **5** 28 days later

Multiple choice—measurement (page 59)

1 A, C **2** B, D **3** B **4** A **5** D **6** A

7 A, C **8** A, C, D **9** B, C **10** A **11** B **12** B

Measurement problems (page 60)

1 A **2** A **3** B **4** D **5** B **6** D

7 C **8** A **9** A **10** D **11** A

Packing cans (page 61)

1 a l = 30 cm w = 20 cm h = 12 cm

b l = 30 cm w = 30 cm h = 24 cm

2 24 There are 2 layers of cans. In each layer there are 4 cans in the length and 3 cans in the width, altogether 12.

3

Ways that cans are packed	length	width	height
12 × 1 × 1	120	10	12
6 × 2 × 1	60	20	12
4 × 3 × 1	40	30	12
3 × 2 × 2	30	20	24
2 × 2 × 3	20	20	36
4 × 1 × 3	40	10	36
1 × 3 × 4	10	30	48
1 × 2 × 6	10	20	72
1 × 1 × 12	10	10	144

Speed challenges (page 62)

1 a 300 m = 25 × 3 × 4
Note that in 1 minute Sarah walks 25 × 3 metres.

b 4500 m = 4.5 km **c** 2 minutes

d 40 minutes

2 a 400 metres **b** 6 km

c 120 seconds = 2 minutes **d** 10 minutes

3 a 225 km **b** 30 km

c 4 hours **d** 1 hour 40 minutes

4 a 126 km **b** 21 km

c 2 hours **d** 45 minutes

Rate problems (page 63)

1 a 30 L **b** 400 L **c** 370 L

2 a 4200 times **b** 40 minutes

3 a 400 metres **b** 40 metres

c 15 seconds as $2\frac{1}{2}$ minutes = 150 seconds and we need $\frac{1}{10}$

4 a $3 **b** 30 c **c** $90 **d** 20 hours

5 a 8 hours **b** $2\frac{1}{2}$ hours **c** 10 m^2 **d** 2 m^2

Water flow (page 64)

1 1200 litres

2 a i 100 litres **ii** 300 litres

b 30 seconds **c** 150 litres **d** 400 litres

3 a 20 litres **b** 120 litres

4 a 180 mL **b** 10800 mL = 180 × 60 mL = 10.8 L

5 1500 mL = 1.5 L = 5 × 5 × 60 mL

Cough medicine (page 65)

1 5 times **2** 5 mL **3** 25 mL **4** 50 doses

5 10 days **6** 150 mL **7** Friday 2 p.m. **8** $1.50

Travel graphs (page 66)

1 Amie **2** Carla **3** Ben and Amie

4 3 hours **5** 25 km/h

Travel graphs (page 66) ***continued...***

6 Ben's speed = $\frac{30}{4}$ = 7.5 km/h and
Amie's speed = $\frac{70}{4}$ = 17.5 km/h. Amie's speed is greater.

7

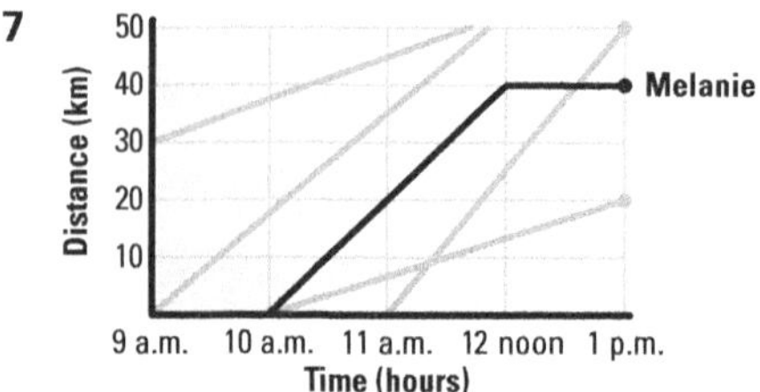

Marbles (page 67)

1 1 marble 15 mL, water 195 mL
2 1 marble 8 mL, water 224 mL (2 marbles = 16 mL)
3 1 marble 8 mL, water 224 mL (4 marbles = 32 mL)
4 1 marble 20 mL, water 280 mL (6 marbles = 120 mL)

Fuel tank (page 68)

1 before 20 L after 10 L used 10 L
2 before 40 L after 20 L used 20 L
3 before 45 L after 15 L used 30 L
4 before 24 L after 8 L used 16 L
5 before 24 L after 12 L used 12 L
6 before 28 L after 14 L used 14 L
7 before 36 L after 9 L used 27 L
8 before 24 L after 6 L used 18 L

Mass (page 69)

1 1430 g = 1.43 kg **2** 2125 g = 2.125 g **3** 24
4 140 kg **5** 20 **6** 20 **7** 48 **8** 10
9 3.2 kg **10** 6
11 a 96 ÷ 3 = 32 kg **b** 40 kg Total weight of 3 = 180 kg
12 43 kg **13 a** $8 **b** $12 **c** $24 **d** $20

How many? (page 70)

Students should be supplied with diagram masters.

1 a

small	7
large	2
total	9

b

small	4
medium	1
large	1
total	6

medium

c

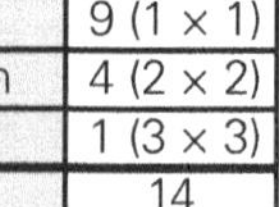

small	9 (1 × 1)
medium	4 (2 × 2)
large	1 (3 × 3)
total	14

medium

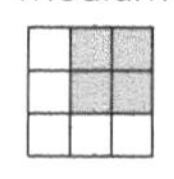

d

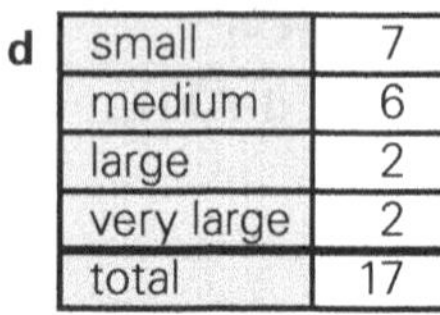

small	7
medium	6
large	2
very large	2
total	17

small medium

large very large

2 a

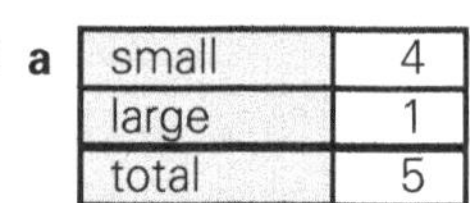

small	4
large	1
total	5

b

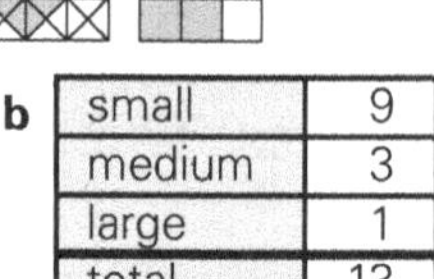

small	9
medium	3
large	1
total	13

medium

c

small	10
medium	4
total	14

medium

d

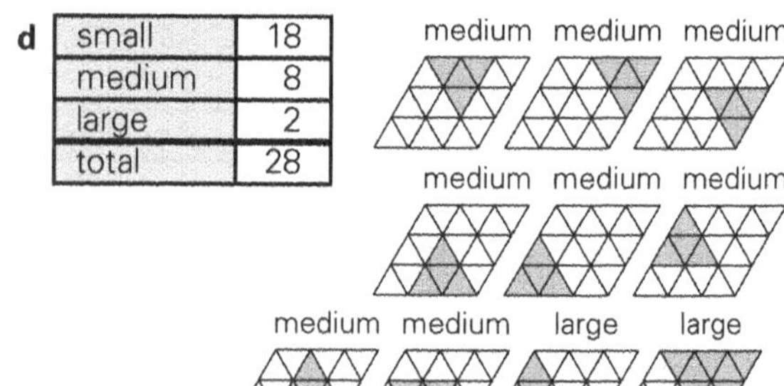

small	18
medium	8
large	2
total	28

Amazing graphs (page 71)

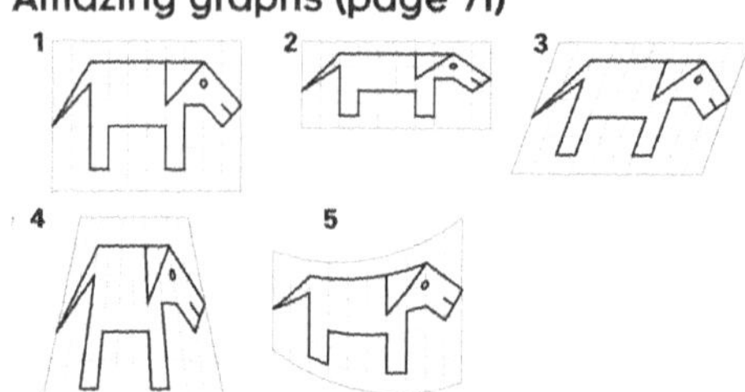

Rolling boxes (page 72)

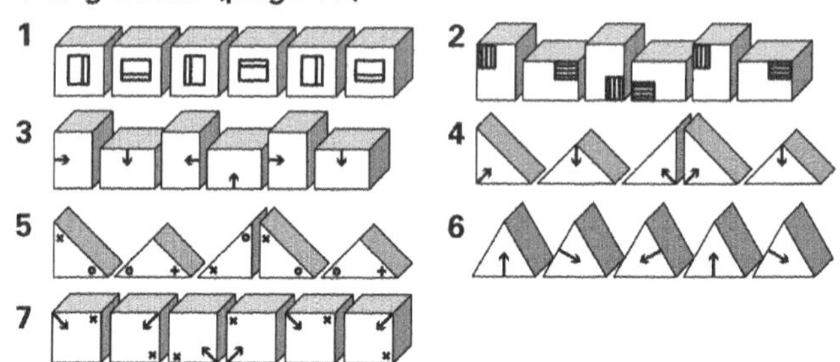

Geometry and dots (page 73)

1 a 5 squares

b 8 squares

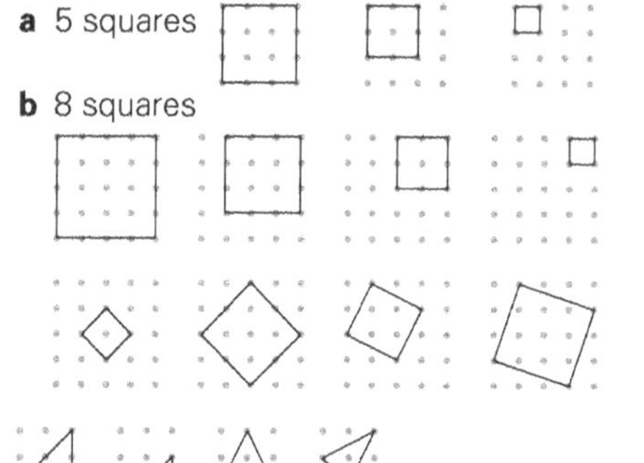

2

3 a 4 isosceles triangles

b 4 isosceles triangles

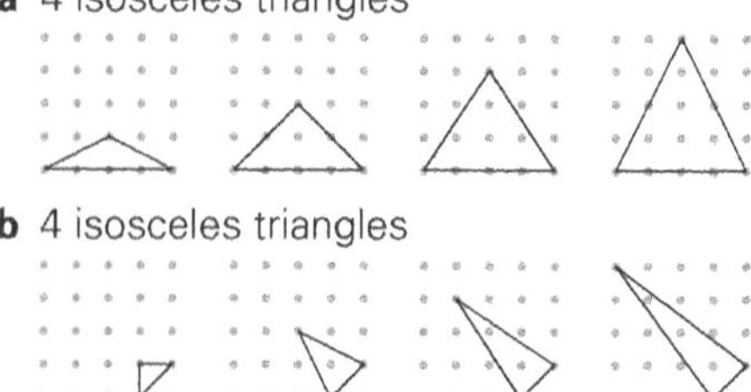

Challenges using cubes (page 74)

1 Students' attention should be drawn to the fact that each cube has six faces, so if a cube is joined to only one other cube, it will have five red faces and one yellow face. If a cube is joined to two other cubes, it will have four red faces and two yellow faces. Note that in figures **b**, **c** and **e** there are cubes with only three red faces, as they are each joined to three others.

It is essential that the students build the solids for themselves. The solutions show cubes with five red faces as shaded, and those with four red faces as marked with an 'X'.

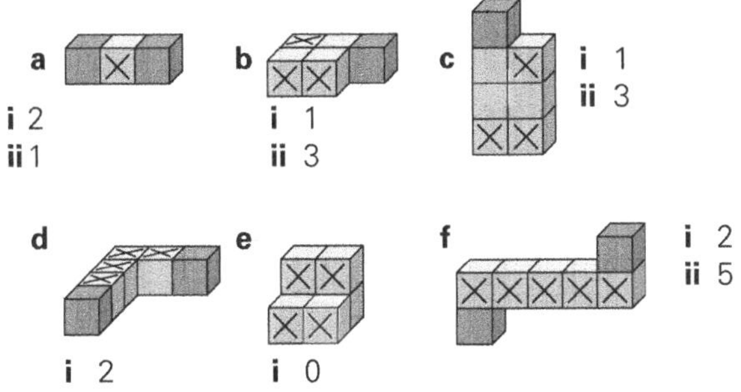

2 The blue cube will be 3 × 3 × 3, and there will be 27 cubes altogether. They will all be blue except for the centre yellow one; therefore 26 blue cubes are needed.

It is essential that students build this large cube themselves.

3 15 more cubes: 4 + 5 + 6
Altogether 21 cubes

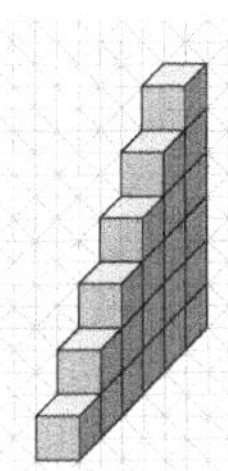

Puzzle with shapes (page 75)

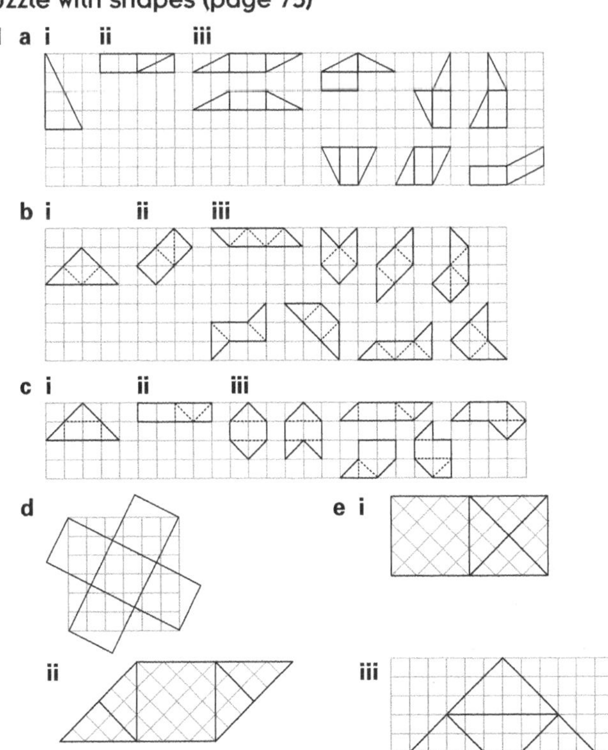

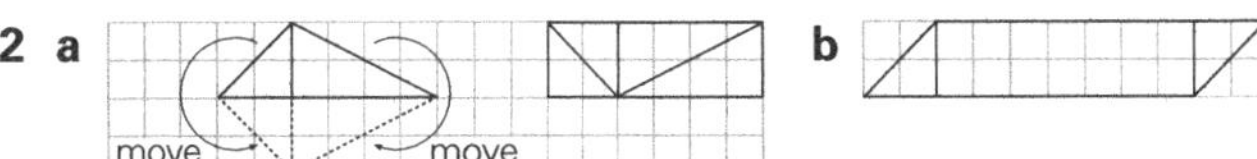

Polyominoes (page 76)

1 Students should check carefully that all their shapes are different; they should circle identical examples. A class discussion about identical shapes is worthwhile.

These are the four tetrominoes.

2 The pentominoes can be identified by letters.

Finding all the shapes will require systematic working and concentration. Again, students will need to check that all their shapes are different.

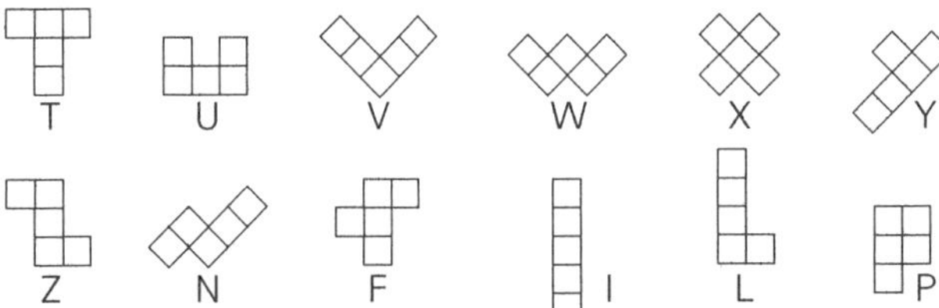

Tessellations (page 77)

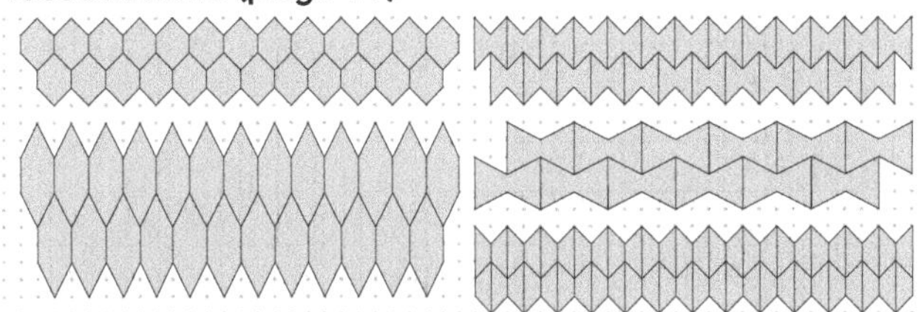

Extending tessellations (page 78)

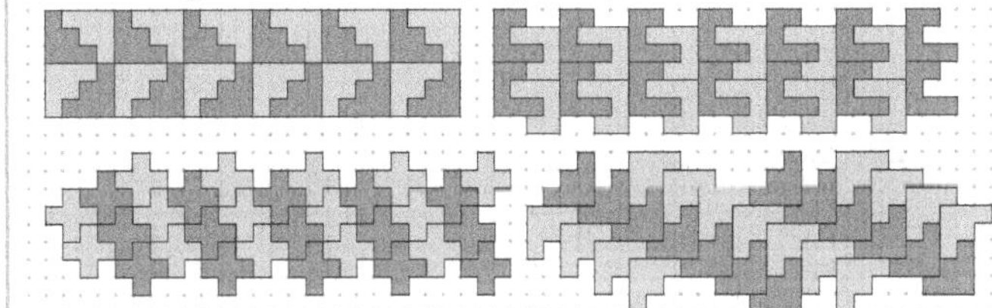

Cutting cakes (page 79)

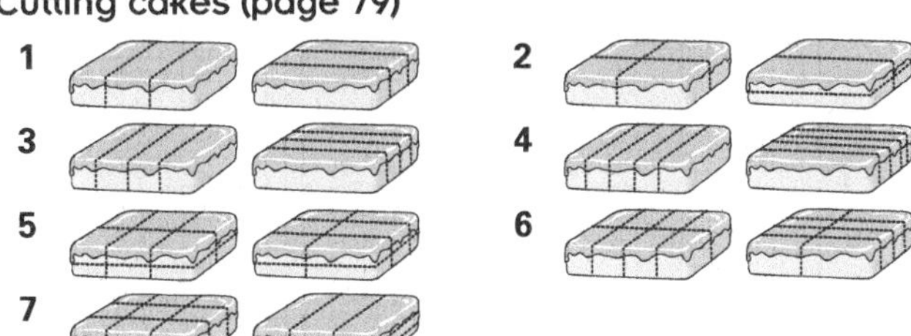

Knight's tour (page 80)

A great deal of perseverance is required to find a solution for all these puzzles, and students should be given several weeks in which to complete them. Possible solutions are given for each one, all of which are the work of students. There are many other solutions.

1 5 × 5 square

5	20	11	16	25
12	17	6	21	10
7	4	19	24	15
18	13	2	9	22
3	8	23	14	1

Knight's tour (page 80) *continued...*

2 a **i** 6 × 6 square

19	28	15	6	21	30
14	5	20	29	16	7
27	18	35	8	31	22
4	13	26	17	36	9
25	34	11	2	23	32
12	3	24	33	10	1

ii 3 × 4 grid

8	11	6	3
5	2	9	12
10	7	4	1

iii 4 × 3 grid

12	5	10
9	2	7
6	11	4
3	8	1

iv There is no solution to the 4 × 4 square.

v 5 × 4 grid

18	13	4	9
3	8	19	14
12	17	10	5
7	2	15	20
16	11	6	1

vi 4 × 5 grid

12	9	18	5	14
17	4	13	10	19
8	11	2	15	6
3	16	7	20	1

b 8 × 8 square

7	18	35	46	9	20	37	48
34	45	8	19	36	47	10	21
17	6	55	58	63	60	49	38
44	33	64	61	54	57	22	11
5	16	43	56	59	62	39	50
30	27	32	53	42	25	12	23
15	4	29	26	13	2	51	40
28	31	14	3	52	41	24	1

Completing the 9 x 9 square and the 12 x 12 square Knight's tour is a difficult challenge for those students who have completed the above successfully.

9 × 9 square

9	38	11	24	7	36	13	22	5
54	25	8	37	12	23	6	35	14
39	10	55	70	65	60	79	4	21
26	53	64	61	80	71	66	15	34
63	40	81	56	69	78	59	20	3
52	27	62	77	58	67	72	33	16
41	76	57	68	73	44	19	2	47
28	51	74	43	30	49	46	17	32
75	42	29	50	45	18	31	48	1

9	50	37	24	11	52	39	26	13
36	23	10	51	38	25	12	53	40
49	8	61	78	73	68	63	14	27
22	35	72	67	62	79	74	41	54
7	48	77	60	81	64	69	28	15
34	21	66	71	58	75	80	55	42
47	6	59	76	65	70	57	16	29
20	33	4	45	18	31	2	43	56
5	46	19	32	3	44	17	30	1

12 × 12 square

11	74	31	52	13	76	33	54	15	78	35	56
30	51	12	75	32	53	14	77	34	55	16	79
73	10	113	102	125	90	115	104	127	92	57	36
50	29	124	89	114	103	126	91	116	105	80	17
9	72	101	112	135	132	139	142	93	128	37	58
28	49	88	123	138	141	136	133	106	117	18	81
71	8	111	100	131	134	143	140	129	94	59	38
48	27	122	87	144	137	130	95	118	107	82	19
7	70	99	110	85	120	97	108	83	62	39	60
26	47	86	121	98	109	84	119	96	41	20	63
69	6	45	24	67	4	43	22	65	2	61	40
46	25	68	5	44	23	66	3	42	21	64	1

Amazing mathematics (page 81)

If the whole class follows the instructions and then you ask the students to call out their answers one at a time, they will be amazed at the result. The reasoning given below is for teachers' benefit, but it may be shown to an exceptionally gifted child.

1 Your result is always 1.

Let the original number be ▲. Now, working through the problem:

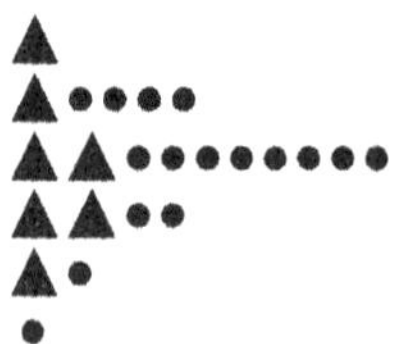

2 Your result is always the number you first thought of.

Let the original number be ◯.

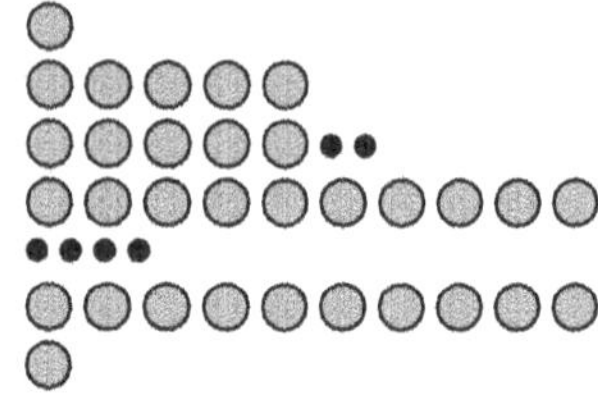

3 Yes. It is known that if you follow these instructions, starting with any number less than 1 million, you will eventually reach 1.

This and Goldbach's conjecture are two instances in which all examples seem to work and yet mathematicians are not able to prove the statement for all numbers.

Note that from the example, any starting number from the chain (11 → 34 → 17 → 52 → 26 → 13 → 40 → 20 → 10 → 5 → 16 → 8 → 4 → 2 → 1) will reach 1.

In fact, any even multiple from the chain will also reach 1:
since 11 → … → 1
so does 22 → … → 1 and 44 → … → 1

a 12 → 6 → 3 → 3 × 3 + 1 = 10 → 5 → 5 × 3 + 1 = 16 → 8 → 4 → 2 → 1
Since 12 reaches 1, so do 24 and 48.

b 7 → 7 × 3 + 1 = 22 → 11, and we have shown in the example that 11 will reach 1.
Since 7 reaches 1, so do 14 and 28.

c We already know the following numbers reach 1, as they occur in the chains above:
2, 3, 4, 5, 6, 7, 8, _, 10, 11, 12, _, 14, _ 16, _, _, _, 20, _, 22, _, 24, (_, 26, _, 28, _, _, _, 32, _, 34, _, _, _, _, _, 40, _, _, _, 44)
Now show that each of the missing numbers will reach 1.
9 → 28 → 14 → 7 → … → 1
13 → 40 → … → 1
15 → 46 → 23 → 70 → 35 → 106 → 53 → 160 → 80 → 40 → … → 1
17 → 52 → 26 → 13 → … → 1
18 → 9 → … → 1
19 → 58 → 29 → 88 → 44 → … → 1
21 → 64 → 32 → … → 1
23 → … → 1 (see 15 →)

Logic using scales (page 82)

1  

Replace ◯ with ★★ on the first scale.

2 

Replace ◯◯ with ★ on the first scale.

3 Replace ○○○ with ★ on the first scale, then remove ★ from each side.

4 Replace ○○ with ★ on the first scale.

5 Replace ○ with ★★ on the first scale, then remove ★★ from both sides.

6 Replace ○○ with ★ on the first scale, remove ★★★ from both sides, then halve.

7 Replace ○ with ★★ on the first scale, then halve.

8  Replace ○○ with ★ on the first scale.

Puzzles with scales (page 83)

1 Students may find this easier if they express the problem orally: If an A box balances two B boxes, how many B boxes balance a certain number of A boxes?

a 6 **b** 5 **c** 6 **d** 10 **e** 9 **f** 9

2 **a** 15 kg **b** 18 kg **c** 9 kg **d** 6 kg **e** 12 kg **f** 6 kg
g 4 kg **h** 3 kg **i** 2 kg **j** 3 kg **k** 3 kg **l** 4 kg

Scale puzzles (page 84)

1 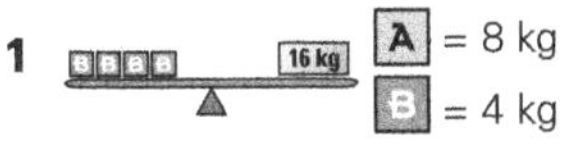 A = 8 kg, B = 4 kg

2 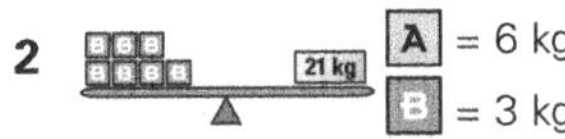 A = 6 kg, B = 3 kg

3 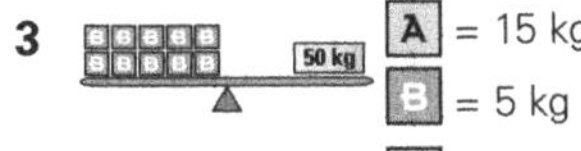 A = 15 kg, B = 5 kg

4 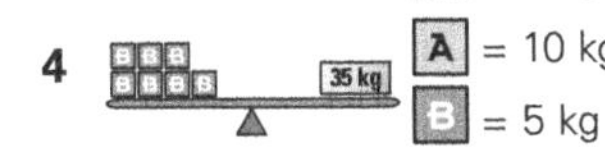 A = 10 kg, B = 5 kg

5 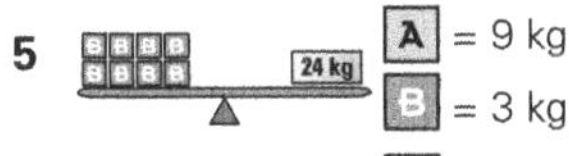 A = 9 kg, B = 3 kg

6 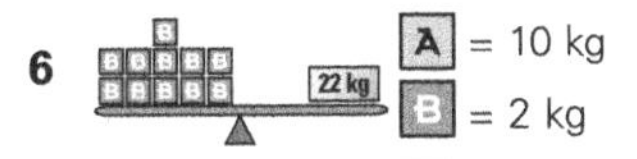 A = 10 kg, B = 2 kg

7 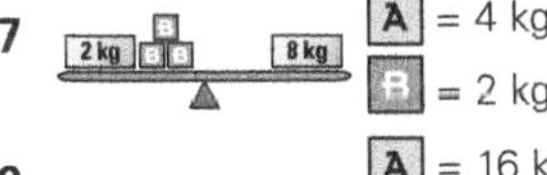 A = 4 kg, B = 2 kg

8 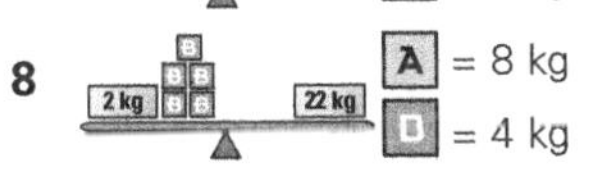 A = 8 kg, B = 4 kg

9  A = 16 kg, B = 4 kg

10 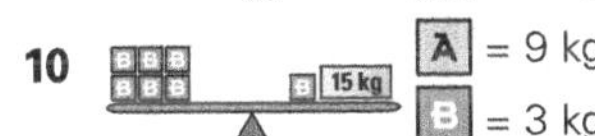 A = 9 kg, B = 3 kg

Shapes and symbols (page 85)

1  2  3 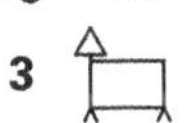 4  5 

6  7  8  9 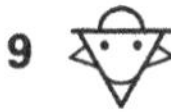

Puzzles with matches (page 86)

1 **a** 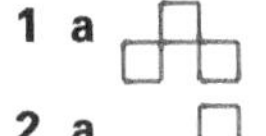 **b** 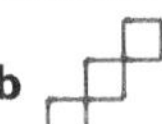

2 **a** 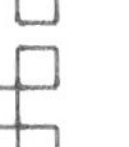 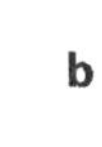  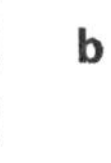 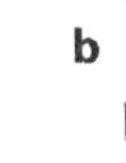 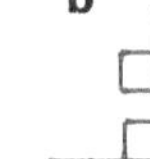 **b** 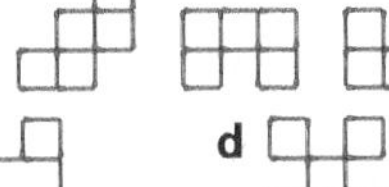 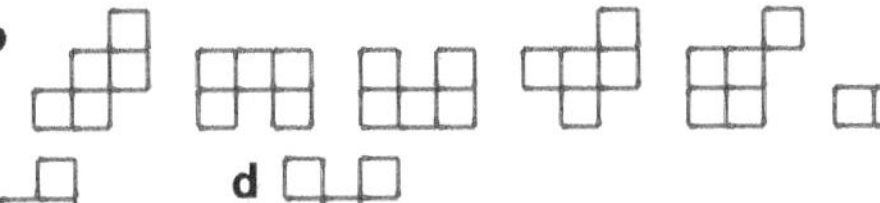

c **d**

3  *Note*: This is a difficult puzzle, as students are not expecting to make squares of different sizes. It may be necessary to give a hint.

4 **a**  **b**  or  5 

6  This is a tricky question, as not all the triangles are the same size. If the students become frustrated, this can be given as a hint.

Matchstick puzzles (page 87)

Encourage students to express each rule in words; in the rule formulas, □ stands for the number of original shapes.

1 The number of matches is twice the number of triangles plus 1. Many students will say that the number of matches is going up (or increasing) by 2, but this is not really a rule and they should be led to see the relationship between the figures and the number of matches. If they cannot predict the number for the tenth diagram, they will have to complete the table with figures 6 to 10.

								Rule
1	Triangles	1	2	3	4	5	→ 10	2 × □ + 1
	Matches	3	5	7	9	11	→ 21	
2	Squares	1	2	3	4	5	→ 10	3 × □ + 1
	Matches	4	7	10	13	16	→ 31	
3	Hexagons	1	2	3	4	5	→ 10	5 × □ + 1
	Matches	6	11	16	21	26	→ 51	
4	Houses	1	2	3	4	5	→ 10	4 × □ + 1
	Matches	5	9	13	17	21	→ 41	
5	Double squares	1	2	3	4	5	→ 10	5 × □ + 2
	Matches	7	12	17	22	27	→ 52	
6	Diamonds	1	2	3	4	5	→ 10	5 × □ − 1
	Matches	4	9	14	19	24	→ 49	

Prisoners in cells (page 88)

1 To solve this puzzle students should be given or asked to construct a board and be provided with counters.

2 Be sure that all students' solutions are different and one is not a rotation or reflection of another.

3

STUDENT BOOK Solutions

BOOK 5

What's the value? (page 1)

 = 6 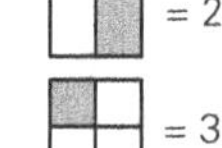 = 8 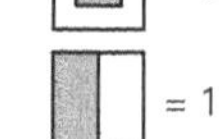 = 2 = 5 = 7

= 4 = 9 = 3 = 1

1 $1 \times 1 = 1$ **2** $1 + 1 = 2$ **3** $2 + 2 = 2 \times 2$

4 $4 \div 2 = 2$ **5** $4 + 1 = 5$ **6** $9 - 4 = 5$

7 $5 + 2 = 7$ **8** $7 + 1 = 8$ **9** $8 + 1 = 9$

10 $9 \div 3 = 3$ **11** $2 \times 3 = 6$

Find my rule (page 2)

	Rule	Missing Nos		Rule	Missing Nos
1	× 4, + 7	12, 19	**2**	− 7, × 6	3, 18
3	− 3, × 9	6, 54	**4**	÷ 2, + 5	3, 8
5	x 8, − 11	48, 37	**6**	x 9, − 5	45, 40
7	− 4, × 7	2, 14	**8**	÷ 3, × 8	2, 16
9	÷ 4, × 5	4, 20	**10**	x 6, ÷ 2	48, 24

Missing numbers (page 3)

In questions 1 and 2, students need to find the rule for each shape. In questions 3, 4 and 5 they are expected to apply the rule, starting with the first given number. In questions 6–12 they should work backwards as well.

For example, in question 12: 22 was reached by dividing a number by 3, so to find that number by working backwards we must multiply 22 by 3, making the number 66. Again working backwards: 66 was reached by subtracting 10 from a number; to find it we need to add 10 to 66, giving us 76. Complete the exercise in the same way.

1 **Rule:** + 8, × 2, − 10, ÷ 3 **2** **Rule:** + 8, × 2, − 10, ÷ 3

3 17, 34, 24 **4** 11, 22, 12, 4

5 20, 40, 30, 10 **6** 36, 88, 78, 26

7 24, 32, 54, 18 **8** 21, 29, 58, 16

9 18, 26, 52, 14 **10** 15, 23, 46, 12

11 27, 35, 70, 60 **12** 30, 38, 76, 66

Number squares (page 4)

1 A = 3, B = 2, C = 1, D = 6, E = 4, F = 5

6	1	3	10
3	2	4	9
2	3	5	10
11	6	12	

2 A = 3, B = 1, C = 4, D = 5, E = 2

5	3	4	12
2	4	5	11
1	5	2	10
8	12	11	

2 In the second row and the third column C + D = 9, therefore A = 3 (in the first row). Now use trial and error to find the value of C and D: if D = 4 and C = 5, then B = 2, which is impossible, since we know that E = 2. Therefore, D = 5 and C = 4. Alternatively, we could work from the bottom row, where B + D = 6, and then use trial and error.

3 A = 1, B = 4, C = 5, D = 3, E = 2

3	1	2	6
4	5	2	11
1	3	4	8
8	9	8	

Start: C + D = 8 or E + D = 5 or D + B = 7

5 A = 5, B = 2, C = 6, D = 1, E = 3, F = 4

5	1	3	9
4	2	6	12
3	4	2	9
12	7	11	

4 A = 3, B = 5, C = 1, D = 2 or C = 2, D = 1, E = 4

4	5	3	12
2 / 1	3	1 / 2	6
1 / 2	4	2 / 1	7
7	12	6	

Start: D + C= 3 or E + B = 9.

Start: F + E = 7 or D + E = 4.

Find the value (page 5)

In solving these puzzles you can use 'guess and check' and/or the given procedure and starting points.

1 A = 2, B = 5, C = 1, D = 4, E = 3

2	2	1	5	10
5	3	5	4	17
3	3	1	4	11
5	4	2	1	12
15	12	9	14	

From the first row C + B = 6. Now, find the value of D in the fourth row or the fourth column, where B + D + A + C = 12
6 + D + 2 = 12, since A = 2

Therefore D = 4
Since B + C = 6, we can work out C in the third column:
C + B + C + A = 9
6 + C + 2 = 9
Therefore C = 1

2 A = 1, B = 3, C = 4, D = 2, E = 5

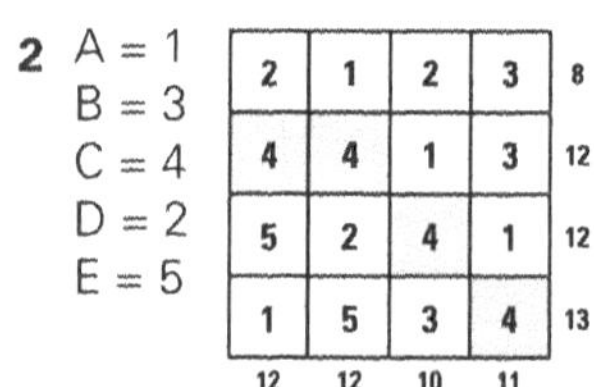

2	1	2	3	8
4	4	1	3	12
5	2	4	1	12
1	5	3	4	13
12	12	10	11	

A + B = 4 from row 2. Therefore B = 3 from column 4, A = 1, E = 5 from row 4 and D = 2 was obvious after we found A and B.

3 A = 2, B = 4, C = 1, D = 5, E = 3

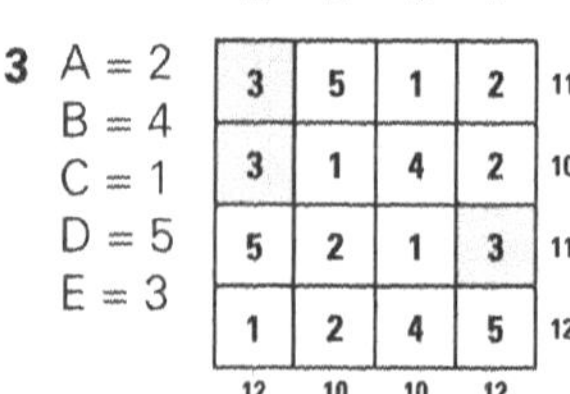

3	5	1	2	11
3	1	4	2	10
5	2	1	3	11
1	2	4	5	12
12	10	10	12	

D + C = 6 from column 1. Therefore A = 2 from row 1 or row 3 or column 2. B = 4 from row 4 as $\underbrace{C + D}_{6} + B + 2 = 12$.

Also, D = 5 from column 4, therefore C = 1.

4 A = 1, B = 4, C = 2, D = 5, E = 3

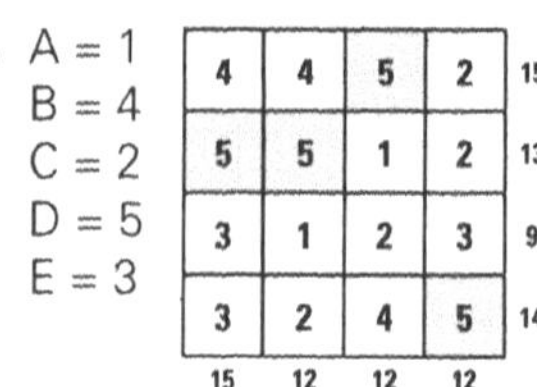

4	4	5	2	15
5	5	1	2	13
3	1	2	3	9
3	2	4	5	14
15	12	12	12	

Start A + C = 3 from row 2. Therefore E = 3 from row 3, B = 4 from column 3 or column 1, so C = 2 and A = 1.

Investigating magic squares (page 6)

1 **a** Yes; the sum is 21. **b** Yes; the sum is 45.

2 **a** Yes; the sum is 12.

b Yes; the sums are 30, 45 and 75, respectively.

16	2	12
6	10	14
8	18	4

(The magic square when you multiply by 2)

c Yes; the sum is $7\frac{1}{2}$

4	$\frac{1}{2}$	3
$\frac{3}{2}$	$\frac{5}{2}$	$\frac{7}{2}$
2	$\frac{9}{2}$	1

(The magic square when you divide by 2)

3 The centre number is one-third of the magic sum.

4 The magic sum of 4 × 4 square is 34.

a When you add 1, 3, 10 or any number (*k*, say) the square remains magic with magic sums of:
$34 + 4 \times 1 = 38$
$34 + 4 \times 3 = 46$
$34 + 4 \times 10 = 74$
$34 + 4 \times k = 34 + 4 \times k$, respectively.

b Magic sum = $34 - 4 \times 1 = 30$ when we subtract 1
magic sum = $34 - 4 \times \frac{1}{2} = 32$ when we subtract $\frac{1}{2}$

c Magic sum = $2 \times 34 = 68$ when multiplied by 2
= $10 \times 34 = 340$ when multiplied by 10
= $k \times 34$ when multiplied by any number k.

d Magic sum = $\frac{1}{2} \times 34 = 17$ when divided by 2.

In all these cases square is still magic.

5 The shaded lines of each square provide the magic sums. In **a** the numbers circled should be worked out first.

a

8	11	2	13
10	5	(16)	3
15	4	9	6
1	(14)	7	12

Magic sum: 34

b

8	15	14	5
9	11	18	4
12	6	7	17
13	10	3	16

Magic sum: 42

c

33	5	7	27
11	23	21	17
19	15	13	25
9	29	31	3

Magic sum: 72

Patterns and sequences (page 7)

		Rule
1	45, 54, 63	Add 9.
2	41, 35, 29	Subtract 6.
3	16, 32, 64	Double.
4	31, 43, 57	Add 2, 4, 6, 8, 10, 12, 14.
5	63, 127, 255	Add 2, 4, 8, 16, 32, 64, 128.
6	1000, 10000, 100000	Multiply by 10.
7	125, 625, 3125	Multiply by 5.
8	25, 36, 49	Square numbers or add 3, 5, 7, 9, 11, 13.
9	4, 2, 1	Halve (divide by 2).
10	1, $\frac{1}{3}$, $\frac{1}{9}$	Divide by 3.
11	5, $\frac{5}{2}$, $\frac{5}{4}$	Halve.
12	15, 21, 28	Triangular numbers: add 2, 3, 4, 5, 6, 7.
13	$\frac{3}{2}$, $\frac{3}{4}$, $\frac{3}{8}$	Halve.
14	0·10, 0·12, 0·14	Add 0·02.
15	0·23, 0·28, 0·33	Add 0·05.
16	0·2, 0·1, 0·05	Halve.
17	26, 37, 50	Add 1, 3, 5, 7, 9, 11, 13.
18	26, 28, 35	Add 7, 2, 7, 2.
19	31, 40, 41	Add 1, 9, 1, 9.
20	15, $7\frac{1}{2}$, $11\frac{1}{2}$	Halve, then add 4.
21	$\frac{2}{81}$, $\frac{1}{243}$, $\frac{1}{1458}$	Divide by 6.
22	13, 21, 34	Fibonacci sequence: each term is the sum of the two previous terms.

What's my rule? (page 8)

1	12	(× 3)	**2**	16	(× 4)	**3**	9	(× 2 + 1)
4	11	(× 3 – 1)	**5**	39	(× 10 – 1)	**6**	21	(× 5 + 1)
7	36	(square number)				**8**	4	(÷ 4)
9	3	(÷ 2)	**10**	66	(× 11)	**11**	22	(× 3 + 1)
12	50	(square + 1)	**13**	27	(× 4 – 1)	**14**	78	(× 11 + 1)
15	3	(÷ 5)	**16**	41	(× 10 + 1)	**17**	15	(square – 1)
18	25	(× 6 + 1)	**19**	55	(× 9 + 1)	**20**	51	(× 7 + 2)

Number sentences (page 9)

These are some possible answers.

1 15 ÷ 5 – 3 = 0 *or* 12 ÷ 4 – 3 = 0 *or* 12 ÷ 3 – 4 = 0

2 4 × 3 – 12 = 0 *or* 3 × 5 – 15 = 0

3 (12 + 3) ÷ 15 = 1

4 (15 – 12) ÷ 3 = 1 *or* (15 – 3) ÷ 12 = 1

5 (5 – 4) × 3 = 3

6 (5 + 4) ÷ 3 = 3 *or* (12 + 3) ÷ 5 = 3

7 (15 + 5) ÷ 4 = 5

8 15 ÷ 5 + 3 = 6 *or* 12 ÷ 4 + 3 = 6

9 15 – 12 + 4 = 7

10 (5 – 3) × 4 = 8

11 12 – (15 ÷ 5) = 9

12 (15 – 12) × 3 = 9

13 15 ÷ 3 + 5 = 10

14 15 – (12 ÷ 3) = 11

15 3 + 4 + 5 = 12

16 5 + 4 × 3 = 17

17 4 × 5 + 3 = 23

18 (12 – 4) × 3 = 24 *or* (5 – 3) × 12 = 24

19 (5 + 4) × 3 = 27

20 (12 – 4) × 5 = 40 *or* (15 – 5) × 4 = 40

Letter values (page 10)

1 B = 5 $\begin{array}{r} 51 \\ \times 5 \\ \hline 255 \end{array}$

2 C = 6 $\begin{array}{r} 61 \\ \times 6 \\ \hline 366 \end{array}$

3 A = 6 $\begin{array}{r} 46 \\ \times 6 \\ \hline 276 \end{array}$

4 M = 2, N = 3 $\begin{array}{r} 23 \\ \times 3 \\ \hline 69 \end{array}$

5 U = 2, V = 4, W = 7 $\begin{array}{r} 247 \\ \times 6 \\ \hline 1482 \end{array}$

6 Since RR must equal 11 (as P + Q must be less than 22), the possible solutions are:

$\begin{array}{r} 9 \\ +2 \\ \hline 11 \end{array}$ $\begin{array}{r} 8 \\ +3 \\ \hline 11 \end{array}$ $\begin{array}{r} 7 \\ +4 \\ \hline 11 \end{array}$ $\begin{array}{r} 6 \\ +5 \\ \hline 11 \end{array}$

$\begin{array}{r} 5 \\ +6 \\ \hline 11 \end{array}$ $\begin{array}{r} 4 \\ +7 \\ \hline 11 \end{array}$ $\begin{array}{r} 3 \\ +8 \\ \hline 11 \end{array}$ $\begin{array}{r} 2 \\ +9 \\ \hline 11 \end{array}$

7 Since the units digits of D + E + F is F, the units digit of D + E is 0, so D + E = 10. Therefore, D = 1 and E = 9. F is any other one-digit number: 2, 3, 4, 5, 6, 7 or 8.

$\begin{array}{r} 1 \\ 9 \\ +F \\ \hline 1F \end{array}$

Letter values (page 10) ***continued...***

8 J = 8, K = 9: 89 + 9 = 98

9 G = 1, H = 0, L = 9: 91 + 9 = 100

10 C = 2, D = 3, E = 9: 23 + 23 + 23 + 23 = 92

C must be an even number and C + C + C + C ← is less than 10, therefore C = 2. D = 3, as D + D + D + D ends in 2, and E = 9.

11 W = 1, X = 9, Y = 8: 88 + 11 + 99 = 198

Since the units digit of Y + W + X is Y, the units digit of W + X is 0, therefore W + X = 10. From the second column, using a similar argument and carrying 1,1 + W + Y = 10. Now W can only be 1 or 2, as the sum of the three two-digit numbers is less than 300, also note X is 1 more than Y. By trial and error it is shown that W = 1, X = 9 and Y = 8.

12 S = 1, P = 3, Q = 4: 34 + 34 + 43 = 111 or S = 2, P = 6, Q = 8: 68 + 68 + 86 = 222

13 W = 1, T = 0, V = 9: 101 − 91 = 10

14 N = 5, L = 1, M = 8: 185 + 185 + 185 = 555 Only 3 × 5 ends in a 5

15 B = 1, A = 9, D = 0: 991 + 119 = 1110

Who am I? (page 11)

There are many different ways to solve these conundrums. One possible method is given for questions 3 to 14.

1 2 **2** 1

3 15 List the two-digit odd numbers less than 20: 11, 13, 15, 17, 19. Now eliminate by crossing out all the numbers that are not composite: ~~11~~, ~~13~~, 15, ~~17~~, ~~19~~.

4 36 This is the only square number greater than 29 and less than 42.

5 30 List the two-digit numbers less than 40 and divisible by 10: 10, 20, 30. Now eliminate those not divisible by 3.

6 12 List the numbers divisible by 4: 4, 8, 12, 16 and 3: 3, 6, 9, 12 and 2: 2, 4, 6, 8, 10, 12 until one number occurs in all three cases.

7 25 List the two-digit square numbers: 16, 25, 36, 49, 64, 81; circle (16) and (25), in which the sum of the digits is 7, and underline the one that is divisible by 5: 25.

8 18 List the even two-digit numbers less than 25: 12, 14, 16, 18, 20, 22, 24. Now underline the number the sum of whose digits is 9: 18. Check that this number is divisible by 3.

9 30 List the two-digit numbers the sum of whose digits is 3: 12, 21, 30. Now underline the number that is divisible by 5 and 2: 30.

10 24 List the two-digit numbers less than 30 that are divisible by 8: 16, 24. Now underline the number that is also divisible by 6: 24.

11 60: List the multiples of 20: 20, 40, 60, 80, ... and check which leaves a remainder of 6 when divided by 9.

12 35 or 63: List the even square numbers: 16, 36, 64. One less gives odd numbers 15, 35, 63. Note the two solutions.

What's my number? (page 12)

1 26 List the numbers between 12 and 32 which leave a remainder of 1 when divided by 5: 16, 21, 26, 31. Now write down numbers, starting from 12, that leave a remainder of 2 when divided by 4: 14, 18, 22, 26, 30. The number that occurs in both lists is 26. Alternatively, from the first list (multiples of 5 plus 1), find the number that leaves a remainder of 2 when divided by 4.

2 49 13, 25, 37, 49, 61 and 73 leave a remainder of 1 when divided by 3 or 4. The number that leaves no remainder when divided by 7 is 49. Alternatively, list the multiples of 7 and eliminate those which do not leave a remainder of 1 when divided by 3 or 4.

3 61 Again it is possible to use several methods. Students can list numbers greater than 50 that when divided by 5, leave a remainder of 1 (51, 56, 61 ...). Eliminate those whose digits do not add to 7 and those that are not prime: my number is 61.

4 77

5 55 List two-digit triangular numbers: 10, 15, 21, 28, 36, 45, 55, 66, 78, 91. Eliminate those not divisible by 5 and those whose digits do not add to 10: my number is 55.

6 23 **7** 8 Four lots of number = 32. Therefore, number is 8.

8 7 Five lots of number = 35. Therefore, number is 7.

9 There are 8 solutions: 12, 23, 34, 45, 56, 67, 78, 89.

10 24 **11** 24 **12** 18

Golbach's conjectures (page 13)

Only some of the possible solutions are listed for each exercise.

1 a 3 + 13, 5 + 11
b 5 + 19, 7 + 17, 11 + 13
c 3 + 43, 5 + 41, 17 + 29, 23 + 23
d 5 + 79, 11 + 73, 13 + 71, 17 + 67, 23 + 61, 31 + 53, 37 + 47, 41 + 43
e 3 + 89, 13 + 79, 19 + 73, 31 + 61
f 5 + 67, 11 + 61, 13 + 59, 19 + 53, 29 + 43, 31 + 41
g 7 + 31, 19 + 19
h 19 + 79, 31 + 67, 37 + 61

2 a 9 = 2 + 2 + 5 = 3 + 3 + 3
b 23 = 5 + 7 + 11 = 5 + 5 + 13
c 29 = 5 + 11 + 13 = 7 + 11 + 11 = 3 + 13 + 13
d 33 = 3 + 13 + 17 = 5 + 11 + 17 = 7 + 7 + 19 = 3 + 11 + 19
e 35 = 5 + 11 + 19 = 7 + 11 + 17 = 3 + 13 + 19
f 53 = 23 + 11 + 19 = 7 + 23 + 23 = 17 + 17 + 19
g 83 = 61 + 19 + 3 = 61 + 11 + 11 = 5 + 19 + 59
h 69 = 53 + 13 + 3 = 53 + 11 + 5 = 3 + 7 + 59
i 45 = 23 + 17 + 5 = 23 + 19 + 3 = 13 + 19 + 13
j 25 = 13 + 7 + 5 = 11 + 11 + 3 = 7 + 7 + 11
k 47 = 29 + 13 + 5 = 29 + 11 + 7 = 5 + 11 + 31
l 101 = 29 + 29 + 43 = 79 + 11 + 11 = 3 + 37 + 61

Digits and numbers (page 14)

1 **a** 1000 **b** 1023 **c** 1011
d 1001 **e** 1111 **f** 235

2 **a** 9999 **b** 9995 **c** 9876
d 7532 **e** 9872 **f** 9870

3 **a i** 9×8 **ii** $(7 + 8) \times 9$ **iii** 87×9
iv 87×96 **v** $98 \div 1$ **vi** $(9 + 8) \div 1$
vii $(6 + 9) \div (1 + 2)$ or $(7 + 8) \div (1 + 2)$ **viii** $96 \div (2 + 1)$

Note that is important to work systematically.

b i $(2 + 3) \times 1$ **ii** 24×13

iii $12 \div 6$, $14 \div 7$, $16 \div 8$, $18 \div 9$

iv $(1 + 2) \div 3$ $(1 + 3) \div 4$ $(1 + 4) \div 5$ $(1 + 5) \div 6$
$(1 + 6) \div 7$ $(1 + 7) \div 8$ $(1 + 8) \div 9$
$(2 + 3) \div 5$ $(2 + 4) \div 6$ $(2 + 5) \div 7$ $(2 + 6) \div 8$
$(2 + 7) \div 9$
$(3 + 4) \div 7$ $(3 + 5) \div 8$ $(3 + 6) \div 9$
$(4 + 5) \div 9$

v $(4 + 5) \div (3 + 6)$, $(3 + 6) \div (4 + 5)$
$(4 + 5) \div (2 + 7)$, $(2 + 7) \div (4 + 5)$
$(4 + 5) \div (1 + 8)$, $(1 + 8) \div (4 + 5)$
$(3 + 6) \div (2 + 7)$, $(2 + 7) \div (3 + 6)$
$(3 + 6) \div (1 + 8)$, $(1 + 8) \div (3 + 6)$
$(2 + 7) \div (1 + 8)$, $(1 + 8) \div (2 + 7)$ } $= 9 \div 9$
$(3 + 5) \div (2 + 6)$, $(2 + 6) \div (3 + 5)$
$(3 + 5) \div (1 + 7)$, $(1 + 7) \div (3 + 5)$
$(2 + 6) \div (1 + 7)$, $(1 + 7) \div (2 + 6)$ } $= 8 \div 8$
$(3 + 4) \div (1 + 6)$, $(1 + 6) \div (3 + 4)$
$(3 + 4) \div (2 + 5)$, $(2 + 5) \div (3 + 4)$ } $= 7 \div 7$
$(2 + 4) \div (1 + 5)$, $(1 + 5) \div (2 + 4)$ $= 6 \div 6$
$(2 + 3) \div (1 + 4)$, $(1 + 4) \div (2 + 3)$ $= 5 \div 5$

vi $17 \div (9 + 8)$ $16 \div (9 + 7)$
$15 \div (9 + 6)$ $15 \div (8 + 7)$
$14 \div (9 + 5)$ $14 \div (8 + 6)$
$13 \div (9 + 4)$ $13 \div (8 + 5)$ $13 \div (7 + 6)$
$12 \div (9 + 3)$ $12 \div (8 + 4)$ $12 \div (7 + 5)$

Missing digits (page 15)

1
```
  4 3 7
+ 3 8 5
  8 2 2
```
2
```
  3 5 8
- 1 7 3
  1 8 5
```
3
```
  3 4 2
- 1 9 7
  1 4 5
```
4
```
    5 8
×     7
  4 0 6
```
5
```
    2 5 2
×       6
  1 5 1 2
```
or
```
    2 5 7
×       6
  1 5 4 2
```
6
```
      3 5
      2 2
      7 0
×   7 0 0
    7 7 0
```
7
```
        7 8
        6 1
        7 8
×   4 6 8 0
    4 7 5 8
```
8
```
        7 4
        3 2
      1 4 8
×   2 2 2 0
    2 3 6 8
```
9
```
      2 9
      1 9
    2 6 1
×   2 9 0
    5 5 1
```
10
```
      1 7
      2 5
      8 5
×   3 4 0
    4 2 5
```
11
```
      1 7
      3 1
      1 7
×   5 1 0
    5 2 7
```
12
```
        7 5
        3 7
      5 2 5
×   2 2 5 0
    2 7 7 5
```
13
```
        6 7 1
          5 1
        6 7 1
×   3 3 5 5 0
    3 4 2 2 1
```
It is obvious that the missing unit digit is a 1, since the answer is a three-digit number when multiplied by 671. The digit in the tens' place must be a 5, since 5 x 6 = 30 and the addition, with carrying, is 34. Then the multiplication is completed to work out the remaining digits.

14
```
   3 2 5
3)9 7 5
```
15
```
   9 6
3)2 8 8
```
16
```
    9 2
6)5 5 2
```
or
```
    9 7
6)5 8 2
```
It might be easier to do this question as a multiplication.

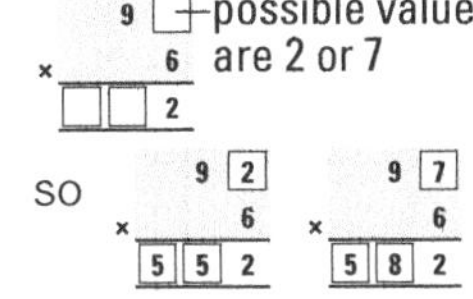

so
```
    9 2        9 7
×     6    ×     6
  5 5 2      5 8 2
```

17
```
  1 4 2
7)9 9 4
```
A very challenging question. Again, consider the equivalent multiplication.

18
```
  1 2 2
8)9 7 6
```
Easier to consider the multiplication.

19
```
          5 7
   12)6 8 4
      6 0
        8 4
        8 4
          0
```
20
```
            3 6      — 6 × 5□ = 312
   52)1 8 7 2        — and now □ = 2
      1 5 6
        3 1 2        — (since no remainder)
        3 1 2
            0
```
must be 3 as 3 × 5 = 15

Multiplication table squares (page 16)

1

×	3	5	4	7
2	6	10	8	14
8	24	40	32	56
3	9	15	12	21
9	27	45	36	63

2

×	4	6	2	7
5	20	30	10	35
3	12	18	6	21
8	32	48	16	56
6	24	36	12	42

3

×	2	3	9	5
2	4	6	18	10
8	16	24	72	40
7	14	21	63	35
4	8	12	36	20

4

×	7	4	11	5
2	14	8	22	10
3	21	12	33	15
9	63	36	99	45
4	28	16	44	20

5

×	3	8	5	10
4	12	32	20	40
5	15	40	25	50
3	9	24	15	30
9	27	72	45	90

6

×	9	2	7	8
3	27	6	21	24
8	72	16	56	64
6	54	12	42	48
2	18	4	14	16

7

×	5	2	10	8
5	25	10	50	40
8	40	16	80	64
3	15	6	30	24
10	50	20	100	80

8

×	5	4	6	8
7	35	28	42	56
6	30	24	36	48
9	45	36	54	72
4	20	16	24	32

Mathematical shortcuts (page 17)

1. a $1400 \div 2 = 700$ b $3600 \div 2 = 1800$
   c $4200 \div 2 = 2100$ d $9800 \div 2 = 4900$
   e $5500 \div 2 = 2750$ f $5900 \div 2 = 2950$
   g $1200 \div 4 = 300$ h $1600 \div 4 = 400$
   i $3600 \div 4 = 900$ j $900 \div 4 = 225$
   k $1800 \div 4 = 450$ l $2100 \div 4 = 525$
2. a 2×3 (hundred) $+ 25 = 625$
   b 3×4 (hundred) $+ 25 = 1225$
   c 5×6 (hundred) $+ 25 = 3025$
   d 7×8 (hundred) $+ 25 = 5625$
   e 10×11 (hundred) $+ 24 = 11\,025$
3. a $5 \times 6 + \frac{1}{4} = 30\frac{1}{4}$ b $7 \times 8 + \frac{1}{4} = 56\frac{1}{4}$
   c $9 \times 10 + \frac{1}{4} = 90\frac{1}{4}$ d $11 \times 12 + \frac{1}{4} = 132\frac{1}{4}$
   e $99 \times 100 + \frac{1}{4} = 9900\frac{1}{4}$

Taking shortcuts (page 18)

1. a $(15 + 5) \times (15 - 5) = 20 \times 10 = 200$
   b $(17 + 7) \times (17 - 7) = 24 \times 10 = 240$
   c $(83 + 17) \times (83 - 17) = 100 \times 66 = 6600$
   d $(58 + 42) \times (58 - 42) = 100 \times 16 = 1600$
2. a $12 \times (88 + 12) = 1200$
   b $23 \times (97 + 3) = 2300$
   c $23 \times (35 - 15) = 460$
   d $23 \times (104 - 4) = 2300$
3. a 20 terms: 10 pairs of numbers whose sum is 21 $= 10 \times 21 = 210$
   b 100 terms: 50 pairs of numbers whose sum is 101 $= 50 \times 101 = 5050$
   c 9 terms: 4 pairs of numbers whose sum is 110, plus the number in the middle $= 4 \times 110 + 55 = 495$
   d 50 terms: 25 pairs of numbers whose sum is 102 $= 25 \times 102 = 2550$
   e 50 terms: 25 pairs of numbers whose sum is 100 $= 25 \times 100 = 2500$

Find three numbers (page 19)

1 □2, △4, ○3 2 □9, △3, ○6 3 □3, △6, ○2
4 □9, △4, ○8 5 □5, △10, ○13 6 □3, △6, ○2
7 □4, △9, ○6 8 □6, △3, ○7 9 □9, △3, ○6
10 □2, △7, ○5

Shopping (page 20)

Students can draw up a table in order to answer the questions.

Lettuces	Tomatoes	Apples	Potatoes
1	2	5	10
2	4	(**3a**) 10	20
3	(**1a**) 6	(**1b**) 15	(**1c**) 30
(**2b**) 4	8	20	40
(**2a**) 5	10	(**3c**) 25	50
(**2c**) 6	12	(**3d**) 30	60
7	14	35	70
8	16	40	80

1. a 6 b 15 c 30
2. a 5 b 4 c 6
3. a 10 b 50 c 25 d 30
4. a 5 b 10 c 50
5. 10 tomatoes = 5 lettuces
   10 apples = 2 lettuces
   Therefore total trade = 7 lettuces
6. a 8 lettuces b 40 apples c 80 potatoes

   Using the table is one way to solve these problems. Another method uses ratios or proportion.

   For **2a**: since 2 tomatoes are worth 1 lettuce, 10 tomatoes (which is 5 lots of 2) are worth 5 lettuces (5 lots of 1).

Special space maths (page 21)

Students could act out some of these problems by making models. They could stick labels on coins.

1. a 1 nive + 4 sives b 2 nives + 2 sives
   c 4 nives + 3 sives d 10 nives or 50 sives
   e 11 nives or 55 sives f 19 nives or 95 sives
2. a 15 sives b 17 sives c 21 sives
   d 25 sives (this is very important) e 75 sives
   f 30 sives
3. a 3 sives b 3 nives
   c 2 nives + 2 sives, or 12 sives d 3 nives
   e 2 nives + 2 sives, or 12 sives
4. a 1 kive, 2 nives and 4 sives b 2 kives, 3 nives
   c 1 kive, 2 nives, 3 sives d 3 kives, 4 nives

What's my question? (page 22)

Students should be given these exercises without any prior preparation or discussion, and the one with the most solutions rewarded. They should discuss their methods with one another. Ideally, the factors should be tested systematically; however, most probably only the better students will do it this way.

1. a $60 \times 2 = 120$, $30 \times 4 = 120$, $15 \times 8 = 120$, $40 \times 3 = 120$, $20 \times 6 = 120$, $24 \times 5 = 120$
   b $70 \times 2 = 140$, $35 \times 4 = 140$, $20 \times 7 = 140$, $28 \times 5 = 140$
   c $75 \times 2 = 150$, $50 \times 3 = 150$, $30 \times 5 = 150$, $25 \times 6 = 150$
   d $72 \times 2 = 144$, $36 \times 4 = 144$, $18 \times 8 = 144$, $48 \times 3 = 144$, $24 \times 6 = 144$, $16 \times 9 = 144$
   e $50 \times 5 = 250$
   f $66 \times 2 = 132$, $33 \times 4 = 132$, $44 \times 3 = 132$, $22 \times 6 = 132$
   g $68 \times 2 = 136$, $34 \times 4 = 136$, $17 \times 8 = 136$
   h $50 \times 4 = 200$, $25 \times 8 = 200$, $40 \times 5 = 200$
   i $36 \times 6 = 216$, $24 \times 9 = 216$, $54 \times 4 = 216$, $27 \times 8 = 216$, $72 \times 3 = 216$

2 a $\begin{array}{r} 87 \\ \times 9 \\ \hline 783 \end{array}$ **b** $\begin{array}{r} 87 \\ \times 94 \\ \hline 8178 \end{array}$ **c** $\begin{array}{r} 874 \\ \times 9 \\ \hline 7866 \end{array}$

The missing sum (page 23)

1

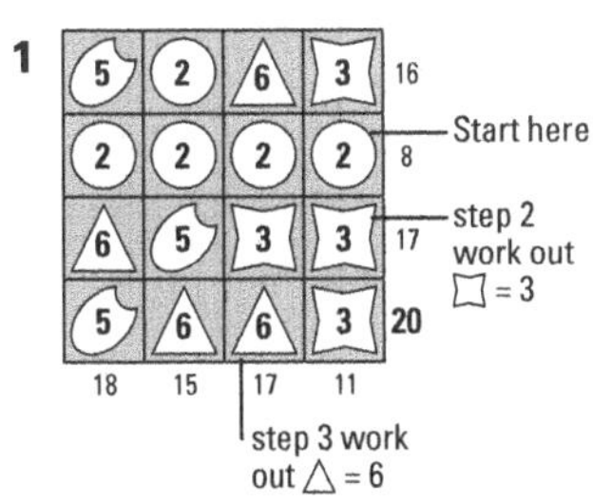

2

Start

4	3	7	1	15
1	3	3	1	8 step 2
4	3	1	4	12
7	3	4	3	**17**
16	12	15	9	

3

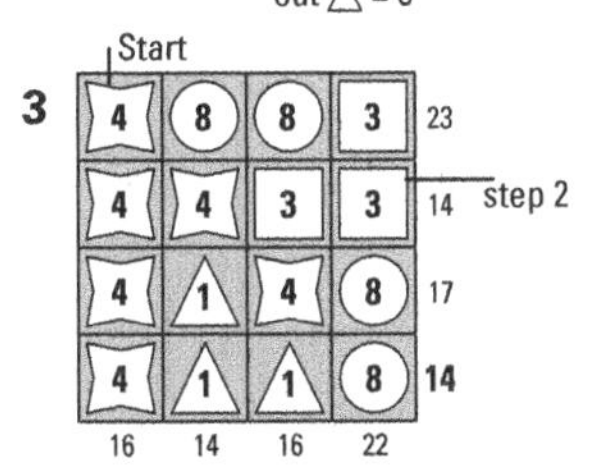

4

step 2

3	2	3	2	10
3	5	3	6	17
5	5	5	5	20 Start
6	2	3	5	**16**
17	14	14	18	

5

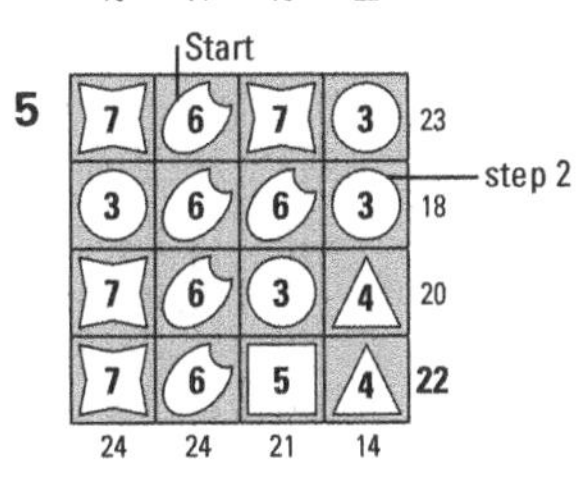

6

Start

7	5	5	7	24 — step 2
9	3	4	7	23
4	4	5	7	20
3	3	9	7	**22**
23	15	23	28	

Shaded fractions (page 24)

1 $\frac{1}{2}$	**2** $\frac{3}{4}$	**3** $\frac{1}{2}$	**4** $\frac{1}{4}$	**5** $\frac{1}{2}$	**6** $\frac{1}{2}$
7 $\frac{1}{2}$	**8** $\frac{1}{2}$	**9** $\frac{1}{2}$	**10** $\frac{1}{2}$	**11** $\frac{1}{4}$	**12** $\frac{1}{2}$
13 $\frac{1}{4}$	**14** $\frac{1}{4}$	**15** $\frac{3}{4}$	**16** $\frac{1}{2}$	**17** $\frac{1}{8}$	**18** $\frac{7}{8}$
19 $\frac{3}{4}$	**20** $\frac{1}{2}$	**21** $\frac{3}{8}$	**22** $\frac{1}{4}$	**23** $\frac{3}{4}$	**24** $\frac{5}{8}$
25 $\frac{1}{2}$	**26** $\frac{1}{3}$	**27** $\frac{1}{6}$	**28** $\frac{1}{2}$	**29** $\frac{1}{3}$	**30** $\frac{5}{9}$

Fraction fun (page 25)

Students should have counters—or some other material—available if they want to work out the problems with the use of some tangible aid.

Students should ask:

1 10Half of what number is 5?
2 $\frac{1}{2}$ What fraction of 8 is 4?
3 12One-quarter of what number is 3?
4 20 One-quarter of what number is 5?
5 $\frac{1}{2}$ What fraction of 18 is 9?

6 $\frac{1}{2}$	**7** $\frac{1}{4}$	**8** $\frac{1}{4}$	**9** 30	**10** $\frac{1}{10}$	**11** $\frac{1}{4}$
12 $\frac{1}{2}$	**13** $\frac{1}{5}$	**14** 40	**15** 50	**16** $\frac{1}{2}$	**17** 27
18 36	**19** 20	**20** 90	**21** $\frac{1}{3}$	**22** $\frac{3}{4}$	**23** 32
24 $\frac{1}{5}$	**25** $\frac{1}{3}$	**26** $\frac{1}{5}$	**27** $\frac{2}{3}$	**28** $\frac{1}{6}$	**29** 15
30 24					

Clock arithmetic (page 26)

1

+	0	1	2	3
0	0	1	2	3
1	1	2	3	0
2	2	3	0	1
3	3	0	1	2

×	0	1	2	3
0	0	0	0	0
1	0	1	2	3
2	0	2	0	2
3	0	3	2	1

2	**a** 1	**b** 2	**c** 3			
3	**a** 1	**b** 3	**c** 1	**d** 3	**e** 4	**f** 2
4	**a** 1	**b** 3	**c** 4	**d** 2	**e** 3	**f** 4
5	**a** 4	**b** 0	**c** 8	**d** 9	**e** 3	**f** 9

Tests of divisibility (page 27)

1 9 or 0 **2** 2 or 5 or 8 **3** 0
4 0 or 5 **5** 1, 3, 5, 7 or 9 **6** 3
7 2 or 5 or 8 **8** 0, 4 or 8 **9** 0 or 9
10 0, 2, 4, 6 or 8 **11** Any whole number from 0–9
12 2 or 6 **13** 2, 8 (must be divisible by 2 and 3)
14 2 or 8 (must be divisible by 3 and 4)
15 11, 02, or 20 (the sum of the missing digits is 2), or 29, 92, 38, 83, 47, 74, 56 or 65 (sum of digits is 11)
16 A number is divisible by 36 if it is divisible by 4 and 9. For this number to be divisible by 4, the last digit must be 0, 4 or 8, and for it to be divisible by 9, the corresponding first digit must be 5, 1 or 6. Therefore the number is:

<u>5</u>3 82<u>0</u> or <u>1</u>3 82<u>4</u> or <u>6</u>3 82<u>8</u>.

17 8 + 9 – (4 + 0 + _) = 11, so missing number is 2.
18 Number is divisible by 4 so ends in 2, or 6. If 2, then missing number is 6. If 6, then missing number is 2. So number is either 846<u>6</u>9<u>2</u> or 846<u>2</u>9<u>6</u>.

Four 5s challenge (page 28)

Students must take special care with the order of operations. It would be a good idea to let them check one another's work before showing it to the teacher.

These are some possible solutions. There are many others. Note: $\sqrt[5]{5^5} \div 5$ (see question 2) might seem a very difficult solution to come from such young mathematicians, but in fact it was an answer given to the author by a 10-year-old. ($\sqrt[5]{5^5}$ = the fifth root of 5 to the power of 5. Since $\sqrt[5]{5^5} = 5$, the student concluded the above.)

1 $5 + 5 - 5 - 5 = 0$ $(5 - 5) \times (5 - 5) = 0$
$(5 + 5) \times (5 - 5) = 0$ $55 - 55 = 0$
2 $(5 \div 5) \times (5 \div 5) = 1$ $\sqrt[5]{5^5} \div 5 = 1$
$55 \div 55 = 1$ $\cdot5 + \cdot5 + \cdot5 - \cdot5 = 1$
3 $5 \div 5 + 5 \div 5 = 2$ $(\sqrt{5 \times 5} + 5) \div 5 = 2$
$\cdot5 + \cdot5 + \cdot5 + \cdot5 = 2$
4 $(5 + 5 + 5) \div 5 = 3$
5 $(5 \times 5 - 5) \div 5 = 4$ $\sqrt{5 \times 5} - (5 \div 5) = 4$
6 $\sqrt{5 \times 5} + 5 - 5 = 5$ $5 \times 5 \div \sqrt{5 \times 5} = 5$
$\sqrt{5 \times 5} \div \cdot5 - 5 = 5$ $5 \times \cdot5 + 5 \times \cdot5 = 5$
7 $\sqrt{5 \times 5} + 5 \div 5 = 6$ $(5 \times 5 + 5) \div 5 = 6$
$55 \div 5 - 5 = 6$
8 $\sqrt{5 \times 5} + \sqrt{5 \times 5} = 10$ $(5 + 5) \times 5 \div 5 = 10$
$(55 - 5) \div 5 = 10$ $5 \div \cdot5 + 5 - 5 = 10$

Four 5s challenge (page 28) ***continued...***

9 $\sqrt{5 \times 5} + 5 + 5 = 15$ $5 \times 5 - 5 - 5 = 15$ $5 \times 5 - (5 + 5) = 15$ $\sqrt{5 \times 5} \div \cdot 5 + 5 = 15$

10 $5 \times 5 - (5 \div 5) = 24$

11 $5 \times 5 \times (5 \div 5) = 25$ $\sqrt{5 \times 5} \times \sqrt{5 \times 5} = 25$ $5 \times 5 + 5 - 5 = 25$

12 $5 \times 5 + 5 \div 5 = 26$

13 $5 \times 5 + \sqrt{5 \times 5} = 30$ $5 + 5 \times \sqrt{5 \times 5} = 30$

14 $(5 + 5) \times 5 - 5 = 45$ $5 \times 5 \div \cdot 5 - 5 = 45$

15 $5 \times 5 + 5 \times 5 = 50$ $5! \times \cdot 5 - 5 - 5 = 50$ $55 - \sqrt{5 \times 5} = 50$

16 $(5 + 5) \times 5 + 5 = 55$ $55 \times 5 \div 5 = 55$ $55 + 5 - 5 = 55$ $5 \times 5 \div \cdot 5 + 5 = 55$

17 $(5 + 5 + 5) \times 5 = 75$

18 $(5 \times 5 - 5) \times 5 = 100$ $5! + 5 - (5 \times 5) = 100$

19 $5 \times 5 \times 5 - 5 = 120$ $\sqrt{5 \times 5} - 5 + 5! = 120$

20 $5! + 5 + 5 - 5 = 125$ $5^5 \div 5 \div 5 = 125$ $5 \times 5 \times \sqrt{5 \times 5} = 125$

21 $5 \times 5 \times 5 + 5 = 130$ $5! + 5 + \sqrt{5 \times 5} = 130$

22 $(5 \times 5 + 5) \times 5 = 150$ $5! + 5 \times 5 + 5 = 150$

Number crunching (page 29)

Only some possible answers are given.

1 a $21 = 7 + 7 + 7$

b $7 = 7 + 7 - 7$ or $7 \times 7 \div 7$ or $7 \times \frac{\sqrt{7}}{\sqrt{7}}$ or $7 \times \frac{7!}{7!}$

c $8 = 7 + 7 \div 7$ or $7 + \frac{\sqrt{7}}{\sqrt{7}}$ or $7 \times \frac{7!}{7!}$

d $84 = 77 + 7$

e $70 = 77 - 7$ or $\frac{7 \times 7}{\cdot 7}$ **f** $11 = \frac{77}{7}$

g $1 = \sqrt{7} \times \sqrt{7} \div 7$ **h** $110 = \frac{77}{\cdot 7}$

i $2 = (7 + 7) \div 7$ or $(7! + 7!) \div 7!$

j $20 = (7 + 7) \div \cdot 7$

2 a $1 = \frac{77}{77}$ or $\frac{7 \times 7}{7 \times 7}$ or $\frac{7 + 7}{7 + 7}$ or $\frac{7 \div 7}{7 \div 7}$

b $2 = 7 \div 7 + 7 \div 7$ or $(\sqrt{7} \times \sqrt{7} + 7) \div 7$

c $3 = (7 + 7 + 7) \div 7$ or $(\sqrt{7} \times \sqrt{7}) \div \cdot 7 - 7$

d $100 = \frac{77}{\cdot 77}$ or $\frac{7}{\cdot 7} \times \frac{7}{\cdot 7}$

e $30 = (7 + 7 + 7) \div \cdot 7$ **f** $14 = \sqrt{7} \times \sqrt{7} + \sqrt{7} \times \sqrt{7}$

g $49 = \sqrt{7} \times \sqrt{7} \times \sqrt{7} \times \sqrt{7}$ **h** $154 = 77 + 77$

i $11 = 77 \div (\sqrt{7} \times \sqrt{7})$ **j** $20 = (\sqrt{7} \times \sqrt{7} + 7) \div \cdot 7$

k $78 = 77 + 7 \div 7$ or $77 + 7! \div 7!$ or $77 + \sqrt{7} \div \sqrt{7}$

l $70 = \sqrt{7} \times \sqrt{7} \times 7 \div \cdot 7$

3 a $1 = 7 \div 7 + 7 - 7 + 7 - 7$ or $\frac{77}{77} + 7 - 7$ or $\frac{7}{7} - \frac{7}{7} + \frac{7}{7}$ or $\frac{7}{7} - \frac{7}{7} + \frac{\sqrt{7}}{\sqrt{7}}$

b $2 = 77 \div 77 + 7 \div 7$ or $\frac{7 \times 7}{7 \times 7} + \frac{7}{7}$ or $\frac{7}{7} + \frac{7}{7} + 7 - 7$

c $3 = 7 \div 7 + 7 \div 7 + 7 \div 7$

d $4 = (7 + 7 + 7 + 7) \div (\sqrt{7} \times \sqrt{7})$

e $78 = 77 + \frac{77}{77}$ or $77 + \frac{\sqrt{7} \times \sqrt{7}}{\sqrt{7} \times \sqrt{7}}$ or $77 + \frac{\sqrt{77}}{\sqrt{77}}$

Calculator challenges (page 30)

1 a 15, 16

Most students will use trial and error. However, another good strategy for approximation is to find the square root of the number; $\sqrt{240} \div 15{\cdot}49$, so the two consecutive numbers are 15 and 16. Alternatively, you can use a different method. Since 10 x 10 = 100 and 20 x 20 = 400, the numbers are between 10 and 20, and since the product ends with 0, possibilities are 14 x 15 or 15 x 16 (10 x 11 is too low and 19 x 20 too high); 15 x 16 gives the correct answer.

b 21, 22 **c** 43, 44 **d** 53, 54

c Since 40 x 40 = 1600 and 50 x 50 = 2500, the numbers are between 40 and 50. As the last digit is a 2, the possibilities for the consecutive integers are 41 and 42, 43 and 44, 46 and 47 or 48 and 49. 48 x 49 will be too high. Use the calculator to check which give a product of 1892.

2 a 8, 9, 10 **b** 14, 15, 16

3 a To test if a number is prime, test to see whether any of the prime numbers 2, 3, 5, 7, …, up to the square root of the number, are factors. Students will not be aware of this method, and it should be explained to them. To find out whether 371 is prime, see if any of the numbers 2, 3, 5, 6, 11, 13, 17 or 19 are factors. As 7 is a factor (7 x 53 = 371), 371 is not prime.

b 631 is prime. You need only test primes up to 25 for factors.

4 The answer will consist of the original number repeated: 273 × 11 × 91 = 273 273. This will always be so, since 11 × 91 = 1001, and to multiply any three-digit number by 1001 will produce this result: *abc* × 1001 = *abc abc*.

5 a i 253 **ii** 594 **iii** 671 **iv** 792 **v** 891 **vi** 308 **vii** 429 **viii** 616

b i 187 **ii** 572 **iii** 517

c i 2541 **ii** 5742 **iii** 26 741 **iv** 46 574 **v** 685 751 **vi** 4037

The rule: consider the number being multiplied by 11. Write down the units digit, then 'add the neighbour digits', each time placing the *units digit only* of this answer to the left of the first one. Finally, again the left-hand digit of the number (being multiplied) will be the left-hand digit of the answer, or one more if you had to carry. In **a**, first write down 3, then write the sum of 2 and 3 (5) in front of it (so far we have 53). Now place the 2 in front of the 5: the answer is 253.

d i 6963 **ii** 89 353 **iii** 5247

6 In fact, all four-digit palindromic numbers are divisible by 11.

We can prove this by algebra; the proof is included here for teachers' interest (and for exceptionally bright students):

All four-digit palindromic numbers are of the form *abba* (e.g. 2992), which in fact is:

$1000a + 100b + 10b + a = 1001a + 110b$

$= 11(91a + 10b)$

$= 11$ x a number

Therefore, *abba* is divisible by 11.

Money challenges (page 31)

1 $4.50 Start with the amount Scott has now and work backwards:

Scott has $9.20. Before he bought the drink, he had $10.00. Before he was paid for the gardening, he had $2.00. Before he bought the present, he had $4.50, which is the amount of his pocket money.

2 $10 (binoculars cost $110)

The 'guess and check' strategy is one method that can be used.

Allot $100 to the binoculars then divide equally the remaining $20.

Binoculars	Case
$100	
$10	$10
$110	$10

3 **a** 20c (Karen had 10c) **b** $12 (Elliot had $36)

4

Adult ($15)	Children ($8)
3	5

5 $31

Mindy	Kaitlin	Lauren	Total
$ 31	$ 25	$ 19	$ 75

6 $27 The family's tickets are equivalent to 7 children's tickets. Therefore, the price of a child's ticket = $94.50 ÷ 7 = $13.50

So the price of an adult ticket = $27

7 $3.50

3R + 4G = $11.50 (writing R for ride and G for game)
4R + 3G = $13

7R + 7G = $24.50 (adding)
Therefore, R + G = $3.50.

To lead students to this method ask, 'What would be the cost of 7 rides and 7 games?'

Money problems (page 32)

1 Total profit = $200 (she made $100 profit each time)

2 After Ivan gives Joseph $15 both have $150, so before this Joseph had $135 and Ivan $165.

3 8, 16, 32 and 64 pengo

Ticks show how each amount of money is expressed in single coins. (For convenience, pengo is abbreviated to 'p'.)

Amount	1p coin	2p coin	4p coin	8p coin
3p	✓	✓		
4p			✓	
5p	✓		✓	
6p		✓	✓	
7p	✓	✓	✓	
*8p				

*For 8 pengo, we need a new coin—the 8 pengo—so we add it to the headings and continue the table, finding that the additional coins required are 8, 16, 32 and 64 pengo.

4 1, 8, 16 and 57 austral

First, 1 austral is needed, then 8 and 16 austral, and the table up to 31 austral will be the same as in exercise 2. (*Note*: a = austral)

Amount	1a	2a	4a	8a	16a	25a
32a	✓	✓	✓			✓
33a				✓		✓
34a	✓			✓		✓
35a		✓		✓		✓

The next coin needed is 57. Note that 1 + 2 + 4 + 8 + 16 + 25 = 56, so that with coins of 1, 2, 4, 8, 16 and 25 austral, every amount up to 56 can be expressed.

5 **a** 3 royals = $36 **b** 2 euros = $36 (∴ 1 euro = $18)

c 3 euros = $18 × 3 = $54 **d** 4 zacs = $54

e 1 zac = $$\frac{54}{4}$ = $13.50

6 **a** 4 postcards, 2 letters

b 3 postcards, 9 letters (use 'guess and check')

P (85c)	L ($1.20)	Total ($)
5	7	$12.65 (too low)
3	9	($2.55 + $10.80) $13.35

Coin puzzles (page 33)

Students can answer these questions either by using 'guess and check' or by acting out the problems, as illustrated in **1a** and **b** and in **2a**.

1

	1st pile	2nd pile
a	27	20
b	21	13
c	8	15
d	36	12
e	24	6

In **a** take 47 counters. Use 7 to begin the first pile. Now divide the remaining 40 equally between the two piles.
In **b** take the counters. Use 8 to begin the first pile. Now divide the remaining 26 equally between the two piles.

2

	1st pile	2nd pile	3rd pile
a	6	9	12
b	20	11	10
c	8	16	32
d	6	12	3
e	3	6	12
f	21	10	7

In **a** take 27 counters. Use 3 to begin the second pile. Now, since the third pile has 3 more than the second, in fact it has 6 more than the first, so place 6 counters in the third pile. Now 18 counters remain, which are divided equally between the three.

3

	5c coins	10c coins	Total coins	Total amount
a	4	2	6	40c
b	6	5	11	80c
c	3	3	6	45c
d	2	6	8	70c
e	6	2	8	50c

Amazing number patterns (page 34)

1 37 × 12 = 444
37 × 15 = 555

2 91 × 4 = 364
91 × 5 = 455

3 9109 × 4 = 36 436
9109 × 5 = 45 545

4 143 × 28 = 4004
143 × 35 = 5005

5 37 × 3333 = 123 321
37 × 33 333 = 1 233 321

6 1234 × 9 + 5 = 11 111
12 345 × 9 + 6 = 111 111

7 9876 × 9 + 4 = 88 888
98 765 × 9 + 3 = 888 888

8 1234 × 8 + 4 = 9876
12 345 × 8 + 5 = 98 765

9 $11\ 111^2$ = 123 454 321
$111\ 111^2$ = 12 345 654 321

10 3367 × 132 = 444 444
3367 × 165 = 555 555

11 37 037 × 12 = 444 444
37 037 × 15 = 555 555

12 999 999 × 5 = 4 999 995
999 999 × 6 = 5 999 994

Strange maths symbols (page 35)

1 **a** 5 + 1 = 6 **b** 5 + 5 = 10

c 2 ★ (12 + 4) = 2 ★ 16
= 32 + 16
= 48

d (6 + 3) ★ 4 = 9 ★ 4
= 36 + 4
= 40

e 8 − 3 = 5 **f** 12 − 2 = 10

Strange maths symbols (page 35) *continued...*

g 3 ⊙ (8 – 4) = 3 ⊙ 4
= 12 – 4
= 8

h (12 – 2) ⊙ 4 = 10 ⊙ 4
= 40 – 4
= 36

2 a Add the two numbers and add 2.
a ⬡ b = a + b + 2
1 ⬡ 3 = 6

b Add the two numbers and square.
a △ b = (a + b) x (a + b)
4 △ 5 = 81

c Multiply the two numbers and add 1.
a ▽ 6 = (a x b) + 1
4 ▽ 7 = 29

d Add the two numbers and halve.
a ◯ b = (a + b) ÷ 2
4 ◯ 6 = 5

e Square each number and add.
a ⌑ b = a x a + b x b
5 ⌑ 1 = 26

f Double the second number and add the first.
a ● b = a + 2 x b
5 ● 2 = 9

Tree diagrams (page 36)

1 H H H
H H T
H T H
H T T
T H H
T H T
T T H
T T T

Note the systematic way in which the possibilities are written out.

2

Bread	Filling
brown	honey
brown	cheese
brown	Vegemite
brown	tuna
brown	salad
white	honey
white	cheese
white	Vegemite
white	tuna
white	salad

3 Reading along all the branches, there are nine possibilities:

train/walking | ferry/walking | car/walking
train/monorail | ferry/monorail | car/monorail
train/taxi | ferry/taxi | car/taxi

Pie graphs (page 37)

1 a i 3 **ii** 6 **iii** 9 **b** $\frac{1}{8}, \frac{3}{8}$

c i 60 **ii** 144 **iii** $\frac{1}{3}, \frac{1}{12}$

$30° = \frac{1}{12} \times 360°$
Therefore, $\frac{1}{12}$ of students = 12 tennis
There are 144 students altogether.
Cricket has 150° at centre so 5 × 12 = 60.

2 a 6 (note $\frac{6}{8}$ = 75%) **b** 4 (note $\frac{4}{6} = \frac{2}{3}$)

c 4 (note $\frac{4}{10}$ = 40%) **d** 4 (note $\frac{4}{12} = \frac{1}{3}$)

Spinners and chance (page 38)

1 $\frac{3}{36} = \frac{1}{12}$ **2** $\frac{4}{36} = \frac{1}{9}$ **3** 7 **4** 2 or 12

5 $\frac{18}{36} = \frac{1}{2}$ **6** 2 and 12 or 3 and 11 or 4 and 10 or 5 and 9 or 6 and 8

7 0 **8** $\frac{15}{36} = \frac{5}{12}$ **9** $\frac{3}{36} = \frac{1}{12}$ **10** $\frac{9}{36} = \frac{1}{4}$

Interpreting graphs (page 39)

1 a i Tessa **ii** Brian **b i** Paul **ii** Marc

c Bob and Kevin are the same age. Kevin is heavier than Bob.

d David and John weigh the same. David is younger than John (or John is older than David).

e Ivy and Rose weigh the same. Ivy is older than Rose.

2 a (mass/height: Anne, Claire) or (height/mass: Anne, Claire)

b (age/mass: Mary, Ruth) or

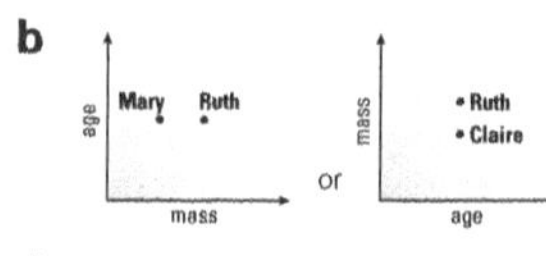

c

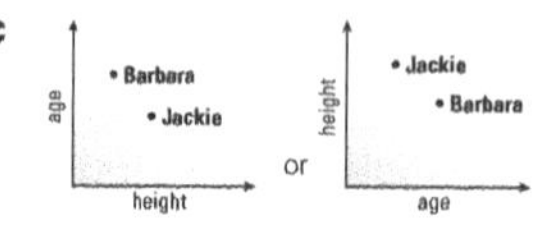

d

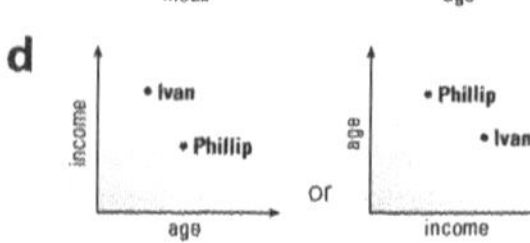

3 a Car A, as it travels further in the same time

b Car D; in fact, car C is stationary

c Neither; cars E and F travel the same distance in the same time. Their speed is the same

d Car G, since car H takes the same time to cover a shorter distance.

Listing possibilities (page 40)

1 a 3 presents **b** 12 presents

2 20 presents

3 a 6 code words
A B C
A C B
B A C
B C A
C A B
C B A

b 27 code words

A A A | B A A | C A A
A A B | B A B | C A B
A A C | B A C | C A C
A B A | B B A | C B A
A B B | B B B | C B B
A B C | B B C | C B C
A C A | B C A | C C A
A C B | B C B | C C B
A C C | B C C | C C C

4

1st	2nd	3rd
A	M	S
A	S	M
M	A	S
M	S	A
S	A	M
S	M	A

5 2 g, 5 g, 7 g, 10 g, 12 g, 15 g, 17 g

6 You should have 19 solutions: 10c, 20c, 30c, 40c, 50c, 60c, 70c, 80c, 90c, $1, $1.10, $1.20, $1.30, $1.40, $1.50, $1.60, $1.70, $1.80, $1.90.

7 29 altogether: 9 two-digit numbers: 11, 22, 33, 44, 55, 66, 77, 88, 99; 20 three-digit numbers: 101, 111, 121, 131, 141, 151, 161, 171, 181, 191, 202, 212, 222, 232, 242, 252, 262, 272, 282, 292.

How many ways? (page 41)

Students should be shown how to set out their work using tree diagrams.

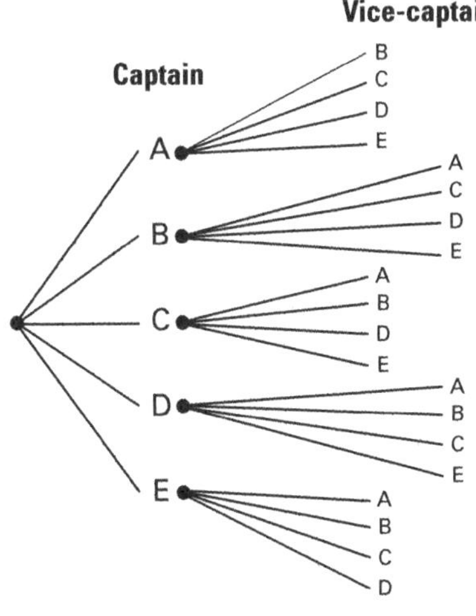

Reading along the branches, there are 20 possibilities.

1 a 20 ways

A B	B A	C A	D A	E A
A C	B C	C B	D B	E B
A D	B D	C D	D C	E C
A E	B E	C E	D E	E D

b i 12 ways

A B	B A	D A	E A
A D	B D	D B	E B
A E	B E	D E	E D

ii 30 possibilities

Using the above patterns, it can be seen that there are 6 ways of choosing the captain and 5 ways of choosing the vice-captain. The tree diagram will have 6 branches then 5 branches, altogether 30 branches so 30 possibilities.

iii $10 \times 9 = 90$ possibilities

2 a 9 different pairs:

N D	I D	K D
N V	I V	K V
N R	I R	K R

b 6 different pairs:

N D	K D
N V	K V
N R	K R

c 8 different pairs:

N D	N R	K D	K R
N V	N J	K V	K J

d

Students auditioning		Queen–king pairs
Girls	Boys	
1	2	2
2	2	4
2	3	6
3	2	6
3	3	9
3	4	12
4	3	12
4	4	16
4	5	20
5	5	25
5	6	30
6	6	36

e The number of queen–king pairs is the product of the numbers of girls times the number of boys auditioning.

Possibilities (page 42)

1

Rings	Friends arriving	Friends present
1	1	1
2	3	4 (= 2^2)
3	5	9 (= 3^2)
4	7	16 (= 4^2)
5	9	25 (= 5^2)
6	11	36 (= 6^2)
10	19	100 (= 10^2)

a 9 **b** 19

c i 16 = (4^2) **ii** 25 = (5^2) **iii** 36 = (6^2)

d 100 = (10^2) The total number of friends present after each ring is the square of the progressive total of rings.

2 a 5 Each group arrriving has three more students than the previous one.

4th group = 14
3rd group = 11
2nd group = 8
∴ 1st group = 5.

b 6 Use 'guess and check' or the result of **a**, where if 5 students are in the first group, when 3 groups have arrived there will be 24 students altogether. Since this is too low, try 6 students in the 1st group, 9 in the 2nd and 12 in the 3rd, which gives 27 altogether.

c 7
12th group = 40 students
11th group = 37 students
10th group = 34 ($40 - 2 \times 3$)
9th group = 31 ($40 - 3 \times 3$)
1st group = 7 ($40 - 11 \times 3$)

3 a There are 2 ways.

Consider the envelopes as A, B, C and the letters as a, b, c. The possible arrangements are: Cross off all groups in which at least one envelope and one letter correspond. Place ticks beside the remaining groups.

A B C
~~a b c~~
~~a c b~~
~~b a c~~
b c a ✓
c a b ✓
~~c b a~~

b There are 9 ways.

If the envelopes are A, B, C, D and the letters a, b, c, d, there are 24 possible arrangements. If all groups in which an envelope and a letter correspond are crossed off, 9 remain:

badc	cadb	dabc
bdac	cdab	dcab
bcad	cdba	dcba.

Time challenges (page 43)

1 a When it is 4 p.m. in Sydney it is 2:30 p.m. in Darwin. Since the flight takes 5 hours it will land at 7:30 p.m.

b When it is 9:20 a.m. in Darwin it is 10:50 a.m. in Sydney. This flight takes 5 hours and 50 minutes.

2 a London is 9 hours behind Melbourne. He must ring 9 hours after 1:30 p.m. Saturday; that is, 10:30 p.m.

b 9 p.m. Saturday in London is 6 a.m. Sunday in Melbourne.

3 120 We need to find LCM of 8, 6 and 15 which is 120.

4 11 p.m. Wednesday
We need to find LCM of 8, 5 and 10, which is 40.
Forty hours after 7 a.m. Tuesday is 11 p.m. Wednesday.

Time challenges (page 43) *continued...*

5 a 10:35 a.m. **b** 10:24 a.m. **c** 11 a.m. **d** 10:48 a.m.

In all of these cases we have to find the LCM of the numbers.

Lighthouse problems (page 44)

1 a 9:10 p.m.

X will flash at 9:02, 9:04, 9:06, 9:08, 9:10.
Y will flash at 9:05, 9:10.

Or, since 2 and 5 are prime numbers and 2 x 5 = 10, they will flash together 10 minutes later.

b

	Time between flashes		Next flash together
	Light X	Light Y	
i	3 min	5 min	11:15 p.m.
ii	3 min	7 min	11:21 p.m.
iii	6 min	7 min	11:42 p.m.
iv	6 min	10 min	11:30 p.m.
v	4 min	11 min	11:44 p.m.
vi	10 min	12 min	12:00 midnight
vii	5 min	15 min	11:15 p.m.
viii	6 min	15 min	11:30 p.m.
ix	10 min	15 min	11:30 p.m.
x	8 min	12 min	11:24 p.m.

Note: if numbers are prime, multiply; otherwise, find LCM. *Or* list when lights flash and find when both occur.

2 7:30 p.m. Since 2, 3 and 5 are prime numbers, we multiply so 2 x 3 x 5 = 30. They flash together every 30 minutes.

3 9 p.m., 10 p.m., 11 p.m.

They will flash together every 60 minutes. Students can arrive at this result in two different ways:

1 By listing the times after 8 p.m. that each of the lights will flash and then finding the time when they all flash at the same time.

2 By finding the lowest common multiple (LCM) of 4, 3 and 10, which is 60.

Area (page 45)

This is an excellent spatial exercise. Most students will need a diagram to understand how to place cards, but some will be capable of visualising the problem.

1 a 4 (2 × 2) **b** 9 (3 × 3) **c** 24 (4 × 6)

2 a 50 (5 × 10) **b** 200 (20 × 10)

3 a 12 (4 × 3) **b** 24 (4 × 6)

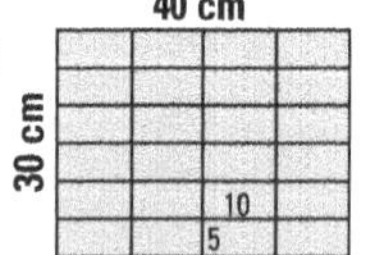

Note: the cardboard pieces can be cut in two different ways.

c 48 (8 × 6)

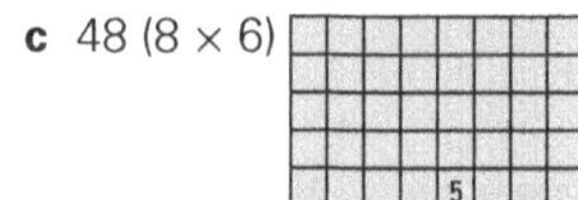

d 60 (10 × 6)

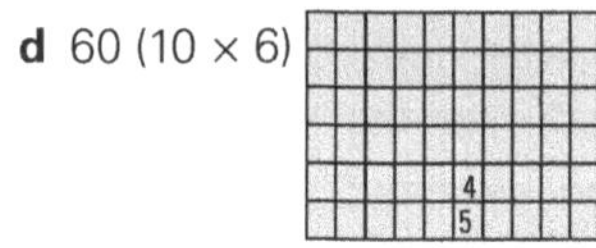

To cut out the maximum number of cardboard pieces, they can be placed only in one way (i.e. 10 columns and 6 rows).

4 a 240 (12 × 20) **b** 480 (24 × 20)

Area and perimeter (page 46)

1 Rectangles with perimeter 16 m

Length (m)	Width (m)	Area (m²)
7	1	7
6	2	12
5	3	15
4	4	16

For a given perimeter, the square has the greatest area.

It is important to discuss with students the fact that the square belongs to the family of rectangles, but the rectangle does not belong to the family of squares.

2 Rectangles with perimeter 20 m

Length (m)	Width (m)	Area (m²)
9	1	9
8	2	16
7	3	21
6	4	24
5	5	25

Largest area: 25 m^2

3 Rectangles with area 24 m^2

Length (m)	Width (m)	Perimeter (m)
24	1	50
12	2	28
8	3	22
6	4	20

Shape with smallest perimeter: 6 m × 4 m

Four rectangles are possible. (In fact, the shape with the smallest perimeter to give this area is the square, with a side of approximately 4.9 m.)

4 Rectangles with area 36 m^2

Length (m)	Width (m)	Perimeter (m)
36	1	74
18	2	40
12	3	30
9	4	26
6	6	24

Shape with smallest perimeter: 6 m × 6 m

Area of triangles (page 47)

1 a 4 cm^2 **b** 6 cm^2 **c** 6 cm^2 **d** 9 cm^2

2 Area of the triangles: A = 5 cm^2, B = 5 cm^2, C = 10 cm^2, D = 10 cm^2, ∴ first two statements are true

Area of the triangle (B + C) = 15 cm^2, area of the triangle (A + D) = 15 cm^2 = shaded area (B + C)
Area of rectangle = 30 cm^2
∴ area of triangle (B + C) = $\frac{1}{2}$ area of rectangle

3 Area of shaded triangle = 12 cm^2 in each case
= $\frac{1}{2}$ area of rectangle
= $\frac{1}{2}$ base × perpendicular height

∴ if a triangle is inscribed within a rectangle so that its base is the length of the rectangle and its height is the width (or the other way around), the area of the triangle will be half the area of the rectangle.

4 Any triangle in which $\frac{1}{2}$ base × perpendicular height = 12 cm^2 will be correct. These are some of the possible figures.

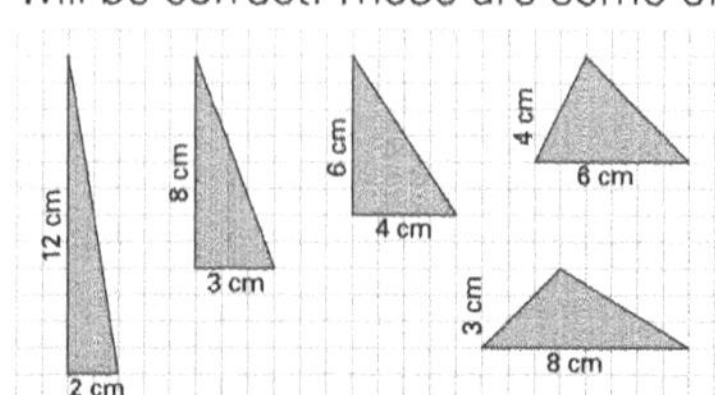

Squares and triangles (page 48)

1 a

	Side (cm)	Perimeter (cm)	Area (cm^2)
	1	4	1
	2	8	4
	3	12	9
	4	16	16
	5	20	25
b i	6	24	36
ii	10	40	100

2 a

	Side (units)	Perimeter (units)	Area (triangular grids)
	1	3	1
	2	6	4
	3	9	9
	4	12	16
	5	15	25
b i	6	18	36
ii	10	30	100

The perimeter and the area can in each case be predicted by following the patterns given in **a**.

Area investigations (page 49)

Graph paper and isometric graph paper are necessary teaching aids when considering the concept of area.

1

Original side (cm)	Original area (cm^2)	Enlarged side (cm)	Enlarged area (cm^2)
3	9	6	36
4	16	8	64
5	25	10	100
6	36	12	144
7	49	14	196

The enlarged area is four times the original area.

2

Original length & breadth (cm)		Original area (cm^2)	Enlarged length & breadth (cm)		Enlarged area (cm^2)
2	1	2	4	2	8
3	2	6	6	4	24
4	3	12	8	6	48
5	2	10	10	4	40

The new area is four times the original area.

3

	Original length & height (cm)		Original area (cm^2)	Enlarged base & height (cm)		Enlarged area (cm^2)
a	1	2	1	2	4	4
b	1	3	$1\frac{1}{2}$	2	6	6
c	2	3	3	4	6	12
d	1	1	$\frac{1}{2}$	2	2	2

The area of an enlarged triangle is four times the original area.

4 Irrespective of shape, if the length of sides are doubled, the new area will be four times the area of the original shape.

Exploring the area (page 50)

1 For squares, rectangles or any other shape, if you treble the dimensions of the sides, the new enlarged area will be nine times the original area.

2 Dimensions × 2, area will be 4 times as large as original.
Dimensions × 3, area will be 9 times as large as original.
Dimensions × 4, area will be 16 times as large as original.

Perimeter and area (page 51)

1 length 18 m, width 6 m

This perimeter (48 m) has eight equal sections, therefore the width (one section) is 6 m and the length (three sections) is 18 m. A simpler problem can be solved using a perimeter of 8 m and drawing the figure on squared paper.

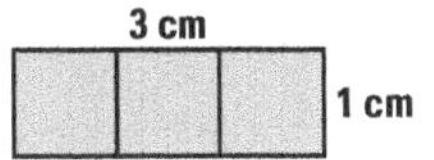

Next use a perimeter of 16m:

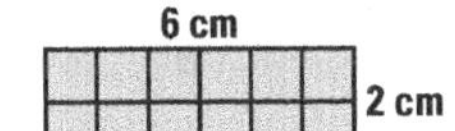

2 8 m The large table has 10 equal sections to total 20 m. Therefore, the side of a square is 2 m and the perimeter of a square is 8 m.

3 60 cm Since two perimeters are equal, 4 equal parts = 3 equal parts + 15. Therefore, 1 equal part = 15 cm.

4 58 m 2 lengths + 1 width = 47 m
1 length + 2 widths = 40 m; adding these two,
3 lengths + 3 widths = 87 m
1 length + 1 width = 29 m
2 lengths + 2 widths = 58 m.

This is a difficult problem and teachers may need to guide students.

5 72 m^2 Note that the four walls have to be painted, not the floor or base.

Area that needs painting
= 6 x 2 x 2 + 12 x 2 x 2 = 72 m^2

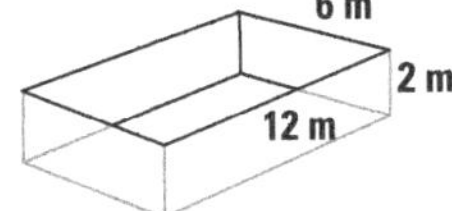

6 52 m

Length x width = 120 m^2

We can use 'guess and check' to work out this problem.

Original length	Original width	New length	New width	New area
12	10	4	14	56
15	8	7	12	84
24	5	7	12	144
30	4	22	8	176
20	6	12	10	120 correct

Perimeter of the original rectangle = 2 x 20 + 2 x 6 = 52 m

7 49 cm^2

A = 12 cm
B = 10 cm
X + 5 = 12
Therefore, X = 7 cm

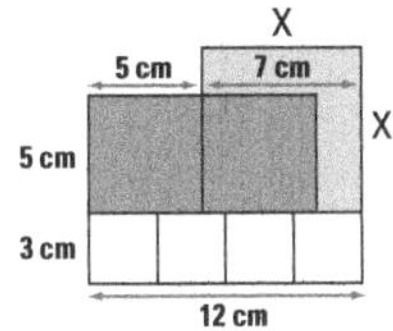

Finding areas (page 52)

1 96 m^2

Method 1:

Length of AB = 16 m
Area of ABIH = 32 m^2
Area of KJCD = 32 m^2
Length of QR = 8 m
Length of QIJR = 16 m^2
Area of HPSK = 16 m^2
Therefore area of path = 96 m^2

Method 2:

Length of AB = 16 m
Length of BC = 12 m
Area of ABCD = 192 m^2
Area of PQRS = 96 m^2
Therefore area of path
= 192 m^2 – 96 m^2 = 96 m^2

2 a Area A + Area B = 20 x 6 + 13 x 5
= 120 + 65 = 185 m^2

b Area C + Area D = 11 x 13 + 7 x 6
= 143 + 42 = 185 m^2

c Area of large rectangle – Area E = 11 x 20 – 7 x 5
= 220 – 35 = 185 m^2

Following these exercises, there should be a discussion about the methods used to find the areas of these shapes.

3 Area of square EFGH = $\frac{1}{2}$ area of square ABCD

= $\frac{1}{2}$ × 36
= 18 m^2

It is easy to see that the total shaded area of the 4 triangles is equal to the area of square EFG.

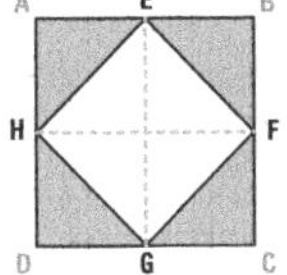

Finding areas (page 52) ***continued...***

4 a Rearrange the shape.
Area = 2 x 4 cm²
= 8 cm²

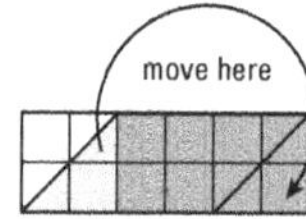

b The shaded area is half the rectangle, *or*

Rearrange the shape to form a rectangle, as shown.
Area = 3 x 4 cm²
= 12 cm²

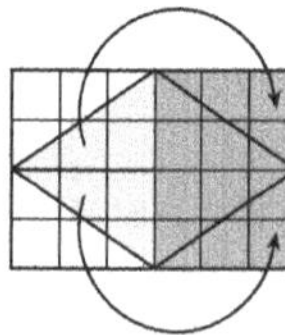

The next shapes can be all rearranged to form rectangles by cutting and fitting shaded figures into dotted sections. This can be done in several ways.

c Area = 4 x 4 cm²
= 16 cm²

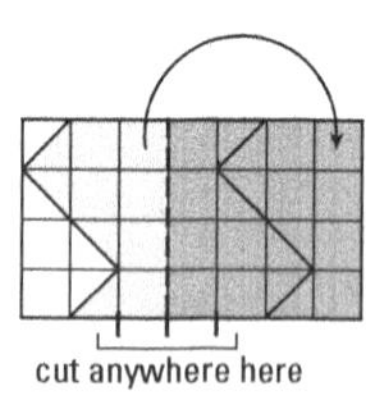

d Area = 2 x 7 cm²
= 14 cm²

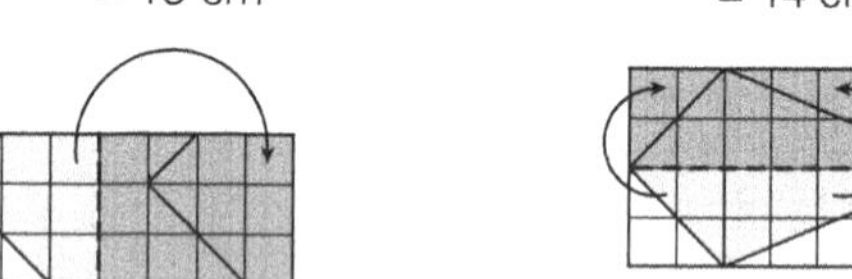

Surface area (page 53)

1 a 148 cm² **b** 104 cm² **c** 80 cm² **d** 88 cm²
i 1 × 2 × 24 (148 cm²) **ii** 3 × 4 × 4 (80 cm²)
Students should build these solids, and should understand the formula:
Surface area = 2 × L × B + 2 × B × H + 2 × L × H

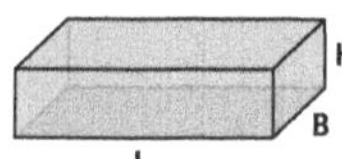

2 2 × (60 × 20 + 60 × 10 + 20 × 10) = 4000 cm²

3 If 6 faces = 486 cm²
1 face = 81 cm²
∴ side = 9 cm

4 a 2 × (1 × 3 + 1 × 5 + 3 × 5) = 46 cm²
b 46 cm² **c** 46 cm² **d** It remains 46 cm².
e Surface area decreases by 2 and becomes 44 cm²

Volume and capacity (page 54)

1 a 10, as 2 L = 2000 mL **b** $\frac{1}{10}$ **c** 6 jugs

2 2 30 x 120 mL = 3600 mL
= 3.6 L
Therefore, need to buy two 2 L ice-cream packs.

3 a He saves in a day, 10 L − $\frac{1}{2}$ L = 9$\frac{1}{2}$ L.
Note that 500 mL = $\frac{1}{2}$ L.

b He saves in a week, 9$\frac{1}{2}$ x 7 = 66$\frac{1}{2}$ L.

c He saves in a year, 52 x 66$\frac{1}{2}$ = 3458 L.

4 14 L In 6 days youngest child drinks 3 L, two older children drink 9 L. Altogether they drink 14 L.

5 32 L The tank had $\frac{3}{4}$ of 64 L = 48 L.
Amount used = $\frac{1}{3}$ of 48 L = 16 L
32 L left in tank, requiring 32 L to fill it

6 1200 cars 60 kL = 60000 L, $\frac{60000}{50}$ = 1200 cars

Solid experiments (page 55)

1 a 6 × 2 × 1, 3 × 4 × 1, 12 × 1 × 1

b i 24 × 1 × 1, 12 × 2 × 1, 8 × 3 × 1, 6 × 4 × 1, 6 × 2 × 2, 4 × 3 × 2
ii 30 × 1 × 1, 15 × 2 × 1, 10 × 3 × 1, 5 × 6 × 1, 5 × 3 × 2
iii 36 × 1 × 1, 18 × 2 × 1, 12 × 3 × 1, 9 × 4 × 1, 6 × 6 × 1, 6 × 3 × 2, 9 × 2 × 2, 4 × 3 × 3

c The height is 3 cm.

2 a 8 cubes

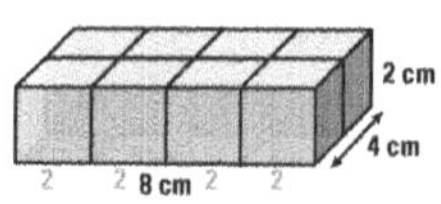

b 90 cubes 6 × 5 = 30 cubes on base with 3 layers, so 90 cubes altogether.

3 a 36 (6 × 3 × 2) **b** 12 (3 × 2 × 2) **c** 300 (10 × 6 × 5)

4 a 7 cm × 5 cm × 4 cm **b** 110 **c** 140 cm²

Measurement puzzles (page 56)

1 a

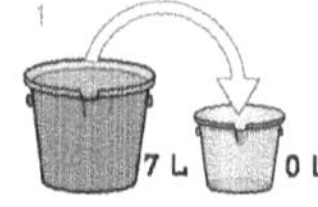

Fill the 7 L bucket. From it fill the smaller bucket, which will take 3 L. 4 L will remain.

b Continuing from **a**,

empty the 3 L bucket. There are 4 L left in the larger bucket; use this to refill the 3 L bucket; 1 L will remain.

c

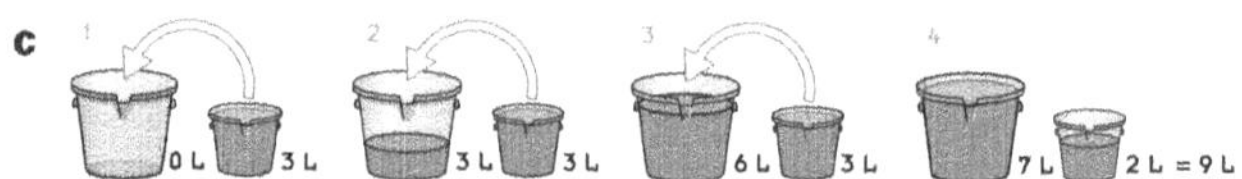

Fill the 3 L bucket and pour the contents into the larger one. Repeat this. The larger bucket now contains 6 L. Fill the 3 L bucket again and from it fill the larger one, which will take 1 L. 2 L will remain.
The last stage of **c** has 7 L in one bucket and 2 L in the other, making a total of 9 L.

d Continuing from **c**,

first empty the larger bucket, then pour into it the water remaining in the small one (2 L). Refill the 3 L bucket and pour its contents into the larger one, which will then contain 5 L.

2 a Fill the 8 L bucket, and from it fill the 5 L bucket; 3 L then remain.

b Fill the 5 L bucket, empty it into the 8 L bucket, refill it, and use this content to fill the 8 L bucket. 2 L remain in the 5 L bucket.

c Follow all the instructions for **b**. Now empty the 8 L bucket, and it pour the 2 L remaining in the 5 L bucket. Refill the 5 L bucket, and add its contents to the 8 L bucket to make 7 L.

3

Fill the 5 L bucket, and from it fill the smaller one (3 L). It now contains 2 L. Empty the smaller bucket, and into it pour the 2 L left in the larger one. Refill the 5 L bucket. The two together now hold 7 L.

or

Fill the 3 L bucket and pour its contents into the 5 L one. Again fill the smaller bucket and use it to fill the larger one. It now contains only 1 L.
Now empty the 5 L bucket, and into it pour the 1 L remaining in the smaller one.
Refill the smaller bucket and pour the contents (3 L) into the larger one, which will then contain 4 L.
Again fill the 3 L bucket. The two together now hold 7 L.

4 It would be ideal to have 2-minute, 3-minute and 5-minute timers available for the puzzles.

a 4 min: 2 min twice (2 + 2)
5 min: 2 min, then 3 min, (2 + 3)
6 min: 3 min, twice or 2 min three times (3 + 3 or 2 + 2 + 2)
7 min: 2 min twice, then 3 min (2 + 2 + 3)
8 min: 2 min four times or 3 min twice, then 2 min (2 + 2 + 2 + 2 or 3 + 3 + 2)
9 min: 3 min three times (3 + 3 + 3)
10 min: 3 min twice, then 2 min twice (3 + 3 + 2 + 2)

b Turn 3 and 5-minute timers over at the same time. When the 3-minute one runs out, turn it over again; when the 5-minute one runs out, 1 min will remain. We will abbreviate this as 1 = 3 + 3 – 5.

2 = 5 – 3 Turn 3 and 5-minute timers over at the same time; when the 3 min runs out, 2 min of the 5 will remain.
3 = 3 4 = 3 + 3 + 3 – 5 (or time 1 min as above, then 3 min)
5 = 5 6 = 3 + 3 7 = 5 + 5 – 3
(2 min above then 5 min) 8 = 5 + 3
9 = 3 + 3 + 3 10 = 5 + 5

Measurement estimation (page 57)

1 a female adult **2** 36.7°C **3** 200°C
4 3 kg **5** 5 pears **6** 190 cm **7** $35
8 120 mL **9** 2000 cm^2 **10** 14 cm **11** 600 cm^2
12 5 g **13** 200 mL

Mass, measurement, time (page 58)

1 True **2** True $\frac{1}{2}$ kg = 500 g
3 False about 4 or 5 average-sized oranges have a mass of 1 kg
4 False **5** True **6** False **7** True **8** False
9 True 1000 g = 1 kg, 1 g = 0·001 kg, 35 g = 0·035 kg
10 True **11** False 6 kg = 6000 g and $\frac{6000}{400}$ = 15 days
12 False It should be much lower. **13** True or a little lower
14 False 100°C is boiling point and you cannot drink it at that temperature.
15 False Temperature should be less than 0°C
16 False Far too high. About 200°C is average. **17** False
18 False It will finish at 10:07 p.m. **19** True 600 ÷ 15 = 40
20 False This is equivalent to 60 km/hour.
21 False 6th April
22 False Time in Melbourne is 30 minutes later than Adelaide

Measurement facts? (page 59)

1 False $\frac{1}{2}$ L to 1 L is more realistic. **2** True
3 False $\frac{1}{8}$ full has 24 L, so full is 192 L. **4** True
5 False It will lose 6 mL in 1 min so 360 mL in 1 hour.
6 True 8 x 250 mL = 2 L
7 True 4 litres will only cover 16 m^2.
8 False 1m = 100 cm cost $2, so 750 cm will cost $15.
9 False It is approximately 2 m high.
10 False This would only give approximate dimensions as 1m x 1m.
11 True Each side is 25 cm.
12 True For $30 you get 25 litres of petrol which will allow you to drive for 225 km.
13 False Brisbane is north.
14 True **15** True 1500 m = $1\frac{1}{2}$ km
16 True 25 ribbons will take 10 min, 5 ribbons will take 2 min. Therefore, 120 = 24 x 5 will take 24 x 2 = 48 min.
17 True **18** False $\frac{1}{2}$ tank = 250 L, now $\frac{250}{75} = 3\frac{1}{3}$ and $3\frac{1}{3}$ hours = 3h 20 min.

Mass problem solving (page 60)

1 a 12 3 kg = 3000 g **b** $\frac{1}{12}$ **c** 4

2 a 5 kg 10 x $1.35 = $13.50
Therefore, 10 x 500 g = 5 kg.

b 15 kg 20 x $2.15 = $43
Therefore, 20 x 750 g = 20 x $\frac{3}{4}$ kg
= 15 kg.

c 12 bags from stand A will cost $16.20.
Either draw up a table and consider the ways that 6 kg can be purchased.

Stand A 500g for $1.35	Stand B 750g for $2.15	Total cost
12 x 500 g = 6 kg	—	$16.20
6 x 500 g = 3 kg	4 x 750 g = 3 kg	$16.70
—	8 x 750 g =	$17.20

∴ cheapest way is buying 12 bags of 500 g for $16.20 altogether.

Or, work out the cost of 250 g for each stand.
For stand A, 250 g cost $67\frac{1}{2}$ cents.
For stand B, 250 g cost $71\frac{2}{3}$ cents.
So stand A is cheaper, where 6 kg cost $16.20.

Mass problem solving (page 60) ***continued...***

3 15 kg Use 'guess and check' and a table. Sheila's weight is 20 kg.

4 30 kg Use the fact there are 3 equal parts.
Jennifer 1 part Carly 2 parts
Total 3 parts = 90 kg
∴ 1 part = 30 kg for Jennifer

5 Leah 65 kg, Andrew 24 kg, Ramah 38 kg

Leah and Andrew weigh 89 kg. Since Leah, Andrew and Ramah weigh 127 kg, Ramah weighs 38 kg. Since Leah and Ramah weigh 103 kg, Leah weighs 65 kg, and so Andrew weighs 24 kg.

6 48 kg If average mass of 3 students = 45 kg, then total mass of 3 students = 135 kg. Therefore total mass of 4 students = 135 + 57 = 192 kg.
∴ average mass of 4 students = 192 ÷ 4 = 48 kg

7 50 kg Use a table like this and guess and check.

Bruce's mass	Natalie's mass	Total
40	45	80 + 135 = 215 (too low)
45	50	90 + 150 = 240 (correct)

Doing jobs (page 61)

1 a 6 hours To do the job, one person will take 3 times as long as 3 people.

b 1 hour 6 people will take half as long as 3 people.

c $1\frac{1}{2}$ hours 4 people will take $\frac{1}{4}$ of the time that 1 person takes.

2 a 54 days 10 people will take 6 times longer than 60 people.

b 30 more people Using the answer to **a**, 90 people take 6 days, so 30 more are needed.

3 a 12 days **b** 180 days (12 x 15)

c 45 days (12 x 15 ÷ 4) **d** 20 days ($\frac{12 \times 15}{9}$)

e 10 days **f** 15 using the answer to **a**

g 5 ($\frac{1}{3}$ of 15, since there are 3 times as many days)

h 3 ($\frac{1}{5}$ of 15, since there are 5 times as many days)

i 3 (6 would do the whole task in 30 days, so 3 will do half in this time.)

It may be helpful to express the information in a table.

Workers	Task	Days
15	$\frac{1}{2}$	6
15	1	12
5	1	36
3	1	60
3	$\frac{1}{2}$	30
9	$\frac{1}{2}$	10
9	1	20
1	1	180
4	1	45

4 a 32 hours **b** 8 hours

5 a 36 days (2 carpenters take 42 x 6 days)

b 28

Ratio problems (page 62)

A table like this is very helpful when teaching ratio.

	Sheep	Cows	Chickens	Animals
	7	8	5	20
a	28	32		
b	35	40	25	100
c		48	30	
d			15	60

1 a 28 **b** 100 **c** 48 **d** 15

2 Either draw up a table, or first work out the total amount of ingredients, which is 150 g. Since the 450 g packet is three times this mass, the amount of each ingredient will be three times that given: 150 g dried fruit, 69 g nuts, 150 g rice bubbles, 81 g bran.

3 a 9 L

b 4 L

Flavouring (L)	Milk (L)	Flavoured milk (L)
$\frac{1}{4}$	2	$2\frac{1}{4}$
1	8	9
4	32	36

4 Students should draw up a table and use logical thinking, being careful to change one variable at a time.

a 6 **b** 12
c 24 **d** 48
e 1 **f** 1
g $\frac{1}{3}$ **h** 6
i 12 **j** 72
k 1 **l** 1
m $\frac{1}{2}$ **n** 48 + 72 = 120

Boys	Notebooks	Weeks
3	3	3
3	6	6
3	12	12
6	24	12
12	48	12
3	1	1
1	1	3
1	$\frac{1}{3}$	1

Girls	Notebooks	Weeks
2	2	2
2	6	6
2	12	12
12	72	12
2	1	1
1	1	2
1	$\frac{1}{2}$	1
Boys + Girls	Notebooks	weeks
12 + 12	48 + 72 = 120	12

Rates and ratio (page 63)

1 a \$2.40 **b** \$7.20

2 a \$22 **b** \$264

3 a 300 L **b** $\frac{5}{60}$ L = $\frac{1}{12}$ L = $83{\cdot}\dot{3}$ mL

4 a 4 barrels **b** 6 barrels **c** 48 barrels
There are 240 barrels.

5

Michael	Danielle	Total	Extra
\$3	\$5	\$8	\$2
× 3 = \$9	× 3 = \$15		
× 10 = \$30	× 10 = \$50		
× 2 = \$6	× 2 = \$10	× 2 = \$16	
× 12 = \$36	× 12 = \$60		
× 7 = \$21	× 7 = \$35		
× 11 = \$33	× 11 = \$55	× 11 = \$88	
× 40 = \$120	× 40 = \$200		× 40 = \$80

a \$15 **b** \$30

c Michael spends \$6 and Danielle spends \$10. **d** \$60

e \$21 **f** Michael spends \$33 and Danielle spends \$55.

g \$200

6 a 90 000 m per hour **b** $\frac{90\,000}{60 \times 60}$m/sec = 25 m/sec

7 a fruit juice, petrol **b** rainfall

c fruit, vegetables **d** tiles

There are many possible answers.

Rolling boxes (page 64)

Students may find it difficult to visualise the position of the symbols as the boxes are rolled. They should carry out the experiment with some concrete objects, e.g. by drawing on an eraser.

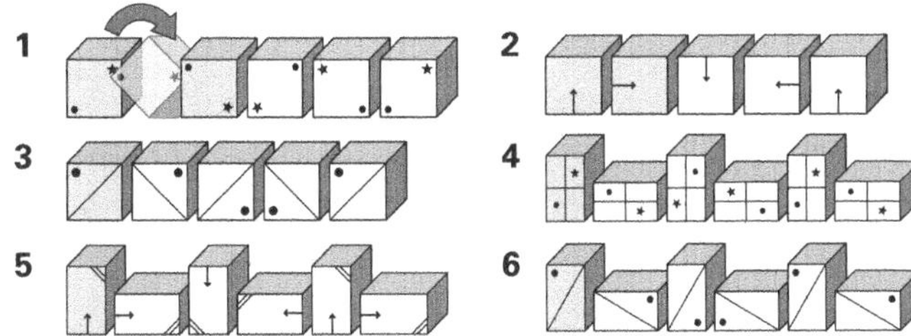

How many? (page 65)

In questions 1, 3 and 5, working systematically and counting all the different-sized shapes is essential. It is also a good idea to set out work in a table for clarity and organisation.

Provide students with copies of diagram masters, since they will need to shade the shapes as they count them, as shown in these middle-sized examples from question 1:

1 20 triangles:
12 single, 6 middle-sized, 2 large 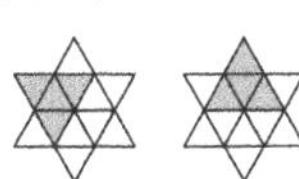

2 18 parallelograms: 6 small, 7 double (illustrated below), 2 treble, 2 quadruple, 1 large.

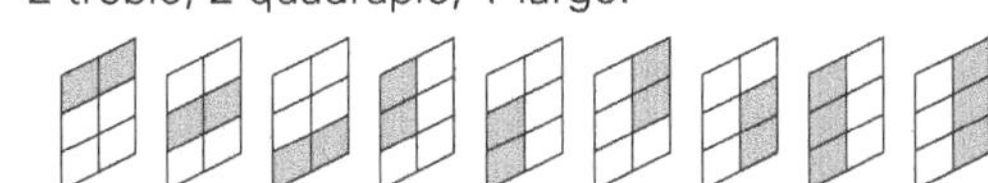

3 18 squares: 8 single, 5 (2 x 2), 4 (3 x 3), 1 (4 x 4)

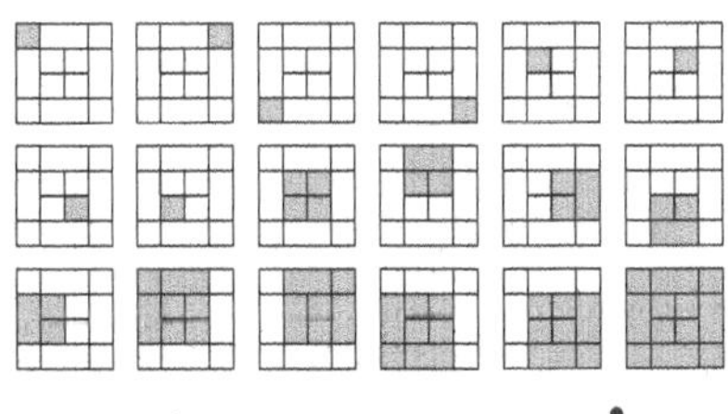

4 **a** 6 **b** 10

5 10 segments: XY, XZ, XU, XW, YZ, YU, YW, ZU, ZW, UW

How many routes? (page 66)

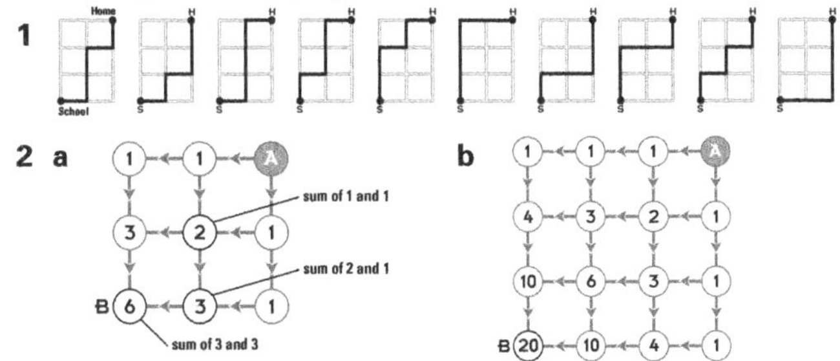

Figure **a** is embedded in figure **b**. For each point that has two arrows pointing to it, the number of routes is the sum of the numbers on the two arms.

A fascinating extension is that the number of routes is part of the pattern of numbers in Pascal's triangle.

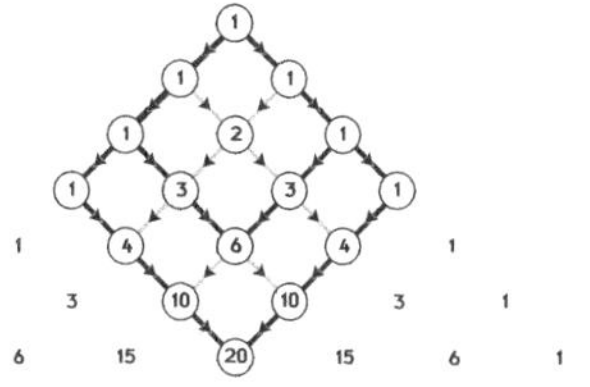

Different routes (page 67)

1 6

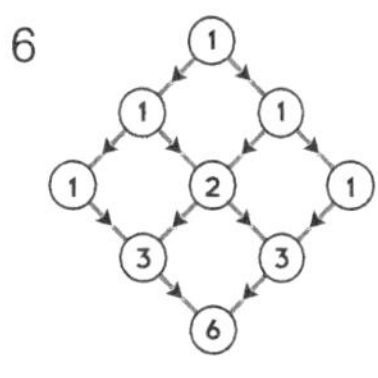

2 **a** 5 **b** 7

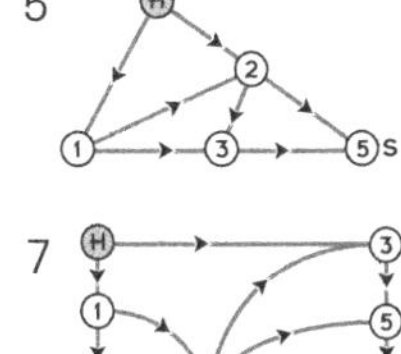

3 **a** 2 **b** 3
c 8 = 4 x 2 **d** 12 = 3 x 4 **e** 24 = 3 x 4 x 2

Polyiamonds (page 68)

For these exercises, students should be provided with isometric grid paper.

1 4 pentiamonds

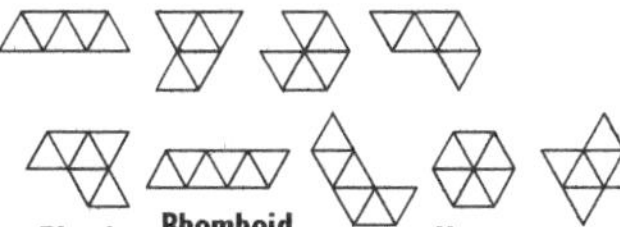

2 12 hexiamonds

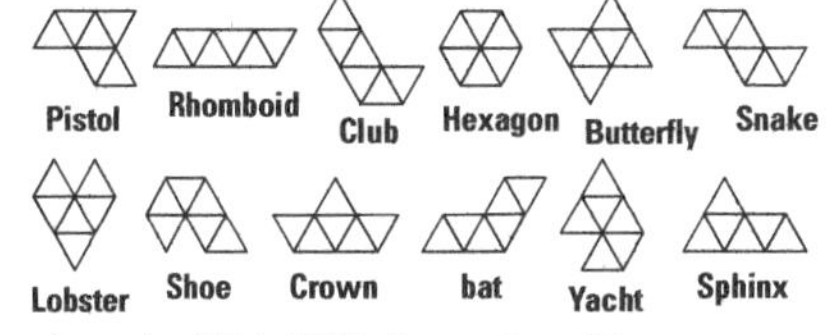

Their names were given by T.H. O'Beirne, the Glasgow mathematician who invented the name 'polyiamonds' in 1965.

Cutting solids (page 69)

Students need draw only the shape of the cross-section formed, not the solid diced.

1 **a** square  **b** square 

2 **a** rectangle 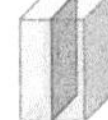 **b** square 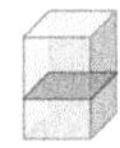

3 **a** circle  **b** circle 

4 **a** rectangle 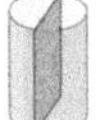 **b** circle 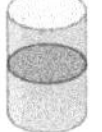

5 **a** triangle 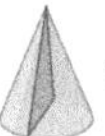 curved shape called hyperbola 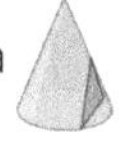

b circle 

6 **a** triangle  (through vertex)

trapezium  (not through vertex)

b square 

Spirolaterals (page 70)

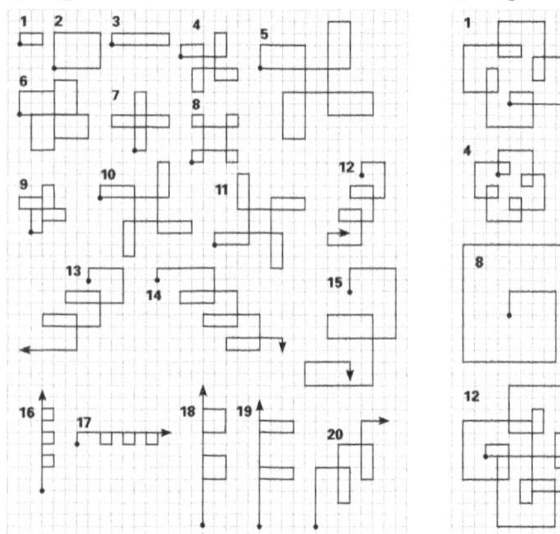

Spirolateral thinking (page 71)

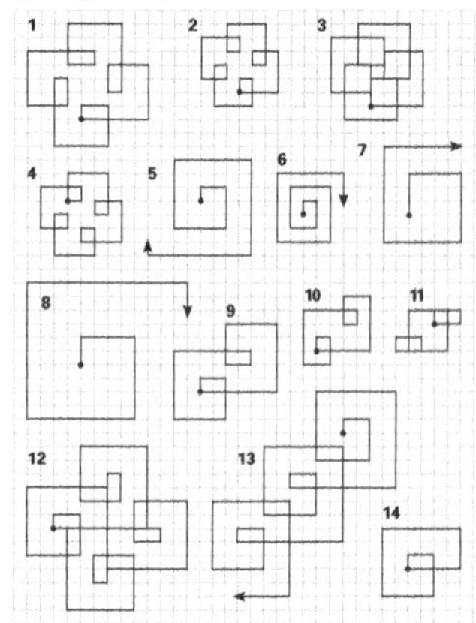

(page 71) Descriptions of certain properties of these figures follow.

a **i** Sequences of 2 terms always form a rectangle (questions 1, 2, 3, page 70).

ii All spirolaterals consisting of three numbers have a 'windmill effect' (questions 4 to 11, page 70) although there are several types.

iii All sequences with an odd number of terms will return to the starting point and will be heading north ready to repeat the pattern after four repetitions (questions 1, 2, 3, 4, 12, page 71). All these form a closed geometrical pattern.

iv All sequences consisting of four numbers—unless the numbers are repeated, as in (3, 4, 3, 4) —will form an open geometrical pattern that will continue indefinitely (questions 12 to 20, page 70)

v All sequences consisting of 6 terms close after only two repetitions (questions 9, 10, 11, page 71).

vi All sequences consisting of 8 terms (or a multiple of 4) will continue indefinitely unless there is a special relationship between the terms, when they will close (questions 13, 14, page 71).

b Multiplying all the terms in a sequence by a constant (number) will create geometrically similar spirolaterals, so that one geometrical pattern will be an enlargement of the other (questions 4 and 5, 12 and 15, 16 and 18, page 70; questions 5 and 8, page 71).

Movements of zero length create interesting patterns.

How many diagonals? (page 72)

1

3 sides 4 sides 5 sides 6 sides 7 sides 8 sides

Number of sides	3	4	5	6	7	8
Number of vertices	3	4	5	6	7	8
Diagonals at each vertex	0	1	2	3	4	5
Total diagonals	0	2	5	9	14	20

Patterns found:

- The number of vertices is the same as the number of sides.
- The number of diagonals at each vertex is three less than the number of sides (or of vertices).
- The difference between the numbers in the total number of diagonals is always increasing by one:

Total diagonals	0	2	5	9	14	20	27	35
Differences		2	3	4	5	6	7	8

- The total number of diagonals = the number of sides x the number of diagonals at each vertex ÷ 2 (halved).

  Note that students are unlikely to discover this formula without some guidance.

2 From the above pattern, the number of diagonals for:

a a nonagon is 27 **b** a decagon is 35.

Solids (page 73)

It is important that students build the solid given in the example and check each view.

They should be told to first lay the base of each one to look the same as the view from the top, then build the front view and check the view from the left.

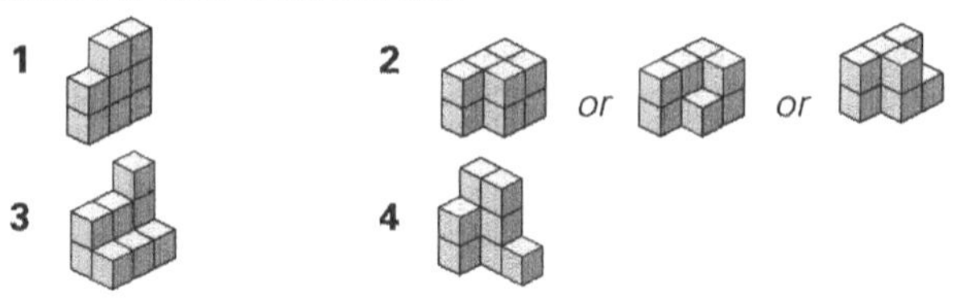

Bearings (page 74)

1 **a** 2S, 1W or 1W, 2S **b** 4E, 1S or 1S, 4E

c 3N, 1E or 1E, 3N **d** 6W, 1N or 1N, 6W

2 **a** S **b** S

c S

d SE **e** 180° clockwise or anticlockwise

Tessellations (page 75)

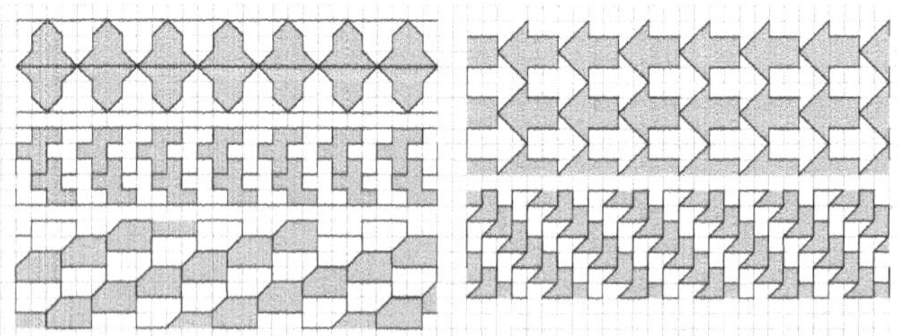

Mathematical curiosities (page 76)

Students will be fascinated with these problems. After trying several different numbers, they can justify the results with a shape for the unknown number as in question 1.

1 The result is the number you first thought of.

2 The result is always 7.

Let the number be □:

□

□□

□□ 17

□□ 14

□ 7

7

3 The result is the number you first thought of.

Let the original number be ○:

○

○○○

○○○ 6

○○○○○○ 12

○○○○○○ 18

○ 3

○

4 The result is the number you first thought of.

Let the original number be △:

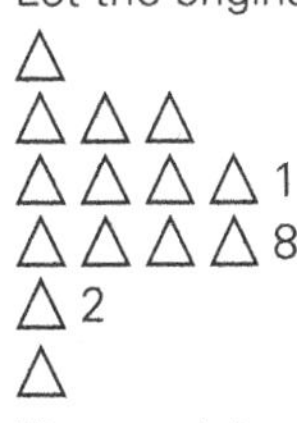

5 The result is always the number you first thought of.

To prove this requires a knowledge of algebra beyond that of students at this level, but the method is nevertheless fascinating for teachers.

Let the original number be a.

Since $(a - 2) \times (a + 2) = a^2 - 4$

$a^2 - 4 + 4 = a^2$

$\sqrt{a^2} = a$

Palindromic numbers (page 77)

1 a

```
   59
 + 95
  154
+ 451
  605
+ 506
 1111
```

b

```
    78
  + 87
   165
 + 561
   726
 + 627
  1353
+ 3531
  4884
```

c

```
     79
   + 97
    176
  + 671
    847
  + 748
   1595
 + 5951
   7546
 + 6457
  14003
+ 30014
  44044
```

2 a 11, 22, 33, 44, 55, 66, 77, 88, 99 are palindromic numbers.

b One-stage palindromic numbers: 10, 12, 13, 14, 15, 16, 17, 18, 20, 21, 23, 24, 25, 26, 27, 29, 30, 31, 32, 34, 35, 36, 38, 40, 41, 42, 43, 45, 47, 50, 51, 52, 53, 54, 56, 60, 61, 62, 63, 65, 70, 71, 72, 74, 80, 81, 83, 90, 92

c Two-stage numbers: 19, 28, 37, 39, 46, 48, 49, 57, 58, 64, 67, 73, 75, 76, 82, 84, 85, 91, 93, 94

d Three-stage numbers: 59, 68, 86, 95

e Four-stage numbers: 69, 78, 87, 96

f Six-stage numbers: 79, 97

Palindromic words include: civic, deed, ewe, eye, gag, gig, kayak, radar, reviver, rotator, level.

Odd one out (page 78)

1 9 (one-digit number)
19 (not a square)

2 4 (one-digit number)
25 (odd number)

3 15 (not prime or composite number)
2 (even number)
62 (not square)

4 9 (one-digit number)
32 (hasn't got 3 as a factor)

5 75 (two-digit number)
245 (not a multiple of 25)

6 38 (not a multiple of 8)
8 (one-digit number)

7 9 (odd number)
4 (not a multiple of 3)

8 12 (not a multiple of 5)
15 (odd number)

9 17 (two-digit number)
8 (even number)

10 12 (two-digit number)
5 (not a factor of 12)

11 24 (even number)
31 (not divisible by 3)

12 18 (even number)
39 (not a multiple of 9)

13 35 (odd number)
48 (not a multiple of 7)

14 100 (three-digit number)
68 (not divisible by 5)

15 128 (three-digit number)
26 (not a multiple of 4 or a power of 2)

16 8 (one-digit number)
74 (not divisible by 4)

17 75 (odd number)
80 (not in the seventies)

18 21 (composite number)
2 (even number, one-digit number)

19 121 (three-digit number)
98 (not a multiple of 11)

20 (2, 5) (pair whose sum is not 10)
(8, 2) (2nd number larger than 1st number)

What's my message? (page 79)

1 ROCK'N'ROLL IS FUN

C = 3 **F** = 6 **I** = 4 **K** = 5 **L** = 2
N = 7 **O** = 8 **R** = 1 **S** = 0 **U** = 9

2 MATHS IS AMAZING

A = 9 **G** = 2 **H** = 8 **I** = 3 **M** = 4
N = 5 **S** = 7 **T** = 1 **Z** = 6

Chessboard squares (page 80)

Note: finding the total number of squares on a chessboard will be very confusing unless:

- simpler boards are considered first
- the results are recorded in a table
- we look for a pattern.

1 a 9 **b** 4 **c** 1 **d** 14

2

Size of board	Number of squares								Total squares	
	1 × 1	2 × 2	3 × 3	4 × 4	5 × 5	6 × 6	7 × 7	8 × 8		
3 × 3	9	4	1	—	—	—	—	—	14	
4 × 4	16	9	4	1	—	—	—	—	30	16
5 × 5	25	16	9	4	1	—	—	—	55	25
6 × 6	36	25	16	9	4	1	—	—	91	36
7 × 7	49	36	25	16	9	4	1	—	140	49
8 × 8	64	49	36	25	16	9	4	1	204	64

3 a part of table above; note the differences in the Total column.

b The number of squares of a particular size is always a perfect square.
The total number of squares is the sum of the squares

Chessboard squares (page 80) ***continued...***

of all the numbers up to the size of the board (e.g. for an 8×8 board, total number of squares = $1^2 + 2^2 + 3^2 + 4^2 + 5^2 + 6^2 + 7^2 + 8^2$).

Chessboard rectangles (page 81)

1 36 rectangles on a 3×3 board

Rectangle size	1×2	2×1	1×3	3×1	2×3	3×2	Subtotal
Number of rectangles	6	6	3	3	2	2	22

∴ with the addition of 14 squares, the total = 22 + 14 = 36

2 100 rectangles on a board 4×4

Rectangle size	1×2	2×1	1×3	3×1	1×4	4×1
Number of rectangles	12	12	8	8	4	4

2×3	3×2	2×4	4×2	3×4	4×3	Subtotal
6	6	3	3	2	2	70

∴ with the addition of 30 squares, the total = 70 + 30 = 100

3

Size of board	2×2	3×3	4×4	5×5	6×6	7×7	8×8
# of rectangles	$9 = 3^2$	$36 = 6^2$	$100 = 10^2$	15^2	21^2	28^2	36^2

These are the squares of the triangular numbers

∴ we can now predict the results for larger boards.

The total number of rectangles on a chessboard is 36^2, which is 1296.

Counting the rectangles systematically is extremely important in this exercise, as is setting out the answers in a table from which a prediction can be made. The importance of looking at simpler cases, drawing up a table and looking for a pattern should also be stressed.

If a student did not consider smaller boards first, to count the numbers of rectangles on a chessboard (even if working systematically) would be an enormous task and one unlikely to be answered correctly.

Working backwards (page 82)

To find all the solutions, we started with the final number and worked backwards. Always do opposite operation.

1 6 (18 divided by 2 gives 9), subtract 3 gives 6
2 5 (Working backwards we get 10, then 5)
3 12 (Working backwards we get 6, then 12)
4 13 (Working backwards we get 6, then 13)
5 10 (Working backwards we get 5, then 10)
6 8 (Working backwards we get 4, then 8)
7 2 (Working backwards we get 8, then 4, then 2)
8 3 (Working backwards we get 12, then 15, then 3)
9 8 **10** 9 **11** 4 **12** 16 **13** 4
14 3 **15** 8 **16** 8 **17** 5 **18** 4
19 10 **20** 2 (Before the result was halved, it was 6; before 2 was added it was 4, and before it was squared it was 2)

Solving problems (page 83)

1 7:24 a.m.: 15 + 12 + 7 + 8 + 3 = 45 minutes, the time needed before the bus leaves at 8:09.

2 50 cupcakes: 18 is $\frac{3}{4}$ of the number left before the gift to the teacher; therefore, 24 cakes were left at the end of the morning. Since this is 1 less than half, the number sold was 25 + 1 = 26 as $\frac{1}{2}$ of 50 = 25

3 40 tickets: 16 was half the number of tickets remaining; therefore 32 were left after the first day's sale of 8

4 Before he sold to the shop he had 16 kg; therefore, he started with 32 kg.

A diagram can be very helpful:

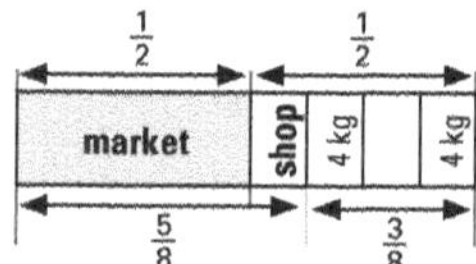

5 Before her last purchase, half Irene's money was \$20. Therefore, she had \$40 when she entered the bookshop. In the record shop she had had \$20 more than that (namely, \$60). Since this was half the money she started with, the original amount was \$120.

6 Farmer Brown had 12 kg before he gave half to charity. Since this 12 kg represented $\frac{1}{4}$ of his potatoes, originally he had 48 kg.

Alternatively, use a diagram:

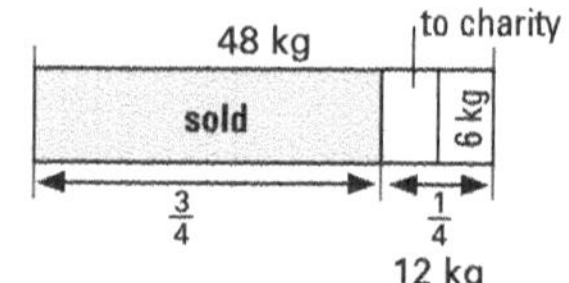

Drawing a diagram (page 84)

Question 1 is very tricky because of the need to consider the corner poles.

1 a 16 A diagram is necessary to show the 16 poles were needed.

b 7 Without a diagram, a student may be tempted to say that 6 poles are needed on each side instead of 7, as illustrated.

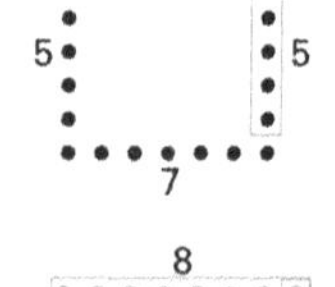

c 4 16 poles are used altogether on the two lengths. Since 20 − 16 = 4, 2 poles are left for each width, so that a shorter side has 4 poles altogether.

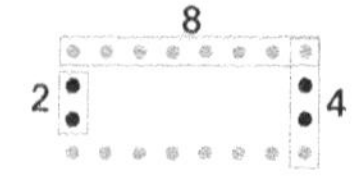

d

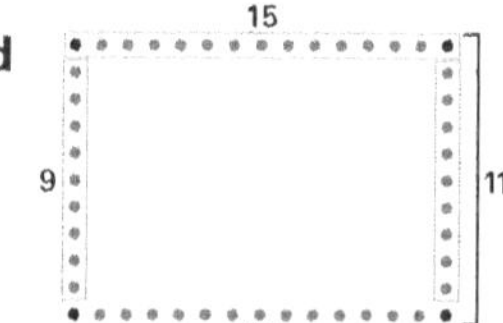

Since corner posts are also part of a shorter side, there are 11 and 15 respectively on the shorter and longer sides, and 48 posts altogether.

2 a 35 km **b** 70 km **c** 35 km

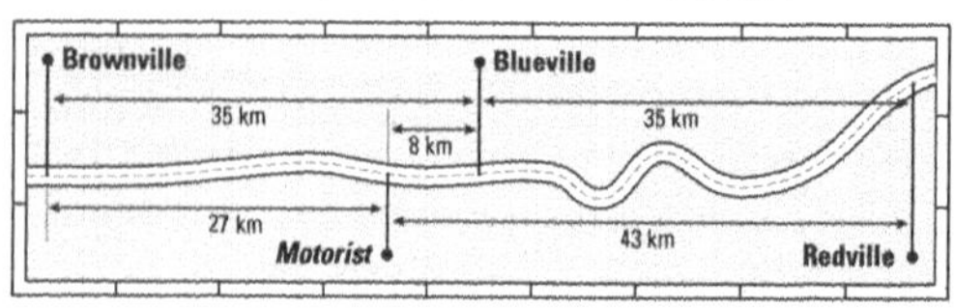

3 a Number of grey tiles = $6 \times 10 = 60$

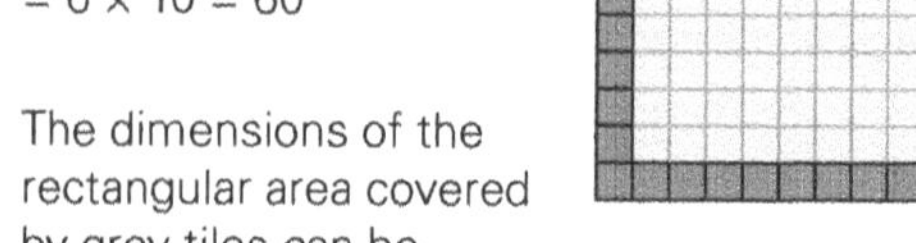

b The dimensions of the rectangular area covered by grey tiles can be

i) 2 × 21, ii) 3 × 14, iii) 6 × 7, and the number of black tiles needed will be
i) 23 + 23 + 2 + 2 = 50 *or*
ii) 16 + 16 + 3 + 3 = 38 *or*
iii) 9 + 9 + 6 + 6 = 30.

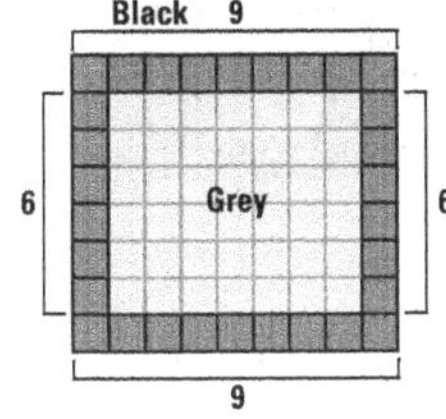

Multiple solutions (page 85)

1 16, 36: List two-digit square numbers (16, 25, 36, 49, 64, 81) and cross out the ones that do not satisfy the other conditions.

2 29, 37: From the first two clues, the possibilities are 11, 13, 17, 19, 23, <u>29</u>, 31, <u>37</u>. The underlined numbers are those compatible with the third clue.

3 19, 37 (referring to the numbers given for question 2)

4 2 4 6
2 3 8
2 2 12 ⎤ Since the question did not specify distinct
3 4 4 ⎦ numbers, these are possibilities as well.

5 16, 61; 27, 72; 38, 83; 49, 94

6 12, 24, 36, 48, 60 (and any further multiple of 12)

7

Blocks	Box 1	Box 2	Box 3	Box 4
	1	2	3	10
	1	2	4	9
	1	2	5	8
	1	2	6	7
	1	3	4	8
	1	3	5	7
	1	4	5	6
	2	3	5	6
	2	3	4	7

Remember to work systematically.

8

3-legged stools	4-legged stools	Total legs
2	9	42
6	6	42
10	3	42

9 The two-digit numbers that satisfy first condition are ⑪, 16, 21, 26, ㉛, 36, ㊶, 46, 51, 56, (61), 66, (71), 76, 81, 86, 91, 96. Circled are the prime numbers.

10 The primes between 100 and 200 are:
101, 103, 107, 109, 113, 127, 131, 137, 139, 149, 151, 157, 163, 167, 173, 179, 181, 191, 193, 197, 199.

To find these numbers and further primes there is a simple rule: test for factors of prime numbers $\leqslant \sqrt{N}$ if testing if N is prime.

It was also obvious that no number ending in 5 or an even number had to be considered. So, for example, to test if 149 is prime, you do the following: $\sqrt{149} \approx 12$.

Primes $\leqslant$ 12 are 2, 3, 5, 7, 11 and in fact you only need to test if 3, 7, or 11 are factors; and since they are not factors 149 is a prime. Now from the above list the numbers for which sum of the digits is 10: <u>109</u>, <u>127</u>, <u>163</u>, <u>181</u>.

Tower of Hanoi (page 86)

The Tower of Hanoi (the puzzle with eight discs) can be bought commercially, but students can easily construct a simple copy by making cardboard discs. Some tangible representation is essential in order to carry out this experiment.

1

Discs	Fewest moves	
2	3	
3	7	4
4	**15**	8
5	**31**	16
6	**63**	32
7	**127**	64
8	**255**	128

Three discs require 7 moves.

With 4 discs, 7 moves are required to transfer the 3 top discs to one of the other two pegs. Now the fourth disc can be moved to the empty peg, and 7 more moved are needed to replace the first 3 discs on top of it.

∴ 15 moves are required.

With 5 discs, 15 moves are needed to transfer the 4 top ones to another peg. The fifth can then be moved to the empty peg, and 15 more moves are needed to replace the first four on top of it.

∴ 5 discs require 31 moves.

2 The smallest number of moves is 2 to the power of the number of discs minus 1.

Each time the number of discs is increased by 1, to find the smallest number of moves you double the previous number and add 1.

3 255 moves

Counting cubes (page 87)

Most students will find it necessary to build towers in order to count the cubes in each layer.

1 a top 1; 2nd 3; **Total 4**
b top 1; 2nd 3; 3rd 6; **Total 10**
c top 1; 2nd 3; 3rd 6; 4th 10; **Total 20**

2 a top 2; 2nd 6; **Total 8**
b top 2; 2nd 6; 3rd 12; **Total 20**
c top 2; 2nd 6; 3rd 12; 4th 20; **Total 40**

3 a top 4; **Total 4**
b top 4; 2nd 16; **Total 20**
c top 4; 2nd 16; 3rd 36; **Total 56**

4 a top 1; 2nd 9; **Total 10**
b top 1; 2nd 9; 3rd 25; **Total 35**
c top 1; 2nd 9; 3rd 25; 4th 49; **Total 84**

The locker problem (page 88)

Lockers numbered 1, 4, 9, 16 and 25 are open.
These locker numbers correspond to square numbers.

Locker numbers (columns); Student numbers (rows)

	1	2	3	4	5	6	7	8	9	10	11	12	13	14	15	16	17	18	19	20	21	22	23	24	25	26	27	28	29	30
1	O	O	O	O	O	O	O	O	O	O	O	O	O	O	O	O	O	O	O	O	O	O	O	O	O	O	O	O	O	O
2		C		C		C		C		C		C		C		C		C		C		C		C		C		C		C
3			C			O			C			O			C			O			C			O			C			O
4				O				O				C				O				O				C				O		
5					C					O					O					C					C					C
6						C						O						C						O						O
7							C							O							O							C		
8								C								C								C						
9									O									O									O			
10										C										O										C
11											C											O								
12												C												O						
13													C													O				
14														C														O		
15															C															O
16																O														
17																	C													
18																		C												
19																			C											
20																				C										
21																					C									
22																						C								
23																							C							
24																								C						
25																									O					
26																										C				
27																											C			
28																												C		
29																													C	
30																														C

STUDENT BOOK Solutions

BOOK 6

Missing numbers (page 1)

From 1 and 2, the rule for each shape can be found:

1 Rule: [+5] (×4) △−8 ◇÷2 **2 Rule:** [+5] (×4) △−8 ◇÷2

Students first have to apply the rule to the starting numbers. In the last ten questions they will have to employ the strategy 'working backwards'. This is an excellent exercise in the use of this method.

3 15, 60, 52 **4** 16, 64, 56, 28 **5** 7, 48, 40, 20
6 6, 11, 36, 18 **7** 3, 8, 32, 12 **8** 4, 9, 36, 14
9 2, 7, 28, 20 **10** 15, 20, 80, 72 **11** 0, 5, 20, 12
12 $\frac{1}{2}$, $5\frac{1}{2}$ or $\frac{11}{2}$, 22, 14 **13** $\frac{3}{4}$, $5\frac{3}{4}$ or $\frac{23}{4}$, 23, 15
14 $\frac{1}{4}$, $5\frac{1}{4}$ or $\frac{21}{4}$, 21, 13

Secret rules (page 2)

In expressing a rule, □ represents the number in the first row.

	Rule	Missing numbers
1	2 x □ – 3: Double the number, subtract 3	17, 13, 7, 19, 23
2	$□^2$ + 1 or □ x □ + 1: Square the number, add 1	10, 5, 82, 122, 65
3	5 x □ + 2: Multiply by 5, add 2	37, 52, 17, 47, 57
4	3 x □ – 2: Multiply by 3, subtract 2	10, 28, 13, 16, 34
5	2 x □ + 5: Multiply by 2, add 5	13, 9, 15, 7, 17
6	$(□ – 1)^2$: Subtract 1, then square	16, 4, 49, 36, 81
7	(9 x □ – 1): Multiply by 9, subtract 1	71, 53, 35, 98, 8
8	(□ + 2) x 2: Add 2, then double *or* 2 x □ + 4: Double, add 4	14, 10, 8, 24, 22
9	(□ – 1) x 3: Subtract 1, multiply by 3 *or* 3 x □ – 3: multiply by 3, subtract 3	9, 0, 3, 27, 18
10	□ x (□ – 1): Multiply the number by the number before it *or* square, then subtract number	12, 0, 2, 90, 132

Square puzzles (page 3)

There are a few ways to do these problems; one method is explained here.

1 A = 5, B = 3, C = 2, D = 1, E = 4

4	1	5	10
3	5	2	10
1	2	3	6
8	8	10	

B + D = 4 (1st column)
C = 2 (3rd row)
and then use 'guess and check'

2 A = 5, B = 3, C = 1, D = 2, E = 4

3	1	2	6
4	5	1	10
1	3	4	8
8	9	7	

3 A = 2, B = 3, C = 5, D = 1, E = 4

5	2	1	8
2	3	4	9
1	5	5	11
8	10	10	

4 A = 2, B = 5, C = 4, D = 3, E = 1

2	5	4	11
3	1	2	6
5	4	3	12
10	10	9	

5 A = 4, B = 2, C = 1, D = 5, E = 3

3	2	5	10
5	4	1	10
1	3	2	6
9	9	8	

Find three numbers (page 4)

1 [5], △10, (9) **2** [8], △11, (1) **3** [6], △12, (9)
4 [12], △4, (8) **5** [24], △4, (6) **6** [24], △8, (3)
7 [36], △4, (6) **8** [36], △9, (4) **9** [4], △5, (20)
10 [27], △3, (8) This is difficult and requires 'guess and check'. If △ = 2, then □ = 8 and ○ = 7.
3rd line becomes 7 × 2 = 8 – 2 which is not correct
so try △ = 3 then □ = 27 and ○ = 8
3rd line becomes 8 × 3 = 27 – 3 (true)
4th line is also true.

What's my pattern? (page 5)

The point of this exercise is to emphasise that 3 terms do not define a pattern and there are many ways to continue it. These are some sample solutions only.

1 27, 81, 243 (× 3) 11, 17, 19 (+ 2, + 6)
27, 33, 99 (× 3, + 6) 11, 33, 35 (+ 2, × 3)
2 9, 14, 23 (each term is the sum of the two previous terms)
20, 21, 84 (× 4, + 1) 12, 13, 28 (× 2, + 2, + 1)
3 10, 15, 21 (triangular numbers, or differences increasing by 1)
8, 11, 13 (+ 2, + 3) 18, 36, 108 (× 3, × 2)
4 8, 12, 17 (differences increasing by 1)
6, 8, 9 (+ 1, + 2) 7, 11, 13 (prime numbers)
5 3, 5, 8 (each term is the sum of the two previous terms)
2, 4, 4 (× 1, × 2) 4, 7, 11 (differences increasing by 1)
6 16, 25, 36 (differences increasing by 2, or square numbers)
12, 17, 20 (+ 3, + 5) 36, 41, 164 (× 4, + 5)
7 8, 10, 12 (+ 2) 12, 14, 28 (× 2, + 2)
10, 16, 26 (each term is the sum of the two previous terms)
8 9, 13, 18, (differences increasing by 1)
7, 9, 10 (+ 1, + 2) 10, 18, 34 (× 2 then – 2)
9 2, $2\frac{1}{2}$, 3 (+ $\frac{1}{2}$) 3, $3\frac{1}{2}$, 7 (× 2, + $\frac{1}{2}$)
$2\frac{1}{2}$, 4, $6\frac{1}{2}$ (each term is the sum of the two previous terms)
10 11, 20, 21 (+ 1, + 9) 11, 110, 111 (+ 1, × 10)
11, 100, 101 (binary numbers: 0, 1, 1, 2, 3, 4, 5, ...)
11 24, 48, 96 (× 2) 15, 30, 33 (+ 3, × 2)
24, 30, 60 (× 2, + 6)
12 125, 625, 3125 (× 5) 29, 145, 149 (+ 4, × 5)
125, 145, 725 (× 5, + 20) 29, 49, 53 (+ 4, + 20)

Magic squares (page 6)

1 a

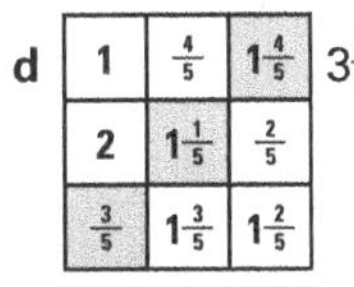

$2\frac{1}{2}$	2	$4\frac{1}{2}$
5	3	1
$1\frac{1}{2}$	4	$3\frac{1}{2}$

9

b

$1\frac{1}{4}$	$1\frac{1}{2}$	$\frac{1}{4}$
0	1	2
$1\frac{3}{4}$	$\frac{1}{2}$	$\frac{3}{4}$

3

c

$\frac{1}{3}$	$2\frac{2}{3}$	1
2	$1\frac{1}{3}$	$\frac{2}{3}$
$1\frac{2}{3}$	0	$2\frac{1}{3}$

4

d

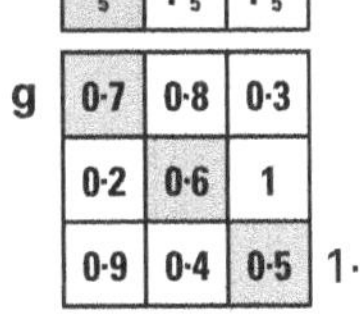

1	$\frac{4}{5}$	$1\frac{4}{5}$
2	$1\frac{1}{5}$	$\frac{2}{5}$
$\frac{3}{5}$	$1\frac{3}{5}$	$1\frac{2}{5}$

$3\frac{3}{5}$

e

$1\frac{1}{2}$	$1\frac{1}{4}$	$2\frac{1}{2}$
$2\frac{3}{4}$	$1\frac{3}{4}$	$\frac{3}{4}$
1	$2\frac{1}{4}$	2

$5\frac{1}{4}$

f

0·6	0·7	0.2
0·1	0·5	0·9
0·8	0·3	0·4

1·5

g

0·7	0·8	0·3
0·2	0·6	1
0·9	0·4	0·5

1·8

h

0·8	0·7	1·2
1·3	0·9	0·5
0·6	1·1	1

2·7

2 a 2·4 and 2·8 **b** 3·5 and 4 **c** 3·7 and 3·8
d 3·25 and 4·25 **e** 2 and 2·5 **f** 2 and 4·5
g 1 and 5·5 **h** 6·4 and 0·1 **i** 2 and 2·2
j 4 and 0·2

Decimal challenges (page 7)

1 C **2** B **3** A **4** D
5 E **6** E **7** C **8** A
9 B **10** A **11** C **12** E

True or false? (page 8)

1 T **2** F **3** T **4** F **5** F **6** F **7** T
8 F **9** F **10** F **11** T **12** T **13** T **14** T
15 T **16** T **17** T **18** T **19** T **20** T **21** F
22 T **23** F **24** T **25** T **26** T **27** T **28** T
29 F **30** T **31** T **32** T **33** T **34** T **35** T
36 T **37** T **38** T **39** T **40** T **41** F **42** T
43 F **44** F **45** T **46** F **47** T **48** F **49** T
50 T **51** T **52** T **53** T **54** F **55** F **56** T
57 T **58** T **59** F **60** T

Fraction fun (page 9)

1 20 (Half of what number is 10?)

2 $\frac{1}{2}$ (What fraction of 40 is 20?)

3 12 **4** $\frac{1}{5}$ **5** 50 **6** $\frac{1}{5}$

7 $\frac{1}{6}$ **8** 36 **9** 25 **10** 16 **11** 24 **12** $\frac{1}{4}$

13 $\frac{1}{3}$ **14** $\frac{1}{2}$ **15** $\frac{1}{3}$ **16** 8 **17** 27 **18** 40

19 16 **20** 20 **21** 2 **22** 3 **23** 2 **24** 3

25 3 **26** 100 **27** 40 **28** 10 **29** 4 **30** 10

It is possible to use 'guess and check' to answer these questions. However, a good understanding of fractions and the factors will lead to a better approach.

Using a diagram would also be very helpful; for example, in **11**: $\frac{1}{4} \times \square = 6$
What is the whole quantity if $\frac{1}{4}$ of it is 6?

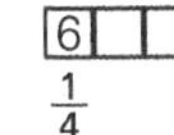

Since each of the blank shapes is also $\frac{1}{4}$ and each represents 6, the total quantity is 4 times 6.

$\therefore \square = 24$.

Fraction challenges (page 10)

1 a 16 **b** 32 **c** 16, since $\frac{1}{4}$ of the number = 4

d 27, since $\frac{1}{3}$ of the number = 9

e 49, since $\frac{1}{7}$ of the number = 7

f 100, since $\frac{1}{10}$ of the number = 10

g 50, since $\frac{1}{5}$ of the number = 10

h 18, since $\frac{1}{3}$ of the number = 6

i 10, using 'guess and check' or logic

j 16, using 'guess and check' or logic

k 12, working backwards: before subtracting 4, it was 6; before halving, it was 12

l $\frac{5}{4}$, working backwards: before subtracting $\frac{1}{2}$, it was $2\frac{1}{2}$ ($\frac{5}{2}$); before doubling, it was $1\frac{1}{4}$ ($\frac{5}{4}$)

m $\frac{3}{4}$, working backwards: before adding $\frac{1}{2}$, it was $1\frac{1}{2}$; before doubling, it was $\frac{3}{4}$

n $\frac{1}{3}$, three times the original number

o $\frac{1}{2}$, working backwards: before adding 5, it was 2; before multiplying by 4, it was $\frac{1}{2}$

2 $\frac{3}{4}, \frac{1}{4}$ **3** $\frac{2}{3}, \frac{1}{3}$ **4** $\frac{3}{2}, \frac{1}{2}$

5 Any fraction greater than 1 (e.g. $\frac{3}{2}$, since $(\frac{3}{2})^2 > \frac{3}{2}$)

6 Any fraction smaller than 1 and greater than 0, e.g. $\frac{1}{2}$ or $\frac{3}{4}$, since $(\frac{1}{2})^2 < \frac{1}{2}$ ($\frac{1}{4} < \frac{1}{2}$) and $(\frac{3}{4})^2 < \frac{3}{4}$ ($\frac{9}{16} < \frac{3}{4}$)

7 a $\frac{1}{2} + \frac{1}{8}$ **b** $\frac{1}{2} + \frac{1}{10}$

c $\frac{1}{2} + \frac{1}{3}$ **d** $\frac{1}{2} + \frac{1}{5}$

e $\frac{1}{2} + \frac{1}{12}$ or $\frac{1}{3} + \frac{1}{4}$

f $\frac{1}{2} + \frac{1}{6} + \frac{1}{15}$

Find the fraction (page 11)

Students must know how to simplify fractions and must understand the concept of equivalent fractions.

1 a □ = 4 **b** ★ = 6 **c** ○ = 6 **d** ◇ = 9
e # = 12 **f** ⬭ = 8 **g** ▯ = 3 **h** ⬠ = 5
i ⊙ = 5, ⊡ = 2 **j** ● = 4, ■ = 7

2 To find the average of a number of fractions, add them and then divide by the number added.

a $(\frac{1}{2} + \frac{1}{3}) \div 2 = \frac{5}{6} \div 2 = \frac{5}{12}$

b $(\frac{1}{2} + \frac{1}{3} + \frac{1}{6}) \div 3 = 1 \div 3 = \frac{1}{3}$

3 a $\sqrt{2\frac{1}{4}} = \sqrt{\frac{9}{4}} = \frac{3}{2}$ **b** $\sqrt{5\frac{1}{16}} = \sqrt{\frac{81}{16}} = \frac{9}{4}$

c $\sqrt{1\frac{7}{9}} = \sqrt{\frac{16}{9}} = \frac{4}{3}$

4 a $1 \div (1 \div \frac{1}{3}) = 1 \div 3 = \frac{1}{3}$ **b** $1 \div (1 \div \frac{1}{5}) = 1 \div 5 = \frac{1}{5}$

c $1 \div (1 \div \frac{3}{5}) = 1 \div \frac{5}{3} = \frac{3}{5}$

d $\dfrac{1}{1 + \frac{1}{2}} = \dfrac{1}{\frac{3}{2}} = 1 \div \frac{3}{2} = \frac{2}{3}$

Find the fraction (page 11) ***continued...***

e $\dfrac{1}{1+\dfrac{1}{1+\frac{1}{2}}} = \dfrac{1}{1+\frac{2}{3}}$ (using the above) $= \dfrac{1}{\frac{5}{3}} = \dfrac{3}{5}$

f $\dfrac{1}{2+\frac{1}{2}} = \dfrac{1}{\frac{5}{2}} = \dfrac{2}{5}$

g $\dfrac{1}{2+\dfrac{1}{2+\frac{1}{2}}} = \dfrac{1}{2+\frac{2}{5}}$ (using the above) $= \dfrac{1}{\frac{12}{5}} = \dfrac{5}{12}$

h $\dfrac{1}{1+\dfrac{1}{1-\frac{1}{4}}} = \dfrac{1}{1+\frac{4}{3}} = \dfrac{3}{7}$

5 **a** $\frac{4}{5}, \frac{5}{6}, \boxed{\frac{6}{7}}, (\frac{1}{2})$ **b** $2, \frac{7}{4}, (\frac{8}{5}), \boxed{4}$ **c** $\frac{4}{9}, \frac{1}{2}, \boxed{\frac{3}{4}}, (\frac{1}{6})$

Spending money (page 12)

Teachers need to stress how much these diagrams help solve these problems.

1 **a** $48

He put $12 in the bank. Present was $24, so he started with $48.

b $72

He put $24 in the bank. Present was also $24.

c $36

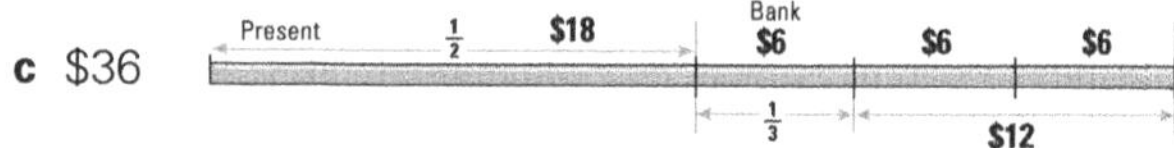

$12 is $\frac{2}{3}$ of remainder. Therefore, he put $6 in bank. Present was $18.

d $48

Movies cost $8 also. Bank $8, present $24.

e $72

$36 is $\frac{3}{4}$ of remainder. Therefore, book cost $12. Present was $24.

2 **a** $8 **b** $48

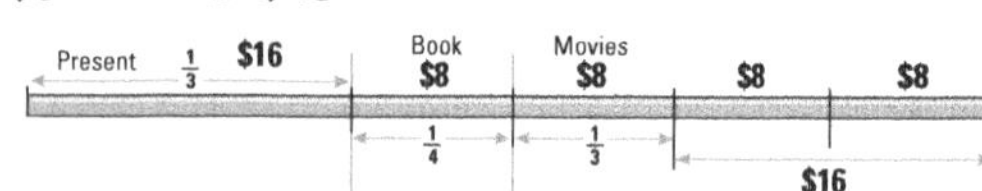

3 **a** $10 **b** $80

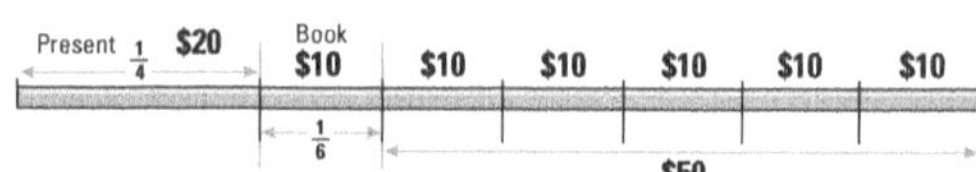

Consecutive numbers (page 13)

There are two methods by which we solve these problems.

(i) Use trial and error, drawing up a table if necessary.

(ii) In **a**, **c** and **f**, if the sum is halved we get an approximation, and the two numbers are the one before and the one after it: $23 \div 2 = 11\frac{1}{2}$, so the numbers are 11 and 12. In **b**, **d**, **e**, **g** and **h**, if we divide the number by 3, we get the middle number.

1 **a** 11, 12 **b** 8, 9, 10 **c** 6, 7 **d** 3, 4, 5
e 16, 18, 20 **f** 19, 21 **g** 24, 25, 26 **h** 23, 25, 27

2 **a** No. The sum of the two consecutive numbers is not divisible by 2, as one of the numbers is always odd and one is always even (e.g. 5 and 6 and 7), and the sum of an even and an odd number is always odd.

b Yes. Since an even number times an odd number is even, the product is divisible by 2.

c Yes. Of 3 consecutive numbers, one will always be a multiple of 3; therefore, the product will be divisible by 3 (e.g. 4 x 5 x 6 or 3 x 4 x 5).

d Yes, since from **c**, above, the product is always divisible by 3; and since the product of any two consecutive numbers is divisible by 2, the product will be divisible by 2 and 3, which is 6.

3 For three consecutive numbers, the square of the middle number minus the product of the other two numbers will always be 1.

For teachers only: If the three numbers are $x-1$, x, $x+1$ then $x^2 - (x-1)(x+1)$
$= x^2 - (x^2 - 1) = x^2 - x^2 + 1$
$= 1$

4 **a, b** For both consecutive even and consecutive odd numbers, the square of the middle number minus the product of the other two numbers will always be 4.

Examples:

4, 6, 8 10, 12, 14 3, 5, 7
36 – 32 = 4 144 – 140 = 4 25 – 21 = 4

11, 13, 15
169 – 165 = 4

For teachers only:

a 3 consecutive even numbers could be $2a-2$, $2a$, $2a+2$
Therefore $(2a)^2 - (2a-2)(2a+2) = 4a^2 - (4a^2 - 4)$
$= 4$

b 3 consecutive odd numbers are
$2a-1$, $2a+1$, $2a+3$
$(2a+1)^2 - (2a-1)(2a+3) = 4a^2 + 4a + 1 - (4a^2 + 4a - 3)$
$= 4$

Factorial notation (page 14)

1 **a** 5

Students will probably expand 5! And 4!, so that $\frac{5!}{4!} = \frac{5 \times 4 \times 3 \times 2 \times 1}{4 \times 3 \times 2 \times 1} = 5$

After a few exercises, they will probably notice the shortcut—that, in fact: $\frac{5!}{4!} = \frac{5 \times 4!}{4!} = 5$

b 6 **c** 7 **d** 10 **e** 20 **f** 20
g 30 **h** 90

2 **a** 4! **b** 7! **c** 9! **d** 8! **e** 7!
f 10! **g** 9! **h** 18! **i** 17! **j** 16!

3 1 x 1! + 2 x 2! + 3 x 3! + 4 x 4! = 5! – 1
1 x 1! + 2 x 2! + 3 x 3! + 4 x 4! + 5 x 5! = 6! – 1

Four 4s problem (page 15)

Some of the possible solutions to this famous problem:

$0 = 4 + 4 - 4 - 4$
$= (4 \div 4) - (4 \div 4)$
$= (4 \times 4) - (4 \times 4)$
$= 4 + 4 - 4 \times \sqrt{4}$

$1 = (4 + 4) \div (4 \times \sqrt{4})$
$= 4 \div 4 + 4 - 4$
$= 4 \times 4 \div (4 \times 4)$
$= \sqrt{4} \div 4 + \sqrt{4} \div 4$
$= (4 + 4) \div (4 + 4)$
$= 4^4 \div 4^4$
$= (4 \times 4!) \div (4 \times 4!)$
$= (4 + 4!) \div (4 + 4!)$
$= 44 \div 44$

$2 = 4 \div 4 + 4 \div 4$
$= 4 \times 4 \div (4 + 4)$
$= (4 + 4) \div (\sqrt{4} \times \sqrt{4})$
$= 4 \div \sqrt{4} \times 4 \div 4$

$3 = 4 \div \sqrt{4} + 4 \div 4$
$= (4 + 4 + 4) \div 4$
$= (4 \times 4 - 4) \div 4$
$= \sqrt{4} + \sqrt{4} - (4 \div 4)$
$= \sqrt{4} \times \sqrt{4} - (4 \div 4)$
$= 4 - \sqrt{4} + 4 \div 4$
$= 4 - \sqrt{4} + \sqrt{4} \div \sqrt{4}$

$4 = 4 + (4 - 4) \times 4$
$= 4 + (4 - 4) \div 4$
$= 4 \times 4 \div (\sqrt{4} \times \sqrt{4})$
$= 4 + 4 - \sqrt{4} - \sqrt{4}$
$= 4! + 4! - 44$
$= 4 \div \sqrt{4} + 4 \div \sqrt{4}$

$5 = (4 + 4 \times 4) \div 4$
$= 4! \div 4 - (4 \div 4)$
$= 4 \div 4 + \sqrt{4} + \sqrt{4}$
$= 4 + 4 \div (\sqrt{4} + \sqrt{4})$
$= 4 \div 4 + (\sqrt{4} \times \sqrt{4})$

$6 = (4 + 4) \div 4 + 4$
$= 4 + 4 \times \sqrt{4} \div 4$
$= 4 + \sqrt{4} + 4 - 4$
$= 4 + 4 - 4 \div \sqrt{4}$
$= 4 + (4 + 4) \div 4$

$7 = 4 + 4 - (4 \div 4)$
$= (4 \times 4 - \sqrt{4}) \div \sqrt{4}$
$= 4 + \sqrt{4} + 4 \div 4$
$= 4 \times \sqrt{4} - (4 \div 4)$
$= 4 \div \sqrt{.\dot{4}} + 4 \div 4$

$8 = 4 + 4 + 4 - 4$
$= (4 + 4) \times (4 \div 4)$
$= (4 + 4) \times (\sqrt{4} \div \sqrt{4})$
$= 4! \div 4 + 4 \div \sqrt{4}$
$= 4! \times \sqrt{.\dot{4}} - 4 - 4$
$= 4! \times \sqrt{.\dot{4}} - 4 \times \sqrt{4}$

$9 = 4 + 4 + 4 \div 4$
$= 4 + 4 + \sqrt{4} \div \sqrt{4}$
$= 4 + 4 + 4! \div 4!$
$= 4 \times \sqrt{4} + 4 \div 4$
$= (4! - 4) \div 4 + 4$

$10 = 4 + 4 + 4 - \sqrt{4}$
$= 4 + 4 + 4 \div \sqrt{4}$
$= 4 \times 4 - 4! \div 4$
$= 4 + \sqrt{4} + \sqrt{4} + \sqrt{4}$
$= 4 \times 4 - 4 \div \sqrt{.\dot{4}}$
$= (4! - 4) \div (4 \div \sqrt{4})$

$11 = 44 \div (\sqrt{4} + \sqrt{4})$
$= 44 \div (\sqrt{4} \times \sqrt{4})$
$= 4! \div \sqrt{4} - (4 \div 4)$
$= (4! + \sqrt{4} - 4) \div \sqrt{4}$
$= (4! + 4) \div 4 + 4$
$= (4! + 4! - 4) \div 4$

$12 = 4 \times 4 - \sqrt{4} - \sqrt{4}$
$= 4 + 4 + \sqrt{4} + \sqrt{4}$
$= 4 + 4 + \sqrt{4} \times \sqrt{4}$
$= 4! \div \sqrt{4} \times (4 \div 4)$
$= 4! \times \sqrt{.\dot{4}} - \sqrt{4} \times \sqrt{4}$

$13 = 44 \div 4 + \sqrt{4}$
$= 4! \div \sqrt{4} + (4 \div 4)$
$= (4! + 4! + 4) \div 4$
$= 4.4 \div .4 + \sqrt{4}$

$14 = 4 \times 4 - 4 + \sqrt{4}$
$= 4 + 4 + 4 + \sqrt{4}$
$= 4 \times 4 - 4 \div \sqrt{4}$
$= 4 \times \sqrt{4} + 4! \div 4$

$15 = 4 \times 4 - 4 \div 4$
$= 44 \div 4 + 4$
$= 4! \times \sqrt{.\dot{4}} - 4 \div 4$
$= 4.4 \div .4 + 4$

$16 = 4 \times 4 + 4 - 4$
$= 4 \times 4 \times 4 \div 4$
$= 4 + 4 + 4 + 4$
$= \sqrt{4} \times \sqrt{4} \times \sqrt{4} \times \sqrt{4}$
$= (\sqrt{4})^4 \times 4 \div 4$
$= (\sqrt{4})^4 + 4 - 4$
$= 4! \times \sqrt{.\dot{4}} + 4 - 4$

$17 = 4 \times 4 + 4 \div 4$
$= (44 + 4!) \div 4$
$= 4! \times \sqrt{.\dot{4}} + 4 \div 4$

$18 = 4 \times 4 + 4 \div \sqrt{4}$
$= 4 \times 4 + 4 - \sqrt{4}$
$= (\sqrt{4} + \sqrt{4}) \times 4 + \sqrt{4}$
$= 4! \times \sqrt{.\dot{4}} + 4 - \sqrt{4}$
$= 4! \times \sqrt{.\dot{4}} + 4 \div \sqrt{4}$

$19 = 4! - 4 - (4 \div 4)$

$20 = 4! \div \sqrt{4} + 4 + 4$
$= 4! + 4 - 4 - 4$
$= 4! \times 4 \div 4 - 4$
$= 4 \times 4 + \sqrt{4} + \sqrt{4}$
$= 4! \times \sqrt{.\dot{4}} + \sqrt{4} \times \sqrt{4}$
$= 4 \times 4 \div \sqrt{.\dot{4}} - 4$

$21 = 4! - 4 + 4 \div 4$

$22 = 4! - 4 + 4 \div \sqrt{4}$
$= 4 \times 4 + 4 + \sqrt{4}$
$= 44 \div (4 \div \sqrt{4})$
$= 4 \times 4 + 4 \div \sqrt{.\dot{4}}$
$= (4.4 \div .4) \times \sqrt{4}$

$23 = (\sqrt{4} \times 4! - \sqrt{4}) \div \sqrt{4}$
$= (4 \times 4! - 4) \div 4$
$= 4 \div 4 + 4! - \sqrt{4}$
$= (44 + \sqrt{4}) \div \sqrt{4}$

$24 = 4! \div \sqrt{4} + 4! \div \sqrt{4}$
$= 4 \times 4 + 4 + 4$
$= 4! \times \sqrt{.\dot{4}} + 4 + 4$
$= 4! + \sqrt{.\dot{4}} + 4 \times \sqrt{4}$

$25 = (\sqrt{4} \times 4! + \sqrt{4}) \div \sqrt{4}$
$= (4 \times 4! + 4) \div 4$
$= 4! + \sqrt{4} - (4 \div 4)$

$26 = (\sqrt{4} \times 4! + 4) \div \sqrt{4}$
$= 4! + \sqrt{4} + 4 - 4$
$= (4! \times 4) \div 4 + \sqrt{4}$
$= (4! \times \sqrt{4}) \div \sqrt{4} + \sqrt{4}$
$= (4! + \sqrt{4}) \times 4 \div 4$

$27 = 4! + 4 - 4 \div 4$
$= 4! + (4 + \sqrt{4}) \div \sqrt{4}$

$28 = 4! + 4 \times 4 \div 4$
$= 4! + 4 + 4 - 4$
$= 44 - 4! \times \sqrt{.\dot{4}}$
$= 4 \times 4 \div \sqrt{.\dot{4}} + 4$

$29 = 4! + 4 + 4 \div 4$

$30 = 4! + 4 + 4 \div \sqrt{4}$
$= 4! + 4 + 4 - \sqrt{4}$
$= 4 \times 4 \times \sqrt{4} - \sqrt{4}$
$= 4! + 4 + 4 - \sqrt{4}$

$31 = 4! + (4! + 4) \div 4$
$= 4! + \sqrt{4} + \sqrt{4} \div .4$

$32 = 4 \times 4 + 4 \times 4$
$= \sqrt{4} \times 4! - 4 \times 4$
$= 4! + 4 + \sqrt{4} + \sqrt{4}$
$= 4 \times 4 \times 4 \div \sqrt{4}$
$= 4 \times 4 \times (4 - \sqrt{4})$
$= (4^4 \div 4) \div \sqrt{4}$
$= 4! \times \sqrt{.\dot{4}} + 4 \times 4$

$33 = \dfrac{\sqrt{4} \times \sqrt{4}}{.\dot{4}} + 4!$
$= \dfrac{4}{.\dot{4}} + (\sqrt{4} + \sqrt{4})!$
$= \dfrac{4 - .4}{.4} + 4!$

$34 = \dfrac{4 \times 4}{.\dot{4}} - \sqrt{4}$
$= 4! + 4 + 4 + \sqrt{4}$
$= \dfrac{.4 \times (4!) + 4}{.4}$
$= 4 \times 4 \times \sqrt{4} + \sqrt{4}$
$= 4 \times (4 + 4) + \sqrt{4}$

$35 = 44 - \dfrac{4}{.\dot{4}}$
$= 44 \div 4 + 4!$
$= \dfrac{4 \times 4 - .\dot{4}}{.\dot{4}}$

$36 = \dfrac{4 \times 4}{.4} - 4$
$= 4! + 4 + 4 + 4$
$= \dfrac{4! \times \sqrt{4}}{4} + 4!$
$= (4 + \sqrt{4}) \times (4 + \sqrt{4})$
$= 4 \times (4 + 4) + 4$
$= 4 \times 4 \times \sqrt{4} + 4$

$37 = \dfrac{4! + \sqrt{4}}{\sqrt{4}} + 4!$
$= \dfrac{4 \times 4 + .\dot{4}}{.\dot{4}}$

$38 = 44 - 4 - \sqrt{4}$
$= \dfrac{4 \times 4}{.\dot{4}} - \sqrt{4}$
$= 4! - \sqrt{4} + 4 \times 4$
$= \dfrac{4 \times 4}{.\dot{4}} + \sqrt{4}$

$39 = 44 - (\sqrt{4} \div .4)$
$= \dfrac{4 \times 4 - .4}{.4}$
$= \sqrt{4} \times 4! - \dfrac{4}{.\dot{4}}$

$40 = (4 + 4 + \sqrt{4}) \times 4$
$= 44 - (\sqrt{4} + \sqrt{4})$
$= 4 \times 4 \times 4 - 4!$
$= \dfrac{4 \times 4}{.\dot{4}} + 4$
$= \sqrt{4} \times 4! - 4 - 4$
$= 4! \div \sqrt{.\dot{4}} + \sqrt{4} \times \sqrt{4}$

$41 = \dfrac{4 \times 4 + .4}{.4}$

$42 = 4 \times 4 + 4! + \sqrt{4}$
$= 44 - 4 + \sqrt{4}$
$= \dfrac{4 \times 4}{.4} + \sqrt{4}$
$= (4! - \sqrt{4}) \times \sqrt{4} - \sqrt{4}$
$= 44 - (4 - \sqrt{4})$
$= 4! \div \sqrt{.\dot{4}} + 4! \div 4$

$43 = 44 - (4 \div 4)$

$44 = 44 + 4 - 4$
$= 44 \times (4 \div 4)$
$= 4! \times 4 \div \sqrt{4} - 4$
$= 4! \div \sqrt{.\dot{4}} + 4 + 4$
$= 4! \div \sqrt{.\dot{4}} + 4 \times \sqrt{4}$
$= (4.4 \div .4) \times 4$

$45 = 44 + (4 \div 4)$
$= \dfrac{4 \times 4 + \sqrt{4}}{.4}$

$46 = 44 + 4 - \sqrt{4}$
$= 4! \times \sqrt{4} + \sqrt{4} - 4$
$= (4! - \sqrt{4}) \times \sqrt{4} + \sqrt{4}$

Four 4s problem (page 15) *continued...*

$47 = 4! \times \sqrt{4} - (4 \div 4)$
$= 4! + 4! - (4 \div 4)$

$48 = (4 + 4 + 4) \times 4$
$= 44 + \sqrt{4} + \sqrt{4}$
$= 4! \times \sqrt{4} + 4 - 4$

$49 = 4! \times \sqrt{4} + 4 \div 4$
$= 4! + 4! + 4 \div 4$

$50 = 44 + 4 + \sqrt{4}$
$= \sqrt{4} \times 4! + 4 - \sqrt{4}$
$= (4 \times 4! + 4) \div \sqrt{4}$
$= 44 + 4 \div \sqrt{.\dot{4}}$

$51 = \frac{4!}{.4} - \frac{4}{.\dot{4}}$
$= \frac{4! - \sqrt{4}}{.4} - 4$

$52 = 44 + 4 + 4$
$= 4! + 4! + \sqrt{4} + \sqrt{4}$
$= 4! \times \sqrt{4} + \sqrt{4} + \sqrt{4}$
$= 4! \div \sqrt{.\dot{4}} + 4 \times 4$

$53 = 44 + \frac{4}{.\dot{4}}$
$= \frac{4! - \sqrt{4}}{.4} - \sqrt{4}$

$54 = 4! \times \sqrt{4} + \sqrt{4} + 4$
$= 4! + 4! + \sqrt{4} + 4$
$= 44 + \frac{4}{.4}$

$55 = \frac{4!}{.\dot{4}} + \frac{4}{4}$
$= (44 \div .4) \div \sqrt{4}$

$56 = 4! + 4! + 4 + 4$
$= 4 + 4! \times \sqrt{4} + 4$
$= \frac{4!}{.\dot{4}} + \frac{4}{\sqrt{4}}$

$57 = 4! \times \sqrt{4} + \frac{4}{.\dot{4}}$
$= \frac{4! - \sqrt{4}}{.4} + \sqrt{4}$

$58 = \frac{4!}{.4} - 4 + \sqrt{4}$
$= \frac{4!}{.4} - \frac{4}{\sqrt{4}}$

$59 = \frac{4!}{.4} - \frac{4}{4}$

$60 = 4^4 \div 4 - 4$
$= 4! \div .4 + 4 - 4$
$= \frac{4 \times 4! + 4!}{\sqrt{4}}$
$= 4 \times 4 \times 4 - 4$
$= 4! \times \sqrt{.\dot{4}} + 44$

$61 = \frac{4!}{.4} + \frac{4}{4}$
$= \frac{4! + \sqrt{4}}{.4} - 4$

$62 = 4 \times 4 \times 4 - \sqrt{4}$
$= 4^4 \div 4 - \sqrt{4}$
$= \frac{4!}{.4} + \frac{4}{\sqrt{4}}$
$= 4! \div .4 + 4 - \sqrt{4}$

$63 = \frac{4! + \sqrt{4} + \sqrt{4}}{.\dot{4}}$
$= \frac{4! + \sqrt{4}}{.4} - \sqrt{4}$

$64 = (4 + 4) \times 4 \times \sqrt{4}$
$= 4 \times 4 \times \sqrt{4} \times \sqrt{4}$
$= 4! + 4! + 4 \times 4$
$= 4! \times \sqrt{4} + 4 \times 4$
$= 4^{4 - \frac{4}{4}}$
$= \frac{4!}{.\dot{4}} + \frac{4}{.4}$

$65 = \frac{4! + 4}{.\dot{4}} + \sqrt{4}$
$= \frac{4! + 4 - \sqrt{4}}{.4}$

$66 = 4 \times 4 \times 4 + \sqrt{4}$
$= 4^4 \div 4 + \sqrt{4}$
$= \frac{4! + 4}{.4} - 4$

$67 = \frac{4! + 4}{.\dot{4}} + 4$
$= \frac{4! + \sqrt{4}}{.4} + \sqrt{4}$

$68 = 4 \times 4 \times 4 + 4$
$= \frac{4! + 4}{.4} - \sqrt{4}$
$= 4^4 \div 4 + 4$

$69 = \frac{4!}{.4} + \frac{4}{.\dot{4}}$
$= \frac{4! + \sqrt{4}}{.4} + 4$

$70 = 44 + 4! + \sqrt{4}$
$= 4! + 4! + 4! - \sqrt{4}$
$= 4! \times 4 - 4! - \sqrt{4}$
$= \frac{4! + \sqrt{4} + \sqrt{4}}{.4}$

$71 = \frac{4! + 4}{.4} + [.4]$
$= (4! + 4.4) \div .4$

$72 = \frac{4! + 4}{.4} + \sqrt{4}$
$= \frac{4}{.\dot{4}} \times 4 \times \sqrt{4}$
$= 44 + 4! + 4$
$= (4 - \frac{4}{4}) \times 4!$

$73 = \frac{\sqrt[.\dot{4}]{4} + .\dot{4}}{.\dot{4}}$
$= (\text{Note } \sqrt[.\dot{4}]{4} = 4^{\frac{5}{2}}$
$= 2^5$
$= 32$

$74 = 4! + 4! + 4! + \sqrt{4}$
$= \frac{4! + 4}{.4} + 4$

$75 = \frac{4!}{.4 \times .4} \div \sqrt{4}$
$= (4! + [.4])(4 - [.4])$

$76 = 4! + 4! + 4! + 4$
$= \frac{4!}{.\dot{4}} + 4! - \sqrt{4}$
$= (4! - 4) \times 4 - 4$

$77 = \frac{4!}{.\dot{4}} + 4! - [.4]$
$= \frac{4!}{.\dot{4} \times \sqrt{.\dot{4}}} - 4$

$78 = (4! - 4) \times 4 - \sqrt{4}$
$= \frac{4!}{.4\dot{4}} + 4!$

$79 = \frac{4! - \sqrt{4}}{.4} + 4!$
$= \frac{4!}{.\dot{4} \times \sqrt{.\dot{4}}} - \sqrt{4}$

$80 = 4 \times 4! - 4 \times 4$
$= \frac{4!}{.\dot{4}} + 4! + \sqrt{4}$
$= (4 \times 4 + 4) \times 4$
$= 4! \div \sqrt{.\dot{4}} + 44$

$81 = (4! - 4) \times 4 + [.4]$

$82 = \frac{4!}{.\dot{4}} + 4! + 4$
$= (4! - 4) \times 4 + \sqrt{4}$

$83 = \frac{4!}{.4} + 4! - [.4]$

$84 = 44 \times \sqrt{4} - 4$
$= 4! \times 4 - (4! \div \sqrt{4})$
$= (4! - 4) \times 4 + 4$
$= \frac{4!}{.4} + 4! \times [.4]$

$85 = \frac{4!}{.4} + 4! + [.4]$
$= \frac{4! + .4}{.4} + 4!$
$= 4 + (\frac{4}{.\dot{4}})^{\sqrt{4}}$
$= \frac{4!}{.\dot{4} \times \sqrt{.\dot{4}}} + 4$

$86 = 44 \times \sqrt{4} - \sqrt{4}$
$= \frac{4!}{.4} + 4! + \sqrt{4}$

$87 = \frac{4! + 4}{.\dot{4}} + 4!$

$88 = 4 \times 4 \times 4 + 4!$
$= 44 + 44$
$= 4! \times 4 - 4 - 4$
$= 44 \times 4 \div \sqrt{4}$
$= 4^4 \div 4 + 4!$

$89 = (4! - \sqrt{4}) \times 4 + [.4]$
$= \frac{4! + \sqrt{4}}{.4} + 4!$

$90 = 4! \times 4 - 4 - \sqrt{4}$
$= 44 \times \sqrt{4} + \sqrt{4}$
$= \frac{4 \times 4}{.\dot{4} \times .4}$

$91 = 4! \times 4 - 4 - [.4]$

$92 = 4! \times \sqrt{4} \times \sqrt{4} - 4$
$= 4! \times 4 - \sqrt{4} - \sqrt{4}$
$= 44 + 4! + 4!$

$93 = 4 \times 4! - \sqrt{4} - [.4]$

$94 = 4! \times 4 + \sqrt{4} - 4$
$= \frac{4! + 4}{.4} + 4!$
$= 4! \times \sqrt{4} \times \sqrt{4} - \sqrt{4}$

$95 = 4! \times 4 - \frac{4}{4}$
$= 4! \times 4 - [.44]$

$96 = 4! + 4! + 4! + 4!$
$= (4 \times 4) \times (4 + \sqrt{4})$
$= 4! \times 4 + 4 - 4$
$= 4! \times 4 \times 4 \div 4$

$97 = 4! \times 4 + \frac{4}{4}$
$= 4! \times 4 + [.44]$

$98 = 4! \times \sqrt{4} \times \sqrt{4} + \sqrt{4}$
$= 4 \times 4! + 4 - \sqrt{4}$

$99 = 4! \times 4 + 4 - [.4]$

$100 = 4! \times \sqrt{4} \times \sqrt{4} + 4$
$= 4! \times 4 + \sqrt{4} + \sqrt{4}$
$= 4! \times 4 + \sqrt{4} \times \sqrt{4}$
$= \frac{44}{.44}$
$= \left(4! + \frac{4}{4}\right) \times 4$
$= (4! \times \sqrt{4} + \sqrt{4}) \times \sqrt{4}$

Some calculations that have been useful include:

$\sqrt{\cdot 4} = \sqrt{\frac{4}{9}} = \frac{2}{3}$ $\qquad 4! \times \sqrt{\cdot 4} = 24 \times \frac{2}{3} = 16$

$4! \div \sqrt{\cdot 4} = 24 \times \frac{3}{2} = 36$ $\qquad 4! \div \sqrt{\cdot 4} = 4 \times \frac{3}{2} = 6$

$\frac{4 - \cdot 4}{\cdot 4} = \frac{40 - 4}{4} = 9$ $\qquad \frac{4! - \sqrt{4}}{\cdot 4} = 22 \times \frac{10}{4} = 55$

$\frac{4!}{\cdot 4 \times \cdot 4} = 24 \times \frac{10}{4} \times \frac{10}{4} = 150$ $\qquad \frac{4!}{\cdot 4 \times \sqrt{\cdot 4}} = 24 \times \frac{9}{4} \times \frac{3}{2} = 81$

$\frac{4! + \cdot 4}{\cdot 4} = \frac{240 + 4}{4} = 61$ $\qquad \frac{4 \times 4}{\cdot 4 \times \cdot 4} = 16 \times \frac{9}{4} \times \frac{10}{4} = 90$

Sequences (page 16)

1 a 35 and 74 Each term is 3 more than the previous one.
12th term = 2 + 11 × 3 = 35 *or* 12 × 3 − 1 = 35
25th term = 2 + 24 × 3 = 74 *or* 25 × 3 − 1 = 74

b 58 and 148
12th term = 3 + 11 × 5 = 58 *or* 12 × 5 – 2 = 58
30th term = 3 + 29 × 5 = 148 *or* 30 × 5 – 2 = 148

c 85 and 295
15th term = 1 + 14 × 6 = 85 *or* 15 × 6 – 5 = 85
50th term = 1 + 49 × 6 = 295 *or* 50 × 6 – 5 = 295

d 65 and 184
8th term = 16 + 7 × 7 = 65 *or* 8 × 7 + 9 = 65
25th term = 16 + 24 × 7 = 184 *or* 25 × 7 + 9 = 184

e 92 and 272
11th term = 2 + 10 × 9 = 92 *or* 11 × 9 – 7 = 92
31st term = 2 + 30 × 9 = 272 *or* 31 × 9 – 7 = 272

f 43 and 167
9th term = 11 + 8 × 4 = 43 *or* 9 × 4 + 7 = 43
40th term = 11 + 39 × 4 = 167 or 40 × 4 + 7 = 167

g 147 and 97
11th term = 197 – 10 × 5 = 147 *or* continuing the pattern:
21st term = 197 – 20 × 5 = 97 each time subtracting 5

h 135 and 83
13th term = 183 – 12 × 4 = 135 *or* continuing the pattern:
26th term = 183 – 25 × 4 = 83 each time subtracting 4

i 16 and 128
5th term = 8 × 2 = 16 *or* continuing the pattern:
8th term = 16 × 2 × 2 × 2 = 128 1, 2, 4, 8, 16, 32, 64, 128

j 27 and 243
6th term = 3 × 3 × 3 = 27 *or* continuing the pattern:
8th term = 27 × 3 × 3 = 243 each time multiplying by 3: $\frac{1}{9}$, $\frac{1}{3}$, 1 3, 9, 27, 81, 243

Obviously, simply to continue the pattern could in some questions be very tedious; using the method shown above can be much quicker.

2 In each question, the sequence is formed by starting with a number, □, adding (or subtracting) another number, △, and continuing to add △ for each term.

First = □, Second = □ + △, Third = □ + 2△,
Fourth = □ + 3△, Fifth = □ + 4△, Sixth = □ + 5△,
Seventh = □ + 6△, Eighth = □ + 7△

To solve these problems, students can use logic or this pattern, or the guide given in the first three solutions.

a 2, 4, 6, 8, 10 Fourth term = 2 + 3△ = 8 so △ = 2.

b 1, 4, 7, 10, 13
1 + 3 x △ = 10, therefore △ = 3.

c 3, 5, 7, 9, 11 Second term = □ + △
Fifth term = □ + 4△
∴ 11 = 5 + 3△
∴ △ = 2

d 2, 7, 12, 17, 22 **e** 2, 6, 10, 14, 18

f 3, 10, 17, 24, 31 **g** 2, 8, 14, 20, 26

h 1, 9, 17, 25, 33

i 51, 47, 43, 39, 35 This question is difficult, since 4 must be subtracted from each term to get the next one.

j 48, 43, 38, 33, 28 This time 5 is subtracted from a term to get the next term.

Bartering (page 17)

It is suggested that students draw up this table and use it to answer the questions.

Carrots	Bananas	Oranges	Apples	Pineapples
12	6	3	4	1
24	12	6	8	2
36	18	9	12	3
48	24	12	16	4
60	30	15	20	5
×2 72	36	18	24	6
120	60	30	40	10
		*36		*12

1 **a** 2 pineapples **b** 3 pineapples **c** 3 pineapples
d 5 pineapples **e** 12 pineapples

2 **a** 30 oranges **b** 40 apples **c** 60 bananas

3 **a** 20 apples **b** 20 apples **c** 8 apples
d 2 apples **e** 12 apples

4 30 bananas or 15 oranges or 20 apples or 5 pineapples

5 12 oranges = 24 bananas
12 apples = 18 bananas
Total = 42 bananas

6 9 oranges = 3 pineapples
24 bananas = 4 pineapples
8 apples = 2 pineapples
Total = 9 pineapples

7 18 carrots = $1\frac{1}{2}$ pineapples
3 bananas = $\frac{1}{2}$ pineapple
Total = 2 pineapples

Percentage challenges (page 18)

1 90 (60 increased by 30)

2 16
24
original number

3 30 (60 decreased by 30)

4 36
50% 18 | 50%

5 24
25%
18

6 60 20% of 50 is 10, so the number = 50 + 10

7 25 since 120% of number = 30
∴ 20% of number = 5 and
100% of number = 25

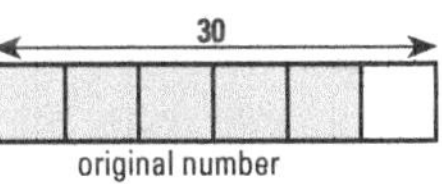

8 40

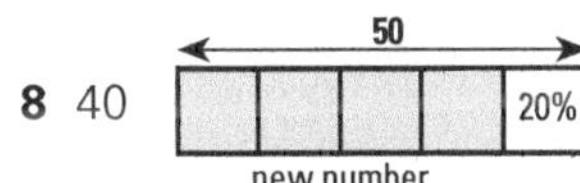

9 40, since 80% of number = 32
∴ 20% of number = 8 and
100% of number = 40
20%
32 (80% of number)

10 32

11 36, since $133\frac{1}{3}$% of number = 48
∴ $33\frac{1}{3}$% of number = 12 and 100% of number = 36

12 72, since $66\frac{2}{3}$% of number = 48
∴ $33\frac{1}{3}$% of number = 24 and 100% of number = 72%

13 60, since 20% = $\frac{1}{5}$ **14** 60, since 5% = $\frac{1}{20}$

15 140, since 5% of number is 7 **16** 250, since 2% = $\frac{1}{50}$

17 40, since 10% of number = 4

18 20, since 10% of number = 2

Percentage challenges (page 18) *continued...*

19 60, since 20% of words = 12 words

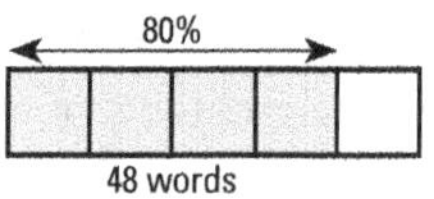

20 $9600

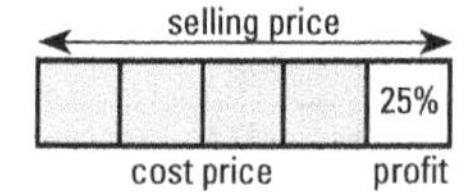

selling price = 125% cost
125% cost = $12 000
25% cost = $2400
100% of cost = $9600

21 $16 000, 75 % of cost = $12 000, 25% of cost = $4000,
100% cost = $16 000

cost price | 25% | selling price | loss

22 $400, if 80% of value = $320
Then 20% of value = $80
∴ 100% value = $400

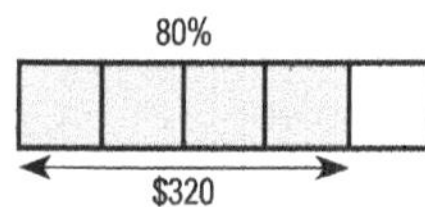

Percentage problems (page 19)

1 No; with a 20% discount the price would be $120. However, with a 10% discount the price is $135 and a further 10% discount reduces it to $121.50.

The 10% employee discount is calculated on the sale price, which is discounted by 10%. This is equivalent not to 20%, but to 19%.

2 If a 20% discount is offered then price is 80% of original price. If 15% tax is added, then we calculate 115% of price.

If the 20% discount is calculated before the 15% tax is added:

$\frac{80}{100} \times \frac{100}{1}$ = $80 (after discount)

$\frac{115}{100} \times 80$ = $92 (after tax)

If the 15% tax is added before the 20% is discounted:

$\frac{115}{100} \times \frac{100}{1}$ = $115 (after tax)

$\frac{80}{100} \times 115$ = $92 (after discount)

In either case, the item will cost $92.

3 The dealer lost $1500.

Car 1 (20% profit)	*Car 2 (20% loss)*
Selling price: $18 000	Selling price: $18 000
120% of cost price = $18 000	80% of cost = $18 000
10% of cost = $1500	10% of cost = $2250
100% of cost = $15 000	100% of cost = $2250
∴ profit = $3000	∴ loss = $4500

4 Price to customer = $1000 × $\frac{125}{100}$ = $1250
Customer selling price = $1250 × $\frac{75}{100}$ = $937.50
Joseph's resale price = $1100
∴ Joseph's profit = $250 + $162.50 = $412.50

5 If Edgar paid $270, then Beverly paid $180, as she made 50% profit. Now, $180 is 120% of the amount Glenn paid.
∴ 20% of amount = $30 and 100% of amount = $150

Fruit arithmetic (page 20)

Use both trial and error and logic to solve these problems.

1 [apple] = 1

```
  1 1
x 1 1
1 2 1
```

2 Since [banana] × [banana][cherries] = 105
and [cherries] × [banana][cherries] = 175,
[banana] = 3 and [cherries] = 5.

```
    3 5
  x 5 3
  1 0 5
1 7 5 0
1 8 5 5
```

3 Since [strawberry] × [strawberry][orange] = 828
and [orange] × [strawberry][orange] = 184
[orange] = 2 and [strawberry] = 9.

```
    9 2
  x 2 9
  8 2 8
1 8 4 0
2 6 6 8
```

4 [banana] = 6 [cherries] = 5
[orange] = 4 [pear] = 3

6	6	6	6	24
6	6	5	5	22
5	3	4	6	18
4	4	3	5	16
21	19	18	22	

5 [cherries] = 2 [strawberry] = 4 [apple] = 6
[grapes] = 7 [banana] = 3 [pear] = 5

6	3	3	6	18
4	7	5	2	18
2	4	5	2	13
3	2	6	2	13
15	16	19	12	

6 [grapes] is either 3 or 7. If it is 3, then [pear] is 7, since 3 x 7 ends in 1. Now test 73 x 73; this gives an incorrect answer, so try [grapes] = 7.
[grapes] = 7 [pear] = 3

```
    3 7
  x 3 7
  2 5 9
1 1 1 0
1 3 6 9
```

7 [apple] = 4 [banana] = 3

```
    4 3
  x 4 3
  1 2 9
1 7 2 0
1 8 4 9
```

8 [pear] = 2 or 3; if it is 2, then [apple] = 4,
[banana] = 1, [strawberry] = 3, [plum] = 7 and [cherries] = 6.
Note if [plum] = 9, then [strawberry] = 4
but [apple] = 4 so contradiction.
or [pear] = 3 [apple] = 9 [strawberry] = 2 [plum] = 7 [cherries] = 6 [banana] = 1

```
   2 3           3 2
2)4 7    or    3)9 7
  4              9
    7              7
    6              6
    1              1
```

9 [orange] = 3 [grapes] = 8 [banana] = 2
[strawberry] = 9 [pear] = 6

```
   3 2
3)9 8
  9
    8
    6
    2
```

10 [apple] = 2 [pear] = 9 [grapes] = 1 [strawberry] = 8
[banana] = 4 [cherries] = 5 [plum] = 7

```
   4 7
2)9 5
  8
  1 5
  1 4
    1
```

Divisibility number puzzles (page 21)

1 36

2 a 60 is divisible by 2, 3, 4 and 5. If it is divisible by 4 and 5, it is a multiple of 20 (20, 40, 60, 80, 100, 120) and the smallest number of these also divisible by 2 and 3 is 60. Alternatively, we could have considered multiples of 3 and 4 (12, 24, 36, 48, 60) and looked for the first one that is divisible by 5.

b 60 is divisible by 2, 3, 4, 5 and 6.

c 420 The number must be divisible by 2, 3, 4, 5, 6 and 7. Since 7 is a prime number, the answer is 60 x 7 = 420.

d 840 The number must be divisible by 2, 3, 4, 5, 6, 7 and 8. Since 420 is not divisible by 8, smallest possible number is 420 x 2 = 840.

3 a 5

b 13 It is possible to do this question by listing the numbers
2: 3, 5, 7, 9, 11, 13, 15, …
3: 4, 7, 10, 13, 16, …
4: 5, 9, 13, …

and find that 13 is the smallest number.

However, a nicer, more sophisticated method is to use the fact that 12 is the smallest number into which 2, 3 and 4 divide, leaving no remainder. Therefore, 13 is the smallest number that will leave a remainder of 1.

d 61, as 60 is the smallest number into which 2, 3, 4 and 5 divide, leaving no remainder. Note that 60 is the LCM (lowest common multiple).

d 61

4 121, using the fact that 120 is the smallest number exactly divisible by 2, 3, 4, 5, 6 or 8.

5 24; the number must be divisible by 2, 3, 4 and 8, so the possibilities are 24, 48, 72, 96, 120, 144 and so on. Both 24 and 144 leave a remainder of 4 when divided by 5, but 24 is correct as the party is unlikely to have consisted of 144 children or more.

6

Rabbits	Hutches	1	2	3	4	5	6	7	8
	5 in a hutch (2 over)	7	12	17	22	27	32	37	42
	6 in a hutch	6	12	18	24	30	36	42	

Therefore, there are 42 rabbits and 8 hutches.

7 a B= 0 A = 3, 6 or 9 *or* B = 4, A = 2, 5 or 8 *or* B = 8, A = 1, 4 or 7

For a number to be divisible by 12 it must be divisible by 4 and 3. To be divisible by 4, the last 2 digits must be divisible by 2.

∴ B = 0, 4 or 8
To be divisible by 3, sum of digits must be divisible by 3
∴ if B = 0, A = 3, 6 or 9
if B = 4, A = 2, 5 or 8
and if B = 8, A = 1, 4 or 7

b A= 7 and B = 2

For a number to be divisible by 8, the last 3 digits must be divisible by 8. Therefore B = 2 as 432 is divisible by 8. For a number to be divisible by 11, sum of digits in odd places – sum of digits in even places = 0 or a multiple of 11.

∴ 9 + 4 + 2 – (A + 5 + 3) = 0
∴ A = 7

8 72 We need to find the LCM of 6, 8, 9 and 12.

$$\left.\begin{array}{l}6 = 2\times 3\\ 8 = 2\times 2\times 2\\ 9 = 3\times 3\\ 12 = 2\times 2\times 3\end{array}\right] \therefore \text{LCM} = 2\times 2\times 2\times 3\times 3 = 72$$

Money problems (page 22)

1 Cost of TV set = $850
deposit = $130
amount owing = $720
instalment = $60

2 Deposit = $600 – 12 x 40
= $120

3 a The first 6 words cost $2.40 and the next 15 cost $4.50. The 21-word telegram cost $6.90.

b The first 6 words cost $2.40 and the rest $3.30, which is the cost of 11 words. The telegram contained 17 words.

4 a 10 kg = $32
15 (3 x 5) = $24
3 kg = $8
28 kg = $64

b $80 less $32 = $48, and since 48 = 6 x 8, the parcel's mass = 10 + 6 x 5 kg = 40 kg. However, the mass could be any figure greater than 35 kg and less than 40 kg.
35 kg < mass of parcel ⩽ 40 kg

5

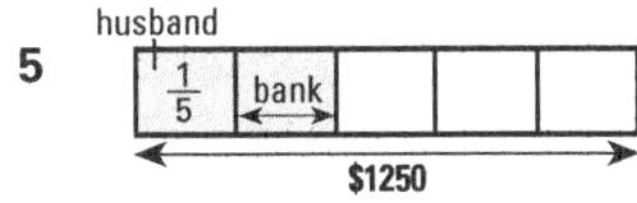

∴ Mrs King gave $250 to her husband and she put $250 in the bank. She had $750 left.

6 7 lots of money = $56
∴ 1 lot of money = $8
∴ Evelyn has $8 more than Georgina.

Use your calculator (page 23)

1 $26 000 **2** $600 **3 a** $128 **b** $640

4 a $230 (cost = 50 + 12 x 15 = $230 **b** $30

5 a $1080 **b** $230

6 a $740 **b** $742 **c** Nicky **d** $2

7 a $6560 (800 + 160 x 36 = $6560) **b** $1560

Tessa's world trip (page 24)

1 a A$2.50 = £1
A$10 = £4
A$800 = £320

b A$200 = £80
c $280 = £112 (112 = 4 x 28)

2 a A$5 = 3 Euros
A$400 = 240 Euros

b A$60 = 36 Euros
c A$200 = 120 Euros

3 a A$1 = 6 kronor
A$300 = 1800 kronor

b A$40 = 240 kronor
c A$250 = 1500 kronor

4 a A$1 = 150 forints
A$200 = 30 000 forints

b A$66 = 9900 forints

5 a A$1 = 30 rupees
A$200 = 6000 rupees

b A$12.50 = 375 rupees

6 a A$1 = US$0.75 = US$$\frac{3}{4}$
A$$\frac{4}{3}$ = US$1
A$1.33 = US$1

b A$800 = US$$\frac{3}{4}$ x 800
= US$600

c A$$\frac{4}{3}$ x 60 = US$60
=A$80

d A$480 = US$$\frac{3}{4}$ x 480
= US$360

e A$400 = US$300

Hiring a bus (page 25)

1 a $300
($250 + 25 x 2 = $300)

b $12
($300 ÷ 25 = $12)

Hiring a bus (page 25) *continued...*

c $280
($250 + 15 x 2 = $280)

d $340
($250 + 45 x 2 = $340)

2 a $275
($150 + 25 x 5 = $275)

b $11
($275 ÷ 25 = $11)

c $225
($150 + 15 x 5 = $225)

d $375
($150 + 45 x 5 = $375)

3 a $275
($25 x 11 = $275)

b $495
($45 x 11 = $495)

4 a The Economy bus company and Travelsafe bus company are the same price.

b The Friendly bus company

c The Friendly bus company
(Friendly would be $320, Economy would be $325, Travelsafe would be $385)

Matchstick puzzles (page 26)

1

Matches on each side	1	2	3	4	5	→ 10
Total No. of matches	3	6	9	12	15	→ 30

The number of matches required is 3 times the number on each side of a triangle.

2

Number of L's	1	2	3	4	5	→ 10
Total No. of matches	2	6	12	20	30	→ 110

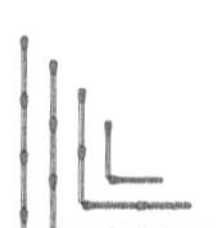

The number of matches required is the number of L's times the next highest number.

3

Number of steps	1	2	3	4	5	→ 10
Total No. of matches	4	10	18	28	40	→ 130

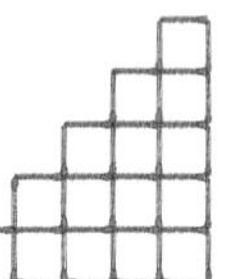

The number of matches required is the number of the step times the number 3 more than the height.

4

Number of hexagons	1	2	3	4	5	→ 10
Total No. of matches	6	10	14	18	22	→ 44

The number of matches required is 4 times the number of hexagons plus 2.

5

Connected diamonds	1	2	3	4	5	→ 10
Total No. of matches	4	10	16	22	28	→ 58

The number of matches required is 6 times the number of diamonds, less 2.

6

No. of pentagon patterns	1	2	3	4	5	→ 10
Total No. of matches	5	12	19	26	33	→ 68

The number of matches required is 2 less than 7 times the number of pentagons.

Patterns with squares (page 27)

1 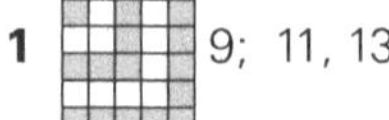 9; 11, 13

2 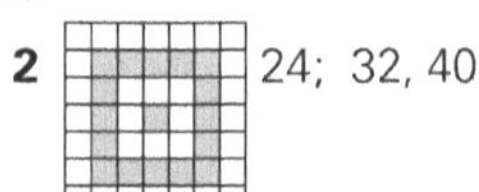 24; 32, 40

The pattern may not be immediately obvious in this case. To help them, students could be asked to draw the next figure.

3 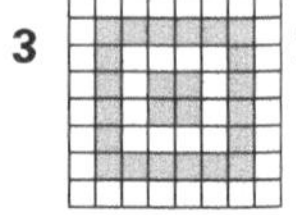 28; 36, 44

4 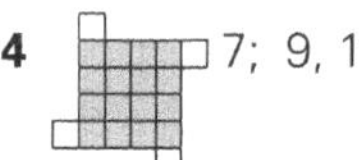 7; 9, 11

Counting techniques (page 28)

Some of these questions can be done by drawing a tree diagram or by listing all possibilities systematically or by considering the number of possible ways of choosing each digit.

1 a 3 x 2 = 6 possibilities
They are 34, 35, 43, 45, 53, 54.

b 3 x 3 = 9 possibilities. Now beside the possibilities above, we can have 33, 44, 55 as well.

2 a 12: There are 4 ways of choosing the first digit and 3 ways of choosing the second digit. Students should list the possibilities, working systematically.

35,	37,	38
53,	57,	58
73,	75,	78
83,	85,	87

b 3: There is only 1 way of choosing the second digit to ensure it is even and there are 3 ways of choosing the first digit; namely, we can have 38 or 58 or 78.

c 24: There are 4 ways of choosing the first digit, 3 ways of choosing the second digit and 2 ways of choosing the third digit.

Students should list the possibilities, working systematically:

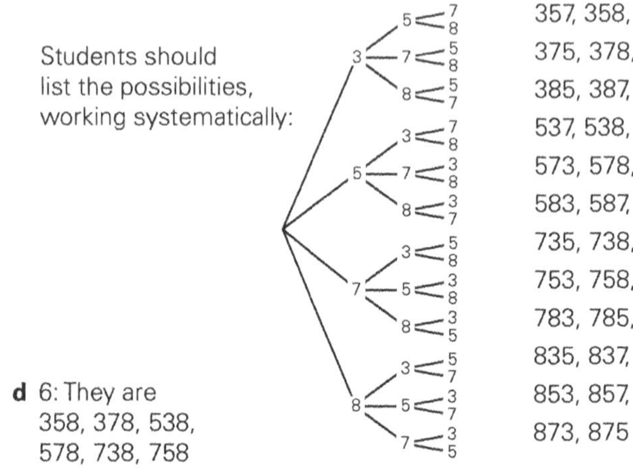

357, 358,
375, 378,
385, 387,
537, 538,
573, 578,
583, 587,
735, 738,
753, 758,
783, 785,
835, 837,
853, 857,
873, 875

d 6: They are 358, 378, 538, 578, 738, 758

e 18: There are 3 ways of choosing the third digit, 3 ways of choosing the first digit and 2 ways of choosing the second digit. Note that the odd and even 3-digit numbers give the same answer as the number of 3-digit numbers. Listing the possibilities: 375, 385, 387, 583, 587, 783, 785, 357, 537, 573, 735, 753, 835, 853, 837, 873, 857, 875.

f 24: There are 4 ways of choosing the first digit, 3 ways of choosing the second digit, 2 ways of choosing the third digit and 1 way of choosing the last digit. Students should list the possibilities, working systematically:
3578, 3587, 3758, 3785, 3857, 3875
5378, 5387, 5738, 5783, 5837, 5873
7358, 7385, 7538, 7583, 7835, 7853
8357, 8375, 8537, 8573, 8735, 8753

g 6: They are 3578, 3758, 5378, 5738, 7358, 7538.

h 18: From the answers in **f**, cross off the even numbers.

i 6 The number must end in 5, so possibilities are: 375, 385, 735, 785, 835, 875.

j 4: A number is divisible by 3 if the sum of the digits is divisible by 3: 5 and 7 so 57 or 75 or 7 and 8 so 78 or 87.

k 12: A number is divisible by 3 if the sum of the digits is divisible by 3:

3 + 5 + 7 so 357, 375, 537, 573, 735, 753
3 + 7 + 8 so 378, 387, 738, 783, 837, 873

3 8 possibilities

aAbB, aABb, AabB, AaBb

bBaA, bBAa, BbaA, BbAa

Note that each couple can be seated in 2 different ways (i.e. 2 × 2) ways and then the couples can be swapped.

4 a Note that each couple can be seated in 2 different ways (i.e. 2 × 2 × 2) ways and then couples can be arranged in 2 different ways as Aa is fixed. This gives a total of 16 possibilities. Students can list these.

b There are 6 ways of arranging the women and 6 ways of arranging the men, giving 6 x 6 = 36 possibilities.

Number plates (page 29)

1 100 cars: the number plates used are 00, 01 → 99.

2 1000 cars: the numbered 000, 001 → 999.

3 a 26, as any letter can be followed by 00 (A00, B00, C00 → Z00).

b 26 again, as in **a**. **c** 100 (B00, B01, B02 → B99)

d 100, as in **c**. **e** 10 (P80, P81,→ P89).

f 2600 (26 x 100) since as seen in **c** and **d**, each letter of the alphabet gives 100 possibilities.

4 a 26 (AA00, BA00, CA00 → ZA00)

b 676 (26 x 26)

26	26	26		26
AA99	BA99	CA99		ZA99
AB99	BB99	CB99	→	ZB99
AC99	BC99	CC99		ZC99
↓	↓	↓		↓
AZ99	BZ99	CZ99		ZZ99

← 26 columns →

c 100 (XX00, XX01, XX02 → XX99)

d There are 26 × 26 possibilities for the two letters and 100 for the two numbers, so total possibilities = 26 × 26 × 100 = 676 000.

5 a 676 (26 x 26 possibilities for the first two letters)

b 17 576 (26 × 26 × 26 possibilities for the three letters)

c 1000 (DAD000, DAD001 → DAD999)

d 17 576 000 (26 × 26 × 26 × 1000)

Listing possibilities (page 30)

1

50c	20c	10c
2		
1	2	1
1	1	3
1		5
	5	
	4	2
	3	4
	2	6
	1	8
		10

2

20c	10c	5c
2	1	
2	0	2
1	3	
1	2	2
1	1	4
1		6
	5	
	4	2
	3	4
	2	6
	1	8
		10

3

Bread	Filling
brown	egg
brown	cheese
brown	tuna
brown	honey
white	egg
white	cheese
white	tuna
white	honey

or use a tree diagram

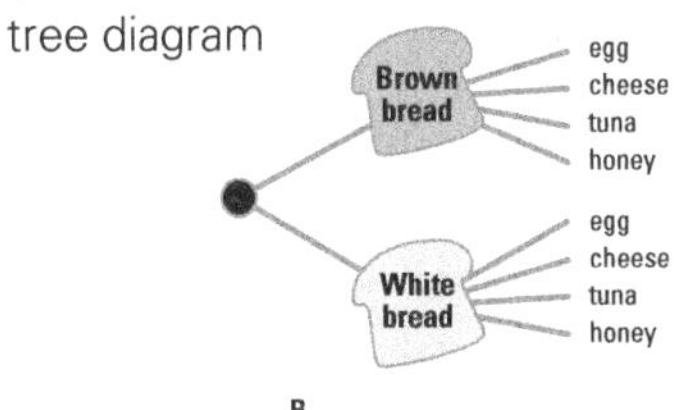

4 a 3

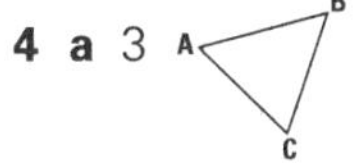

b 6 **c** 10

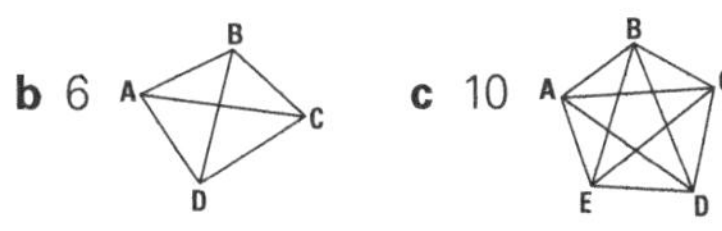

d A pattern is obvious from the above solutions: 3, 6 and 10 are the triangular numbers, so if 6 girls enter, 15 games will be played.

e 45 games will be played: 3 → 6 → 10 → 15 → 21 → 28 → 36 → 45

5 a 6 cards (each person sent 2)

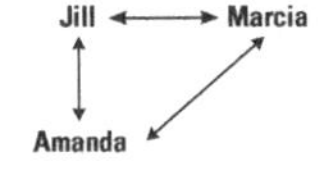

b 12 cards (each person sent 3)

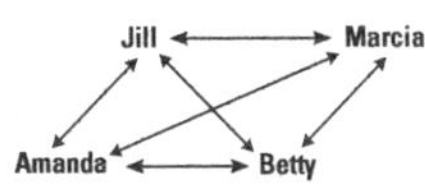

c 20 cards (each person sent 4)

d

Friends	3	4	5	6	7	8	→20
Cards sent	(3 × 2) 6	(4 × 3) 12	(5 × 4) 20	(6 × 5) 30	(7 × 6) 42	(8 × 7) 56	(20 × 19) 380

The table above is very useful as the sequence in the number of cards sent is that of the triangular numbers doubled, the triangular numbers being 1, 3, 6, 10, 15, 21, 28.

6 a 2, by rolling (1, 1)

b i (2, 6), (3, 5), (4, 4), (5, 3), (6, 2) 5 combinations

ii (6, 3), (5, 4), (4, 5), (3, 6) 4 combinations

iii (6, 4), (5, 5), (4, 6) 3 combinations

iv (6, 5), (5, 6) 2 combinations

v (6, 6) 1 combination

Possibilities (page 31)

1

50c	20c	10c	5c
1	1	1	4
1		4	2
	4	1	2
	3	4	

2 31 possibilities

$1	50c	20c	10c
1	1	1	3
1	1	2	1
1		5	
1		4	2
1		3	4
1		2	6
1		1	8
	3	1	3
	3	2	1
	2	5	0
	2	4	2
	2	3	4
	2	2	6
	2	1	8
	1	7	1
	1	6	3
	1	5	5
	1	4	7
	1	3	9
	1	2	11
	1	1	13
		10	
		9	2
		8	4
		7	6
		6	8
		5	10
		4	12
		3	14
		2	16
		1	18

3 a

	0	1	2	3	4	5	6
0	✓						
1	✓	✓					
2	✓	✓	✓				
3	✓	✓	✓	✓			
4	✓	✓	✓	✓	✓		
5	✓	✓	✓	✓	✓	✓	
6	✓	✓	✓	✓	✓	✓	✓

Remember that there is only one domino of each combination.
∴ total number of dominoes
= 1 + 2 + 3 + 4 + 5 + 6 + 7
= 28

b 7

c Using the table in **a** and counting down the columns:

Dots on 1st half	Dots on 2nd half	Total dots
0 (on 7 dominoes)	0 + 1 + 2 + 3 + 4 + 5 + 6 = 21	21
1 (on 6)	1 + 2 + 3 + 4 + 5 + 6 = 21	27
2 (on 5)	2 + 3 + 4 + 5 + 6 = 20	30
3 (on 4)	3 + 4 + 5 + 6 = 18	30
4 (on 3)	4 + 5 + 6 = 15	27
5 (on 2)	5 + 6 = 11	21
6 (on 1)	6 = 6	12 168

Or each number from 0 to 6 occurs 8 times, making 168 dots (on 7 dominoes with 1 double).

4 63

Pages	Digits
1 to 9	9
10 to 19	20
20 to 29	20
30 to 36	14
Total number	**63**

5 a Pages 1 to 9 need 9 digits; since 129 digits are used altogether, 120 remain to be written. When writing the numbers from 10 to 19, 20 digits are needed, and each 10 numbers require a further 20. The 120 digits will cover numbers 10 to 69.
∴ if 129 digits are used, the book has 69 pages.

b Pages 1 to 9 need 9 digits. Pages 10 to 99 need 90 x 2 (= 180) digits. However, since 204 digits are used, a further 15 will cover numbers 100 to 104.
∴ the book has 104 pages.

6 a 39 palindromic numbers:
11, 22 → 99 (9) 101, 111 → 191 (10)
202, 212 → 292 (10) 303, 313 → 393 (10)

b 99 numbers

7 a 81 There are 90 numbers between 10 and 99 and of these, 9 are palindromic, i.e. they have the same digit (11, 22 … 99).
Or we can do this question another way! The tens digit can be chosen in 9 different ways from the numbers 1 to 9 while the unit digit can also be chosen in 9 different ways as we must not choose the digit selected already; however, we can now include 0. Altogether 9 x 9 = 81 possibilities.

b 41 There are 45 even two-digit numbers and, of these, 4 are palindromic (22, 44, 66, 88).

c 40 Either 81 – 41 = 40 as the number is either even or odd.
Or there are 5 ways of selecting the odd unit digit and since the tens digit cannot use this digit, nor 0, there are 8 ways of choosing the tens digit, giving 5 x 8 = 40 possibilities.

d 72 Since the three-digit number starts with 1, the tens digit can be any one of 9 digits and the unit digit can be any one of 8 digits, giving 9 x 8 = 72 possibilities.

8 a 8 The sum of numbers is even if both are even or both are odd:
2 ways of selecting from 4 and 6
6 ways of selecting from 3 and 5 or 3 and 7 or 5 and 7.

b 12 Odd number = even and odd or odd and even
2 x 3 + 3 x 2 = 12 possibilities

What's the chance? (page 32)

1 a 64 **b** 2 **c** 16

d i $\frac{1}{64}$ **ii** $\frac{3}{64}$ **iii** $\frac{8}{64} = \frac{1}{8}$

e 2 and 16, 3 and 15, 4 and 14, 5 and 13, 6 and 12, 7 and 11, 8 and 10

+	1	2	3	4	5	6	7	8
1	2	3	4	5	6	7	8	9
2	3	4	5	6	7	8	9	10
3	4	5	6	7	8	9	10	11
4	5	6	7	8	9	10	11	12
5	6	7	8	9	10	11	12	13
6	7	8	9	10	11	12	13	14
7	8	9	10	11	12	13	14	15
8	9	10	11	12	13	14	15	16

2 a i a $\frac{1}{4}$, b $\frac{1}{2}$, c $\frac{1}{8}$, d $\frac{3}{4}$

ii a $\frac{5}{12}$, b $\frac{1}{12}$, c $\frac{1}{4}$, d $\frac{1}{2}$

iii a $\frac{1}{4}$, b $\frac{1}{6}$, c $\frac{1}{6}$, d $\frac{5}{12}$

iv a $\frac{1}{4}$, b $\frac{1}{8}$, c $\frac{1}{2}$, d $\frac{3}{4}$

v a $\frac{1}{5}$, b $\frac{2}{5}$, c $\frac{1}{5}$, d $\frac{3}{5}$

Chance and data (page 33)

1 a A person turns on the tap for 5 minutes, filling the bathtub to a height of about 15 cm. He then sits in the bath, raising the water level by another 5 cm. He relaxes and plays in the bath for 5 minutes before turning the tap on again for about 5 minutes to raise the level to 40 cm, with him in the bathtub. Possibly the water became cold and he put in some hot water. The person then relaxed in the bathtub for another 5 minutes before getting out of the bath and pulling out the plug. The water takes 5 minutes to drain from the bath.

b

Height (centimetres)
40
30
20
10
5 10 15 20 25
Time (minutes)

2 a

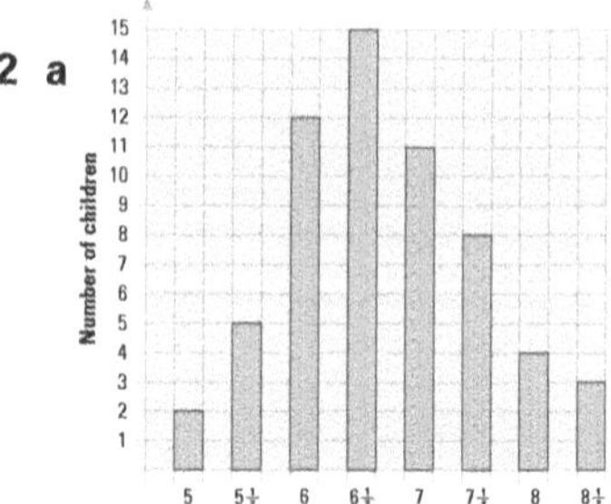

b 60 **c** $\frac{12}{60} = \frac{1}{5}$

d 20% **e** $\frac{15}{60} = \frac{1}{4}$ or 25%

f 25% **g** $\frac{5}{60} = \frac{1}{12} = 8\frac{1}{3}$ %

Interpreting graphs (page 34)

1 **a** Marc **b** Bob

2 1: George 2: Claire 3: Faye 4: Tim 5: Marc

3 **a** 1: Jessica 2: Alice 3: Doug 4: Monica 5: Harry

b 1: Jessica 2: Alice 3: Harry 4: Monica 5: Doug

c 1: Jessica 2: Alice 3: Harry 4: Monica 5: Doug

4 **a** A person travels at the same speed for 2 hours, stays in one place (possibly resting) for an hour, then returns at a slower speed taking 3 hours.

b A person travels 10 km in 10 minutes ($S = \frac{D}{T}$, which $= \frac{10}{\frac{1}{6}} = 60$ km/h, since 10 min $= \frac{1}{6}$h), then travels 20 km in 10 min, which is 120 km/h.

5 Daisy spoke for 30 min, the longest time for the least cost, so her call was likely to be local—or at least nearer than Margaret's or Eva's. Eva called the longest distance— a 10 min call cost her the most. Margaret's call was further than Daisy's but not as far as Eva's.

Graphs (page 35)

1

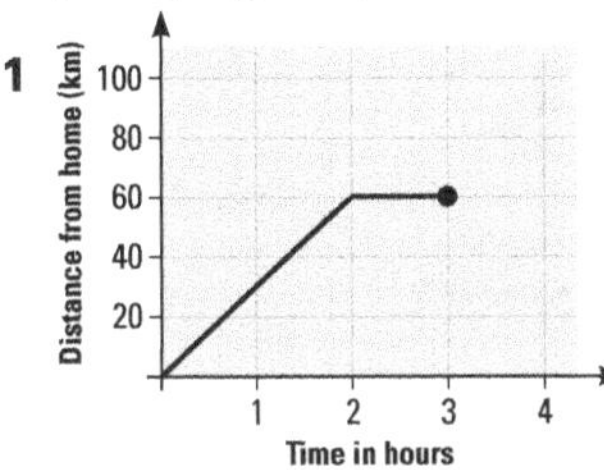

2 **a** A

b C and D are the same age. However, C is also taller than D. A and C are the same height. However, A is younger than C.

3 **a** 150 L **b** 10 seconds **c** 120 L

d 450 L (Every 10 secs 90 L flows from hose B. In 50 secs 450 L will flow.)

Venn diagrams (page 36)

1 7 do only ballet

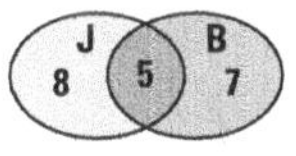

2 3 do both jazz and ballet

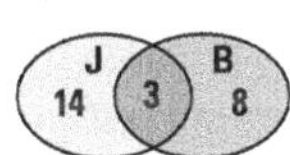

3 3 do both

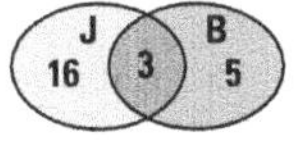

Note 19 + 8 = 27, so 3 must do both.

4 **a** 7 do both

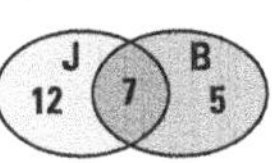

Note 19 + 12 = 31, so 7 must do both.

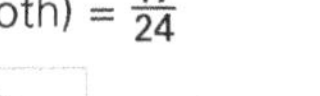

b **i** P (only ballet) $\frac{5}{24}$ **ii** P (only jazz) $\frac{12}{24} = \frac{1}{2}$

iii P (not both) = $\frac{17}{24}$

5 **a**

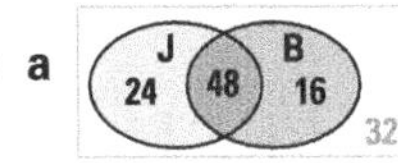

72 jazz, 48 both, 24 only jazz

now $48 = \frac{3}{4}$ B

B = 64

∴ 16 only do ballet

120 – 88 = 32 do neither

b **i** P (B) = $\frac{64}{120} = \frac{8}{15}$ **ii** P (neither) = $\frac{32}{120} = \frac{4}{15}$

Odd one out (page 37)

1 205 cents **2** 403c **3** \$3.00 ÷ 100 **4** 1 L 45 mL

5 5000 mL **6** 5·8 L **7** 0·131 m **8** 8·52 cm

9 10·001 km **10** 2500 mm **11** 0·385 cm **12** 0·26 km

13 0·254 kg **14** 0·075 kg **15** thirty minutes to 5

16 10:11 p.m. **17** 62 s **18** 104 hours

Time zones (page 38)

Note: Daylight saving time differences have not been mentioned here.

1 8 ahead **2** 10 behind **3** 5 behind **4** 9 ahead

5 14 ahead **6** 8 ahead **7** 4 ahead **8** 19 behind

World times (page 39)

1 **a** 4:30 p.m. Tuesday **b** 5 p.m. Tuesday **c** 7 a.m. Tuesday

d 3 p.m. Tuesday **e** 2:30 p.m. Tuesday **f** 7 p.m. Tuesday

g 3 p.m. Tuesday **h** 8 a.m. Tuesday **i** 10 a.m. Tuesday

2 **a** 9 a.m. Monday **b** 12 noon Monday **c** 1 p.m. Monday

d 7 a.m. Monday **e** 2 a.m. Tuesday **f** 3 a.m. Tuesday

g 1 a.m. Tuesday **h** 8 p.m. Monday **i** 7 p.m. Monday

3 6 a.m next day **4** 9 p.m. previous day

Mass puzzles (page 40)

Note: All table measurements are in kilograms.

1

Left side	Right side	
Mass kg	Mass kg	Carrots kg
3 + 3	5	1
5	3	2
3		3
3 + 3 + 3 or 5 + 5	5 or 3 + 3	4
5		5
3 + 3		6
5 + 5	3	7
5 + 3		8
3 + 3 + 3		9
5 + 5		10
5 + 3 + 3		11
5 + 5 + 5	3	12
5 + 5 + 3		13
5 + 3 + 3 + 3		14
5 + 5 + 5		15
5 + 5 + 3 + 3		16

2

Left side	Right side	
Mass kg	Mass kg	Carrots kg
8 + 8	5 + 5 + 5	1
5 + 5	8	2
8	5	3
8 + 8 + 8	5 + 5 + 5 + 5	4
5		5
8 + 8	5 + 5	6

3 24 g

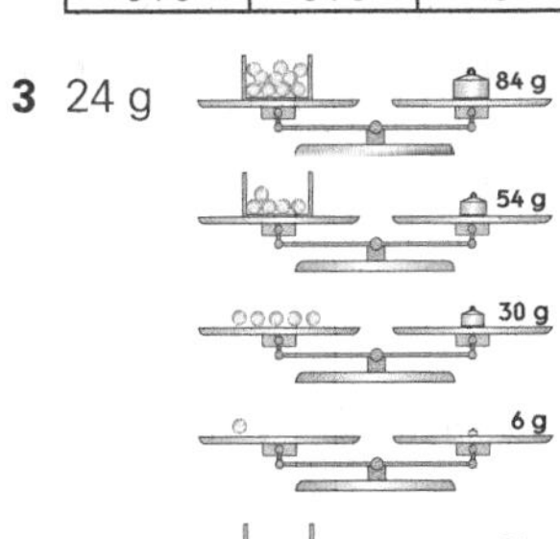

4 12 kg

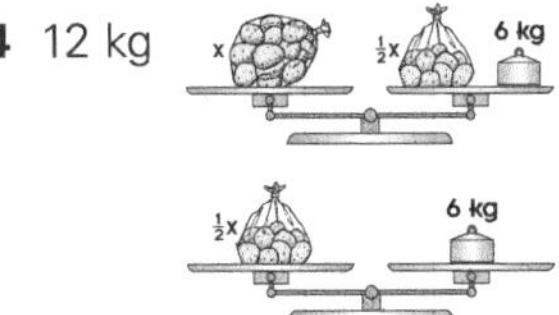

5 4 kg

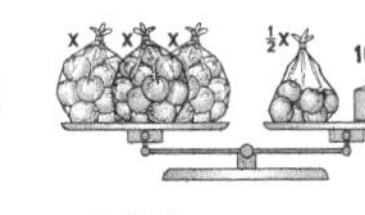

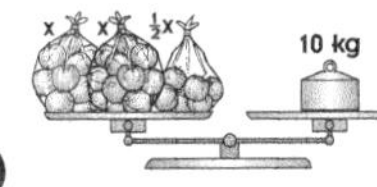

Fractions and measurement (page 41)

1 **a** 90 min **b** 45 min

c 40 min **d** 80 min

2 **a** 16 **b** 18

3 4 kg **4** 40 kg

5 $1 + \frac{3}{4} = \frac{7}{4}$

∴ 7 equal parts = 105 kg, 1 part = 15 kg, 3 parts = 45 kg

Fractions and measurement (page 41) ***continued...***

6 240 L $\frac{1}{12}$ of capacity = 20 L

7 $\frac{1}{6}$ case of apples (without the mass of case) = 3 kg
∴ $\frac{1}{2}$ case of apples (without mass of case) = 9 kg
∴ mass of empty case = 1 kg.

8 $\frac{1}{6}$ case of pears (without the mass of case) = 5 kg
∴ $\frac{1}{2}$ case of pears (without mass of case) = 15 kg
∴ mass of empty case = 3 kg.

Measuring with fractions (page 42)

1 80 s = 1 min 20 s
2 lengths x 50 m = 100 m
$\frac{5}{4}$ m in 1 sec
5 m in 4 sec
100 m in 80 s = 1 min 20 s

2 L = 20 cm
W = 15 cm

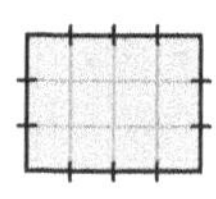

If the length is 4 units, then width is 3 units and perimeter = 14 units.
∴ 14 units = 70 cm.
∴ Therefore 1 unit = 5 cm.

3 6 cm, 8 cm, 12 cm

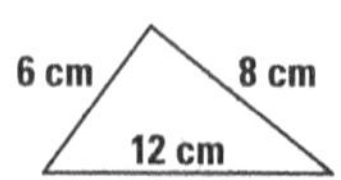

4 32 L There is $\frac{3}{4}$ x 64 = 48 L of fuel in the tank. Now $\frac{1}{3}$ is used, so 32 L remains and 32 L are needed to fill the tank.

5 240 mL

120 mL is $\frac{3}{4}$ of remainder ∴ 40 mL is $\frac{1}{4}$ of remainder and 40 mL was spilled. 80 mL is $\frac{1}{3}$ of container.
240 mL filled the container.

6 480 mL

$\frac{1}{4}$ of container is 120 mL ∴ full container 480 mL.

7 504 kg, from the following steps:

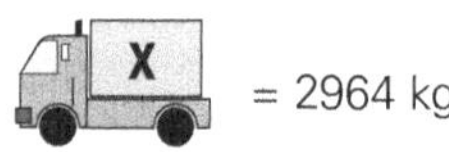

= 2964 kg

= 2712 kg

½X = 252 kg

X = 504 kg

8 48 hours

$\frac{3}{4}$ of remaining time = 24 hours.
∴ $\frac{1}{4}$ of remaining time = 8 hours (8 hours playing sport, 16 hours sleeping, so camp was 48 hours.)

9 12 km

Perimeter and area (page 43)

1 length = 24 cm, width = 6 cm
A clear diagram drawn on squared paper is helpful: . In the perimeter of 60 cm there are altogether 10 equal parts, so each part is 6 cm.
∴ length of rectangle = 4 parts = 24 cm
width of rectangle = 1 part = 6 cm

2 36 cm² Perimeter of square = 24 cm = 4 x side
∴ side = 6 cm; area = 36 cm²

3 32 cm Area of square = 64 cm² ∴ each side = 8 cm
Perimeter = 32 cm

4 13 cm

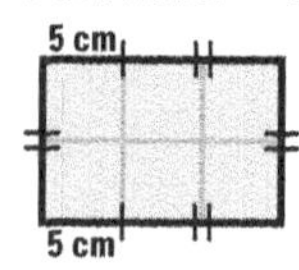

From the diagram, the perimeter of 42 cm is equal to 4 equal parts + 10 cm.

∴ 4 equal parts = 32 cm
1 part = width of rectangle = 8 cm
length of rectangle = 5 + 8 = 13 cm
Alternatively, use trial and error, drawing up a table.

5 40 cm²

Width (cm)	Length (cm)	Perimeter (cm)
5 cm	8 cm	26 cm

∴ area = 40 cm²

6 **a** 3 cm, 6 cm, 6 cm **b** 7 cm, 14 cm, 14 cm

7 **a** 80 cm² For the perimeter of 48 cm, 12 equal sides
∴ side of each square = 4 cm
area of each square = 16 cm²
area of figure = 5 x 16
= 80 cm²

b 36 cm Area of 5 squares = 45 cm²
∴ area of each square = 9 cm²
side of each square = 3 cm
perimeter of figure = 36 cm

c Area of 5 desks = 32 000 cm²
∴ area of 1 desk = 6400 cm²
side of each desk = 80 cm
∴ perimeter of figure = 12 x 80 cm
= 960 cm

8 **a**

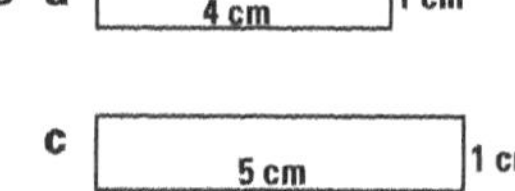

b

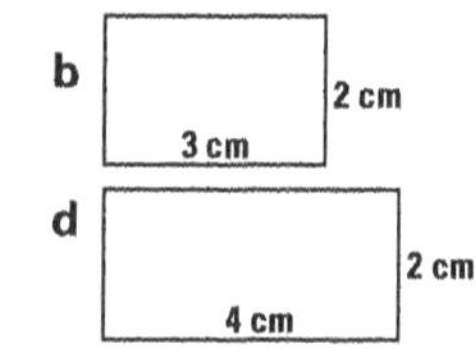

c **d**

Perimeter challenges (page 44)

1 Since a square was folded in half, the breadth of the rectangle is half its length
∴ perimeter of rectangle
= 3 x length = 24 cm, length of rectangle
= 8 cm, perimeter of square = 32 cm

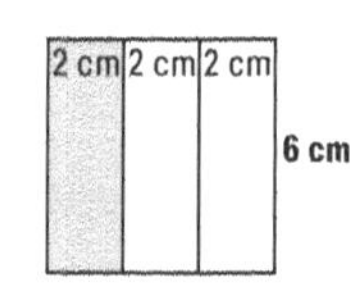

2 The perimeter of one rectangle (shaded)
= 16 cm = 8 equal parts
∴ width of rectangle = 2 cm and length of rectangle = 6 cm (so perimeter of square = 24 cm)

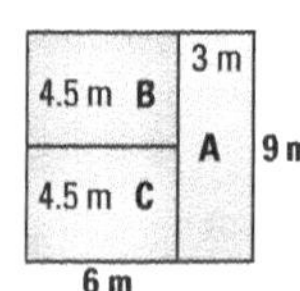

3 The area of the square = 81 m².
∴ area of each rectangle
= 81 ÷ 3 = 27 m²

Method 1
Since the length of the rectangle A is 9 m, its width is 3 m.
Therefore length of rectangle B = 6 m, width of rectangle B = $4\frac{1}{2}$ m

Method 2

Since rectangles B and C are identical, the width of each is $9 \div 2 = 4\frac{1}{2}$ m. Now use the fact that the area of B = 27 m^2.

4 Each side of the large square is 6 cm.

We divide square into nine congruent squares.

a Area of small square = 4 cm^2
side of small square = 2 cm
∴ perimeter of small square = 8 cm

or side of large square = 6 cm
side of small square = 2 cm
∴ perimeter of small square = 8 cm

b Area of rectangle = 6 cm^2 = 3 cm x 2 cm
∴ perimeter of rectangle = 10 cm

5 a Each side of the large square is 12 cm, and the area of each rectangle is 12 cm^2 in both cases. Therefore:

i each rectangle = 4 cm x 3 cm
perimeter = 14 cm

ii each rectangle = 6 cm x 2 cm
perimeter = 16 cm

b The only measurement for these congruent rectangles is 6 cm x 3 cm.
There are 8 rectangles.

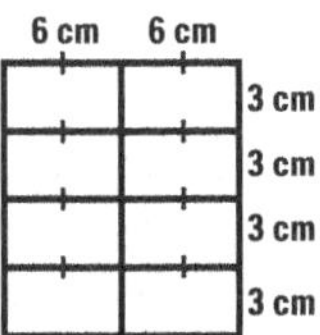

Area challenges (page 45)

1 110 11 cards can be cut in one row, and there will be 10 rows, so 110 cards can be cut.

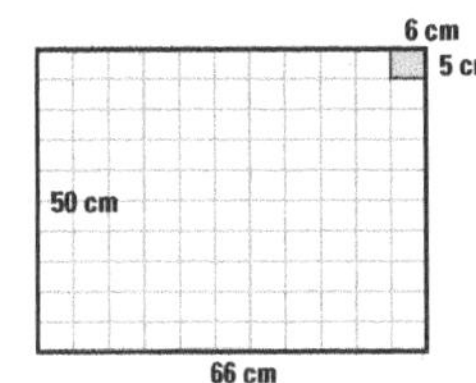

2 15 x 30 = 450 tiles

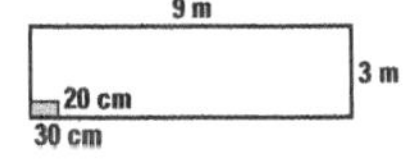

3 a Area of 6 squares = 294 cm^2
∴ area of 1 square = 49 cm^2
∴ side of 1 square = 7 cm
∴ perimeter of figure = 7 x 14
= 98 cm

b Perimeter = 42 cm
∴ side of 1 square = 3 cm
∴ area of 1 square = 9 cm^2
∴ are of figure = 9 x 6 cm^2
= 54 cm^2

4 The side of square A is 8 cm. Therefore:

Side of square B = 8 + 1 = 9 cm
Side of square D = 8 – 1 = 7 cm
Side of square C = 9 + 1 = 10 cm
Side of square E = 7 + 8 = 15 cm
Side of square F = 10 + 9 – 15 = 4 cm
Side of square G = 4 + 10 = 14 cm
Side of square H = 14 + 4 = 18 cm

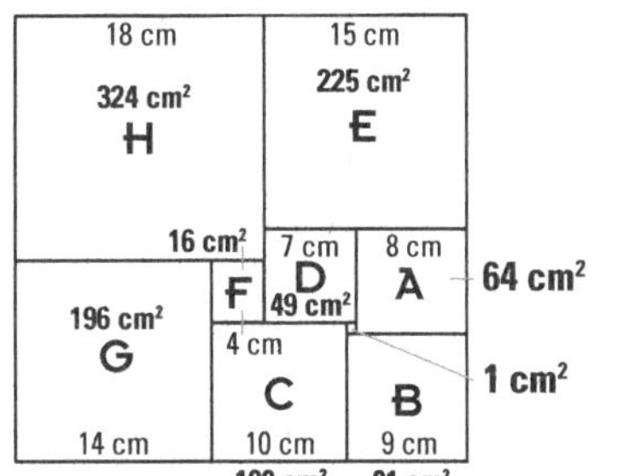

a 324 cm^2

b B = 81 cm^2
C = 100 cm^2
D = 49 cm^2
E = 225 cm^2
F = 16 cm^2
G = 196 cm^2.

5 The height is 10 cm. Since the total length or ribbon was 2 m (200 cm), the part going around the box = 150 cm.

Let the height of the box be △.
Then 65 x 2 + 2 x △ = 150
Therefore (by trial and error) △ = 10 cm

The width of the box was not required in solving the problem.

6

length		height		width		bow	
40	x 2 +	10	x 4 +	25	x 2 +	20	= 190 cm

7 a length = 4 cm, width = 3 cm, height = 2 cm

b length = 9 cm, width = 3 cm, height = 4 cm

Paving stones (page 46)

1 a

Length of side garden	1	2	3	4
Number of paving stones	8	12	16	20

b *Method 1*

From the table the pattern suggested, number of paving stones = 4 x (length of side of garden + 1)
Therefore = 4 x 11 when side length is 10.
= 44

Method 2

Continue the pattern.

Length of side garden	5	6	7	8	9	10
Number of paving stones	24	28	32	36	40	44

Method 3

Number of paving stones
= 4 x 10 + 4 (corners)

Method 4

Number of paving stones
= 2 x A + 2 x B
= 2 x 12 x 1 + 2 x 10 x 1
= 44

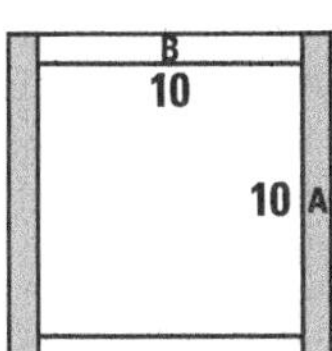

c Number of paving stones
Rule 1 = 4 x (length of side of garden increased by 1)

Rule 2 = 4 x length of side of garden plus 4 (for the corners)

d 4 x 30 + 4 = 124

e $\frac{84 - 4}{4} = 20$

Paving stones (page 46) *continued...*

2 **a** Area = 9 x 8 – 5 x 4
= 72 – 20
= 52 m^2

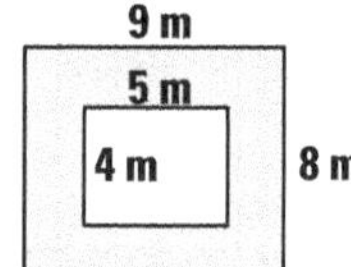

b 52 **c** 11 x 10 – 9 x 8 = 110 – 72 = 38

3 60 black, 180 brown

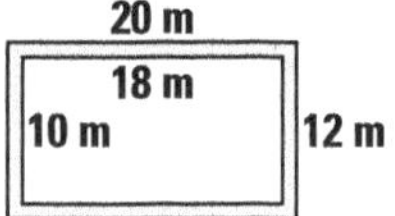

18 x 10 = 180 brown paving stones
20 + 20 + 10 + 10
= 60 black paving stones
or 20 x 12 – 18 x 10 = 240 – 180
= 60 black paving stones

Area explorations (page 47)

In order to work systematically, students need to consider rectangular shapes in which the breadth is first 1 m, then 2 m, 3 m and so on; they next find the corresponding length, and then the area.

1

Length (m)	Breadth (m)	Area (m^2)
46	1	46
44	2	88
42	3	126
40	4	160
38	5	190
36	6	216
34	7	238
32	8	256
30	9	270
28	10	280
26	11	286
*24	12	288
22	13	286
20	14	280
18	15	270
16	16	256
14	17	238
12	18	216
10	19	190

*From the table, it seems that the garden of largest area measures 24 m x 12 m.

2 **a** Area of square A = 9 cm^2
area of square B = 16 cm^2
area of square C = 25 cm^2
∴ side of square C = 5 cm

b Area of square A = 25 cm^2
area of square B = 144 cm^2
∴ area of square C = 169 cm^2
∴ side of square C = 13 cm

c Area of square A = 36 cm^2
area of square B = 64 cm^2
area of square C = 100 cm^2
∴ side of square C = 10 cm

3 33 cm^2 Using the areas of the two rectangles, the following diagrams are possible (not drawn to scale):

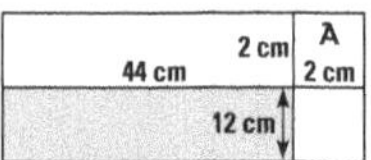

Perimeter of shaded rectangle:
2 x 88 + 2 x 24 = 224 cm

Perimeter of shaded rectangle:
2 x 44 + 2 x 12 = 112 cm

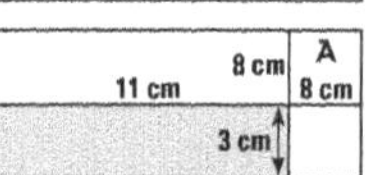

Perimeter of shaded rectangle:
2 x 22 + 2 x 6 = 56 cm

Perimeter of shaded rectangle:
2 x 11 + 2 x 3 = 28 cm

Since this is the required perimeter, the area of the shaded rectangle is 33 cm^2. Students need not list all these possibilities to find the correct solution. However, some interesting discussions can be held about the diagrams.

Cough medicine (page 48)

1 **a** 4 times (at 8 a.m., 12 noon, 4 p.m., 8 p.m.)
b 20 mL (5 mL each time) **c** 40 **d** 10 days

2 **a** 5 times (7 a.m., 11 a.m., 3 p.m., 7 p.m., 11 p.m.)
b 50 mL **c** 20 **d** 4 days
e Tuesday 11 p.m.

3 **a** 5 times (6 a.m., 10 a.m., 2 p.m., 6 p.m., 10 p.m.)
b 12·5 mL (2·5 x 5 = 12·5 mL) **c** 80
d 16 days = 2 weeks 2 days ($\frac{80}{5}$ = 16 or $\frac{200}{12{\cdot}5}$ = 16)
e Sunday 10 p.m., 2 weeks 2 days later

Volume & surface area (page 49)

1

Length of side (cm)	1	2	3	4	5	6
Total edge length (cm)	12	24	36	48	60	72
Surface area (cm^2)	6	24	54	96	150	216
Volume (cm^3)	1	8	27	64	125	216

Each cube has 12 edges and 6 faces.

To find the total edge length, multiply the length of one side by 12.
To find the surface area, multiply the area of one face by 6.
To find the volume, the formula is side x side x side.

2 A cube has 12 edges. Therefore:
12 edges = 36 cm
edge = 3 cm
volume = 3 x 3 x 3 cm^2
= 27 cm^2

Area of each face = 3 x 3 cm^2
total surface area = 6 x 9 cm^2
= 54 cm^2

These measurements can also be found in the column 3 of the table exercise 1.

3

Length	Width	Height	Total surface area	
24	1	1	98 cm^2	2 x (24 + 24 + 1)
12	1	2	76 cm^2	2 x (12 + 24 + 2)
8	1	3	70 cm^2	2 x (8 + 24 + 3)
6	2	2	56 cm^2	2 x (12 + 12 + 4)
4	3	2	52 cm^2	2 x (12 + 8 + 6)

4 **a** 2 cm Each side = 8 cm as $\sqrt[3]{512}$ = 8. Each ball has a diameter = 4, therefore radius = 2 cm.

b 1728 cm^3 If radius = 3 cm,
diameter = 6 cm
side = 12 cm
volume = 12 cm^3
= 1728 cm^3

Volume and capacity (page 50)

1 24 4 rows of 3 will fit in the bottom layer and there will be 2 layers giving 24 boxes

2 a 50 5 × 5 bottom layer, 2 layers

b 200 5 × 10 bottom layer, 4 layers

c 1000 10 × 10 bottom layer, 10 layers

d 10 000 50 × 20 bottom layer, 10 layers

3 125 5 × 5 boxes bottom layer, 5 layers = 125 boxes

4 a i The original cube is sliced in half in the 3 directions, giving 2 × 2 × 2 = 8 smaller cubes.

ii The original cube is sliced into 3 sections in the 3 directions, giving 3 × 3 × 3 = 27 smaller cubes.

b i 64 = 4 × 4 × 4 so the length of each smaller cube is 3 cm

ii 216 = 6 × 6 × 6 so the length of each smaller cube is 2 cm

5 Volume of rectangular block is equal to volume of new block.
Therefore 12 × 8 × 6 = 9 × 4 × height
Therefore height = 16 cm

6 Volume of 6 cm cube is 27 times volume of 2 cm cube
mass is 27 times mass of 2 cm cube = 27 × 16 g
= 432 g

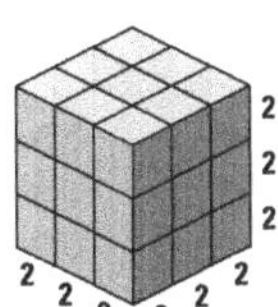

Volume investigation (page 51)

Paper size (cm)	Size of cut square (cm)	Size of box (cm)	Volume of box (cm³)
10 × 10	1 × 1	1 × 8 × 8	64
	2 × 2	2 × 6 × 6	72
	3 × 3	3 × 4 × 4	48
	4 × 4	4 × 2 × 2	16
12 × 12	1 × 1	1 × 10 × 10	100
	2 × 2	2 × 8 × 8	128
	3 × 3	3 × 6 × 6	108
	4 × 4	4 × 4 × 4	64
	5 × 5	5 × 2 × 2	20
14 × 14	1 × 1	1 × 12 × 12	144
	2 × 2	2 × 10 × 10	200
	3 × 3	3 × 8 × 8	192
	4 × 4	4 × 6 × 6	144
	5 × 5	5 × 4 × 4	80
	6 × 6	6 × 2 × 2	24
15 × 15	1 × 1	1 × 13 × 13	169
	2 × 2	2 × 11 × 11	242
	3 × 3	3 × 9 × 9	243
	4 × 4	4 × 7 × 7	196
	5 × 5	5 × 5 × 5	125
	6 × 6	6 × 3 × 3	54
	7 × 7	7 × 1 × 1	7

Volume (page 52)

1 5 cm 1 litre = 1000 cm^3
∴ 10 × 20 × depth = 1000, so depth = 5 cm

2 a 3 cm Water level will rise by 120 ÷ (8 x 15) = 1 cm,
∴ new water level = 3 cm

b 240 cm^3 We want level to rise by 2 cm. So add stone to double the volume of first stone = 240 cm^3

3 a The stone has a volume of 70 cm^3, and since the base of the tank is 5 cm by 7 cm = 35 cm^2, the water level will rise by 2 cm.

b 140 cm^3

4 15 cm Amount of fluid = 6 x 5 x 8
= 240 cm^3
= 4 × 4 × height in new container
∴ height = $\frac{240}{16}$ = 15 cm

5 a 200 mL = 200 cm^3

b Volume of prism = s × s × 8 = 200
side of square = 5 cm

Scale drawings (page 53)

1 a i If the actual length of the room is 30 m, then the width is 18 m and the actual perimeter of the room is 96 m.

ii If the actual width of the room is 9 m, then the length is 15 m, and the actual perimeter of the room is 48 m.

b i If the actual width of the room is 12 m, then the length is 18 m and the actual perimeter of the room is 60 m.

ii If the actual length of the room is 30 m, then the width is 20 m and the actual perimeter of the room is 100 m.

2 a i 15 m **ii** 30 cm or 0·3 m **iii** 4·5 m

b i 11 cm **ii** 1·2 cm **iii** 50 cm

3 7 cm

4 a 150 km **b** 350 km

Map	Actual distance
5 cm	250 km
1 cm	50 km
3 cm	150 km
7 cm	350 km

5 15 m = 1500 cm
Model = 15 cm

Ratio problems (page 54)

Note that * in the tables indicates the figures given in the exercises.

1 Drawing up a table is a very good strategy. Otherwise, students can use logic.

a 60 **b** 200 **c** 240

Grey tiles	Yellow tiles	Total tiles	
4	3	7	
*80 (4 x 20)	60 (3 x 20)	140 (7 x 20)	60 yellow
200 (4 x 50)	*150 (3 x 50)	350 (7 x 50)	200 grey
240 (80 x 3)	180 (60 x 3)	*420 (140 x 3)	240 grey

2 a 6 cricket balls **b** $20 **c** 24 tennis balls

d 16 cricket balls

	Cricket balls	Tennis balls	Cost	
	2	3	$4	
	4	6	$8	
a	6	9	*$12	6 cricket balls
b	8	*15	$20	$20
c, d	16	24	*$32	24 tennis balls 16 cricket balls

3 Note that table measurements are in cubic metres.

Metal (4 parts)	Sand (3 parts)	Cement (1 part)	Concrete mixture (8 parts)
4	3	1	*8
8	6	2	16
20 (4 x 5)	15 (3 x 5)	5 (1 x 5)	*40 (8 x 5)

a 8 m^3 of metal; 3 m^3 of sand; 1 m^3 of cement

b 20 m^3 of metal; 15 m^3 of sand; 5 m^3 of cement

Ratio problems (page 54) *continued...*

4 a Building shadow = 3 x tree shadow
∴ building height = 3 x tree height
= 18 m

b Pole height = 2 x tree height
∴ pole shadow = 2 x tree shadow
= 8 m

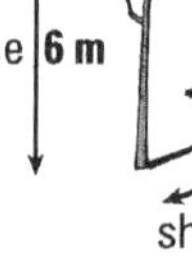

c building = $3\frac{1}{2}$ x 6
shadow = $3\frac{1}{2}$ x 4 = 14 m

5 a 24 m Shadow = 10 m = 2·5 x 4 Building = 6 x 4 = 24 m

b 9 m

6 a i 9 x 72 = 648 km **ii** 24 km **iii** 96 km

b i 3 L
ii 2 L
iii 9 L

Litres	Kilometres
6	72
1	12
2	24
3	36
8 (= 2 x 4)	96 (= 2 x 48)
9	108
54 (= 6 x 9)	648 (= 6 x 108 or 9 x 72)

Ratio and proportion (page 55)

Students can use 'guess and check' together with tables. A better way is to consider parts and shares as shown in these solutions, in which case diagrams can be very helpful.

1 a 3 shares = $24
∴ 1 share = $8
1 person gets $8, the other $16

b 4 shares = $24
∴ 1 share = $6
1 person gets $6, the other $18.

c 6 shares = $24

∴ 1 share = $4
1 person gets $4, the other $20.

2 a 3 shares = $60
∴ 1 share = $20
1 person gets $20, the other $40.

b 4 shares = $60
∴ 1 share = $15
1 person gets $15, the other $45.

c 5 shares = $60
∴ 1 shares = $12
1 person gets $12, the other $48.

3 M:S:C = 4:3:2
Concrete mixture is made up of 9 parts. All measurements are in cubic metres.

Metal	Sand	Cement	Concrete
4	3	2	9
12	**9**	**6**	**27**

4 5 parts = 15
∴ 1 part = 3

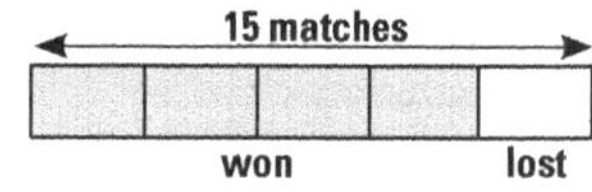

The team lost 3 matches and won 12.

5 K:M:I = 1:2:4.

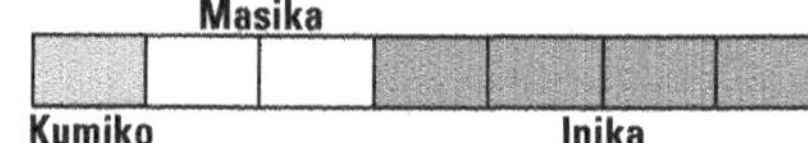

Note that Inika receives four times as much as Kumiko.
7 shares (total) = $28
∴ 1 share = $4
Kumiko receives $4, Masika $8 and Inika $16.

6 If the prize were $1000, Gilah would receive $500, Jahinger $300 and Olga $200. Since it is $2000, these amounts are doubled. Gilah receives $1000, Jahinger $600 and Olga $400.

7 6 children You can solve this problem by trial and error. If there were 5 children, all would be given $48 and when the eldest distributed his share among 4 brothers and sisters each would have $12 extra, which is too high.

Try 6 children, who would all be given $40; the distributed share of the eldest would give 5 brothers and sisters each $8 extra, which is correct.

8 Since there is 3 metres between the dog's rate and the rat's rate, in 3 lots of 3 he will catch up the 9 metres. The dog has to run 21 metres.

Rate problems (page 56)

1 15 kg

Boarders	Potoatoes (kg)
24	20
12	10
6	5
18	15

2 a 43 tables **b** 129 tables (43 x 3)

c 10 hours, using the answer to **a**.

3 It is most important to draw a diagram.
Four cuts are needed to make 5 pieces, so each cut takes 5 minutes. To get 6 pieces will require 5 cuts, taking 25 minutes.

4 Together they worked 7 hours for $84.
∴ 1 hour of work = $12
Dimitri gets $36. Lindsay gets $48.

5 a If 4 cans of food will feed 3 dogs for 1 day,
8 cans are needed to feed 6 dogs for 1 day.
40 cans are needed to feed 6 dogs for 5 days.

b Since 8 cans are needed to feed 6 dogs for 1 day,
48 cans will feed 6 dogs for 6 days.

6 Each tap fills 15 bottles in 1 min, so 20 taps will fill 300 bottles (20 × 15). *Or* each tap fills 5 bottles every 20 s, so 20 taps will fill 100 bottles every 20 s and 300 bottles every minute.

7 a 25 + 11 × 4 = 69c (4 additional minutes)

b Subtract 25c (for 1 minute) from the total charge, to leave 88c. Therefore, the person spoke for another 8 minutes. and the call lasted 9 minutes.

$$\begin{array}{r} \$1.13 \\ -\ \$0.25 \\ \hline \$0.88 \end{array}$$

8 a $45 × 5 + $0.12 × 800 = $321

b First subtract the rent from the total cost:
$132 – $90 = $42

Now calculate how many lots of 12c there are in $42:
$42 ÷ $0.12 = 350
Mr Campbell travelled 350 km.

9 **a** 300 km **b** 15 km
c 40 L **d** $12 (need 10 L)
e 1125 km

Litres	km	
30	450	
10	150	
∴ 20	300	a
1	15	b
40	600	c
75	1125	e

Dripping tap (page 57)

1 **a** 120 drops
b 24 mL 2 x 60 drops in 1 min
$\frac{120}{5}$ mL = 24 mL

2 **a** 8 mL **b** 480 mL **c** 4·8 L **d** 10 min
e 25 min. **f** 10 h 25 min. 5 L = 5000 mL
= 25 x 25= 625 min
= 10 h 25 min

3 **a** 4 mL **b** 1200 drops **c** 5 drops
d 0·2 mL **e** 5·76 L

4 **a** 50 cm^3 The tap is dripping at a rate of 5 drops every minute, so 300 drops in an hour = 50 cm^3 in an hour.
b 10 hours Total volume if height is 5 cm = 10 x 10 x 5 = 500 cm^3. It will take 10 hours to get a volume of 500 cm^3.

Speed challenges (page 58)

1 **a** 90 km/h = 90 × 1000 m/h = 90 000 m/h
b 90 km/h = $\frac{90 \times 1000}{60}$ = 1500 m/min
c 90 km/h = $\frac{90 \times 1000}{60 \times 60}$ m/sec. = 25 m/s
d 4 hours
e 1 h 20 min, since it takes 20 min to travel 30 km
f 135 km **g** 15 km (90 ÷ 6)
h 1500 m, using that fact from **b** that the speed = 1500 m/min.

2 **a** 20 km 72 km/h = $\frac{72 \times 1000}{60 \times 60}$ m/sec
b 54 km 15 m/sec = $\frac{15 \times 60 \times 60}{1000}$ km/h *or* using **a**:
20m/sec = 72km/h
5m/sec = 18km/h
∴ 15m/sec = 54 km/h

3 Speed = $\frac{\text{Distance}}{\text{Time}}$ $(S = \frac{D}{T})$
If S = 60 km/h, 60 = $\frac{50}{T}$
∴ T = $\frac{5}{6}$ h = 50 min
If S = 50 km/h, T = 1 h
At 50 km/h it will be 10 min late.

4 Average speed = $\frac{75}{5}$ = 15 km/h

5 Adam first takes 1 hour and 20 minutes, and then he takes 40 minutes for the rest of the journey.
Average speed = $\frac{\text{Total distance}}{\text{Total time}}$ = $\frac{110}{2}$ = 55 km/h

6 60 km/h In $1\frac{1}{3}$ hours motorist travelled 64 km, then in $\frac{2}{3}$ hour the motorist travelled 56 km.
Average speed = $\frac{120}{2}$ km/h = 60 km/h

7 Students can draw up a table to solve this problem, remembering that when Janelle starts, Angelica has travelled 15 km.

Time (hrs)	Distance travelled by Angelica	Distance travelled by Janelle
3	15	
4	20	11
5	25	22
6	30	33 (so Janelle has passed Angelica)
$5\frac{1}{2}$	27·5	27·5

So Janelle overtakes Angelica after she has been cycling for $2\frac{1}{2}$ hours (and Angelica has been walking for $5\frac{1}{2}$ hours).

Or, a more sophisticated method:

We can use the fact that Distance = Speed × Time.

Let *t* hours be the time taken for Angelica to overtake Janelle: $d = 11t = 5(t + 3)$. Students can use guess and check to find the value of *t*. $t = 2\frac{1}{2}$.

Speed problems (page 59)

1 After walking for $1\frac{1}{2}$ hours at a speed of $5\frac{1}{2}$ km/h, Malcolm walked $8\frac{1}{4}$ km.
This leaves $3\frac{3}{4}$ to walk at a speed of 3 km/h.
3 km will take 1 hour, 1 km will take 20 minutes,
so $\frac{3}{4}$ km will take 15 minutes.
Therefore $3\frac{3}{4}$ km will take 1 hour and 15 minutes, giving total of 2 hours and 45 minutes.

2 Nicole travels 80 x 81 cm each minute, so the school is 80 x 81 x 12 cm = 777·60 metres away.
a Ivan travels 90 x 72 each minute, so he takes $\frac{80 \times 81 \times 12}{90 \times 72}$ = 12 minutes to get to school.
b $\frac{80 \times 81 \times 12}{90 \times 16}$ = 54 cm

3 **a** David takes 25 minutes, while Michelle takes 20 minutes to drive 25 km. So it takes Michelle 5 minutes less than David.
b Jason takes 20 minutes to drive 15 km, while Paul takes $22\frac{1}{2}$ minutes.
Paul's time $T = \frac{D}{S} = \frac{15}{40} = \frac{3}{8}$ h $= \frac{3}{8} \times 60 = 22\frac{1}{2}$ min
Or Paul takes 15 minutes to drive 10 km at 40 km/h and he takes $7\frac{1}{2}$ minutes to drive 5 km at 40 km/h, so he takes $22\frac{1}{2}$ minutes to drive 15 km.

4 **a** $1\frac{1}{2}$ hours after leaving home he is not cycling for $\frac{1}{2}$ hour.
b 2 hours after leaving home. He takes 1 hour to travel 20 km, so his speed is 20 km/h. Between D and E.
c 40 km **d** Average speed = $\frac{40}{3}$ = $13\frac{1}{3}$ km/h
e From D to E she is travelling at 20 km/h and from A to B she is travelling at 15 km/h, so she is travelling 5 km/h faster.

Cutting rectangles (page 60)

Students should be given 1 cm squared paper for questions 1, 2, 3 and 4 and 0·5 cm paper for question 5.

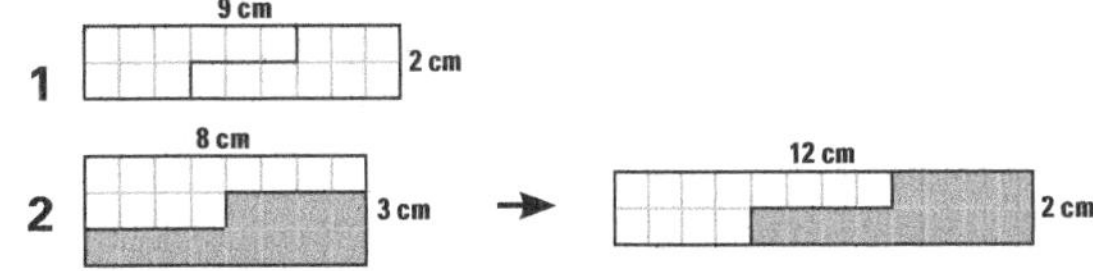

Cutting rectangles (page 60) ***continued...***

3 Cut in half, then rearrange pieces.

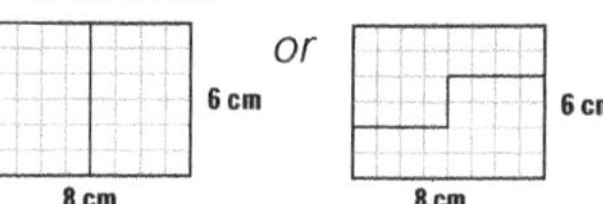

4

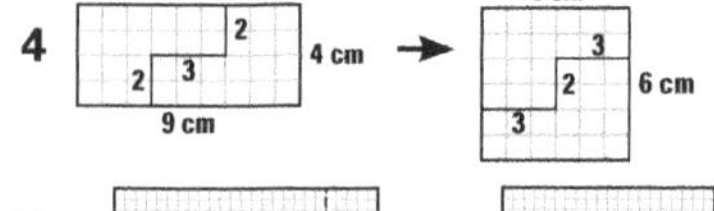

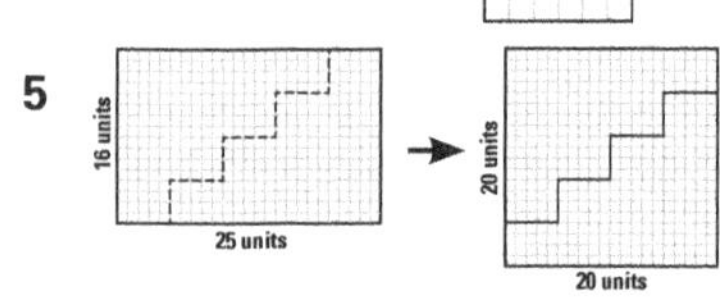

5

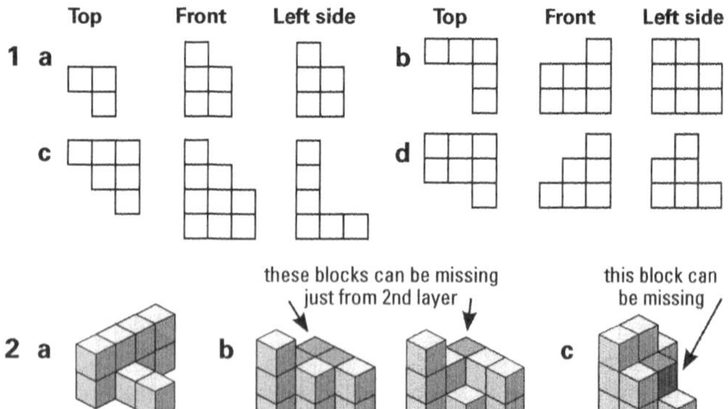

Cut along the stepped lines, then rearrange the shapes. Note that measurements are in units, not centimetres.

Solids (page 61)

Top Front Left side Top Front Left side

1 a b

c d

2 a b c

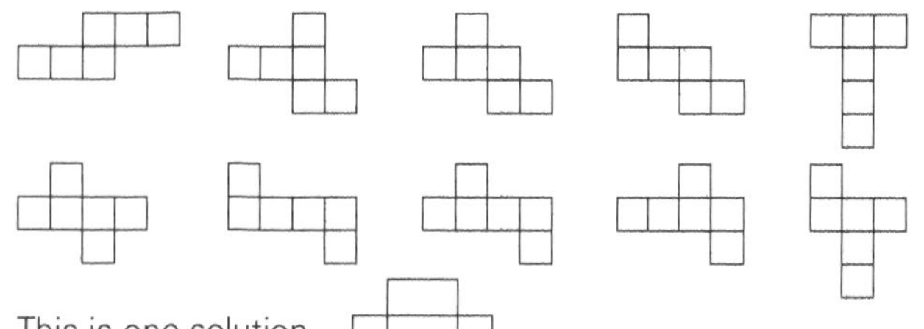

Nets (page 62)

1 **a** square pyramid **b** triangular pyramid

c triangular prism **d** triangular prism

2 Finding all the nets for a cube is difficult and students are not expected to discover them all. Their attention should be drawn to the question of symmetries, as this reduces the number of nets of a cube. Check carefully that no two drawings are actually the same by rotating or reflecting each one.

3 This is one solution.

4 **a**, **c** and **e**.

Polygons (page 63)

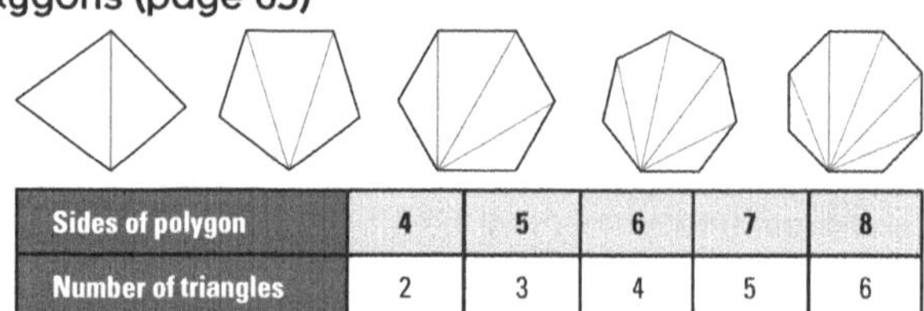

1

Sides of polygon	4	5	6	7	8
Number of triangles	2	3	4	5	6
Angle sum (degrees)	2 × 180 = 360°	3 × 180 = 540°	4 × 180 = 720°	5 × 180 = 900°	6 × 180 = 1080°

2 Angle sum of an '*n*' sided polygon is (*n* – 2) lots of 180°, i.e. subtract 2 from the number of sides of the polygon and then multiply it by 180°.

3 **a** 8 x 180 = 1440° **b** 10 x 180 = 1800°

What's my shape? (page 64)

1 **a** rhombus or square

b any isosceles triangle

c square or rectangle

d square, rectangle, rhombus or parallelogram

e rectangle or square

f square **g** circle

h isosceles triangle **i** equilateral triangle

j i square **ii** rhombus or rectangle

iii isosceles trapezium or kite

2 **a** cube **b** rectangular prism

c triangular pyramid (tetrahedron)

d triangular prism **e** sphere

f square pyramid **g** cylinder

h cone **i** square pyramid

j pentagonal pyramid

Billiards (page 65)

1 *Note*: the dimensions of figures **p**, **q**, **r**, **s** and **t** can be reduced.

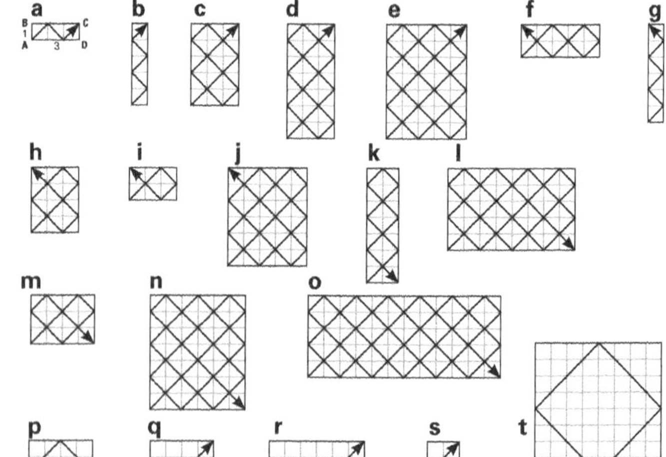

2

	Length	Width	Corner pocket
a	odd	odd	upper right
b	even	odd	upper left
c	odd	even	lower right

d If the table is even by even, the dimensions can always be reduced; and the reduced form should be used to work out into which pocket the ball will fall. For example, 4 × 4 can be reduced to 1 × 1, which is odd by odd, so the ball will fall in the upper right pocket; 12 × 8 can be reduced to 3 × 2, which is odd by even, so it will fall in the lower right pocket.

3 Expressing some of the results in a table, it is possible to observe the following rule: if we consider the reduced dimensions, in each case the number of times the ball hits the side of the table will equal the sum of the length and the width, minus 2.

Table	Hits	Length + width
1 × 3	2	4
5 × 1	4	6
5 × 3	6	8
7 × 3	8	10
2 × 5	5	7
4 × 3	5	7
6 × 5	9	11
7 × 2	7	9
3 × 4	5	7
7 × 6	11	13
5 × 12	15	17
1 × 2	1	3
1 × 1	0	2
3 × 2	3	5

Cube explorations (page 66)

Students should build these solids to check their answers.

1 a 27 (This large cube is 3 × 3 × 3.)

b 19 (students may become confused by the cubes with two or three faces showing.)

c 8 (There are 27 cubes altogether of which 19 can be seen, so 8 cannot be seen.)

2 a 64 (This cube is 4 × 4 × 4.) b 37 c 27 (64 – 37)

3 Students can carry out experiments to solve these problems. To set up an experiment, use centicubes: start with a black cube, say, and around it construct a 3 × 3 × 3 solid of yellow cubes; around that build a 5 × 5 × 5 large red cube. Now remove in turn the cubes of each colour.

If you start with a 2 × 2 × 2 cube and follow the same method, you will first build a 4 × 4 × 4 cube and then a 6 x 6 x 6 cube.

Size of cheese	Cubes cut	Cubes with wax	Cubes without wax
3 cm × 3 cm × 3 cm	27	26	1 (1 × 1 × 1)
4 cm × 4 cm × 4 cm	64	56	8 (2 × 2 × 2)
5 cm × 5 cm × 5 cm	125	98	27 (3 × 3 × 3)
6 cm × 6 cm × 6 cm	216	152	64 (4 × 4 × 4)
10 cm × 10 cm × 10 cm	1000	488	512 (5 × 5 × 5)
8 cm × 6 cm × 4 cm	192	144	48 (6 × 6 × 6)
7 cm × 5 cm × 4 cm	140	110	30 (7 × 7 × 7)

Experiments with cubes (page 67)

1 a 27

b i 8  ii 12 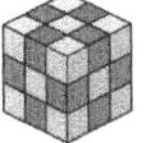 iii 6  iv 1

Obviously the shaded cubes are only those that can be seen from one angle.

c 27 This is the same answer as for **a** above.

2 a 64

b i 8 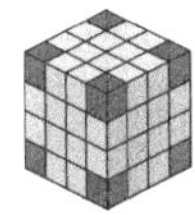 ii 24  iii 24 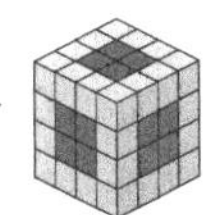
iv 8

Note that, as in question 1, the sum of the answers to **b**, **i–iv** equals the answer to **a**.

3 The following points can be noted:

- Irrespective of the size of a larger cube, the number of small cubes with paint on three faces is always 8.
- The total number of small cubes with paint on three faces, two faces, one face and none is equal to the total number that makes up the large one.

4 a The blue cube's dimensions will be 3 × 3 × 3.
∴ cubes needed = 27 – 1 = 26

b The yellow cube's dimensions will be 5 × 5 × 5.
∴ cubes needed = 125 – 27 = 98

5 a The blue cube will measure 4 × 4 × 4
∴ cubes needed = 64 – 8 = 56

b The yellow cube will measure 6 × 6 × 6.
∴ cubes needed = 216 – 64 = 152

Curves from lines (page 68)

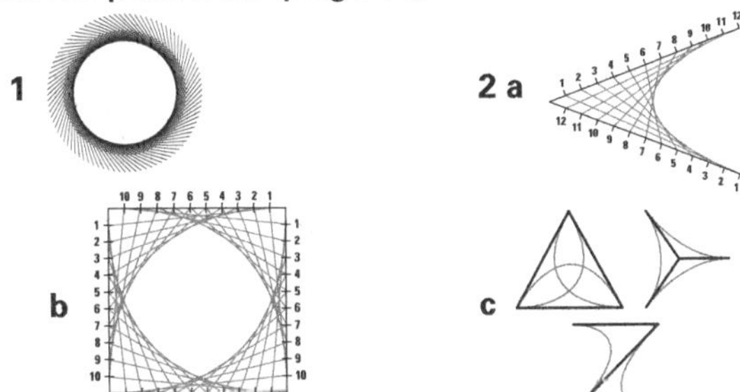

More curves (page 69)

Students should be provided with copies of the circle.

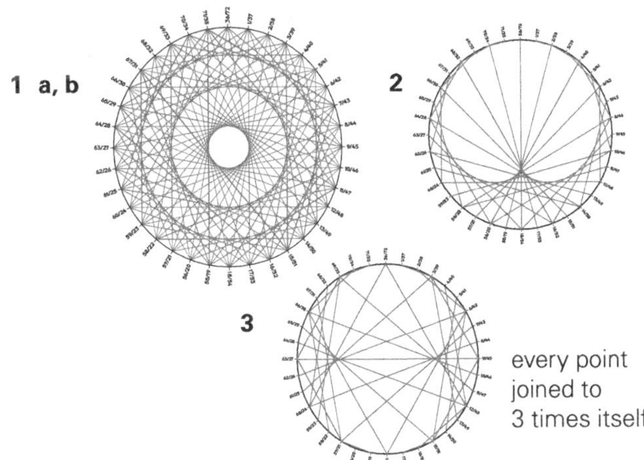

Euler's formula (page 70)

1

Solid	Number of faces (F)	Number of vertices (V)	Number of edges (E)	F + V – E
Cube	6	8	12	2
Rectangular prism	6	8	12	2
Triangular prism	5	6	9	2
Square pyramid	5	5	8	2
Triangular pyramid	4	4	6	2

2 Euler proved that F + V – E = 2 for simple closed surfaces. Students can test the formula on many different figures.

For figure **a**:
F + V – E = 8 + 12 – 18 = 2
Therefore formula holds.

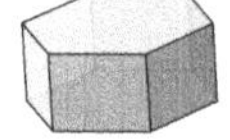

For figure **b**:
F + V – E = 7 + 10 – 15 = 2
Therefore formula holds.

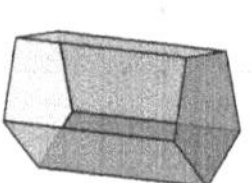

Locus (page 71)

It is important to note that students experiment by tying one end of a string to a pencil and, except for problems **2** and **3**, holding the other end fixed. Make sure that the string measures 5 cm *after* being tied. For these experiments, 1 cm paper can be used.

1

2

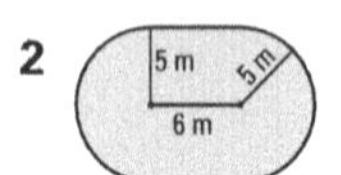

3

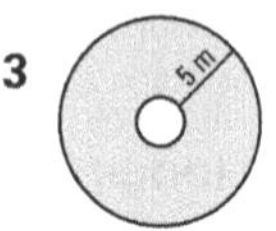

4

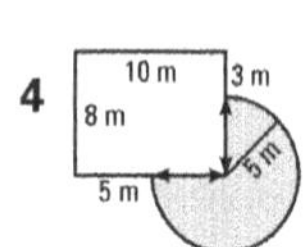

5 a

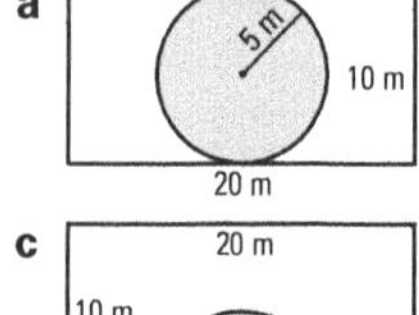

b

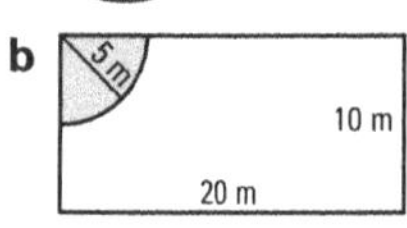

c

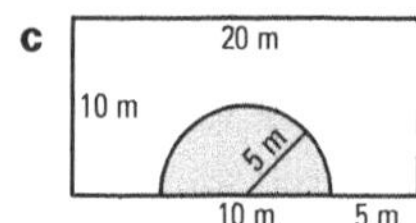

d

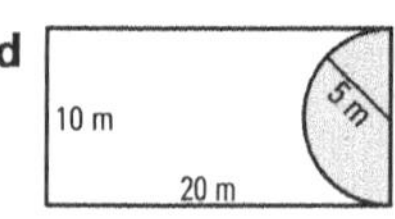

What's my message? (page 72)

1 DO THE RIGHT THING

Code	H	I	O	G	D	N	E	R	T
	1	2	3	4	5	6	7	8	9

$1 \times 7 = 7$, $3^2 = 9$, $2^2 = 4$,
$3 + 3 + 3 = 9$, $9^2 = 81$,
$14 \div 2 = 7$, $5 \times 5 = 25$

$$\begin{array}{r} 9126 \\ +\,5289 \\ \hline 14415 \end{array} \qquad \begin{array}{r} 129 \\ -\,43 \\ \hline 86 \end{array}$$

2 I LOVE DANCING

Code	A	C	N	V	I	E	D	L	O	G
	1	2	3	4	5	6	7	8	9	0

$7 - 7 = 0$, $3 \times 1 = 3$,
$3 \times 2 = 6$, $6^2 = 36$
$40 \div 4 = 10$, $\sqrt{4} = 2$
$1 + 2 = 3$, $10 - 1 = 9$

$$\begin{array}{r} 790 \\ +\,790 \\ \hline 1580 \end{array} \qquad \begin{array}{r} 8137 \\ -\,7618 \\ \hline 519 \end{array} \qquad \begin{array}{r} 25 \\ +\,25 \\ \hline 50 \end{array}$$

Simpler problems first (page 73)

1 66 connections; the answers to **a**, **b**, **c** and **d** are indicated on the table.

Students should draw diagrams such as these:

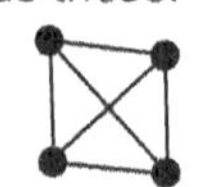

They should also recognise the pattern in the table as that of triangular numbers, *or* notice a pattern in the differences.

	Couples	Connections	
a	2	1	
b	3	3	2
c	4	6	3
d	5	10	4
	6	15	5
	7	21	6
	8	28	7
	9	36	8
	10	45	9
	11	55	10
	12	**66**	11
	13	78	12
	14	91	13

2 a This solution uses exactly the same pattern as question **1**, extended to two further lines. There were 91 handshakes.

b 10 people

3 Students can nominate which person wins each match.

a 7 matches

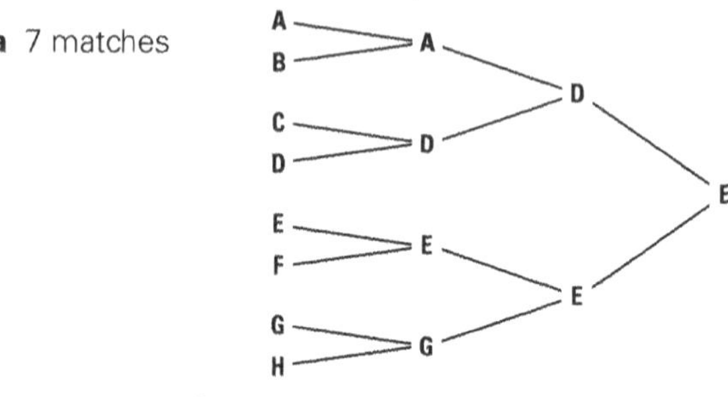

8 matches

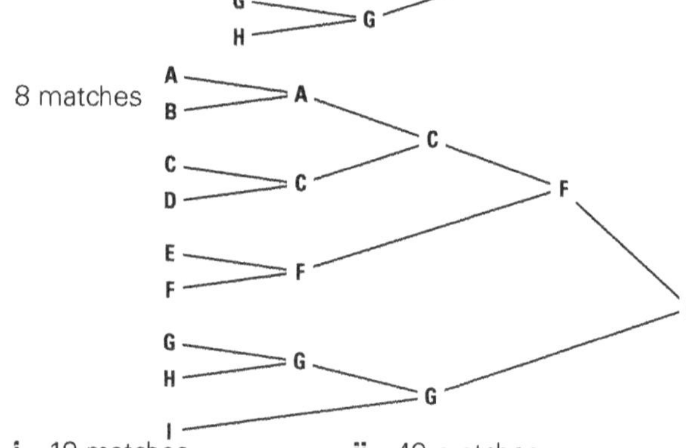

b i 19 matches **ii** 49 matches

Number of students	6	7	8	9	20	50
Matches needed to find a winner	5	6	7	8	19	49

Since the number of matches required to find a winner is one less than the number of students in the competition, predicting the result for 20 and 50 students in the competition is easy.

Logic using scales (page 74)

1

2 Using scale 2, replace □□ on scale 1.

3 Using scale 2, replace □□ on scale 1.

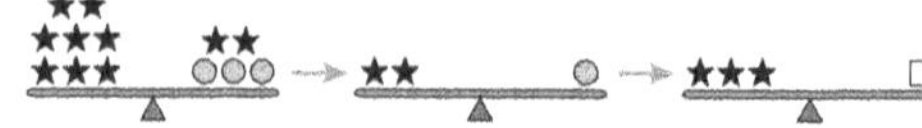

4 Double scale 1, replace □□, using scale 2.

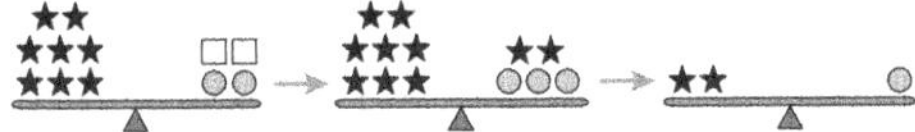

5 Replace ◯◯ on scale 1.

6 Replace ◯ on scale 1.

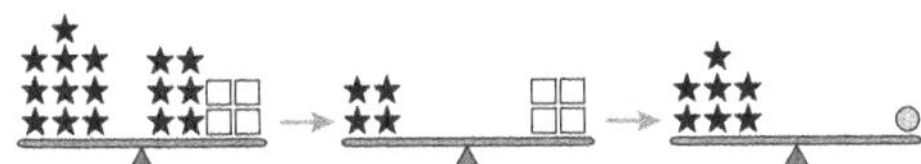

7 Replace □ on scale 1.

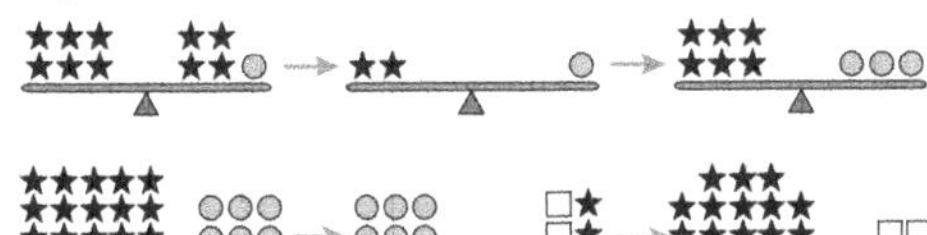

8

9 Double scale 1, add □ to each pair, then replace ◯◯□ using scale 2.

10 Double scale 1, then use scale 2.

Challenge on averages (page 75)

1 **a** Average of 1, 2, 3, 4, 5 is 3

b Average of 1,2 3, 4, 5, 6, 7 is 4

c Average of 1, 2, 3, 4, 5, 6, 7, 8, 9 is 5

d Average of 2, 4, 6, is 4. (It is the middle number.)

e Average of 2, 4, 6, 8, 10, 12, 14, 16, 18 is 10. (It is the middle number.)

f Average of 1, 2, 3 → 15 is the middle number, which is 8

2 The mass of three friends is 35 kg, 36 kg and 37 kg, so the lightest person has a mass of 35 kg.

3 The temperatures from Monday to Friday are 6°C , 7°C, 8°C, 9°C, 10°C, respectively, so the average temperature from Monday to Wednesday is 7°C.

4 The total mass of the five cases is 5 × 13 = 65 kg. With the sixth case, the total mass becomes 72 kg. Therefore, the average mass of the six cases is 72 ÷ 6 = 12 kg.

5 Total age of 5 children = 60
total age with teacher = 84
∴ average age of the six = 14 years

6 The sum of the temperatures on the five days is 5 × 6 = 30°C and since the sum of the temperatures on the first four days is 20°C, on the fifth day it was 10°C.

7 If the average for four tests is 80, the total score is 320. Therefore, 90 marks are still needed.

8 Total mass of the 8 rowing crew = 576 kg
total mass of crew plus cox = 9 × 70
= 630 kg
∴ mass of cox = 54 kg

Working together (page 76)

1 If 15 painters take 12 days, 5 painters take 36 days, and 20 painters take 9 days.

2 John paints 1 room in 3 hrs and $\frac{1}{3}$ room in 1 hour.
Apprentice paints 1 room in 6 hrs and $\frac{1}{6}$ room in 1 hour
∴ Together they paint $\frac{1}{3} + \frac{1}{6} = \frac{1}{2}$ room in 1 hour
∴ Together they can paint 1 room in 2 hours.

3 Michael cleans the whole house in 3 hrs, and $\frac{1}{3}$ in 1 h. His mother cleans the whole house in 2 hrs, and $\frac{1}{2}$ in 1 h.
Together they clean $\frac{1}{2} + \frac{1}{3} = \frac{5}{6}$ of the house in 1h
∴ Together they clean the whole house in $\frac{6}{5}$ h, or 1 h 12 min

4 Lachlan needs 1 hour. His dad needs $\frac{1}{2}$ hour.
His dad in 1 hour would do 2 equivalent jobs.
Together in 1 hour they would do 3 equivalent jobs.
Therefore, together they will do the job in $\frac{1}{3}$ h = 20 min.

5 Eddie needs 4 days to mow the lawn and 1 day to mow $\frac{1}{4}$ of it. Mark and David need 3 days to mow the lawn and 1 day to mow $\frac{1}{3}$ of it. The three together need 1 day to mow $\frac{1}{4} + \frac{1}{3}$ (or $\frac{7}{12}$) of the lawn. ∴ Together to mow the whole lawn they need $\frac{12}{7}$ days, or 1 day 17 hours $8\frac{1}{2}$ minutes.
Note that $1\frac{5}{7}$ of a day = 1 day + $\frac{5}{7}$ × 24 hours
= 1 day + 17 hours + $\frac{1}{7}$ hour
= 1 day + 17 hours + $8\frac{1}{2}$ minutes.

Counting frame mathematics (page 77)

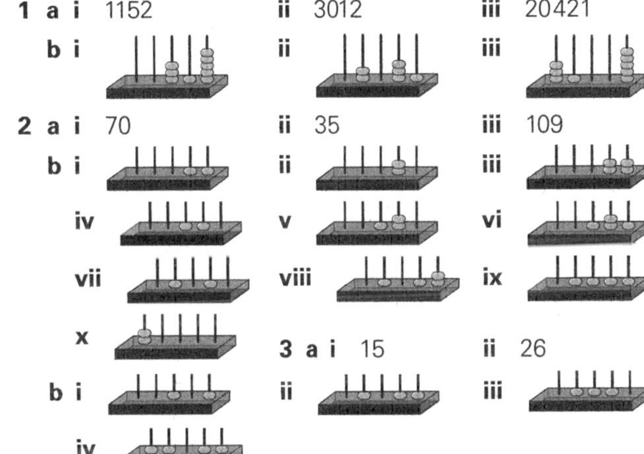

1 a i 1152 ii 3012 iii 20421
b i ii iii

2 a i 70 ii 35 iii 109
b i ii iii
iv v vi
vii viii ix
x

3 a i 15 ii 26
b i ii iii
iv

Triangular numbers (page 78)

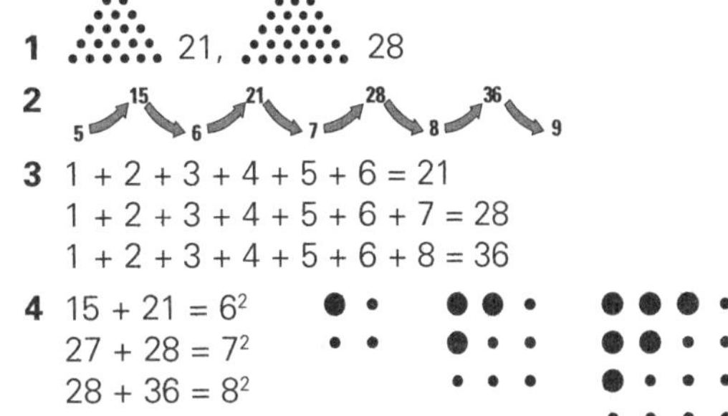

1 21, 28

2 5 → 15 → 6 → 21 → 7 → 28 → 8 → 36 → 9

3 1 + 2 + 3 + 4 + 5 + 6 = 21
1 + 2 + 3 + 4 + 5 + 6 + 7 = 28
1 + 2 + 3 + 4 + 5 + 6 + 8 = 36

4 15 + 21 = 6^2
27 + 28 = 7^2
28 + 36 = 8^2

If one triangle number is added to the next, the result is a square number.

Triangular numbers (page 78) *continued...*

5 $11 \times 11 - 1 = 120 = 8 \times 15$ **6**$8 \times 10 + 1 = 9^2$
$13 \times 13 - 1 = 168 = 8 \times 21$ $8 \times 15 + 1 = 11^2$
$8 \times 21 + 1 = 13^2$
$8 \times 28 + 1 = 15^2$

7 19 = 3 + 6 + 10　　21 = 6 + 15　　29 = 28 + 1
35 = 28 + 6 + 1　　36 = triangular number
37 = 36 + 1　　38 = 28 + 10
46 = 45 + 1 or 36 + 10 or 28 + 15 + 3

Pascal's triangle (page 79)

1

Row	1	Sum of numbers
1	1 1	2 $= 2^1$
2	1 2 1 Triangular numbers	4 $= 2^2$
3	1 3 3 1	8 $= 2^3$
4	1 4 6 4 1	16 $= 2^4$
5	1 5 10 10 5 1	32 $= 2^5$
6	1 6 15 20 15 6 1	64 $= 2^6$
7	1 7 21 35 35 21 7 1	128 $= 2^7$
8	1 8 28 56 70 56 28 8 1	256 $= 2^8$
9	1 9 36 84 126 126 84 36 9 1	512 $= 2^9$

2 121, 1331 and 14 641, which are the numbers in the second, third and fourth rows

3 a 6　**b** 9　**c** 15

4 a 3　**b** 6　**c** 10

d They are the triangular numbers.

e 45: the triangular number next in line indicated on the triangle.

5 a 2, 3, 4, 5, 6, 7, …　**b** 3, 6, 10, 15 … (triangular numbers)

6 3, 6, 10, 15, … (again triangular numbers)

7 a 2, 4, 8, 16, 32, 64

b They are the powers of 2, and the powers are the same as the row numbers.

8 Third, seventh, eleventh, …

9 In every odd row, the middle numbers are the same.

In every row, the second and the second last numbers are always also the row number.

The number of terms in each row is one more than the row number.

How many routes? (page 80)

1 a There are six paths from P to Q.

b　**c**

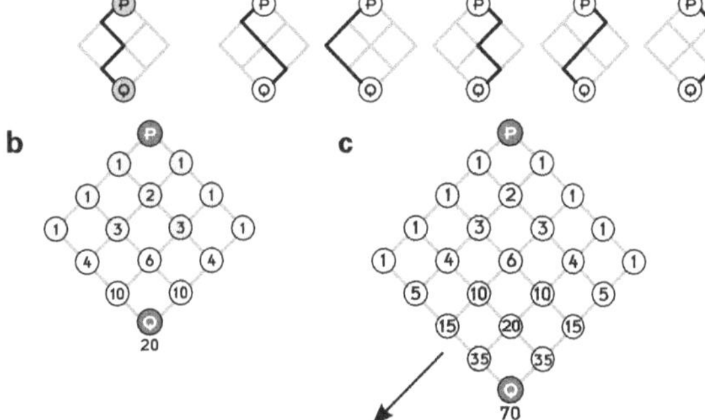

The symmetry figure diagram in **1 b** is embedded in this one; the numbers of paths follow the numbers in Pascal's triangle.

2 a

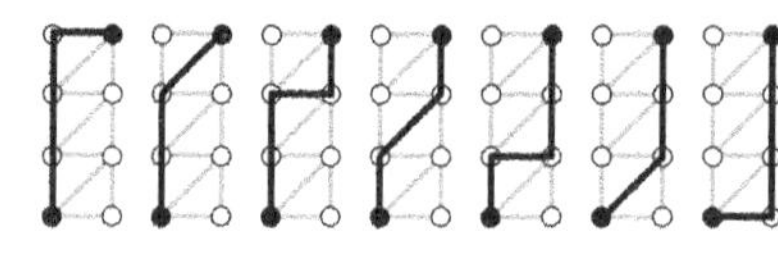

b

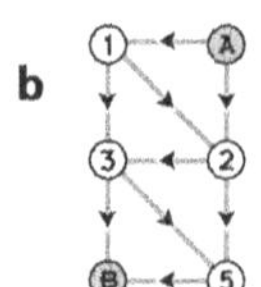

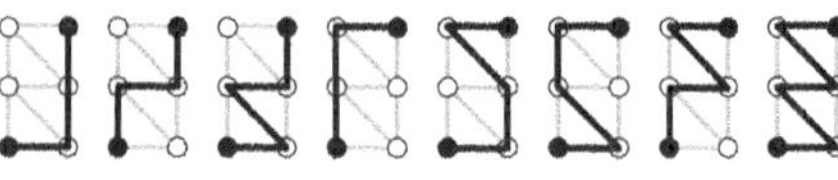

Fibonacci numbers (page 81)

1 55, 89, 144, 233, 377, 610

2 a 20: one less than the eighth term

b 33: one less than the ninth term

c 143: one less than the twelfth term

3 If we consider three consecutive Fibonacci numbers, the square of the middle number is either one more *or* one less than the product of the first and last terms.

4 The product of the first and last of these numbers and the product of the middle two always differ by 1.

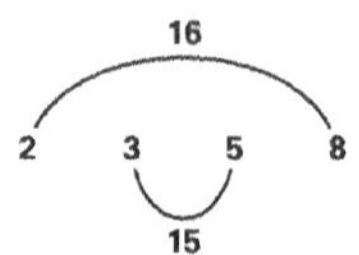

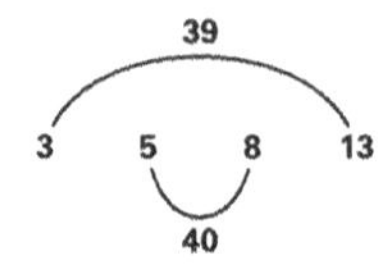

5 a $1^2 + 12 + 2^2 + 3^2 + 5^2 = 5 \times 8$
$1^2 + 1^2 + 2^2 + 3^2 + 5^2 + 8^2 = 8 \times 13$
$1^2 + 1^2 + 2^2 + 3^2 + 5^2 + 8^2 + 13^2 = 13 \times 21$
$1^2 + 1^2 + 2^2 + 3^2 + 5^2 + 8^2 + 13^2 + 21^2 = 21 \times 34$

b $3^3 + 5^3 - 2^3 = 144$
$5^3 + 8^3 - 3^3 = 610$
$8^3 + 13^3 - 5^3 = 2584$

6 Some examples for which this is true:
1 + 1 + 2 + 3 + 5 + 8 + 13 + 21 + 34 + 55 = 143 = 11 x 13
1 + 2 + 3 + 5 + 8 + 13 + 21 + 34 + 55 + 89 = 231 = 11 x 21

7 a Divisible by 2 (an even number)

b Divisible by 3 (a multiple of 3)

c Divisible by 5 (a multiple of 5)

d Divisible by 8 (a multiple of 8)

Repeating cycles (page 82)

1 a

b

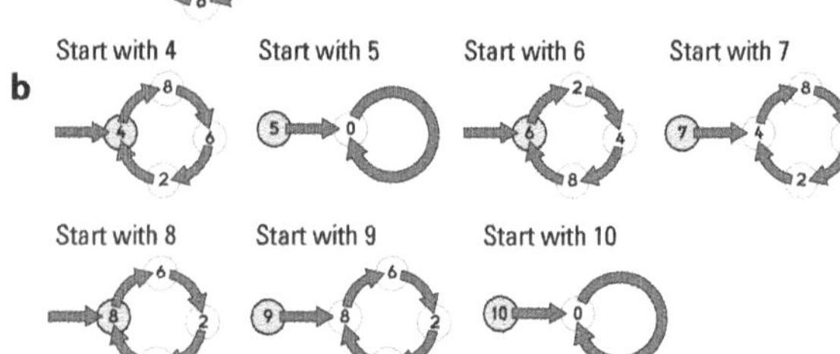

These can be summarised like this:

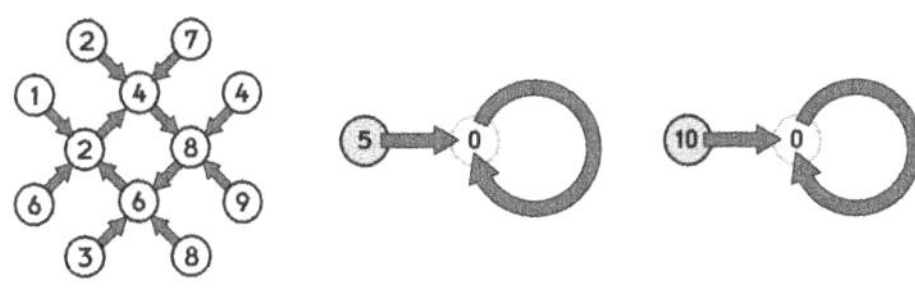

2 a

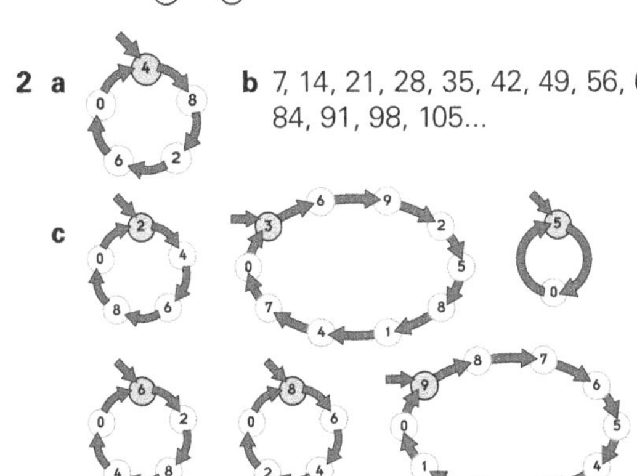

b 7, 14, 21, 28, 35, 42, 49, 56, 63, 70, 77, 84, 91, 98, 105...

c

Power cycling (page 83)

1 $2^7 = 128$, $2^8 = 25\underline{6}$, $2^9 = 51\underline{2}$, $2^{10} = 102\underline{4}$.
The unit digits, which are the numbers underlined, show that these numbers fit the cycle.

2 powers of 3: 3 9 7 1 3 9 7 1 ...
powers of 4: 4 6 4 6 4 6 4 6 ...
powers of 6: 6 6 6 6 6 6 6 6 ...
powers of 7: 7 9 3 1 7 9 3 1 ...
powers of 8: 8 4 2 6 8 4 2 6 ...
powers of 9: 9 1 9 1 9 1 9 1 ...
powers of 10: 0 0 0 0 0 0 0 ...
powers of 11: 1 1 1 1 1 1 1 ...
powers of 12: 2 4 8 6 2 4 8 6 ...
} note for interest

Following these patterns, we can express repeating cycles:

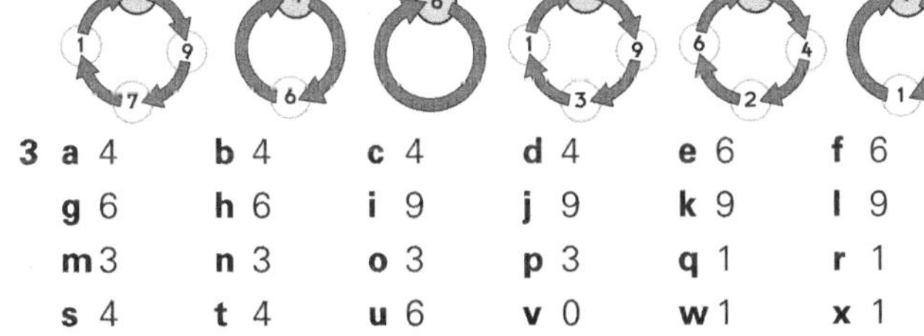

3

a 4	**b** 4	**c** 4	**d** 4	**e** 6	**f** 6
g 6	**h** 6	**i** 9	**j** 9	**k** 9	**l** 9
m 3	**n** 3	**o** 3	**p** 3	**q** 1	**r** 1
s 4	**t** 4	**u** 6	**v** 0	**w** 1	**x** 1

Since there are four numbers in the pattern of repeating cycles for powers of 2, 3 and 7; every fourth power in these cycles ends with the same unit digit.

In a canoe (page 84)

The use of tangible materials is essential, so students should use discs, coins or buttons to represent the children in the canoe. They may need some guidance with the first few moves.

1 8 moves

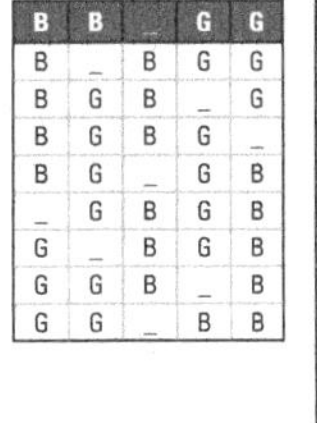

B	B	_	G	G
B	_	B	G	G
B	G	B	_	G
B	G	B	G	_
B	G	_	G	B
_	G	B	G	B
G	_	B	G	B
G	G	B	_	B
G	G	_	B	B

2 15 moves

B	B	B	_	G	G	G
B	B	B	G	_	G	G
B	B	_	G	B	G	G
B	_	B	G	B	G	G
B	G	B	_	B	G	G
B	G	B	G	B	_	G
B	G	B	G	B	G	_
B	G	B	G	_	G	B
B	G	_	G	B	G	B
_	G	B	G	B	G	B
G	_	B	G	B	G	B
G	G	B	_	B	G	B
G	G	B	G	B	_	B
G	G	B	G	_	B	B
G	G	_	G	B	B	B
G	G	G	_	B	B	B

3 24 moves

B	B	B	B	_	G	G	G	G
B	B	B	B	G	_	G	G	G
B	B	B	_	G	B	G	G	G
B	B	_	B	G	B	G	G	G
B	B	G	B	_	B	G	G	G
B	B	G	B	G	B	_	G	G
B	B	G	B	G	B	G	_	G
B	B	G	B	G	_	G	B	G
B	B	G	_	G	B	G	B	G
B	_	G	B	G	B	G	B	G
_	B	G	B	G	B	G	B	G
G	B	_	B	G	B	G	B	G
G	B	G	B	_	B	G	B	G
G	B	G	B	G	B	_	B	G
G	B	G	B	G	B	G	B	_
G	B	G	B	G	B	G	_	B
G	B	G	B	G	_	G	B	B
G	B	G	_	G	B	G	B	B
G	_	G	B	G	B	G	B	B
G	G	_	B	G	B	G	B	B
G	G	G	B	_	B	G	B	B
G	G	G	B	G	B	_	B	B
G	G	G	B	G	_	B	B	B
G	G	G	_	G	B	B	B	B
G	G	G	G	_	B	B	B	B

4 *Note the pattern*: The smallest number of moves is the number of boys (or girls) multiplied by the number two greater than this.

2 boys x 4 = 8
) 7
3 boys x 5 = 15
) 9
4 boys x 6 = 24
) 11
∴ 5 boys x 7 = 35

Happy numbers (page 85)

Note: Once a number has appeared in a happy or sad chain, it is not listed separately.

1 32 is also happy.

2 Since 13 and 10 also reach 1, they too are happy.

3 2 → 4 → 16 → 37 → 58 → 89 → 145 → 42 → 20 → 4 (all sad)
3 → 9 → 81 → 65 → 61 → 37 → (sad, from the above)
5 → 25 → 29 → 85 → 89 → (sad, from 2 →)
6 → 36 → 45 → 41 → 17 → 50 → 25 → (sad from 5→)
7 → 49 → 97 → 130 → 10 → 1 → (all HAPPY)
8 → 64 → 52 → 29 → (sad)
9 → 81 → (sad)
11 → 2 → (sad)
12 → 5 → (sad)
13 → 10 → 1 (HAPPY)
14 → 17 → (sad)
15 → 26 → 40 → 16 → (sad)
18 (sad, since 81 is sad)
19 → 82 → 68 → 100 → 1 (HAPPY)
21 → 5 (sad)
22 → 8 (sad)
23 → (HAPPY, in students' examples)
24 → (sad, since 42 is sad)
27 → 53 → 34 → 25 (sad)
28 → (HAPPY, since 82 is happy from 19)
30 → (sad, since 3 is sad)
31 → (HAPPY, since 13 is happy)
32 → 13 → (HAPPY)
33 → 18 → (sad)
35 → (sad, since 53 is sad from 27)
38 →73 → (sad, since 37 is sad)
39 → 90 → 81 → (sad)
40 → (sad, since 4 is sad)
43 → (sad, since 34 is sad)

Happy numbers (page 85) ***continued...***

44 → 32 → (HAPPY)

46 → (sad, since 64 is sad)

47 → 65 → (sad)

48 → 80 → 64 → (sad)

4 Since 4 is sad, so is 40; since 3 is sad so is 30. Since 42 is sad, so is 24: any 2-digit numbers consisting of identical digits reversed are the same type. Once we enter a cycle that we know will be sad, it is not necessary to continue it.

Happy numbers

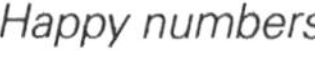

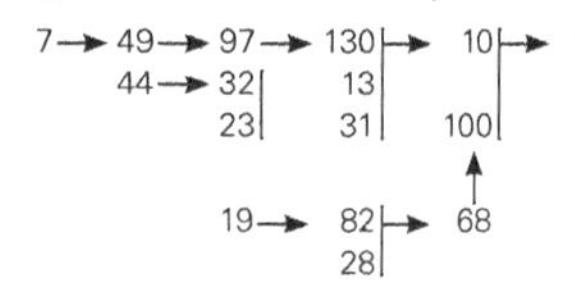

Sad numbers

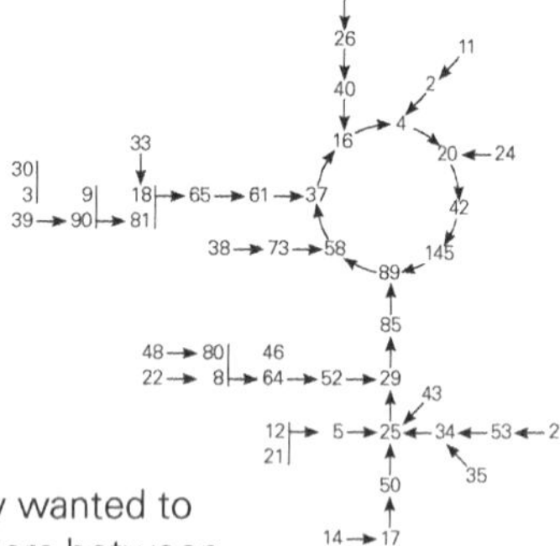

Although we only wanted to investigate numbers between 2 and 50, we had to include some greater than 50 as part of the chains. Students' attention should be drawn to the following closed chain.

58 → 89 → 145 → 42 → 20 → 4 → 16 → 37

Mixed challenges (page 86)

1

3-legged stools	4-legged stools	Total legs
14	15	3 x 14 + 4 x 15 = 102

2

Jolly	Bossy	Sporty	Popular	Helpful
20	2	11	28	14

3

Day	1	2	3	4	5
Jelly beans eaten	8	14 (8 + 6)	20 (8 + 12)	26 (8 + 18)	32 (8 + 24)

Charley ate 20 jelly beans on the third day.

4 K + L = 7, L + M = 17, K + M = 14

It is possible to use 'guess and check' to find the solutions.

Or add these 3 equations:

2K + 2L + 2M = 38

∴ K + L + M = 19

Now using the above, K = 2, L = 5, M = 12

5 a 11

Correct (3)	Incorrect (–1)	Total score
11	4	29 = 33 – 4

b 7

Correct (5)	Incorrect (–1)	Total score
23	7	108 = 115 – 7

Logic puzzles (page 87)

1 Only 9 cuts are needed, so the time taken = 9 x 5 min = 45 min

2 First, 4 verticals are needed, as shown, then 4 more are needed vertically but perpendicular to the first. Finally, 4 horizontal cuts are needed. Altogether, there will be 12 cuts.

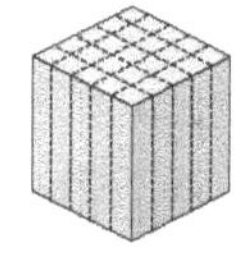

3 When they meet, they are the same distance from Sydney.

4 Since the substance doubles every minute and the container is full at 10:00 a.m., it was:

a half full at 9:59 a.m., 1 min before 10:00 a.m.

b one-quarter full at 9:58 a.m., 1 min before it was half full.

In these questions there are possible hidden assumptions, and finding a solution may be a most frustrating experience. Restrictions that are assumed may be held very strongly and may completely block the way of thinking that leads to an answer.

5 The fifth triangle is the large one!

6  The triangle will be 3-dimensional. This figure is called a triangular pyramid, or tetrahedron.

7 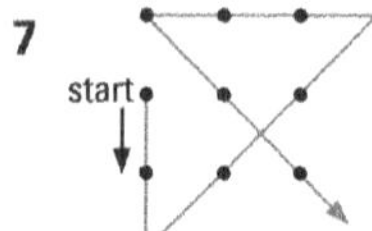

Students may make the assumption that the line segments must not go beyond the dots.

8 Three flags in one horizontal plane form an equilateral triangle, and one flag could be on top of a hill or in a valley. Students may assume that the flags must be in the same horizontal plane.

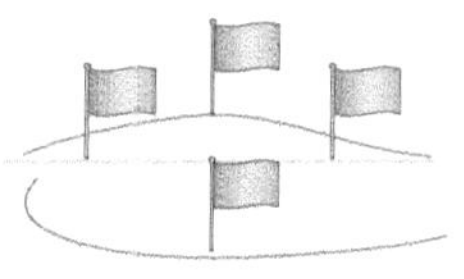

9 Students may assume that the squares are to be the same size.

10 Divide the coins into three piles of three and compare the first two. OOO OOO OOO
A **B** **C**

If A balances B, then C contains the heavier coin. If A does not balance B, the heavier side contains the heavier coin. Now take the heavier pile and compare two of the three coins. As in weighing the three piles, the heavier coin will be obvious.

Models or diagrams (page 88)

1 Only 3 cuts, so each takes 4 minutes.

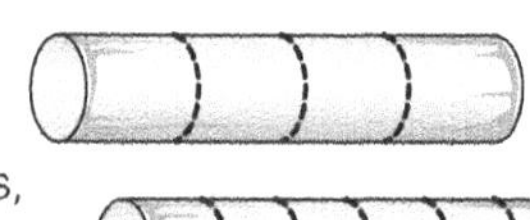

6 pieces require 5 cuts, therefore taking 20 minutes.

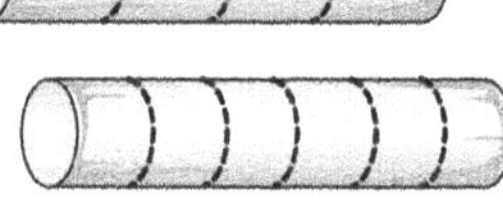

2 a

b

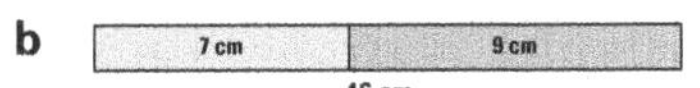

c

7 cm 9 cm
15 cm 1 cm

d

e

9 cm 15 cm
7 cm 17 cm

f

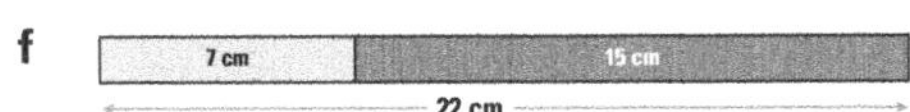

g

7 cm 15 cm
9 cm 13 cm

3 The ball travels the height of each bounce *twice*, up and then down; therefore, the total distance travelled
= 320 + 160 x 2 + 80 x 2 + 40 x 2 + 20
= 900 cm
= 9 m

320 cm
160 cm
80 cm
40 cm
caught 20 cm

4 a 2 h 20 min **b** $4\frac{1}{2}$ h

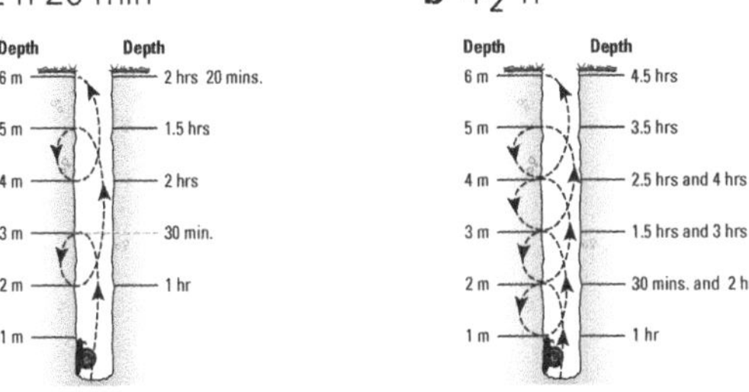

Note: if the snail climbs 3 m in 30 min, it would climb 2 m in 20 min.

5 a 70 min **b** 110 min

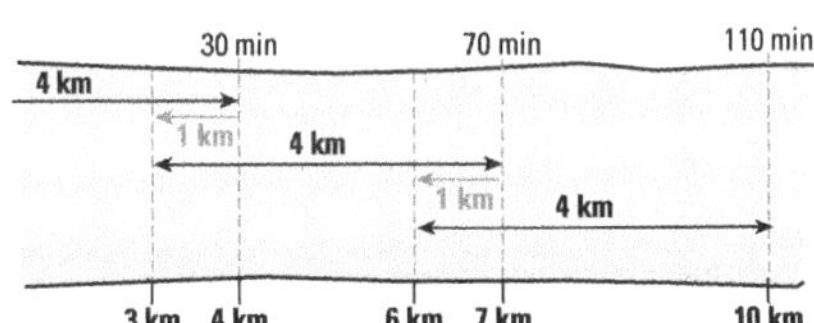